T0310310

The Coloration of Wool and other Keratin Fibres

Current and future titles in the Society of Dyers and Colourists – John Wiley Series

Published

The Coloration of Wool and other Keratin Fibres
David M. Lewis and John A. Rippon

Forthcoming

Natural Dyeing for Textiles: A Guide Book for Professionals
Debanjali Banerjee

Colour for the Design Industry
Vien Cheung

The Coloration of Wool and other Keratin Fibres

Edited by

DAVID M. LEWIS

Department of Colour Science,
University of Leeds, UK

and

JOHN A. RIPPON

CSIRO Materials Science and Engineering, Australia

Published in association with the Society of Dyers and Colourists

Series Editor: Andrew Filarowski

society of dyers
and colourists

WILEY

Library of Congress Cataloging-in-Publication Data

The coloration of wool and other keratin fibres. / edited by David M. Lewis and John A. Rippon.
    pages cm
  Includes bibliographical references and index.
  ISBN 978-1-119-96260-1 (cloth)
  1. Dyes and dyeing–Wool.   I. Lewis, D. M. (David M.), editor of compilation.   II. Rippon, John A., editor of compilation.   III. Title.
   TP899.W66 2013
   667′.2–dc23

2013005805

A catalogue record for this book is available from the British Library.

ISBN: 9781119962601

Set in 10/12pt Times by Aptara Inc., New Delhi, India

Printed and bound in Singapore by Markono Print Media Pte Ltd

Contents

List of Contributors

Peter J. Broadbent, *Colour Chemistry Consultant, UK*

Stephen M. Burkinshaw, *School of Design, University of Leeds, UK*

Robert M. Christie, *School of Textiles & Design, Heriot-Watt University, UK*

Peter A. Duffield, *Retired; Global Textile Associates Ltd, UK*

Paul Hamilton, *Bulmer & Lumb Group Limited, UK*

Jamie A. Hawkes, *Perachem Limited, UK*

David M. Lewis, *Department of Colour Science, University of Leeds, UK*

Keith R. Millington, *CSIRO Materials Science and Engineering, Geelong, Victoria, 3216, Australia*
Co-operative Research Centre for Sheep Industry Innovation, University of New England, NSW, 2800, Australia

Olivier J.X. Morel, *Xennia Technology Ltd., UK*

Muriel L.A. Rigout, *School of Materials, University of Manchester, UK*

John A. Rippon, *CSIRO Materials Science and Engineering, Geelong, Victoria, 3216, Australia*

Arthur C. Welham, *The Dyehouse Doctor Ltd, UK*

Society of Dyers and Colourists

SDC is the world's leading independent, educational charity dedicated to advancing the science and technology of colour. Our mission is to communicate the science of colour in a changing world. We do this by:

- Placing education at the heart of all our operations.
- Providing information and expertise in the fields of colour and the science of coloration.
- Promoting good ethical and environmental working practice within the textile and coloration supply chain.
- Acting as an advocate for the textile and coloration supply chain to promote SDC, good practice and raise the profile of the sector.
- Increasing the reach of SDC to deliver the core message to a wider constituency globally.

We are a global organisation. With our Head Office and trading company based in Bradford, UK, we have members based worldwide and an international network of regions and activities. In India we have an office in Mumbai and regions in Hong Kong, Pakistan and China, with events and training extending far beyond this.

SDC was established in 1884 and became a registered educational charity in 1962. SDC was granted a Royal Charter in 1963 and is the only organisation in the world that can award the Chartered Colourist status, which remains the pinnacle of achievement for coloration professionals.

The dissemination of knowledge and information relating to colour is at the heart of SDC's publications activities. We offer print, electronic and Web-based products. Our publications include over 25 text books covering a full range of dyeing and finishing topics with an ongoing programme of new and revised titles. In addition we publish *Coloration Technology*, the world's leading peer-reviewed journal dealing with the application of colour, and the only journal that covers all aspects of coloration technology. *Coloration Technology*'s scope embraces colorants of all classes, chemicals, application practice, application theory, analysis testing and the theory and practice of ancillary processes.

For further information please email: info@sdc.org.uk, or visit www.sdc.org.uk.

Preface

An important series of books on wool dyeing, edited firstly by C.L. Bird ("Theory and Practice of Wool Dyeing") and later by D.M. Lewis ("Wool Dyeing") was published by The Society of Dyers and Colourists.

This book entitled *The Coloration of Wool and other Keratin Fibres*, has fully updated and expanded the content of the earlier publication "Wool Dyeing". Even though wool has become a minority fibre, it is still cherished by consumers as the basis of clothing and furnishings that provide excellent warmth, comfort and drape. It is the only textile fibre that has been bioengineered over millions of years to be worn next to the animal's skin. No other fibre has such a delightfully complex heterogeneous chemical and morphological structure, consisting of thousands of proteins.

The subjects covered by recognised experts in the field in this book include:

The Structure of Wool, The Chemical and Physical Basis for Wool Dyeing, The Role of Auxiliaries in the Dyeing of Wool and other Keratin Fibres, Ancillary Processes in Wool Dyeing, Bleaching and Whitening of Wool: Photostability of Whites, Wool-dyeing Machinery, Dyeing Wool with Acid and Mordant Dyes, Dyeing Wool with Metal-complex Dyes, Dyeing Wool with Reactive Dyes, Dyeing Wool Blends, The Coloration of Human Hair and Wool Printing.

Compared with the earlier Wool Dyeing books cited above, additional chapters dealing with the bleaching and photostability of whites and the important area of human hair coloration have been included. The chemical and physical natures of wool and of human hair fibres are very close and yet there are remarkable differences in the dyeing methods used.

It is hoped that the new book will continue to provide valuable source material for science and SDC students, especially those studying the science of protein-based materials and those involved in the textile and hair dyeing industries.

The managing editors would like to thank the dedicated John Wiley & Sons team for their copy-editing and support throughout this project. Where appropriate, the assistance of the authors' employers in providing facilities and illustrations for inclusion is also gratefully acknowledged.

David M. Lewis and John A. Rippon

1

The Structure of Wool

John A. Rippon
CSIRO Materials Science and Engineering, Geelong, Victoria, 3216, Australia

1.1 Introduction

The textile industry uses substantial quantities of fibres obtained from animals, of which the wool from sheep is commercially the most important. Natural fibres are biodegradable, so few examples of ancient textiles have survived to the present time; it is thus unclear when wool was first used as a textile material. Archaeological finds suggest, however, that it was probably the first fibre to be used for making cloth, which may have been a wool felt [1]. Early breeds of sheep were covered not in the off-white, continuously growing fleece of the modern animal, but in a brownish coat [2]. This consisted of an outer covering of coarse hairs (kemp fibres) and a finer undercoat. Both the kemp and undercoat fibres were shed annually.

Following domestication of the sheep, selective breeding led to the progressive development of animals with finer wool. The discovery of dyeing probably had an important impact on early sheep breeding, as it would have created a demand for whiter wools. The exact date at which this occurred is again uncertain, but dyed woven cloth made from wool was definitely in use in ancient Egypt several thousand years ago.

The various breeds of sheep produce a wide range of wool types, which are classified according to fibre length and diameter [3,4].

Coarse wools are generally used in interior textiles, such as carpets and upholstery, and fine wools are used to produce fabrics used for apparel. Examples of sheep that produce coarse wools are Corriedale (diameter 28–33 μm), Romney (33–37 μm), Perendale (31–35 μm), Lincoln (39–41 μm), Leicester (37–40 μm), Suffolk (30–34 μm) and Blackface (40–44 μm). The most important breed for producing fine wools is the Merino, which originated in Spain during the Middle Ages. Merino sheep were introduced into Australia

The Coloration of Wool and other Keratin Fibres, First Edition. Edited by David M. Lewis and John A. Rippon.
© 2013 SDC (Society of Dyers and Colourists). Published 2013 by John Wiley & Sons, Ltd.

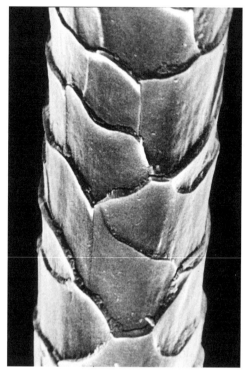

Figure 1.1   *Scanning electron micrograph of a clean Merino wool fibre (courtesy of CSIRO).*

around 200 years ago, where they were developed to produce wool with desirable fineness, length, lustre, crimp and colour. Merino fibres typically range in diameter from 17 to 25 μm. A Merino wool fibre, viewed under the scanning electron microscope (SEM), is shown in Figure 1.1.

1.2   Composition of Wool

Raw wool can contain 25–70% by mass of impurities [3]. These consist of wool grease, suint, dirt and vegetable matter, such as burrs and seeds. Wool grease is a complex mixture of various fatty esters and fatty acids. Suint, which arises from perspiration, is composed mainly of the potassium salts of short-chain acids, plus some sulphate, phosphate and nitrogenous material [5]. Grease, suint and dirt are removed by scouring [6,7]. Vegetable matter is removed in worsted processing by carding and combing [8], or in woollen processing by carbonising [9]. The wool discussed in this chapter is the fibrous material from which the surface contaminants have been removed.

Wool is a member of a group of proteins known as keratins [10], which name is derived from the Greek word for 'horn'. A precise definition of a keratin is not possible because of the diversity of its various forms with respect to both structure and occurrence. Keratins have been classified as 'hard' or 'soft' according to their tactile properties [11]. A characteristic feature of hard keratins, such as wool, hair, hooves, horns, claws, beaks and feathers, is a higher concentration of sulphur (in excess of 3%) than is found in soft keratins, such as those in skin [12]. The sulphur in keratins is mainly present in the form of residues of the amino acid cystine (see Table 1.1).

Wool fibres grow in follicles in the skin of sheep [15]. Cell growth occurs throughout the bulbous base of the follicle and is complete immediately above the bulb, from where the process of keratinisation commences. Keratinisation, which results in hardening of the fibre, involves the formation of residues containing the disulphide crosslinks of the amino acid cystine (**1**) from pairs of cystine (**2**) residues.

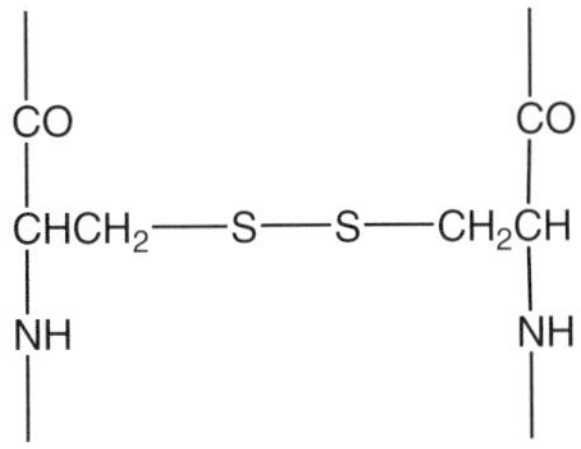

$$
\begin{array}{ccc}
| & & | \\
CO & & CO \\
| & & | \\
CHCH_2-S-S-CH_2CH & & \\
| & & | \\
NH & & NH \\
| & & | \\
\end{array}
$$

1

$$
\begin{array}{c}
| \\
CO \\
| \\
HC-CH_2-SH \\
| \\
NH \\
| \\
\end{array}
$$

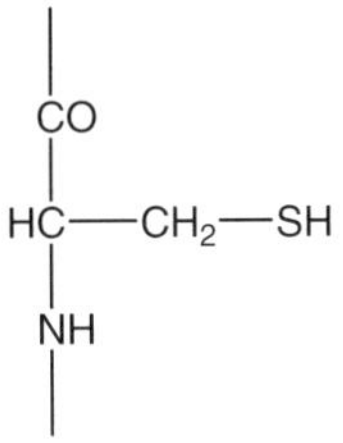

2

Keratinisation is complete before the fibre emerges from the skin.

Keratins have also been classified as α- or β-types, according to their x-ray diffraction patterns [11,12]. Unstretched wool fibres give a pattern characteristic of α-keratin, while stretched wool fibres give a pattern which closely resembles that of β-keratin [16].

Although classified as a keratin, clean wool contains only (approximately) 82% keratinous proteins, which are characterised by a high concentration of cystine (see Section 1.4). Approximately 17% of wool is composed of proteins termed 'nonkeratinous', because of their relatively low cystine content [17–20].

Wool fibre also contains approximately 1% by mass of non-proteinaceous material. This consists mainly of waxy lipids, plus a small amount of polysaccharide material. The nonkeratinous proteins and lipids are not uniformly distributed throughout the fibre but are concentrated in specific regions of the structure. Their location and their importance in determining the behaviour of wool are discussed later.

Table 1.1   *Structure and amount of major amino acids in wool (mol%).*

Amino Acid	Structure[a]	mol% [13]	mol% [14]	Nature of Side Chain
Glycine	HCHCOOH NH$_2$	8.6	8.2	Hydrocarbon
Alanine	CH$_3$CHCOOH NH$_2$	5.3	5.4	Hydrocarbon
Phenylalanine	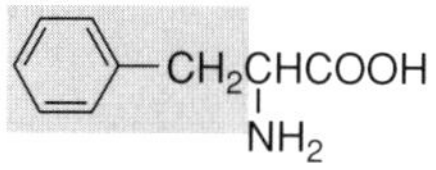 ⬡—CH$_2$CHCOOH NH$_2$	2.9	2.8	Hydrocarbon
Valine	H$_3$CCHCHCOOH H$_3$C NH$_2$	5.5	5.7	Hydrocarbon
Leucine	H$_3$CCHCH$_2$CHCOOH H$_3$C NH$_2$	7.7	7.7	Hydrocarbon
Isoleucine	H$_3$CCH$_2$CHCHCOOH H$_3$C NH$_2$	3.1	3.1	Hydrocarbon
Serine	HOCH$_2$CHCOOH NH$_2$	10.3	10.5	Polar
Threonine	H$_3$CCHCHCOOH HO NH$_2$	6.5	6.3	Polar
Tyrosine	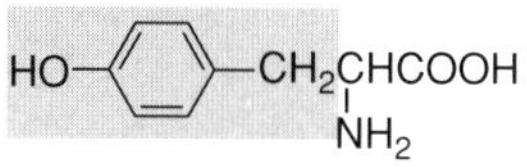 HO—⬡—CH$_2$CHCOOH NH$_2$	4.0	3.7	Polar
Aspartic acid[b]	HOOCCH$_2$CHCOOH NH$_2$	6.4	6.6	Acidic
Glutamic acid[c]	HOOCCH$_2$CH$_2$CHCOOH NH$_2$	11.9	11.9	Acidic
Histidine	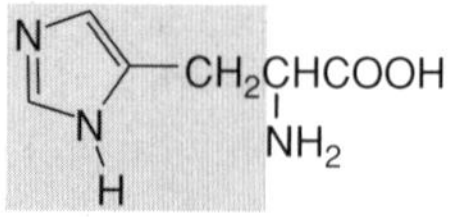 N⬠—CH$_2$CHCOOH N NH$_2$ H	0.9	0.8	Basic
Arginine	H$_2$NCNH(CH$_2$)$_3$CHCOOH HN NH$_2$	6.8	6.9	Basic
Lysine	H$_2$N(CH$_2$)$_4$CHCOOH NH$_2$	3.1	2.8	Basic
Methionine	H$_3$CS(CH$_2$)$_2$CHCOOH NH$_2$	0.5	0.4	Sulphur-containing

Table 1.1   *(Continued)*

Amino Acid	Structure[a]	mol% [13]	mol% [14]	Nature of Side Chain
Cystine[d]	HOOCCHCH$_2$SSCH$_2$CHCOOH 　　｜　　　　　　｜ 　　NH$_2$　　　　　NH$_2$	10.5[e]	10.0[e]	Sulphur-containing
Tryptophan	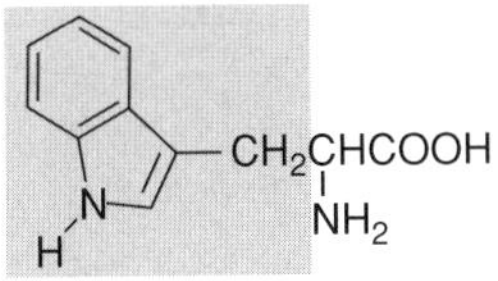 —CH$_2$CHCOOH 　　　　｜ 　　　　NH$_2$	See text		Heterocyclic
Proline	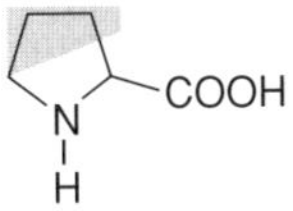 —COOH	5.9	7.2	Heterocyclic

[a] Shading indicates identity of side chain.
[b] Includes asparagine residues (see text).
[c] Includes glutamine residues (see text).
[d] Includes oxidation byproduct, cysteic acid.
[e] Values are for half-cystine (see text).

1.3   Chemical Structure of Wool

The structures of fibrous proteins, in particular that of wool, have been studied extensively over many years. The growth of knowledge in this area has been catalogued in some extensive reviews [4,10,15,17–28].

1.3.1   General Chemical Structure of Proteins

Proteins are natural polymers of high relative molecular mass (r.m.m.). They are very widespread in nature, being essential components of animal and plant tissue. The basic structural units of proteins are α-amino acids, which have the general formula shown in Figure 1.2, where the side chain R can be an aliphatic, aromatic or other cyclic group.

With the exception of glycine (Table 1.1), the amino acids isolated from proteins are optically active, because of the presence of an asymmetric carbon atom. In common with other naturally occurring proteins, the optically active amino acids in wool are laevorotatory. As shown in Figure 1.2, they have a tetrahedral configuration, with the carbon atom at the centre of the tetrahedron. When this schematic diagram is viewed along the C–H bond, the other groups occur in a clockwise direction, in the order R, NH$_2$, COOH.

Proteins are formed by condensation of L-α-amino acids via their carboxyl and amino groups. A dipeptide is formed by the condensation of two amino acid molecules (Scheme 1.1). Condensation of further molecules, of the same or a different amino acid, produces a linear polymer (Scheme 1.2). Such a polymer can be regarded as a polyamide, because each structural unit is joined by an amide group. In the case of proteins, however, the repeat unit (–NHCHRCO–) is referred to as a peptide group, and compounds containing

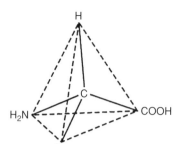

Figure 1.2   *General structure of an amino acid.*

$$NH_2CHRCOOH + NH_2CHRCOOH \xrightarrow{-H_2O} NH_2CHRCONHCHRCOOH$$

Scheme 1.1   *Condensation of two amino acids to produce a dipeptide.*

$$NH_2CHRCONHCHRCOOH + n[NH_2CHRCOOH] \xrightarrow{-nH_2O}$$
$$NH_2CHRCO[NHCHRCO]_n NHCHRCOOH$$

Scheme 1.2   *Formation of a polypeptide by multiple condensation reactions.*

multiples of this group are called polypeptides. The peptide group is also referred to as an 'amino acid residue', because it is the part of the amino acid that remains after the condensation reaction shown in Scheme 1.1.

1.3.2   Amino Acid Composition of Wool

Wool fibres vary in their physical properties, such as length, diameter and crimp. They also vary in chemical composition [29]. Wool can be hydrolysed under acid conditions into its constituent amino acids, with various techniques being used to analyse the hydrolysate [28]. The literature contains details of many amino acid analyses for whole wool and for the various histological components of the fibre [13,21,22,24,30]. Intact wool fibres contain 20 amino acids. Complete acid hydrolysis of wool, however, yields a mixture containing the 18 amino acids shown in Table 1.1. The two not shown in the table are asparagine and glutamine, which during acid hydrolysis are converted into their corresponding acids (Scheme 1.3).

   Thus, the amount of aspartic or glutamic acid in an acid hydrolysate is the sum of the concentration of the original acid plus that derived from the original asparagine or glutamine residues [31]. Hydrolysis of asparagine and glutamine during dissolution can be avoided by using enzymes to digest the wool. This technique has enabled the concentrations of the four residues to be determined separately. The fractions of (aspartic acid + asparagine) and (glutamic acid + glutamine) present as asparagine and glutamine are approximately 60 and 45%, respectively [32,33].

   There is often considerable variation in the values for the concentrations of amino acids obtained by different workers. Although some of the differences may be due to

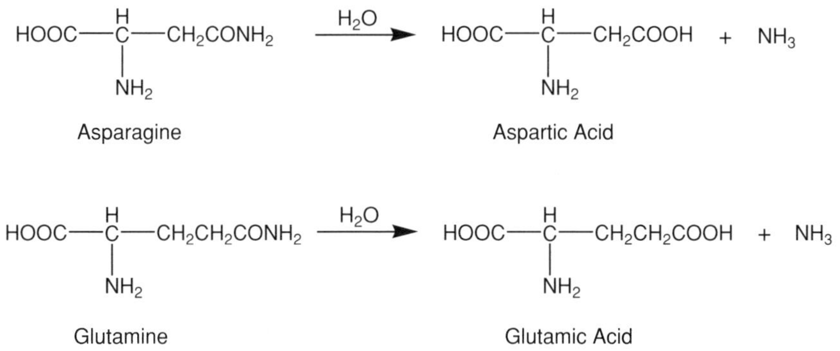

Scheme 1.3   *Hydrolysis of asparagine and glutamine.*

experimental error, others are probably real, and may be caused by several factors [31]. Significant variation in amino acid composition can exist, both between fibres from different individuals of a single species and along the length of single fibres from the same animal [34]. Variation in composition along fibres, in particular the concentration of the thiol groups of cystine residues (**2**), has been shown to affect physical properties, such as stress relaxation from root to tip [35]. These differences tend to decrease on prolonged storage, however, as the thiol groups become oxidised to cysteic acid (**3**).

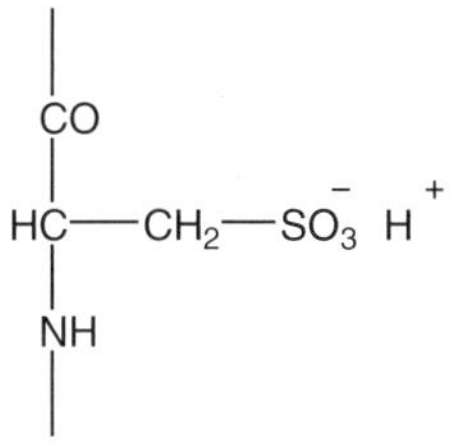

3

The most important factors affecting variations in the chemical structure and physical properties of wool are genetic origin [36], physiological state [34] and nutrition [29,37]. In particular, the cystine content of wool has been shown to be particularly susceptible to changes in diet. Differential weathering of wool while on the back of the sheep can also be responsible for variations in amino acid content; again the most notably is cystine, which is often oxidised to cysteic acid. The method of cleaning the sample before testing may also affect the result, particularly when purification procedures are used which extract labile material from the fibre interior.

In addition to glutamine and asparagine, precise concentrations of some other amino acids cannot be obtained on acid hydrolysates, due to degradation by the hydrolytic procedure. In particular, serine and threonine are progressively degraded during hydrolysis, while tryptophan is completely destroyed. Various techniques have been developed to avoid

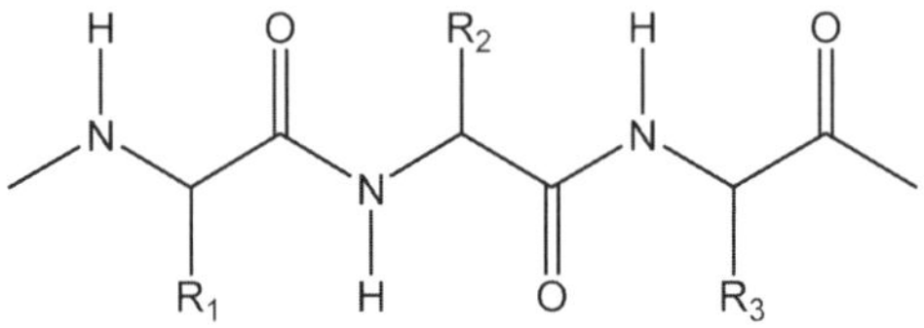

Figure 1.3   *General structure of wool polypeptide.*

this difficulty; for example, correction factors are used for serine and threonine [28]. In the case of tryptophan, a mixture of *p*-toluenesulphonic acid and tryptamine is used as the hydrolysing medium [38]. Alternatively, the wool can be hydrolysed under alkaline conditions or digested with enzymes [31,32]. Values for tryptophan concentration of about 0.5 mol% have been obtained by these methods.

Wool samples usually contain both cystine and – as already discussed – a small amount of its precursor in the reduced state, cystine [31]. Cystine and cystine are partially destroyed by acid hydrolysis, and nonhydrolytic methods have been developed for their determination [39]. In common with many works, in this chapter cystine concentration is expressed in terms of the concentration of its reduction product, cystine (also termed 'half-cystine'). Values for half-cystine usually include both the small amount of cystine that occurs naturally in most samples of wool and also any of the oxidation product, cysteic acid, that is present.

It is worth noting that, in addition to the 20 amino acids already discussed, wool fibres also contain trace amounts of the amino acids orthinine and citrulline [31].

1.3.3   Arrangement of Amino Acids in Wool

The general structure of a wool polypeptide is shown in Figure 1.3, where R_1, R_2, R_3 represent amino acid side chains. A significant proportion of the polypeptide chains in wool are in the form of an α-helix. This ordered arrangement is responsible for the characteristic x-ray diffraction pattern of α-keratin [11,23], discussed in Section 1.2.

The side chains of the various amino acids vary in size and chemical nature (Table 1.1). The nonpolar hydrocarbon side chains of glycine, alanine, phenylalanine, valine, leucine and isoleucine vary in hydrophobic character and have low chemical reactivity. The hydroxyl groups of serine, threonine and tyrosine make their side chains polar in nature. These groups are also chemically more reactive than the hydrocarbon residues, especially under alkaline conditions. The side chains that probably have the most marked overall influence on the properties of wool, including its dyeing properties, are those containing acidic or basic groups. Acidic carboxyl groups are contained in residues of aspartic and glutamic acids, whereas histidine, arginine and lysine residues contain basic side chains: the imidazole, guanidino and amino groups, respectively.

Proline is somewhat unusual in that it is an imino, rather than an amino acid. It does not have a side chain that projects from the main backbone, in the manner of the other amino acids in wool. The bonds linking proline to the polypeptide chain are situated almost at right angles, because of the orientation of the imino and carboxyl groups. Thus, the presence of a proline residue has a marked effect on the conformation of a protein. The frequent occurrence of proline would be expected to result in a highly convoluted structure [23].

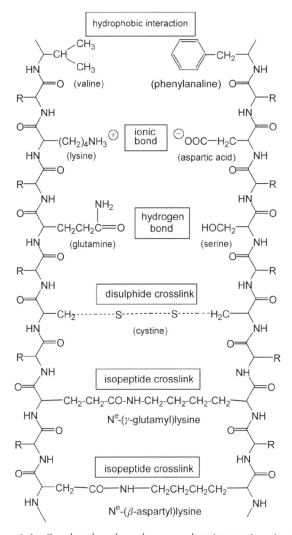

Figure 1.4 *Covalent bonds and noncovalent interactions in wool.*

The individual peptide chains in wool are held together by various types of covalent crosslink and noncovalent interaction (Figure 1.4). In addition to their occurrence between separate polypeptide chains (interchain), these bonds can also occur between different parts of the same chain (intrachain) (Figure 1.5). With respect to the properties and performance of wool, interchain bonds are the more important of the two.

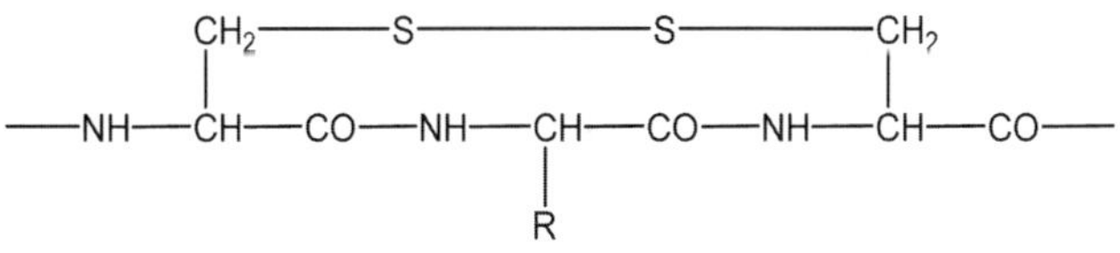

Figure 1.5 *Example of an intrachain crosslink in wool.*

1.3.3.1   Covalent Crosslinks

Except for a small amount of the amino acid methionine, the sulphur content of wool occurs in the form of cystine. As discussed earlier, this is formed within the follicle in the skin of the sheep during keratinisation, or hardening, of the fibre [11,15]. The disulphide bonds of cystine form crosslinks, either between different protein chains (the interchain bonds shown in Figure 1.4) or between different parts of the same protein chain (the intrachain bonds shown in Figure 1.5) [40]. The cystine interchain crosslinks, which have been compared with the rungs in a ladder [40], are the major bonds responsible for stabilising the fibre, particularly in the wet state [10]. Cleavage or rearrangement of the disulphide bonds in wool is involved in important industrial processes such as shrinkproofing and setting [4,41].

Another type of covalent crosslink, the isopeptide bond, has been identified in wool [32,42,43]. Isopeptide crosslinks are formed between the ε-amino groups of lysine and the β- or γ-carboxyl groups of aspartic or glutamic acid, respectively. They are believed to crosslink polypeptide chains, as depicted in Figure 1.4. These bonds are not found in acid hydrolysates of wool because, under the conditions used in the hydrolysis, all peptide bonds are broken, including isopeptide linkages. The pre-analysis digestion of wool by a succession of enzyme treatments enables these bonds to be detected, however. The concentration of N^{ε} (γ-glutamyl)-lysine isopeptide bonds is believed to be much greater than that of N^{ε} (β-aspartyl)-lysine linkages [40].

1.3.3.2   Noncovalent Bonds (Interactions)

In addition to covalent crosslinks, noncovalent bonds or interactions also exist in wool. These secondary bonds, which can occur within a single protein chain or between different chains, act like crosslinks and make an important contribution to fibre properties. The noncovalent bonds in keratin fall into three main groups.

Hydrogen Bonds   The –CO and –NH groups in the peptide chains and the amino and carboxyl groups in the side chains can interact through hydrogen bonds. These bonds can also exist between suitable donor and acceptor groups in the amino acid side chains [44]. A large number of hydrogen bonds in wool occur between suitable groups within an α-helical chain.

Ionic Interactions   The side chains of wool contain approximately equal numbers of basic amino and acidic carboxyl groups [31]. These groups are responsible for the amphoteric nature of the fibre and its ability to combine with large amounts of acids or bases [45].

Scheme 1.4 shows that at neutrality, both types of group are fully ionised, and the net electrical charge carried by the fibre is zero. This condition is known as the isoelectric

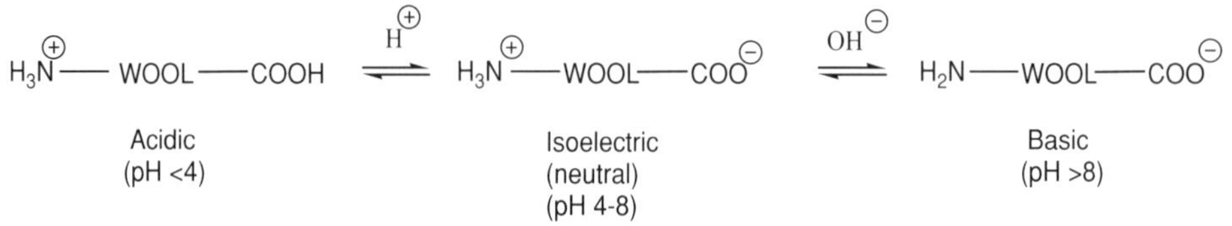

$$\overset{\oplus}{H_3N} — WOOL — COOH \quad \underset{}{\overset{\overset{\oplus}{H}}{\rightleftharpoons}} \quad \overset{\oplus}{H_3N} — WOOL — \overset{\ominus}{COO} \quad \underset{}{\overset{\overset{\ominus}{OH}}{\rightleftharpoons}} \quad H_2N — WOOL — \overset{\ominus}{COO}$$

| | Acidic (pH <4) | | Isoelectric (neutral) (pH 4-8) | | Basic (pH >8) |

Scheme 1.4   *Amphoteric nature of wool.*

state. Strong electrostatic interactions occur between ionised amino and carboxyl groups. An example of such a linkage, between the ionised terminal groups of lysine and aspartic acid, is shown in Figure 1.4. These ionic interactions have also been referred to as 'ionic bonds' or 'salt linkages'. As can be seen from Scheme 1.4, the number of ionic interactions depends on the pH; in fact, their existence in wool was first proposed to explain changes in the mechanical properties of the fibre with varying pH [45]. Both salt linkages and hydrogen bonds contribute markedly to the physical properties of dry wool. Both types of interaction are progressively disrupted as wool absorbs water, but even when the fibre is fully saturated some of these interactions within the protein structure remain undisturbed. The contribution of salt linkages and hydrogen bonds to the physical properties of wet wool is less than it is for dry wool. For this reason, physical tests to determine the effect of chemical treatments (including dyeing) on the covalent bonds in wool are often carried out in the wet state. When tests – such as burst strength – are performed on conditioned-only wool, the considerable contribution of salt linkages and hydrogen bonds tends to mask strength losses caused by fission of peptide and disulphide bonds [46].

Hydrophobic Interactions   Hydrophobic interactions (sometimes called hydrophobic bonds) can occur between the nonpolar groups of alanine, phenylalanine, valine, leucine and isoleucine, with the exclusion of associated water molecules [47]. This type of interaction is believed to contribute to the mechanical strength of keratin, particularly at high water contents [48]. It is important in the setting of wool and contributes to the smooth-drying properties of fabrics [4].

1.3.4   The Structure of Wool Proteins

It has been estimated that wool contains about 170 different types of polypeptide molecules [20]. These are not uniformly distributed throughout the fibre. Despite the overall classification of wool as a keratin, the constituent wool proteins have been termed 'keratinous' or 'nonkeratinous', according to their cystine content. Nonkeratinous proteins contain fewer than one residue in every 33 of half-cystine and have a relatively low concentration of disulphide crosslinks compared with keratinous proteins [18–20]. This makes them more labile and less resistant to chemical attack than the keratinous proteins of wool. Zahn has defined nonkeratins in terms of the material digested from wool by the proteolytic enzyme pronase [19,20]. Nonkeratinous proteins constitute approximately 17% of the total fibre mass, whereas keratinous proteins account for around 82% (see Section 1.4).

Two methods have been used to determine the amino acid sequence of keratinous wool proteins. Both involve solubilisation of the fibre, or its morphological components, followed by separation of the extract into the various protein fractions [24,28]. The extraction procedures involve solubilising the proteins by conversion of disulphide crosslinks into anionic groups. Conditions are employed that avoid fission of peptide bonds. The relatively low cystine content of nonkeratinous proteins precludes their solubilisation by these techniques and results in their separation as a solid residue [20].

The first method involves treatment with peracetic [49] or performic [50] acid, both of which oxidise cystine to cysteic acid residues. The resulting extract, which represents about 85% of the total mass [31], is separated into three fractions on the basis of acid or alkali solubility. These fractions have been designated α-, β- and γ-keratoses. Performic acid is preferred to peracetic acid because it produces less peptide fission. Both reagents

Table 1.2   *Amino acid composition of various protein fractions isolated from wool (mol %).*

Amino Acid	Low-Sulphur Fraction (SCMKA) [52]	High-Sulphur Fraction (SCMKB) [52]	High-Glycine–Tyrosine (HGT Type I) [52]	High-Glycine–Tyrosine (HGT Type II) [52]	Whole Wool[a]
Alanine	6.9	2.9	1.5	1.1	5.4
Arginine	7.3	5.9	5.4	4.7	6.9
Aspartic acid[b]	9.0	3.0	3.3	1.8	6.5
1/2-cystine	6.0	18.9	6.0	9.8	10.3
Glutamic acid[c]	15.7	8.4	0.6	0.7	11.9
Glycine	7.7	6.9	27.6	33.6	8.4
Histidine	0.6	0.8	1.1	0.1	0.9
Isoleucine	3.6	3.6	0.2	0.2	3.1
Leucine	10.2	3.9	5.5	5.3	7.7
Lysine	3.5	0.6	0.4	0.4	3.0
Methionine	0.6	0.0	0.0	0.0	0.5
Phenylalanine	2.5	1.9	10.3	4.5	2.9
Proline	3.8	12.5	5.3	3.0	6.6
Serine	8.2	12.7	11.8	10.9	10.4
Threonine	4.8	10.3	3.3	1.7	6.4
Tyrosine	3.6	2.1	15.0	20.3	3.9
Valine	6.1	5.6	2.1	1.4	5.6

[a]Mean values from Table 1.1.
[b]Includes asparagine residues (see Section 1.3.2).
[c]Includes glutamine residues (see Section 1.3.2).

oxidise tryptophan and methionine residues, however, and the peracid oxidation procedure is now regarded as inferior to the reduction/carboxymethylation method [24]. This technique, which minimises side reactions, involves reduction of the disulphide bonds to cystine residues [51]. These are then converted into the *S*-carboxymethylcystine derivative by alkylation with iodoacetic acid. Subsequent extraction with alkali dissolves the 80% of the fibre that is composed of keratinous proteins [27]. The alkali-soluble *S*-carboxymethylkerateine can then be separated, by either gel electrophoresis or chemical fractionation [24,28], into the following three groups of proteins, each group having a characteristic amino acid composition [23,24,27,52] (Table 1.2):

- low-sulphur proteins (designated SCMKA);
- high-sulphur proteins (designated SCMKB);
- high-glycine, high-tyrosine proteins (HGT).

The HGT proteins have been further divided into two subgroups, known as Type I and Type II, according to their differing cystine contents [52].

These three families of proteins have been characterised according to their r.m.m., with various ranges of values being quoted by different workers [24,27]. Gillespie has placed the low-sulphur proteins in the r.m.m. range 44 000–57 000 Da, the high-sulphur proteins in the range 10 000–30 000 Da and the high-glycine–tyrosine proteins at below 10 000 Da [27].

The amino acid composition of many of the proteins in the three groups have been determined [11,24,52] (Table 1.2). The high-sulphur proteins are rich in cystine, proline, serine and threonine; together, these amino acids constitute more than half of the amino acid residues in the proteins of this group [27]. They contain little aspartic acid, lysine, alanine or leucine, and no methionine. In contrast, the low-sulphur proteins are particularly rich in the amino acids that contribute to α-helix formation [27], namely glutamic and aspartic acids, leucine, lysine and arginine. The two types of high-glycine–tyrosine protein, which are also rich in serine, differ mainly in their contents of phenylalanine and of cystine [52]. Approximately 65–70% of the composition of both Type I and Type II HGT proteins is accounted for by three or four amino acids [27].

Compared with the number of studies carried out on keratinous wool proteins, relatively little work has been published on the composition of the nonkeratinous proteins. These can be isolated from wool by extraction with formic acid [22] or enzymes [19,20]. Analysis of the extracts suggests that proteins rich in glycine, tyrosine, phenylalanine, serine and glutamic acid, but low in cystine, are present in the intercellular regions (see Section 1.4.3).

The specific location within the fibre of the various types of wool protein will be discussed in Section 1.4.

1.3.5 Wool Lipids

Wool contains a small amount (0.8–1.0% by mass) of lipid material [19,22,53,54]. This is believed to be concentrated mainly in the intercellular regions of the fibre (Section 1.4.3). Material extracted from clean wool by solvents has been shown to contain a high proportion of fatty acids, plus cholesterol and lanosterol [55]. The fatty acids have been found to be different from those present in wool grease. The internal lipids extracted from wool have been shown to contain every straight-chain saturated and monounsaturated fatty acid between C_7 and C_{26} [54], sterols [56], triglycerides [57], diglycerides and polar lipids, in particular sphingolipids and phospholipids [58] (Table 1.3).

Table 1.3   *Composition of wool lipids [54].*

Lipid Component	Proportion of Total Lipid (Approx. %)	Major Constituents
Sterols	40	Cholesterol Desmosterol
Polar lipids	30	Cholesterol sulphate Ceramides Glycosphingolipids
Fatty acids	25	Stearic Palmitic Oleic Myristic 18-methyleicosanoic[a]
Phospholipids	Trace	–

[a] This acid is unique in that it is not an internal lipid, but is covalently bound to the fibre surface (see Section 1.4.1).

1.4 Morphological Structure of Wool

In addition to being chemically heterogeneous (Section 1.3.4), wool and other keratin fibres are also physically heterogeneous and can be considered as biological composites [17–19,22–25,53,59].

Three methods have been used to determine the chemical composition of the various morphological components present in keratin fibres [25]:

(1) *Chemical analysis of extracts obtained by digestion of the whole fibre*, followed by assignment of the components to various regions identified by microscopy. Examples of these components are the α-, β- and γ-keratoses and the S-carboxymethylkerateines, discussed in Section 1.3.4. This technique is limited in respect of the amount of information that can be obtained.

(2) *Selective staining of specific chemical groups* with reagents that have a high electron-scattering power, followed by examination of sections of the fibre under the transmission electron microscope (TEM). This procedure enables the morphological components of the fibre to be highlighted [60]. Several techniques involving staining with heavy-metal salts have been used to identify the location of cystine in wool. Absolute specificity for cystine has been established, however, for only three methods [25]. These use either organomercurial compounds [61,62], a mixture of silver nitrate and hexamethylenete-tramine [63] or a uranyl salt followed by post-staining with an alkaline solution of a lead salt [64]. Specific staining procedures using phosphotungstic acid have also been developed to identify amino and other basic groups [65]. Carboxyl groups can be identified by a technique that uses uranyl acetate [64].

(3) *Preferential separation or dissolution of the components*, usually by enzymatic digestion [66,67]. The effect of the treatment is monitored by examination under the electron microscope, in conjunction with chemical analysis of the separate compounds. This procedure has proven to be extremely useful in providing a large amount of information on the composition of the various morphological components of wool [66–73]. The complex morphological structure of fine wool fibres is shown schematically in Figure 1.6. Not shown in this diagram is the hydrophobic F-layer (see Section 1.4.1.2). This consists of fatty acids covalently bound to the surface of the epicuticle.

Fine wools contain two types of cell: the cells of the external *cuticle* and the cells of the internal *cortex*. Together, these constitute the major part of the mass of clean wool. Table 1.4, taken from the data of Bradbury [22], shows the proportions of the cuticle and cortex in fine wools, plus the amounts of the other minor histological components of the fibre.

Coarse keratin fibres (usually of diameters greater than 35 μm) may contain a third type of cell: those of the *medulla* [11,23,25,74]. This is a central core of cells, arranged either continuously or intermittently along the fibre axis and wedged between the cortical cells, often in a ladder-like manner. Air-filled spaces lie between the medullary cells. The function of the medulla in the live animal appears to be to confer maximum thermal insulation and provide economy of weight. The presence of a medulla increases the light-scattering properties of fibres, particularly for blue light [23]. This makes medullated fibres appear whiter than those of unmedullated wools, thus restricting the use of these wools for certain purposes.

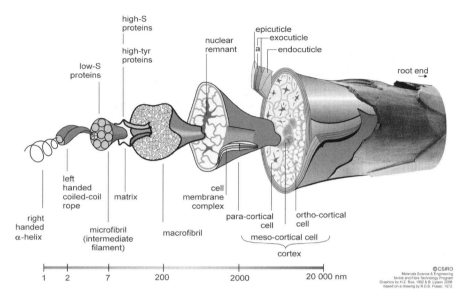

THE STRUCTURE OF A MERINO WOOL FIBRE

Figure 1.6 *Diagram of the morphological components of a fine wool fibre (courtesy of CSIRO). See colour plate section for a full-colour version of this image.*

Table 1.4 *Amounts of various morphological components in fine wool (% o.m.f.).*

Component	Keratinous Proteins	Nonkeratinous Proteins	Nonprotein Matter
Cuticle[a]			
exocuticle	6.4		
endocuticle		3.6	
Cortex[b]			
intermediate filaments	35.6		
matrix	38.5		
nuclear remnants and intermacrofibrillar material		12.6	
Cell membrane complex[c]			
soluble proteins from the cell membrane complex		1.0	
resistant membranes[d]	1.5		
lipids			0.8
Total	**82.0**	**17.2**	**0.8**

[a] Total cuticle 10%.
[b] Total cortex 86.7%.
[c] Total cell membrane complex 3.3%.
[d] Including the epicuticle (0.1%).

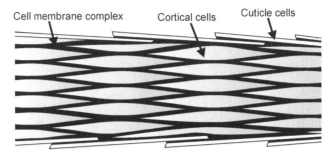

Figure 1.7   *Simplified diagram of the cuticle and cortex of wool (courtesy of CSIRO).*

Cuticle cells are separated from the underlying cortex – and individual cortical cells are separated from each other – by the *cell membrane complex* [17–19,22,53,75,76]. A fine wool fibre can, therefore, be considered an assembly of cuticle and cortical cells held together by the cell membrane complex (Figure 1.7). The cell membrane complex, which has several components, is of particular importance because it constitutes the only continuous phase in wool (see Section 1.4.3).

Each individual cuticle and cortical cell is surrounded by a thin, chemically resistant, proteinaceous membrane [11,22,76,77] (see Section 1.4.3). In fine wools, these *resistant membranes* constitute approximately 1.5% of the total fibre mass (Table 1.4). The term 'resistant membrane' has arisen because this component is the last part of the fibre to dissolve when whole wool fibres, or individual cuticle or cortical cells, are digested by various degradative procedures [76,77]. In cases where there are two adjacent cortical cells, the resistant membrane surrounding each cell is considered to be part of the cell membrane complex (see Section 1.4.3). The *epicuticle* is defined as that part of a cuticle cell resistant membrane that is located on the fibre surface (see Section 1.4.1).

The families of proteins listed in Table 1.2 are not uniformly distributed between the morphological regions of the fibre. This is reflected in a difference in the amino acid composition of the various components (Table 1.5).

1.4.1   The Cuticle and the Fibre Surface

The cuticle cells, or scales, constitute the outermost surface of the wool fibre and are responsible for important properties such as wettability [78,79], tactile properties [79] and felting behaviour [41,79,80]. Approximately 10% of a fine wool fibre consists of cuticle cells, which can be seen clearly in the light microscope or SEM (Figure 1.1). According to Bradbury, Merino cuticle cells range in thickness from 0.3 to 0.5 μm and are about 30 μm in length and 20 μm in width [22]. Other workers, however, have claimed that the dimensions of cuticle cells and their arrangement on the fibre are more complex and varied, with some cells forming a spiral around the fibre [81]. The cells overlap rather like tiles on a roof, with the edge of every scale pointing from the root to the tip of the fibre. The function of cuticle cells in keratin fibres appears to be to anchor them in the follicle on the skin of the animal [23]. A consequence of the ratchet-like arrangement of cuticle cells on the fibre surface is that the coefficient of friction along the fibre is much less in the root-to-tip direction than it is from the tip to the root [41,45,80]. This directional frictional effect (DFE) is believed

Table 1.5 Amino acid composition of the morphological components of wool (mol%).

| Amino Acid | Whole Wool[a] | Cuticle | | | | Resistant Membranes (Total) (22,76,166) | Cortex (22) | | Intercellullar Cement[b] (14,53) | Nuclear Remnants and Intermacrofibrillar Material (66) |
		Whole (67)	Exo- (67)	Endo- (67)	Epi- (96)		Ortho-	Para-		
Alanine	5.4	5.8	6.4	6.7	4.6	6.8	5.6	5.4	5.8	7.5
Arginine	6.9	4.3	4.8	5.0	4.3	4.7	6.8	6.5	6.4	6.2
Aspartic acid[c]	6.5	3.5	2.1	7.4	5.8	6.8	6.7	6.3	7.1	9.9
Citrulline[d]	–	–	–	–	0.9	0.6	–	–	–	–
1/2-cystine[a]	10.3	15.6	19.9	3.1	11.9	9.0	10.3	12.9	1.3	3.1
Glutamic acid[f]	11.9	8.7	8.5	10.3	10.7	10.8	12.1	12.6	8.9	11.2
Glycine	8.4	8.2	8.7	8.2	15.4	11.6	8.6	7.5	16.8	9.4
Histidine	0.9	0.8	0.5	1.1	1.0	1.2	0.7	0.7	1.6	1.7
Isoleucine	3.1	2.7	2.9	3.9	2.5	3.3	3.2	3.3	3.5	5.6
Leucine	7.7	6.1	4.6	9.3	5.5	6.7	8.4	7.3	7.9	8.7
Lysine	3.0	2.7	2.1	4.2	4.8	7.2	2.8	2.3	3.9	6.5
Methionine	0.5	0.3	0.2	0.8	–	–	0.4	0.4	0.9	1.4
Phenylalanine	2.9	1.7	1.2	3.9	1.9	2.3	2.7	2.2	4.4	3.0
Proline	6.6	10.5	12.3	8.9	5.8	6.9	6.3	7.0	3.3	4.8
Serine	10.4	14.3	11.9	10.7	13.7	10.1	10.2	10.5	10.8	7.1
Threonine	6.4	4.4	3.9	5.5	3.6	5.1	6.1	7.0	4.9	4.3
Tyrosine	3.9	2.8	2.0	3.6	2.1	0.6	3.4	2.4	7.4	3.1
Valine	5.6	7.5	8.2	7.5	5.7	6.3	5.7	5.7	5.1	6.6

[a] Mean values from Table 1.1.
[b] Material extracted by formic acid at 20 °C (believed to originate from cell membrane complex).
[c] Includes asparagine residues.
[d] Includes the hydrolysis byproduct ornithine.
[e] Includes the oxidation byproduct cysteic acid.
[f] Includes glutamine residues.

to assist in expelling foreign matter from the fleece [23]. The DFE is also responsible for wool's unique property among textile fibres: the ability to felt [41,45,80].

Felting occurs when individual fibres, either within a loose mass or in a yarn or fabric, move preferentially in one direction. Such movement occurs readily when the fibre assembly is agitated in water. The term 'felting' is used to describe this behaviour when its effect is undesirable, such as in the laundering of knitted garments. Felting is also carried out by the textile industry to produce fabrics in which the structure has been consolidated or closed up. This process of controlled felting, which is called 'milling' or 'fulling', has been described in detail elsewhere [82].

The amount of each cuticle cell visible on the wool surface varies with fibre diameter; for fine wools, the amount of scale overlap is approximately 15% [15,83]. Except where two cells overlap, the cuticle of Merino wool fibres is only one cell thick. The cuticle of coarse keratin fibres, however, consists of up to 15 layers of cells [22]. Shoulders or 'false' scale edges also occur on 25% of the cuticle cells on Merino fibres. False scale edges, which are believed to be the imprint of the serrated inner root sheath of the hair follicle on the fibre, are formed prior to keratinisation [84].

Cuticle cells can be separated from the cortex by ultrasonic disruption [13,85] (Figure 1.8); by shaking in either formic acid [86] or aqueous sodium dodecylsulphate [87]; or by refluxing in 98% formic acid [88].

The substructure of the cuticle has been studied extensively by many workers, using a wide range of physical and chemical methods [22,23]. The cuticle has a higher cystine content than has whole wool [13] (Table 1.5) and contains certain cuticle-specific proteins [89]. Cuticle cells are also rich in cysteic acid, serine, proline, glycine and valine. They are poorer than whole wool in aspartic acid, threonine, glutamic acid, methionine, isoleucine, leucine, tyrosine, phenylalanine and arginine. The former group of amino acids is considered to be generally non-helix-forming, whereas the latter group favours formation of an α-helical structure. Thus, it has been concluded that the cuticle has a more amorphous structure than the rest of the fibre [13]. The cuticle is much less extensible than the cortex,

Figure 1.8 *Light micrograph (phase contrast) of cuticle cells produced from wool by an ultrasonic technique [85]. Reproduced with permission of CSIRO Publishing.*

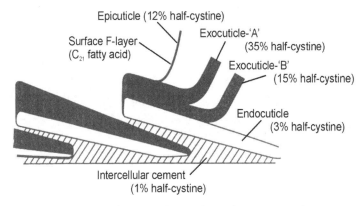

Epicuticle (12% half-cystine)

Surface F-layer
(C$_{21}$ fatty acid)

Exocuticle-'A'
(35% half-cystine)

Exocuticle-'B'
(15% half-cystine)

Endocuticle
(3% half-cystine)

Intercellular cement
(1% half-cystine)

Figure 1.9 *Schematic diagram of a wool cuticle (courtesy of CSIRO) [59].*

presumably because of the higher level of cystine (and hence higher crosslink density). The lower extensibility is shown by cracking of the cuticle cells when wool fibres are stretched [90]. Unlike human hair, however, extension does not lead to detachment of cuticle cells from the cortex [91].

The epicuticle (mentioned in Section 1.4) is the thin membrane covering the surface of the cuticle [92]. It is difficult to detect by microscopy, because of the poor contrast between it and the embedding medium [22,25]. It has been observed on hair by evaporating a thin coating of metal on to the fibre surface, followed by post-staining and examination of a thin section under the TEM [93].

The major part of wool cuticle cells is composed of two distinct major layers [22], identified by heavy-metal staining techniques in conjunction with the electron microscope (see Section 1.4). These layers, namely the outer *exocuticle* and inner *endocuticle* – shown schematically in Figure 1.9 – differ mainly in their cystine content [22,25,67,92]. Staining techniques have also shown that the exocuticle contains two poorly defined subcomponents (A-layer and B-layer), which also differ in cystine content.

1.4.1.1   The Epicuticle and the Allwörden Reaction

The epicuticle is defined as the membrane that is raised as bubbles or sacs along the fibre following immersion in chlorine water [22,53,94]. This phenomenon is called the Allwörden reaction, after its discoverer [95] (Figure 1.10).

The proteinaceous epicuticle membrane is approximately 2–7 nm thick and accounts for around 0.1% of the mass of the fibre [15,22,23,93,96]. The epicuticle is believed to be derived from the plasma membranes of the outer layer of a cuticle cell [11,97].

The Allwörden bubbles produced by treatment of wool with chlorine water occur as a result of reaction with the proteins beneath the chlorine-resistant epicuticle membrane on the surface of the cuticle cells (i.e. in the A-layer). The reactions involve oxidation of the disulphide bonds of cystine [98] and cleavage of peptide bonds at tyrosine residues [31,41,80,99]. These two reactions produce peptide fragments that are water soluble, because they contain anionic sulphonic (cysteic) acid residues. The fragments, which are too large to diffuse

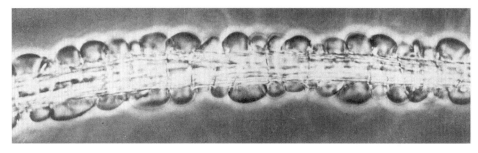

Figure 1.10   *Formation of Allwörden bubbles on wool [94]. Reproduced with permission of Macmillan Publishers Ltd.*

through the semipermeable epicuticle membrane, permit the absorption of large amounts of water, which results in the generation of an osmotic pressure, stretching the epicuticle membrane outwards [97,99]. This mechanism is supported by the finding that the bubbles collapse when exposed to a concentrated salt solution [100]. Chemical modification of the disulphide bonds interferes with the formation of Allwörden bubbles, presumably because chlorine water is incapable of oxidising the modified bonds to sulphonic acid residues [99]. In order to generate sufficient sulphonic acid groups to produce Allwörden bubbles, a high concentration of cystine must be present beneath the epicuticle [99]. As shown in Figure 1.9, the A-layer of the exocuticle contains approximately 35% half-cystine, which is the highest level in the fibre. This concentration, which represents 1 in every 2.7 amino acid residues in the form of half-cystine, is sufficient to produce the concentration of osmotically active oxidation products necessary to swell the resistant membrane [53].

Sacs, similar to those produced by chlorine water, are also produced on wool fibres by immersion in bromine water [101]. The bubbles are, however, of a different nature from those produced by chlorine water, and the surrounding membrane is thicker and appears to include material from layers of the cuticle beneath the epicuticle [102]. This is likely to be because bromine is less reactive than chlorine and, therefore, would be expected to diffuse further into the cuticle before complete reaction occurs.

The epicuticle was originally treated as a unique component of the wool fibre, but it is now considered to be part of the resistant membrane system that surrounds all cuticle and cortical cells [22,53,97] (see Section 1.4). The chemical structure of the epicuticle is discussed, along with those of the other resistant membranes in wool, in Section 1.4.3.

The relationship between the epicuticle and the fibre surface has been the subject of considerable debate. Lindberg *et al.* [92] suggested that the epicuticle is a continuous membrane which surrounds every fibre, like a sausage skin. Some workers, however, believed it to be discontinuous and to encapsulate each separate cuticle cell [103]. This alternative view was disputed, mainly on the grounds that the Allwörden bubbles often cover several scale edges [104,105]. The question remained unresolved until it was demonstrated by Leeder and Bradbury that Allwörden bubbles can be produced on isolated cuticle cells [94,97]. These authors reasoned that the epicuticle must surround each cuticle cell, otherwise the osmotically active, soluble proteins would escape from the edges of isolated cells and Allwörden bubbles would not form [99]. The absence of sacs on the underside of single cuticle cells was explained in terms of the relatively low level of cystine in the adjacent endocuticle (i.e. approximately 3% half-cystine) (Table 1.5 and Figure 1.9). This concentration was presumed to be too low to produce the increase in osmotic pressure required to raise the membrane. The formation of Allwörden bubbles that appear to cover more than one cuticle cell can be explained by the existence of the false scale edges, as previously discussed [84].

1.4.1.2 *The Epicuticle and the Hydrophobic Surface of Wool*

As shown schematically in Figures 1.7 and 1.9, the surface of wool fibres is made up of the epicuticle plus a very small component (approximately 0.05%) consisting of the region between the cuticle cells that extends to the fibre surface [106,107]. Wool fibres from which the surface grease has been removed are hydrophobic. This property, which is difficult to explain if the epicuticle is composed solely of protein, has attracted considerable interest over many years. The wettability of wool is increased dramatically by treatment for a few seconds with an alcoholic solution of potassium hydroxide [92]. Lindberg pointed out the apparent paradox presented by this observation and the marked resistance of the epicuticle to dissolution in alkaline reagents [78]. He proposed that alkali removes a layer of hydrophobic material only a few molecules thick. A hydrophobic fatty acid layer, bound to the surface of wool, has also been proposed by other workers [108,109].

Support for the idea of a waxy component of the epicuticle was provided by King and Bradbury [96], who found lipid material to be associated with the epicuticle. Treatments used to modify the epicuticle on intact fibres, such as the alcoholic potassium hydroxide procedure already discussed, are very degradative, and reaction is not confined to the fibre surface. A technique has, however, been developed which enables the surface of wool fibres to be treated with an alkaline reagent (potassium tert-butoxide dissolved in tert-butanol) under conditions where damage or modification of the fibre interior cannot occur [79,110]. This treatment was carried out on predried wool under strictly anhydrous conditions, which prevents penetration of alkali beyond the fibre surface. It produced changes in a range of properties governed by the fibre surface: in particular a dramatic increase in wettability, an increase in wet and dry friction and improved adhesion of polymers [79]. The increased dry friction led to a substantial harshening of handle and the increased wet friction produced a reduction in felting shrinkage. Despite the changes in properties, there was no visible modification of the fibre in the SEM and there was no measurable change in the mass of the treated wool. Allwörden bubbles could still be raised, showing that the epicuticle was still intact [110] and that the reaction was confined to the fibre surface. The effect of the treatment has been explained in terms of the removal of a very thin fatty layer to expose the 'clean' protein surface of the epicuticle, because lubricants such as cationic softeners reduce the dry friction and improve the handle [79,107]. Leeder and Rippon named this lipid component the 'F-layer', as they considered it to be separate to the proteinaceous epicuticle [79]. Furthermore, in view of the difficulty in effecting its removal, they also suggested that the F-layer is chemically bound to the epicuticle. Analysis of liquors obtained following treatment of wool with anhydrous potassium tert-butoxide in tert-butanol confirmed the presence of fatty acids (approximately 0.025% on mass fabric (o.m.f.)) [111]. The major component was found to be an unusual C_{21} fatty acid, containing a branched chain. Evans *et al.* [111] agreed with the earlier suggestion [79] that the acid was covalently bound to the epicuticle via an ester or thioester bond. It was confirmed later that the C_{21} fatty acid is mainly located in the cuticle [112].

This fatty acid has subsequently been found in human hair [113] and in the hair of other mammals [114]. Wertz and Downing confirmed its structure as 18-methyl-eicosanoic acid (**4**) [113]. They agreed with Evans *et al.* [111] that the most likely form of attachment is via ester or thioester linkages. Subsequently, it was shown that around 60% of the bound fatty acids are released by reaction with aqueous chlorine below pH 3 [115] and by hydroxylamine [116]. These results support the proposal that the fatty acid is mainly bound to wool by a thioester, rather than an ester bond, since aqueous chlorine and hydroxylamine cleave thioester bonds more rapidly than oxygen ester bonds [116,117].

$$CH_3CH_2CH(CH_2)_{16}COOH$$
$$|$$
$$CH_3$$

4

The thickness of the lipid layer on the fibre surface has been estimated by x-ray photoelectron spectroscopy (XPS) to be 0.9 nm [118], which is around 40–50% of the value calculated from the dimensions of a C_{21} fatty acid monolayer [119]. This difference could be due to partial hydrolysis of the thioester during sample preparation, or to an artefact of the anhydrous, high-vacuum conditions used in XPS.

1.4.1.3   The Exocuticle

The exocuticle is the layer of keratinous protein immediately below the epicuticle (Figure 1.9). In Merino wool, the exocuticle, which is approximately 0.3 μm thick, represents around 60% of the total cuticle cell [67] and may extend partly around the edge of the scale [120]. The major part of the cystine content of the cuticle is believed to be in the exocuticle [11,61,62,67,121] (Table 1.5). Chemical analysis has shown that the exocuticle contains one crosslink for every five amino acid residues, which is double the crosslink density for whole wool [67]. As already mentioned, a subcomponent (the A-layer) has also been identified at the surface of the exocuticle [61,105,121,122]; this upper layer is not well defined, but may account for around 30–50% of the total thickness of the exocuticle [22]. The dense A-layer is believed to have a higher cystine content than the underlying B-layer [11,22,61,62,67,121] (Figure 1.9) and its importance in the Allwörden reaction has already been discussed.

1.4.1.4   The Endocuticle

The endocuticle is a well-defined layer that lies below the exocuticle [22,92] (Figure 1.9). It is bounded on the underside by the cell membrane complex, which separates it from other cuticle cells and from the cells of the cortex (Figure 1.7). The endocuticle of Merino wool is around 0.2 μm thick and constitutes approximately 40% of the whole cuticle [22,67]. It is believed to be derived from material left over from the developing cell [61]. Bradbury [22] has noted that, in this respect, it is comparable with the intermacrofibrillar material (see Section 1.4.2), which is residual matter remaining after formation of the cortical cells.

The endocuticle has a relatively low crosslink density, with only one amino acid residue in every 33 taking the form of half-cystine [67] (Table 1.5). It is therefore classified in Table 1.4 as one of the nonkeratinous components of the fibre [19].

A consequence of the low concentration of disulphide crosslinks, together with a total concentration of acidic, basic and polar amino acids similar to that of whole wool (Table 1.5), is that the endocuticle is readily swollen by polar liquids. By using Zahn's 'swelling factor' calculations [17,19], it has been estimated that the endocuticle has a swelling capacity greater than that of whole wool but less than that of the intercellular cement [53,123]. The low cystine content also makes the endocuticle more susceptible than the exocuticle to chemical attack, for example from acids [68] or proteolytic enzymes [66,67,93]. Specific chemical attack on the endocuticle has been used industrially to remove the scales from wool [124,125]. The endocuticle is mechanically a relatively weak region of the fibre and preferential fracture often occurs along this component during carpet wear [126].

1.4.2   The Cortex

The cortex constitutes almost 90% of keratin fibres (Table 1.4) and is largely responsible for their mechanical behaviour. The extremely complex structure of the cortex of fine wool is illustrated by the transmission electron micrograph shown in Figure 1.11 and by the schematic diagram in Figure 1.6.

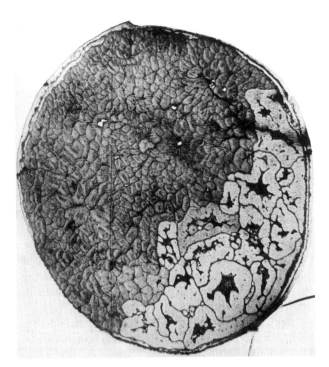

Figure 1.11   *Transmission electron micrograph of a 21 μm Merino wool fibre (courtesy of CSIRO).*

Figure 1.12   *Light micrograph (phase contrast) of cortical cells produced from wool by an ultrasonic technique [85]. Reproduced with permission of CSIRO Publishing.*

The cortex of fine wool consists of closely packed, overlapping cortical cells, arranged parallel to the fibre axis. Cortical cells are approximately 100 μm long and 3–6 μm wide [127,128]. As mentioned before, each cell is surrounded by a cell membrane complex, which is a continuous phase that extends throughout the whole fibre (Figure 1.7).

Cortical cells can be liberated for analysis by treatment with enzymes [66], hydrochloric acid [129], formic acid [86,123] or ultrasonic disruption [13,85], or by techniques involving sequential treatments [130] (Figure 1.12). After liberation from the fibre, a fluorescence-activated cell sorter can be used to separate the different types of cell [131].

Fine wool fibres contain two main types of cortical cell: orthocortical and paracortical [22,25,74,132]. A third type, mesocortical, is sometimes present at the boundary between the orthocortex and the paracortex [133,134]. Mesocortical cells have some of the characteristics of the main cell types [134,135]. Where present, the mesocortex usually accounts for no more than 4% of the fibre [22].

1.4.2.1   Intermediate Filament/Matrix Structure

The cells of the cortex are composed of rod-like elements of crystalline proteins, surrounded by a relatively amorphous matrix [11,23,27,61,74,136]. In the older literature, the rod-like elements were called *microfibrils*, but in accordance with the terminology used for other proteins, they are now called *intermediate filaments* [27,137]. They are approximately 7 nm in diameter [21–23,74]. Their length is known with less certainty, but is believed to be at least 1 μm [22]. When cross-sections of the cortex of keratin fibres are examined in the TEM – following reduction and staining with heavy metals – the intermediate filaments can be seen as lightly stained circular areas set in a more heavily stained surrounding region (the matrix) [11,61,136] (Figure 1.13). The appearance of the intermediate filaments suggests a ring and core structure [11], whereas the matrix is featureless [27].

The three groups of proteins shown in Table 1.2 are all present in the cortex of wool fibres [23,24]. The high-sulphur and high-glycine/high-tyrosine proteins are concentrated in the matrix [138], while the intermediate filaments are relatively rich in low-sulphur proteins [139]. The latter proteins are rich in the amino acids that favour α-helix formation, namely lysine, aspartic and glutamic acids and leucine [27] (Table 1.2). Each intermediate filament

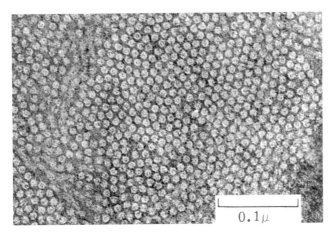

Figure 1.13 *Transmission electron micrograph showing the ring/core structure of the intermediate filaments and matrix of cortical cells. (courtesy of CSIRO).*

consists of a rod-like central domain composed of four lengths of α-helix separated by three segments of nonhelical material (Figure 1.14). The α-helical sections, which are of different lengths, show a heptad repeat and take the form of a two-chain coiled coil [140] (Figure 1.6). The ends of the polypeptide chains consist of nonhelical domains terminated in a carboxyl or *N*-acetyl group [27,141,142].

1.4.2.2 *Macrofibrils, Nuclear Remnants and Intermacrofibrillar Material*

Electron microscopy shows that within a cortical cell the intermediate filaments are grouped together in aggregates, known as macrofibrils [61,136] (Figure 1.6). These are cylindrical units, around 0.3 μm in diameter [68,70], which range in length from 10 μm [70] to the length of an entire cortical cell [25]. Each macrofibril has been estimated to contain an average of 19 intermediate filaments [143]. Macrofibrils of greater diameter are believed to result from fusion of smaller macrofibrils.

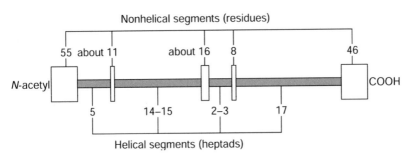

Figure 1.14 *Schematic representation of the structure of an intermediate filament (courtesy of CSIRO).*

The cells of the cortex contain around 13% of nonkeratinous proteins (Table 1.4). These consist of nuclear remnants and intermacrofibrillar material, and are derived from the nucleus and cytoplasm of the once-living cells. The composition of the nonkeratinous material in the cortical cells is believed to be similar in many respects to the endocuticular material described in Section 1.4.1 [22,25] (Table 1.5).

Orthocortical and paracortical cells are identified by the manner in which the nonkeratinous material is distributed within them [61,134–136]. Paracortical cells are generally more clearly outlined than those of the orthocortex, with the nonkeratinous material concentrated in prominent regions of variable size, called nuclear remnants (Figure 1.11); these are also present in mesocortical cells. The macrofibrils in para- and mesocortical cells are not well defined and have a fused appearance.

Nuclear remnants are less apparent in the cells of the orthocortex, because the nonkeratinous material is distributed between the macrofibrils, rather than being concentrated in specific regions. The network of intermacrofibrillar material in the orthocortex clearly delineates the macrofibrils but reduces the definition of the boundaries of the orthocortical cells compared with those of the paracortex.

Orthocortical and paracortical cells also differ in the composition and arrangement of the intermediate filament/matrix system within each macrofibril. In the well-defined macrofibrils of the orthocortex, the intermediate filaments are poorly resolved and are grouped together in a whorl [74,136] or 'fingerprint' [135] pattern. This arrangement is believed to arise from twisting of the intermediate filaments around a central core [74]. The macrofibrils of the mesocortex contain intermediate filaments packed in a hexagonal pattern, whereas the arrangement of the well-defined intermediate filaments in the paracortex is largely random, with an occasional hexagonal pattern [135].

The relative proportion of intermediate filaments and matrix differs between the three types of cell. Paracortical cells contain a higher proportion of matrix [135], and hence a greater proportion of high-sulphur proteins [68,130,144], than do orthocortical cells. The cells of the orthocortex, however, contain a higher proportion of intermediate filaments and are therefore richer in the low-sulphur proteins that favour α-helix formation [130] (Tables 1.2 and 1.5). Furthermore, it has been shown that the proteins of the intermediate filaments are similar in both the ortho- and the paracortex [145]. There appears to be some differences between the high-sulphur matrix proteins in the two types of cell, with the proteins with the highest sulphur content concentrated in the paracortex.

1.4.2.3   Ortho/Para Segmentation of the Cortex

The relative proportions and arrangements of the different types of cortical cell in wool vary with fibre diameter [74] and often within a fibre [106]. In general, for fine Merino wools, the orthocortex usually accounts for over 50% of the fibre cross-section (Figure 1.11). In many wools, the cortex is transversely segmented [74,132,146]. Bilateral segmentation of ortho- and paracortical cells predominates in wool fibres of diameters up to 25 μm. Less distinct segmentation occurs in wool of 25–35 μm diameter and the distribution of cell types is very variable in fibres thicker than 35 μm [74,146]. Some coarse wools, such as Lincoln, have cylindrical asymmetry, usually with a central core of orthocortex surrounded by an annulus of paracortical cells [22]. The bilateral segmentation of fine wools is associated with the highly desirable natural crimp of the fibres [132,135,146]. In these wools, the orthocortex

Orthocortex

Paracortex

Figure 1.15 *Relationship between ortho/para segmentation and fibre crimp (courtesy of CSIRO).*

is always orientated towards the outside radius of the crimp curl. In order to achieve this, the two segments of the cortex twist around the fibre in phase with the crimp (Figure 1.15).

As discussed earlier, the orthocortical and paracortical cells differ in the manner in which the nonkeratinous material is distributed between the macrofibrils. There is also a difference in the amount of crosslinked matrix between the intermediate filaments in each cell. The orthocortex contains a more extensive network of easily swollen intermacrofibrillar material and a smaller amount of matrix between the intermediate filaments than does the paracortex. These differences make the orthocortex generally more accessible to reagents and more chemically reactive than the paracortex [22].

The two cortices were first identified as a result of their differential staining with dyes. Basic dyes [132], cationic surfactants [147] and many high-r.m.m. ions containing heavy metals [65,134] preferentially stain the more accessible orthocortex. Dyes and chemicals reach the cortical cells by diffusing along the network of the cell membrane complex that extends throughout the whole fibre (see Chapter 2). It has been suggested that the bilateral staining of Merino fibres with basic dyes is due to differences in the structure of the nonkeratinous proteins of the cell membrane complex between the orthocortex and the paracortex [148].

The situation for acid dyes is less clear [22], and there was some dispute among early workers as to whether or not these dyes show preferential distribution between the two cortices [132,149]. Examples of both orthocortical [149,150] and paracortical [151] preference have been found. It seems, however, that most acid dyes show little or no preference for either cortex – nor, it is interesting to note, do anionic surfactants [152].

The dissimilarity in distribution of nonkeratinous material and in the intermediate filament/matrix structure gives rise to other differences in the properties of the two segments. The extensive network of readily swollen, nonkeratinous intermacrofibrillar material in the orthocortex makes this segment more wettable [22] and more susceptible to acid hydrolysis [68,153] and extraction by enzymes [18,19,146] than is the paracortex. The lower crosslink density of the orthocortical matrix leads to higher rates of stress relaxation [154] and setting [155] in the orthocortex. Differential stress relaxation in the ortho- and paracortices has been utilised to generate additional crimp in wool fibres [156,157].

1.4.3 The Cell Membrane Complex

As already discussed, the cuticle and cortical cells in wool fibres are separated by the cell membrane complex. This continuous network, which for Merino wool is around

25 nm wide [22], is visible in the light microscope [158]. It provides adhesion between the cells [75] and can be partly dissolved or disrupted by enzymes [66,75,159] or formic acid [53,57,76,86,123]. These treatments eventually lead to separation of the fibre into its constituent cortical and cuticle cells.

The cell membrane complex is believed to originate in the hair follicle, from the two plasma membranes of adjacent living cells [53,136,160]. During keratinisation or hardening of the growing fibre, the membranes around the cells consolidate and the intercellular cement is laid down, providing the adhesive layer between the cells [11,136,160]. Examination under the TEM of fibre sections that have been pretreated with a reducing agent and then stained with a heavy metal salt reveals the structure for the cell membrane complex shown in Figure 1.16.

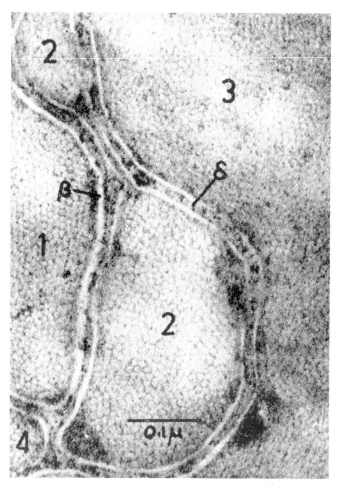

Figure 1.16 *Transmission electron micrograph of a Lincoln wool fibre, showing the cell membrane complex between four cortical cells. Adapted from G E Rogers, Electron Microscopy of Wool, J. Ultrastruc. Res., **2**:3, 309–330 (1959) with permission of Elsevier.*

A densely stained central region (the δ-layer) is sandwiched between two lightly stained segments (the β-layers) [61,136]. Each β-layer is bounded by an inert, chemically resistant membrane. The δ-layer, which is often referred to as 'intercellular cement', is of variable thickness. In some places it is undetectable, with the outer membranes being very close together [25,53], while elsewhere it is 15 nm wide [161]. The thickness of the inert, nonstaining β-layers probably lies in the range 2.5–5.0 nm [25,162].

1.4.3.1   Intercellular Cement

Although its exact composition is not known, the intercellular cement of the δ-layer is believed to consist mainly of lightly crosslinked proteins, plus some lipid material [14,18,19,22,53,163]. The low level of crosslinking of the intercellular cement makes it easily swollen by many reagents. One of these is formic acid, which has been widely used to remove material from this region for analysis [14,18,19,22,53]. It is likely that this reagent removes mainly the most labile material (i.e. that of lowest crosslink density and r.m.m.), with the more resistant components remaining in the cell membrane complex [53]. Analytical data for the composition of the intercellular cement show that the levels of glycine and the aromatic amino acids tyrosine and phenylalanine are higher than in whole wool [14,163,164] (Table 1.5). The concentration of cystine is very low, however, and this qualifies the intercellular cement for classification in Table 1.4 as a nonkeratinous component of wool [17–19,53,123].

1.4.3.2   Lipid Component of the Cell Membrane Complex

The lightly stained β-layers, seen in transmission electron micrographs of stained sections of wool fibres (Figure 1.16), are generally believed to arise from the hydrophobic ends of a lipid bilayer [19,22,93,136]. The composition of lipid extracts isolated from wool was discussed in Section 1.3.5. The β-layers were believed by Rogers [136] to constitute regions of relative weakness in the cell membrane complex, because he observed splitting along these planes during the preparation of fibre cross-sections. This suggestion is consistent with the model of a bimolecular lipid leaflet sandwiching the intercellular cement. This model has been questioned, however, on the grounds that extraction with lipid solvents does not markedly alter the appearance of the β-layers under the TEM [57]. It has been suggested [53,57] that a factor in the appearance of the β-layers may be poor uptake of the histochemical stain by the chemically resistant membranes surrounding each cell. This may mask the ultrastructure of the cell membrane complex. Further work is therefore required before the location of the lipids in the cell membrane complex can be known with certainty. The use of energy-filtered TEM has been shown to give improved definition of fine detail in the cell membrane complex [165]. This technique, in conjunction with modified staining methods, may prove to be useful in determining the ultrafine structure of the cell membrane complex [15].

1.4.3.3   Resistant Membranes

The resistant membrane represents the boundary between a cortical or cuticle cell and the remainder of the cell membrane complex. As previously discussed, the membranes are considered to be an integral part of the cell membrane complex [22,53]. The membranes

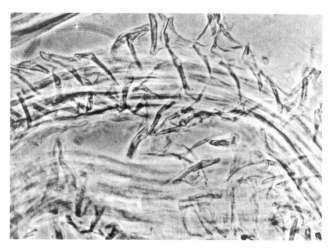

Figure 1.17   *Light micrograph (phase contrast) showing the resistant membranes of wool [77]. Reproduced with permission of SAGE UK.*

are relatively chemically inert and are the last part of the fibre to dissolve when wool is treated with reagents such as acids, alkalis, proteolytic enzymes and oxidising or reducing agents [53,76,77,166,167] (Figure 1.17).

Resistant membranes from cuticle and cortical cells have very similar amino acid compositions, except that those from cuticle cells contain a small amount of citrulline (Table 1.5) [76,86]. Both cuticle and cortical membranes contain approximately the same proportion of cystine crosslinks as whole wool [22,53,76,166], which makes their high chemical inertness difficult to explain. The concentration of lysine in the membranes is, however, approximately two to three times that in whole fibres (Table 1.5) [53]. This observation has led to the suggestion that isopeptide crosslinks, formed between lysine and glutamic or aspartic acid side chains, contribute to the chemical stability of wool membranes (i.e. the N^ε (γ-glutamyl)-lysine and N^ε (β-aspartyl)-lysine crosslinks shown in Figure 1.4 [19,76,168]. The possibility that isopeptide bonds are solely responsible for the high resistance of the membranes to chemical attack has been questioned on the grounds that there is no justification for supposing that this type of peptide linkage should be more stable than the other peptide bonds in wool [53]. Furthermore, because it is known that isopeptide bonds are broken during acid hydrolysis (see Section 1.3.3), there is no apparent reason why this type of bond should be more resistant to chemical attack when located in the membranes than when located in the rest of the fibre.

1.4.3.4   Amount of Cell Membrane Complex in Wool

The total amount of material in the cell membrane complex is not known with certainty. Bradbury quotes a value of 3.3% of the total fibre mass (Table 1.4), whereas Leeder [53] favours a higher value of around 6% o.m.f. The chemical inertness of the resistant membranes has allowed the concentration of this component to be estimated with a reasonable degree of accuracy (1.5% o.m.f.). The concentrations of intercellular cement and lipid material are not exactly known, because the amounts of these components extracted from the fibre depend on the procedure used [53]. It has been suggested that the values in Table 1.4 are too low, and 3.0 and 1.5% o.m.f. have been proposed for the concentrations of lipid and intercellular cement, respectively [53,57].

1.4.3.5   Differences Between Cell Membrane Complexes in Cuticle and Cortex

When observed in the electron microscope, the structure of the cell membrane complex between cuticle and cortical cells appears to be different from that of the cell membrane complex between two cuticle or two cortical cells [136]. The cell membrane complex of the cuticle has been found to be more resistant to modification by formic acid than that of the cortex [76]. Other workers [169,170] have obtained evidence that the intercellular cement has different chemical compositions in the cuticle/cuticle, cuticle/cortical and cortical/cortical intercellular regions. These differences are reflected in a difference in the chemical reactivity of the cell membrane complex between the two cell types, particularly in the ease of separation of cuticle and cortical cells [171,172].

1.4.3.6   The Cell Membrane Complex and Fabric Properties

Although the cell membrane complex accounts for only a small proportion of the total mass of wool (Table 1.4), it has been the subject of a great deal of recent research because it is known to have a large influence on the mechanical and chemical properties of the fibre [17–19,53,59].

When wool worsted fabrics are abraded during wear, breakdown of fibres occurs through fibrillation [53,173]. A similar pattern of fibre fracture is seen in fibres taken from fabrics abraded on the Martindale abrasion tester [173] (Figure 1.18). It appears that application of torsional stress, such as occurs during the abrasion of a fabric in wear, causes fracture to occur mainly along the boundaries between cortical cells and to a lesser extent along the

Figure 1.18   Scanning electron micrograph showing fibre fibrillation obtained in wear or Martindale abrasion testing (courtesy of CSIRO).

intermacrofibrillar regions [126,173,174]. Thus it is now accepted that the cell membrane complex is a region of relatively low mechanical strength in the overall fibre composite [53,59]. As discussed in the previous section, splitting along intercellular boundaries was first noticed by Rogers [136], who identified the fracture planes with the β-layers.

The abrasion resistance of worsted fabrics is decreased by many treatments that modify the cell membrane complex, in particular prolonged dyeing at low pH and chemical finishing operations [18,19,175–177] (see Section 1.6), but is actually increased by extraction with certain organic solvents. Polar solvents, in particular lower alcohols [173,175] and formic acid [175,178], are effective at room temperature. The nonpolar solvent, perchloroethylene, also improves the abrasion resistance of wool at normal regain [59,177,179], but prolonged extraction at a higher temperature is required in order to achieve an improvement similar to that obtained at room temperature with polar solvents. Milder treatment conditions can be used if a small amount of a fibre-swelling solvent, such as methanol, is added to the perchloroethylene [59,180]. The cell membrane complex appears to be modified by some solvents [57], and it is thought that removal of material from this region is responsible for the improved abrasion resistance of solvent-treated wool [173,175,177]. Rippon and Leeder found a correlation between the amount of internal lipid material extracted from a wool fabric by perchloroethylene and the improvement in abrasion resistance [177]. The mechanism by which extraction improves resistance to abrasion is not known, but increased intercellular adhesion following lipid removal may be involved [178].

Fabric abrasion resistance can also be improved by treatment with low levels of crosslinking agents, such as formaldehyde [59,176]. It appears that increasing the crosslink density of the intercellular regions improves the resistance of wool fibres to torsional fatigue. High levels of crosslinking produce a decrease in abrasion resistance, however, presumably because of embrittlement of the fibre [59].

1.5   Chemical Reactivity of Wool

Wool, in common with many other proteins, will react with a large range of chemicals. Wool contains three main types of reactive group: peptide bonds, the side chains of amino acid residues and disulphide crosslinks. The chemical reactions involving these groups have been studied extensively and discussed in various textbooks and reviews [4,31,40,45,181–186].

The highly reactive nature of wool has enabled many industrial treatments to be developed, particularly in the areas of shrinkproofing [4,41,80,181,186,187], dyeing [4,181,187–192], bleaching [4,181,187,193–195], flame-resistance treatment [196,197] and finishing [190,191,198–204].

1.6   Damage in Wool Dyeing

Conventional methods used in wool dyeing involve prolonged periods at or near the boil; this is necessary in order to obtain good levelling and penetration into the fibre. Depending on the dyes and equipment used, wool dyeing is carried out within the pH range 2–7. Under these conditions, wool proteins can be modified in several ways [18,46,187,205]. This modification, or 'damage', often results in unacceptable levels of yellowing [206], reduced

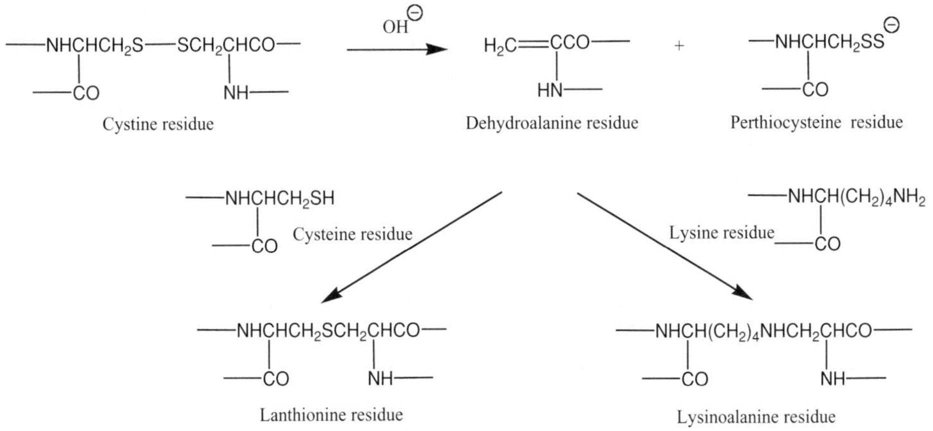

Scheme 1.5 *Formation of lanthionine and lysinoalanine from cystine.*

productivity and yields in processing [207] and an impairment of end-product performance, such as abrasion resistance [59] (see Section 1.4.3).

Damage to wool in hot aqueous, acidic dye liquors occurs mainly through hydrolysis of peptide bonds, particularly at aspartic acid residues [18,21,46,185,205,208] and at the amide group of tryptophan [45]. Chemical attack on wool in alkaline solution is less selective and more rapid than that under acidic conditions. Peptide bonds are broken, but other linkages, notably cystine, are also progressively hydrolysed with increasing liquor pH [18,21,45,46,184–186,205,208]. Under alkaline conditions, cystine can also undergo a β-elimination reaction to produce lanthionine and lysinoalanine crosslinks [40]. These are generated via a dehydroalanine intermediate, as shown in Scheme 1.5. Hydrogen sulphide, produced by hydrolysis of perthiocysteine residues, is believed to catalyse these reactions, leading to a rapid increase in the rate of fibre degradation in boiling dyebaths [192,209,210]. Lanthionine formation is believed to lead to fibre embrittlement and a reduction in abrasion resistance [211].

Both lanthionine and lysinoalanine crosslinks are more stable than disulphide bonds and contribute to the stability of wool fabrics after permanent setting in finishing [40] and in imparting permanent set during dyeing [192].

1.6.1   Nonkeratinous Proteins and Damage in Dyeing

As discussed in Section 1.4.3, the nonkeratinous components of wool – particularly the cell membrane complex – are regions of relative chemical and physical weakness in the overall composite structure of the fibre. Preferential attack on these readily swollen regions is a major factor in the impaired physical performance of wool often found after dyeing [18,53,59,175–178]. When wool is dyed at the boil, soluble proteins, termed 'wool gelatins' [212], are extracted from the fibre. Wool gelatins have a low cystine content, and are therefore believed to originate from the nonkeratinous regions, in particular the cell membrane complex. The yield of wool gelatins is regarded as a measure of the extent of wool damage

[18], because at a given liquor pH, the mass material extracted from the fibre is proportional to the treatment time [213]. This is a clear indication of the penalty, in terms of fibre damage, incurred whenever lengthy dyeing cycles are employed. Although the amount of soluble protein extracted from wool during dyeing is relatively small compared with the total fibre mass, the effect on physical properties can be large. For example, extraction of 2% o.m.f. of wool gelatins resulted in a decrease in wet tensile strength of 25% [18].

1.6.2   Influence of Dyebath pH on Fibre Damage

The role of ionic interaction (salt linkages) in stabilising the structure of wool proteins was discussed in Section 1.3.3. It was shown in Scheme 1.4 that the number of these interactions in a fibre depends on the pH. The importance of dyebath pH on damage during dyeing has been recognised for many years [214,215]. Furthermore, it has been demonstrated that the level of damage is kept to a minimum when wool is dyed at a pH value within the isoelectric region of the fibre. Under these conditions, the concentration of salt linkages is at a maximum level, and hence their stabilising effect on the wool proteins is greatest [46].

Baumann and Möchel have measured the effect of liquor pH on the yield and composition of the soluble proteins extracted when wool is boiled for a fixed time [216]. In the absence of added electrolytes, the yield of wool gelatins appears to be independent of pH in the range pH 3–8. A dramatic increase in yield has been found in liquors below pH 3. In the presence of electrolyte, however, the amount of soluble protein extracted shows a minimum around pH 3.5–5.0, which coincides approximately with the isoelectric region of the fibre, and hence with the maximum concentration of salt linkages. In the absence of electrolyte, the Donnan effect causes the internal pH of wool to lag behind the pH of the external solution [21,45]. Neutral salts decrease the difference between the internal pH of the fibre and that of the external solution [45,217]. Thus, in an acid medium the effect of electrolyte is to decrease the internal pH, whereas in an alkaline solution the internal pH is raised. In both cases, the shift in pH away from the isoelectric region will result in an increase in damage and a concomitant increase in the yield of wool gelatins. In a comprehensive study, Peryman [218] measured the effect on various physical and chemical properties of treatment for 3 hours in boiling aqueous liquors over the pH range 1.5–9.0. This author found that the presence of sodium sulphate in the liquor had little effect on damage within the pH range 1.7–6.8, but above pH 6.8 the electrolyte caused a marked increase in damage. Peryman's results in alkaline liquors are consistent with the data of Baumann and Möchel [216] on the effect of electrolyte on extraction of wool gelatins. The difference in the results obtained under acid conditions by the two groups of workers is difficult to explain. As discussed earlier, however, alkaline damage is less selective than acid hydrolysis and includes extensive fission of disulphide bonds. It may be that the physical tests used by Peryman are more sensitive to disulphide bond fission than they are to the peptide hydrolysis that predominates under acid conditions.

Peryman concluded that minimum fibre damage occurs when wool is dyed at pH 3.0–3.5 [218], whereas Elöd and Reutter suggested that pH 4.5–5.0 gives optimum results [215]. Baumann, from his own studies [213,216,219] and those of other workers [215,218], deduced that the optimum pH for wool dyeing is in the range pH 3.5–4.0 [18]. Following these studies on the effect of pH on fibre damage, ranges of dyes are now marketed that are especially suitable for application at or around the isoelectric region of wool [46,181,187].

1.7 Conclusion

As discussed in this chapter, wool has the most complex chemical and physical structure of all the fibres used by the textile industry. The evolution of wool occurred over many thousands of years to produce a unique material whose insulating properties protect sheep from heat, cold and rain. Its complex structure makes it a very versatile fibre and it is used in a diverse range of products, the most important of which include clothing, carpets and upholstery. Compared with some other fibres, wool can be easily coloured, by either dyeing or printing, with several ranges of dyes, selected according to the end use of the product. These dyes and their methods of application are described in the following chapters.

References

[1] N Hyde, *National Geographic Society*, **173** (1988) 552.

[2] M L Ryder, *Scientific American*, **Jan.** (1987) 100.

[3] D C Teasdale, *Wool Testing and Marketing Handbook* (New South Wales, Australia: University of New South Wales, 1988).

[4] J R Christoe, R J Denning, D J Evans, M G Huson, L N Jones, P R Lamb, K R Millington, D G Phillips, A P Pierlot, J A Rippon and I M Russell, *Encyclopedia of Polymer Science and Technology* online edition http://www.mrw .interscience.wiley.com/epst/index.html (Chichester, UK: John Wiley & Sons Ltd, 2003), last accessed 1 February 2013.

[5] E V Truter, *Wool Wax* (London, UK: Cleaver-Hume Press, 1956).

[6] R G Stewart, *Wool Scouring and Allied Technology*, 2nd Edn (Christchurch, New Zealand: WRONZ, 1985).

[7] J R Christoe and B O Bateup, *Wool Science Review*, **63** (1987) 25.

[8] B V Harrowfield, *Wool Science Review*, **64** (1987) 44.

[9] T E Mozes, *Textile Progress*, **17**(3) (1988).

[10] M Feugelman, *Encyclopedia of Polymer Science and Engineering*, 2nd Edn, **Vol. 8** (New York, USA: John Wiley and Sons, 1997), 566.

[11] R D B Fraser, T P Macrae and G E Rogers, *Keratins: Their Composition, Structure and Biosynthesis* (Springfield, USA: C.C. Thomas, 1972).

[12] H P Lungren and W H Ward, in *Ultrastructure of Protein Fibers*, ed. R Borasky (New York, USA: Academic Press, 1963).

[13] J H Bradbury, G V Chapman and N L R King, *Australian Journal of Biological Sciences*, **18** (1965) 353.

[14] J D Leeder and R C Marshall, *Textile Research Journal*, **52** (1982) 245.

[15] L N Jones, D E Rivett and D J Tucker, in *Handbook of Fiber Chemistry*, ed. M Lewin and E M Pearce (New York, USA: Marcel Dekker, 1998), 355.

[16] A Y Bhoyro, J S Church, D G King, G J O'Loughlin, D G Phillips and J A Rippon, *Proceedings of the Textile Institute 81st World Conference, Melbourne, Australia* (CD ROM, 2001, ISBN 1870372433).

[17] H Zahn, *Lenzinger Berichte*, **4** (1977) 19.

[18] H Baumann, in *Fibrous Proteins: Scientific, Industrial and Medical Aspects*, **Vol. 1**, ed. D A D Parry and L K Creamer (London: Academic Press, 1979) 299.

[19] H Zahn, *Plenary Lecture, 6th International Wool Textile Research Conference, Pretoria*, **1** (1980).

[20] H Zahn and P Kusch, *Melliand Textilber*, English Edn, **10** (1981) 75.

[21] W G Crewther, R D B Fraser, F G Lennox and H Lindley, in *Advances in Protein Chemistry*, **Vol. 20**, ed. C B Anfinsen Jr, M L Anson, J T Edsall and F M Richards (New York, USA: Academic Press, 1965), 191.

[22] J H Bradbury, in *Advances in Protein Chemistry*, **Vol. 27**, ed. C B Anfinsen Jr, J T Edsall and F M Richards (New York, USA: Academic Press, 1973), 111.

[23] R D B Fraser, L N Jones, T P Macrae, E Suzuki and P A Tulloch, *Proceedings of the 6th International Wool Textile Research Conference, Pretoria*, **1** (1980) 1.

[24] H Lindley, in *Chemistry of Natural Protein Fibers*, ed. R S Asquith (New York, USA: Plenum Press, 1977), 147.

[25] J A Swift, in *Chemistry of Natural Protein Fibers*, ed. R S Asquith (New York, USA: Plenum Press, 1977), 81.

[26] D A D Parry, in *Fibrous Proteins: Scientific, Industrial and Medical Aspects*, **Vol. 1**, ed. D A D Parry and L K Creamer (London, UK: Academic Press, 1979), 393.

[27] J M Gillespie, in *Cellular and Molecular Biology of Intermediate Filaments*, ed. R D Goldman and P M Steinert (New York, USA: Plenum Press, 1990), 95.

[28] J C Fletcher and J H Buchanan, in *Chemistry of Natural Protein Fibers*, ed. R S Asquith (New York, USA: Plenum Press, 1977), 1.

[29] D A Ross, *Proceedings of the New Zealand Society of Animal Production*, **21** (1961) 153.

[30] I J O'Donnell and E O P Thompson, *Australian Journal of Biological Sciences*, **15** (1962) 740.

[31] J A Maclaren and B Milligan, *Wool Science: The Chemical Reactivity of the Wool Fibre* (New South Wales, Australia: Science Press, 1981).

[32] M Cole, J C Fletcher, K L Gardner and M C Corfield, *Applied Polymer Symposium*, **18** (1971) 147.

[33] L A Holt, B Milligan and C M Roxburgh, *Australian Journal of Biological Sciences*, **24** (1971) 509.

[34] R C Marshall and J M Gillespie, in *The Biology of Wool and Hair*, ed. G E Rogers, P J Reis, K K A Ward and R C Marshall (London, UK and New York, USA: Chapman and Hall, 1988), 117.

[35] B Rigby, M S Robinson and T W Mitchell, *Journal of the Textile Institute*, **73**(2) (1982) 94.

[36] J M Gillespie and R L Darskus, *Australian Journal of Biological Sciences*, **24** (1971) 1189.

[37] J M Gillespie, A Broad and P J Reis, *Biochemical Journal*, **112** (1969) 41.

[38] T Y Liu and Y H Chang, *Journal of Biological Chemistry*, **246** (1971) 2842.

[39] S J Leach, *Australian Journal of Chemistry*, **13** (1960) 547.

[40] K Ziegler, in *Chemistry of Natural Protein Fibers*, ed. R S Asquith (New York, USA: Plenum Press, 1977), 267.

[41] J A Rippon, in *Friction in Textile Materials*, ed. B S Gupta (Cambridge, UK: Woodhead, 2008), 253.

[42] R S Asquith, M S Otterburn, J H Buchanan, M Cole, J C Fletcher and K L Gardner, *Biochimica et Biophysica Acta*, **221** (1970) 342.

[43] R S Asquith and M S Otterburn, *Applied Polymer Symposium*, **18** (1971) 277.

[44] D Poland and H A Scheraga, in *Poly α-amino Acids*, ed. G D Fasman (New York, USA: Marcel Dekker, 1967).

[45] P Alexander and R F Hudson, in *Wool: Its Chemistry and Physics*, ed. C Earland (London, UK: Chapman and Hall, 1963).

[46] D M Lewis, *Review of Progress in Coloration and Related Topics*, **19** (1989) 49.

[47] G Nemethy and H A Scheraga, *Journal of Physical Chemistry*, **66** (1962) 1773.

[48] H Zahn and G Blankenburg, *Textile Research Journal*, **34** (1964) 176.

[49] P Alexander and C Earland, *Nature*, **166** (1950) 396.

[50] E O P Thompson and I J O'Donnell, *Australian Journal of Biological Sciences*, **12** (1959) 282.

[51] J A Maclaren, D J Kilpatrick and A Kirkpatrick, *Australian Journal of Biological Sciences*, **21** (1968) 805.

[52] J M Gillespie, in *Biochemistry and Physiology of Skin*, ed. L A Goldsmith (Oxford, UK: Oxford University Press, 1983), 475.

[53] J D Leeder, *Wool Science Review*, **63** (1986) 3.

[54] D E Rivett, *Wool Science Review*, **67** (1991) 1.

[55] C A Anderson and J D Leeder, *Textile Research Journal*, **35** (1965) 416.

[56] H E Crabtree, P Nicholls and E V Truter, *Proceedings of the Analytical Division of the Chemical Society*, **16** (1979) 235.

[57] J D Leeder, D G Bishop and L N Jones, *Textile Research Journal*, **53** (1983) 402.

[58] A Schwan, J Herrling and H Zahn, *Colloid Polymer Science*, **264** (1986) 171.

[59] H D Feldtman, J D Leeder and J A Rippon, in *Objective Evaluation of Apparel Fabrics*, ed. R Postle, S Kawabata and M Niwa (Osaka, Japan: Text. Mach. Soc. Japan, 1983), 125.

[60] E Zeitler and G F Bahr, *Experimental Cell Research*, **12** (1957) 44.

[61] G E Rogers, *Annals of the New York Academy of Sciences*, **83** (1959) 378.

[62] M G Dobb, R Murray and J Sikorski, *Journal of Microscopy*, **96** (1972) 285.

[63] J A Swift, *Journal of the Royal Microscopical Society*, **88** (1967) 449.

[64] P Kassenbeck, *Journal of Polymer Science*, **C20** (1967) 49.

[65] P Kassenbeck and R Hagege, *Proceedings of the 3rd International Wool Textile Research Conference, Paris*, **1** (1965) 245.

[66] D E Peters and J H Bradbury, *Australian Journal of Biological Sciences*, **25** (1972) 1225.

[67] J H Bradbury and K F Ley, *Australian Journal of Biological Sciences*, **25** (1972) 1235.

[68] S J Leach, G E Rogers and B K Filshie, *Archives of Biochemistry and Biophysics*, **105** (1964) 270.

[69] R A Dedeurwaeder, M G Dobb, L A Holt and S J Leach, *Archives of Biochemistry and Biophysics*, **120** (1967) 249.

[70] J H Bradbury and D E Peters, *Textile Research Journal*, **42** (1972) 471.

[71] J A Swift and B Bews, *Journal of the Society of Cosmetic Chemists*, **25** (1974) 355.

[72] J A Swift and B Bews, *Journal of the Society of Cosmetic Chemists*, **25** (1974) 13.

[73] R A Dedeurwaeder, M G Dobb and B J Sweetman, *Nature*, **203** (1964) 48.

[74] D F G Orwin, in *Fibrous Proteins: Scientific, Industrial and Medical Aspects*, **Vol. 1**, ed. D A D Parry and L K Creamer (London, UK: Academic Press, 1979), 271.

[75] R Burgess, *Journal of the Textile Institute*, **25** (1934) T289.

[76] D E Peters and J H Bradbury, *Australian Journal of Biological Sciences*, **29** (1976) 43.

[77] K R Makinson, *Textile Research Journal*, **46** (1976) 360.

[78] J Lindberg, *Textile Research Journal*, **23** (1953) 585.

[79] J D Leeder and J A Rippon, *Journal of the Society of Dyers and Colourists*, **101** (1985) 11.

[80] K R Makinson, *Shrinkproofing of Wool* (New York, USA: Marcel Dekker, 1979).

[81] K H Phan, H. Thomas and E. Heine, *Proceedings of the 9th International Wool Textile Research Conference, Biella*, **II** (1995) 19.

[82] J T Marsh, *Introduction to Textile Finishing* (London, UK: Chapman and Hall, 1966).

[83] H M Appleyard and C M Grevelle, *Nature*, **166** (1950) 1031.

[84] J H Bradbury and J D Leeder, *Australian Journal of Biological Sciences*, **23** (1970) 843.

[85] J H Bradbury and G V Chapman, *Australian Journal of Biological Sciences*, **17** (1964) 960.

[86] J H Bradbury, G V Chapman, A N Hambly and N L R King, *Nature*, **210** (1966) 1333.

[87] K F Ley, R C Marshall and W G Crewther, *Proceedings of the 7th International Wool Textile Research Conference*, **1** (1985) 152.

[88] K Stewart, P L Spedding, M S Otterburn and D M Lewis, *Journal of the Society of Dyers and Colourists*, **113** (1997) 32.

[89] K F Ley and W G Crewther, *Proceedings of the 6th International Wool Textile Research Conference*, **3** (1980) 13.

[90] E Lehmann, *Melliand Textilber*, **22** (1941) 145.

[91] S B Ruetsch and H D Weigmann, *Proceedings of the 9th International Wool Textile Research Conference, Biella*, **II** (1995) 44.

[92] J Lindberg, E H Mercer, B Philip and N Gralén, *Textile Research Journal*, **19** (1949) 673.

[93] J A Swift and A W Holmes, *Textile Research Journal*, **35** (1965) 1014.

[94] J D Leeder and J H Bradbury, *Nature*, **218** (1968) 694.

[95] K von Allwörden, *Z Angewandte Chemie*, **29** (1916) 77.

[96] N L R King and J H Bradbury, *Australian Journal of Biological Sciences*, **21** (1968) 375.

[97] J D Leeder and J H Bradbury, *Textile Research Journal*, **41** (1971) 563.

[98] J B Speakman, B Nilssen and G H Elliott, *Nature*, **142** (1938) 1035.

[99] J H Bradbury and J D Leeder, *Australian Journal of Biological Sciences*, **25** (1972) 133.

[100] C W Hock, R C Ramsay and M Harris, *American Dyestuff Reporter*, **30** (1941) 449 and 469.

[101] W Herbig, *Z Angewandte Chemie*, **32** (1919) 120.

[102] R D B Fraser and G E Rogers, *Biochimica et Biophysica Acta*, **16** (1955) 307.

[103] D F O'Reilly, J C Whitwell, R O Steele and J H Wakelin, *Textile Research Journal*, **22** (1952) 441.

[104] E H Mercer and R L Golden, *Textile Research Journal*, **23** (1953) 43.

[105] G Lagermalm, *Textile Research Journal*, **24** (1954) 17.

[106] R C Marshall, *Proceedings of the 8th International Wool Textile Research Conference, Christchurch*, **1** (1990) 169.

[107] J D Leeder, J A Rippon and D E Rivett, *Proceedings of the 7th International Wool Textile Research Conference, Tokyo*, **4** (1985) 312.

[108] R L Elliott and B Manogue, *Journal of the Society of Dyers and Colourists*, **68** (1952) 12.

[109] V Kopke and B Nilssen, *Journal of the Textile Institute*, **51** (1960) T1398.

[110] J D Leeder and J H Bradbury, *Textile Research Journal*, **41** (1971) 215.

[111] D J Evans, J D Leeder, J A Rippon and D E Rivett, *Proceedings of the 7th International Wool Textile Research Conference, Tokyo*, **1** (1985) 135.

[112] U Kalkbrenner, A Körner, H Höcker and D E Rivett, *Proceedings of the 8th International Wool Textile Research Conference, Christchurch*, **1** (1990) 398.

[113] P W Wertz and D T Downing, *Lipids*, **23** (1988) 878.

[114] P W Wertz and D T Downing, *Comparative Biochemistry and Physiology B*, (1989) 759.

[115] A P Negri, H J Cornell and D E Rivett, *Textile Research Journal*, **62** (1992) 381.

[116] D J Evans and M Lanczki, *Textile Research Journal*, **67** (1997) 435.

[117] L N Jones and D E Rivett, *Micron*, **28** (1997) 469.

[118] R J Ward, H A Willis, G A George, G B Guise, R J Denning, D J Evans and R D Short, *Textile Research Journal*, **63** (1993) 362.

[119] D J Peet, R E H Wettenhall and D E Rivett, *Textile Research Journal*, **65** (1995) 58.

[120] J H Bradbury and G E Rogers, *Textile Research Journal*, **33** (1963) 452.

[121] J Sikorski and W S Simpson, *Nature*, **182** (1958) 1235.

[122] M G Dobb, F R Johnston, J A Nott, L Oster, J Sikorski and W S Simpson, *Journal of the Textile Institute*, **52** (1961) T153.

[123] J D Leeder and J A Rippon, *Journal of the Textile Institute*, **73** (1982) 149.

[124] H Hojo, *Proceedings of the 7th International Wool Textile Research Conference, Tokyo*, **4** (1985) 322.

[125] R. Levene and G. Shakkour, *Journal of the Society of Dyers and Colourists*, **111** (1995) 352.

[126] D F G Orwin and R W Thomson, *Proceedings of the 5th International Wool Textile Research Conference, Aachen*, **2** (1975) 173.

[127] L W Lockhart, *Journal of the Textile Institute*, **51** (1960) T295.

[128] R E Chapman and B F Short, *Australian Journal of Biological Sciences*, **17** (1964) 771.

[129] W H Ward and J J Bartulovich, *Journal of Physical Chemistry*, **60** (1956) 1208.

[130] H Ito, H Sakabe, T Miyamoto and H Inagaki, *Proceedings of the 7th International Wool Textile Research Conference, Tokyo*, **1** (1985) 115.

[131] K F Ley, L M Dowling and R D Rossi, *Proceedings of the 8th International Wool Textile Research Conference, Christchurch*, **1** (1990) 215.

[132] M Horio and T Kondo, *Textile Research Journal*, **23** (1953) 373.

[133] T D Brown and W J Onions, *Nature*, **186** (1960) 93.

[134] R M Bones and J Sikorski, *Journal of the Textile Institute*, **58** (1967) 521.

[135] I J Kaplin and K J Whiteley, *Australian Journal of Biological Sciences*, **31** (1978) 231.

[136] G E Rogers, *Journal of Ultrastructural Research*, **2** (1959) 309.

[137] W G Crewther, L M Dowling, A S Inglis, L G Sparrow, P M Strike and E F Woods, *Proceedings of the 7th International Wool Textile Research Conference, Tokyo*, **1** (1985) 85.

[138] R D B Fraser, J M Gillespie and T P Macrae, *Comparative Biochemistry and Physiology*, **44B** (1973) 943.

[139] L N Jones, *Biochimica et Biophysica Acta*, **446** (1976) 515.

[140] R D B Fraser, T P Macrae, G R Millward, D A D Parry, E. Suzuki and P A Tulloch, *Applied Polymer Symposium*, **18** (1971) 65.

[141] L M Dowling, W G Crewther and A S Inglis, *Biochemical Journal*, **236** (1986) 695.

[142] J F Conway, R D B Fraser, T P Macrae and D A D Parry, in *The Biology of Wool and Hair*, ed. G E Rogers, P J Reis, K A Ward and R C Marshall (London, UK: Chapman and Hall, 1988), 127.

[143] R D B Fraser, G E Rogers and A D Parry, *Journal of Structural Biology*, **143** (2003) 85.

[144] V G Kulkarni and J H Bradbury, *Australian Journal of Biological Sciences*, **27** (1974) 383.

[145] L M Dowling, K F Ley and A M Pearce, *Proceedings of the 8th International Wool Textile Research Conference, Christchurch*, **1** (1990) 205.

[146] R D B Fraser and G E Rogers, *Australian Journal of Biological Sciences*, **8** (1955) 288.

[147] L A Holt and I W Stapleton, *Proceedings of the Textile Institute World Conference, Sydney*, (1988) 420.

[148] H Sakabe, H Ito, T Miyamoto and H Inagaki, *Textile Research Journal*, **56** (1986) 635.

[149] J H Dusenbury and A B Coe, *Textile Research Journal*, **25** (1955) 354.

[150] J Menkart and A B Coe, *Textile Research Journal*, **28** (1958) 218.

[151] J D Leeder and J A Rippon, *Proceedings of the International Symposium on Fiber Science and Technology, Hakone*, **II-9** (1985) 203.

[152] L A Holt and I W Stapleton, *Journal of the Society of Dyers and Colourists*, **104** (1988) 387.

[153] R L Elliott and J B Roberts, *Journal of the Society of Dyers and Colourists*, **72** (1956) 370.

[154] P Miro and J A Heuso, *Textile Research Journal*, **38** (1968) 770.

[155] A R Haly and J Griffith, *Textile Research Journal*, **28** (1958) 32.

[156] R Umehara, Y Shibata, Y Masuda, H Ito, T Miyamoto and H Inagaki, *Textile Research Journal*, **58** (1988) 22.

[157] J R Cook and B E Fleischfresser, *Textile Research Journal*, **60** (1990) 77.

[158] J M Appleyard and C M Dymoke, *Journal of the Textile Institute*, **45** (1954) T480.

[159] E H Mercer, J L Farrant and A G L Rees, *Proceedings of the 1st International Wool Textile Research Conference, Australia*, **F** (1955) 120.

[160] D F G Orwin, R W Thomson and N E Flower, *Journal of Ultrastructural Research*, **45** (1973) 1, 15, 30.

[161] G E Rogers, in *The Epidermis*, ed. W Montagna and W C Lobitz (New York, USA: Academic Press, 1964), 179.

[162] R I Logan, L N Jones and D E Rivett, *Proceedings of the 8th International Wool Textile Research Conference, Christchurch*, **1** (1990) 408.

[163] V G Kulkarni and H Baumann, *Textile Research Journal*, **50** (1980) 6.

[164] J H Bradbury, G V Chapman and N L R King, *Proceedings of the 3rd International Wool Textile Research Conference, Paris*, **1** (1965) 359.

[165] L N Jones, T J Horr and I J Kaplin, *Micron*, **25** (1994) 589.

[166] J H Bradbury, J D Leeder and I C Watt, *Applied Polymer Symposium*, **18** (1971) 227.

[167] J H Bradbury and N L R King, *Australian Journal of Chemistry*, **20** (1967) 2803.

[168] A Schwan and H Zahn, *Proceedings of the 6th International Wool Textile Research Conference, Pretoria*, **2** (1980) 29.

[169] Y Nakamura, T Kanoh, T Kondo and H Inagaki, *Proceedings of the 5th International Wool Textile Research Conference, Aachen*, **2** (1975) 23.

[170] Y Nakamura, K Kosaka, M Tada, K Hirota and S Kunugi, *Proceedings of the 7th International Wool Textile Research Conference, Tokyo*, **1** (1985) 171.

[171] A W Holmes, *Textile Research Journal*, **34** (1964) 706.

[172] J A Swift and B Bews, *Journal of the Textile Institute*, **65** (1974) 222.

[173] L A Allen, R E Bacon-Hall, B C Ellis and J D Leeder, *Proceedings of the 6th International Wool Textile Research Conference, Pretoria*, **4** (1980) 185.

[174] D H Tester, *Textile Research Journal*, **54** (1984) 75.

[175] H D Feldtman and J D Leeder, *Textile Research Journal*, **54** (1984) 26.

[176] H D Feldtman and J D Leeder, *Proceedings of the 7th International Wool Textile Research Conference, Tokyo*, **4** (1985) 471.

[177] J A Rippon and J D Leeder, *Journal of the Society of Dyers and Colourists*, **102** (1986) 171.

[178] J D Leeder and J A Rippon, *Journal of the Society of Dyers and Colourists*, **99** (1983) 64.

[179] J A Rippon and J D Leeder, *Textile Chemist and Colorist*, **17**(4) (1985) 74.

[180] J A Rippon and J D Leeder, *Research Disclosures* (UK: K Mason Publications, 1983), 169.

[181] J A Rippon and D J Evans, in *Handbook of Natural Fibres: Vol. 2: Processing and Applications*, ed. R. Kozlowski (Cambridge, UK: Woodhead, 2012), Chapter 3, 63.

[182] E H Hinton, *Textile Research Journal*, **44** (1974) 233.

[183] N H Leon, *Textile Progress*, **7**(1) (1975) 1.

[184] A Robson, *Proceedings of the 5th International Wool Textile Research Conference, Aachen*, **1** (1975) 137.

[185] R S Asquith and N H Leon, in *Chemistry of Natural Protein Fibers*, ed. R S Asquith (New York, USA: Plenum Press, 1977), 193.

[186] R S Asquith, in *Fibrous Proteins: Scientific, Industrial and Medical Aspects*, **Vol. 1**, ed. D A D Parry and L K Creamer (London, UK: Academic Press, 1979), 371.

[187] D M Lewis, *Proceedings of the 8th International Wool Textile Research Conference, Christchurch*, **4** (1990) 1.

[188] H Zollinger, *Proceedings of the 5th International Wool Textile Research Conference, Aachen*, **1** (1975) 167.

[189] D M Lewis, *Review of Progress in Coloration and Related Topics*, **8** (1977) 10.

[190] D S Taylor, *Proceedings of the 6th International Wool Textile Research Conference, Pretoria*, **1** (1980) 93.

[191] D S Taylor, *Proceedings of the 7th International Wool Textile Research Conference, Tokyo*, **1** (1985) 27.

[192] D M Lewis, in *Handbook of Textile and Industrial Dyeing, Part I: Textile Applications*, ed. M Clark (Cambridge, UK: Woodhead, 2011), Chapter 1, 3.

[193] J Cegarra and J Gacen, *Wool Science Review*, **59** (1983) 3.

[194] P A Duffield and D M Lewis, *Review of Progress in Coloration and Related Topics*, **15** (1985) 38.

[195] J Y Cai, *Fibres and Polymers*, **10** (2009) 502.

[196] L Benisek, *Wool Science Review*, **52** (1976) 30.

[197] A R Horrocks, *Review of Progress in Coloration and Related Topics*, **16** (1986) 62.

[198] C S Whewell, *Textile Progress*, **2**(3) (1970) 1.

[199] T Shaw and J Lewis, *Textile Progress*, **4**(3) (1972) 1.

[200] M Lipson, *Proceedings of the 5th International Wool Textile Research Conference, Aachen*, **1** (1975) 209.

[201] C S Whewell, in *Chemistry of Natural Protein Fibers*, ed. R S Asquith (New York, USA: Plenum Press, 1977) 333.

[202] T Shaw and M A White, in *Handbook of Fiber Science and Technology*, **Vol. 2**, Part B, ed. M Lewin and S B Sello (New York, USA: Marcel Dekker, 1984), Chapter 8.

[203] A G De Boos, *Textile Progress*, **20**(1) (1989) 1.

[204] P R Brady, *Finishing and Wool Fabric Properties* (Geelong, Australia: CSIRO Wool Technology, 1997).

[205] R L Hill, in *Advances in Protein Chemistry*, **Vol. 20**, ed. C B Anfinsen Jr, M L Anson, J T Edsall and F M Richards (New York, USA: Academic Press, 1965) 37.

[206] W Beal, K Dickinson and E Bellhouse, *Journal of the Society of Dyers and Colourists*, **76** (1960) 333.

[207] J A Rippon and F J Harrigan, *Proceedings of the 8th International Wool Textile Research Conference, Christchurch*, **4** (1990) 50.

[208] D M Lewis, *Journal of the Society of Dyers and Colourists*, **106** (1990) 257.

[209] H Zahn, *Proceedings of the 9th International Wool Textile Research Conference, Biella*, **1** (1995) 1.

[210] D M Lewis and S M Smith, *Journal of the Society of Dyers and Colourists*, **107** (1991) 351.

[211] P Ponchel and M Bauters, *Proceedings of the 5th International Wool Textile Research Conference, Aachen*, **3** (1975) 213.

[212] H Zahn and J Meienhofer, *Proceedings of the 1st International Wool Textile Research Conference, Australia*, **C** (1955) 62.

[213] H Baumann and H Müller, *Textilveredlung*, **12** (1977) 163.

[214] J B Speakman, *Transactions of the Faraday Society*, **29** (1933) 148.

[215] E Elöd and H Reutter, *Melliand Textilber*, **19** (1938) 67.

[216] H Baumann and L Möchel, *Textil Praxis*, **29** (1974) 507.

[217] J Steinhardt and M Harris, *Journal of Research of the National Bureau of Standards*, **24** (1940) 335.

[218] R V Peryman, *Journal of the Society of Dyers and Colourists*, **70** (1954) 83.

[219] H Baumann and B Potting, *Textilveredlung*, **13** (1978) 74.

2

The Chemical and Physical Basis for Wool Dyeing[1]

John A. Rippon
CSIRO Materials Science and Engineering, Geelong, Victoria, 3216, Australia

2.1 Introduction

The complex structure of wool was described in Chapter 1, where it was shown that wool fibres are biological composites consisting of regions of different chemical and physical composition. The main components that affect dyeing behaviour are the (approximately) 170 different types of proteins; these are not uniformly distributed throughout the fibre. The proteins in wool have been classified as either 'keratinous' or 'nonkeratinous', depending on their cystine content, and hence their crosslink density. Keratinous proteins account for around 82% of the fibre and nonkeratinous proteins for around 17% (Table 1.4). An important minor component of wool is waxy lipids (Tables 1.3 and 1.4). Although composing less than 1% of the fibre mass, the lipids have an important influence on wool properties, including dyeing behaviour.

2.2 The Chemical Basis for Wool Dyeing

Dyes used for the coloration of wool are mostly the sodium salts of aromatic acids, with relative molecular masses (r.m.m.) around 300–900 Da [1–10a]. They usually contain between

[1] Some information in this chapter is based on that published in Chapters 1 and 2 of *Wool Dyeing*, ed. David M. Lewis, published SDC, 1992 and is used with permission from the Society of Dyers and Colourists.

The Coloration of Wool and other Keratin Fibres, First Edition. Edited by David M. Lewis and John A. Rippon.

one and three sulphonic acid groups per molecule, with a few dyes containing four. A simple way to depict their structure is $D(-SO_3^- Na^+)_n$, where D is the chromophore. As an alternative to sulphonic acid groups, carboxyl groups, phenolic groups or hydrophilic, nonionic substituents have also been used to provide water solubility [11a].

Wool dyes are classified according to their method of application. In general, this depends on their structure, in particular their hydrophilic/hydrophobic balance, which varies markedly between the various types [4,12]. Hydrophilic character is conferred by the solubilising groups and hydrophobic properties by aromatic residues and aliphatic hydrocarbon chains. The various types of wool dyes will be discussed further in Section 2.4.

2.2.1   The Wool–Water System

In common with other textile fibres, wool is usually dyed from aqueous dyebaths. The keratin–water system has been described by Feughelman [13]. It has been shown that the primary adsorption sites for water molecules in keratin do not have equal degrees of hydrophilicity [14]. At low relative humidities, water adsorption by proteins is believed to occur on the polar side chains, but at higher humidities it takes place on the peptide linkages; multilayer formation also occurs [15]. Using the theory of multilayer adsorption, it has been calculated that the surface area available to water molecules at 25 °C is 206 m^2 g^{-1} [16]. This surface area is about 200 times greater than that found for nitrogen, leading to the proposal that the additional internal surface area within wool fibres exists only in the presence of the aqueous swelling agent [16].

In wool dyeing, the combined effect of water, temperature and dyeing auxiliaries make the fibre as accessible to dyes as is possible. When wool is immersed in water, the fibres swell about 16% radially and a little over 1% longitudinally [13]. It appears that, compared with the matrix, the microfibrils do not interact to any great extent with water. As water is sorbed by wool, an internal osmotic pressure is developed. This causes the molecular chains to move apart, until a point is reached at which the cohesive forces balance the difference in osmotic pressure between the outside and the inside of the fibre. Some dyeing auxiliaries, such as acids, salts and surfactants, can modify the noncovalent cohesive forces (shown in Figure 1.4). This increases the degree of swelling, but only within the limits allowed by the permanent crosslinks. In addition to reagents, swelling is also affected by temperature.

2.2.2   The Amphoteric Nature of Wool and Dyeing Behaviour

As discussed in Chapter 1 (Scheme 1.4), wool fibres are amphoteric because of the presence of acidic and basic side chains in some of the constituent proteins (Table 1.1). These groups give wool a very large capacity for adsorbing acids and alkalis [17]. This occurs as a result of adsorption or desorption of hydrogen ions by the carboxyl side chains of aspartic and glutamic acid residues, and the amino side chains of lysine, imidazoyl side chains of histidine and guanidinium side chains of arginine residues. As acid is added to the system, hydrogen ions react with the carboxylate anions to form carboxylic acid groups, leaving the positively charged ammonium groups available to act as sites for anions (including dye anions, as discussed later). Under alkaline conditions, hydrogen ions are abstracted from the positively charged amino groups. The carboxyl anions then confer a negative charge on

Table 2.1 *Acidic and basic amino acid residues in wool.*

	Name	Amount (mol%)	Dissociation Constant at 25 °C (pK$_a$)
Acidic residues	Aspartic acid	6.4	4.2
	Glutamic acid	11.9	3.9
Basic residues	Arginine	6.9	12.5
	Lysine	3.0	10.5
	Histidine	0.9	9.3

the substrate. The maximum acid-binding capacity of wool is governed by the number of carboxyl groups present.

The concentrations and dissociation constants of the acidic and basic residues in wool are shown in Table 2.1.

Wool also contains single-chain protein molecules, with an *N*-terminal and a *C*-terminal amino acid at the opposite ends of each [18]. The concentrations of these groups, however, are only around 1% of the total acid and basic groups in the fibre, as shown in Table 2.1; their influence on the overall amphoteric properties of wool is thus very small. A typical acid/base titration curve for wool is given in Figure 2.1. This shows that the maximum bound acid or bound base capacity is around 0.82 mE g^{-1}, which is in close agreement with the number of amino and carboxyl groups in the fibre (Table 2.1).

The titration curve is characterised by a region, between pH 4 and 10, in which very little acid or base is bound to the wool. The explanation for this is that the wool protein segments have a limited mobility within the water-swollen fibre, due to the tendency of the solid nature of the protein and the covalent crosslinks to oppose expansion. This results in the buildup of a high electrical potential in the fibre, which limits the penetration of additional hydrogen ions. In order to maintain electrical neutrality within the fibre, there must be simultaneous adsorption of both hydrogen ions and counter-ions. This makes the entry of hydrogen ions more difficult and shifts the titration curve to lower pHs, compared

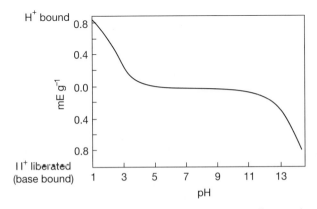

Figure 2.1 *Typical titration curve for wool (mE = milliequivalents).*

with the curves for soluble proteins. The precise effect of the external liquor pH on the charge carried by the fibre depends on whether the external liquor contains an electrolyte. In the absence of salts, because of the Donnan effect [10a], the internal pH of the fibre lags behind the pH of the external solution [10a,17,19]. Addition of a neutral salt to the external liquor swamps the buildup of electrical charge on the wool, which has the effect of decreasing the difference between the internal fibre pH and the pH of the external liquor, decreasing the internal pH in an acidic medium and raising the internal pH in an alkaline solution. This shifts the titration curve for wool so that it agrees more closely with that for soluble proteins [17,20].

In an aqueous dyebath, wool fibres attain equilibrium with the external liquor. As shown in Scheme 1.4, below pH 4 wool carries a positive charge; in the pH range 4–8 the fibre is in the isoelectric state and carries no net charge; and above pH 8 the fibre becomes increasingly negatively charged as the amino groups progressively lose protons. Under acidic conditions, all the cationic groups in wool are potential sites for the attraction of negatively charged, anionic acid dyes to wool. The negative charge on wool under alkaline conditions (above pH 8) makes the fibre substantive to dyes that carry a cationic charge (basic dyes). However, these dyes are not important for wool and will not be discussed in this chapter.

The pH at which the particular groups in the side chains of wool proteins sorb or desorb protons is determined by their pK_a values, shown in Table 2.1. Among the amino acid residues, arginine is particularly important. Its pK_a value of 12.5 means that it is cationic under all normal wool-dyeing conditions (i.e. within the pH range 2–8) [6]. Furthermore, arginine's importance is reinforced by its relatively high concentration in the fibre (6.9 mol%) compared with the other basic amino acids.

As already discussed, ionic attraction between wool and anionic dyes is important in attracting the dyes to the fibre in the early stages of a dyeing cycle. Ionic interactions are also important in wool dyeing because they determine the rate at which a dye is taken up from the dye liquor. The dyeing rate can be controlled by varying the amount of acid added to the dyebath, because this determines the size of the net positive charge on the wool.

2.2.3   Classical Theories of Wool Dyeing

Early theories of wool dyeing were based on the adsorption of acids by wool, as previously discussed. These theories attempted to explain the overall mechanism of wool dyeing under acid conditions solely in terms of the ionic interactions between the positively charged amino groups on the fibres and the negatively charged dye anions [9,10a,11b,17,21]. A typical acidic wool dyebath contains dye anions, hydrogen ions from the acid, electrolytes (e.g. sodium sulphate) and counter-ions from the acid. When wool is immersed in the dyebath, the smallest and most rapidly diffusing ions should be quickly adsorbed, while the larger and more slowly diffusing dye anions will be taken up more slowly. Elöd [22] followed the rate at which a simple, disulphonated dye (Crystal Ponceau) and hydrochloric acid were absorbed from the dyebath. Figure 2.2 shows that, initially, there was a very rapid adsorption of hydrogen ions and chloride anions. With increasing time, the more slowly diffusing dye anions progressively displaced the chloride ions from the wool, as shown by the increase in the concentration of chloride ions in the bath.

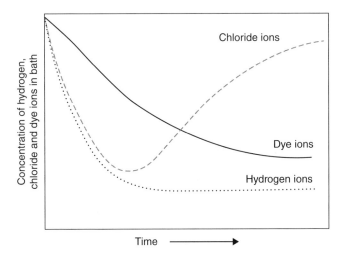

Figure 2.2   *Representation of the rate of adsorption of ions by wool from an acid dyebath* [10a].

Elöd's work led others to develop theories of wool dyeing based on this model [9,10a, 11b,17,21]. Although approaches based on ionic interactions are now considered to over-simplify the very complex wool-dyeing system, the two most important models will be discussed briefly here, due to their historical importance. In both theories, the fibre is considered to be at a uniform electrical potential, different to that of the acid solution.

2.2.3.1   Gilbert–Rideal Theory

In the Gilbert–Rideal theory [23], it was assumed that all the positively charged basic groups in wool interact with anions in an identical manner, and that all the carboxyl groups have the same affinity for protons. It was also assumed that an anion can occupy any positively charged site, irrespective of its location relative to a charged or uncharged carboxyl group. The anions and protons were, thus, regarded as being adsorbed independently of each other, with the only requirement being that the fibre maintained overall electrical neutrality. Using these assumptions, a series of equations were derived. Gilbert and Rideal then adapted the statistically derived dependence of the chemical potential for an uncharged substance, adsorbed at random over a limited number of sites [24], by taking into account the increase in the chemical potential that results from the species being charged. They derived equations for the chemical potential of both protons and chloride ions for a solution of hydrochloric acid in equilibrium with a wool fibre. By combining these, an equation (2.1) [17] was obtained which gave a reasonable fit to the data of Steinhardt and Harris [9,25] for the titration of wool with hydrochloric acid:

$$\log \frac{\theta_H}{1 - \theta_H} = -\text{pH} - \frac{\left(\Delta\mu_H^0 + \Delta\mu_{Cl}^0\right)}{4.6RT} \tag{2.1}$$

where θ_H is the fraction of sites occupied, μ^0 is the chemical potential when θ is 0.5, R is the gas constant ($8.317\,\text{J}\,\text{K}^{-1}\,\text{mol}^{-1}$) and T is the absolute temperature (K).

Gilbert and Rideal also derived an equation (2.2) for the case where an alkali-metal salt, such as sodium chloride, is added to the system, so that titrations can be carried out at constant ionic strength. This equation also fit the data of Steinhardt and Harris [9,25] and of Steinhardt [26], including the dependence of the pH of half-saturation (i.e. half-neutralisation) on ionic strength [17]:

$$(pH_{0.5}) = \log(Cl^-) - \frac{(\Delta\mu_H^0 + \Delta\mu_{Cl}^0)}{2.303RT} \tag{2.2}$$

where $pH_{0.5}$ is the pH of half-saturation.

The value of half-saturation is used as a standard for investigating the effect of salt concentration on the position of the titration curve. Such studies have shown that at higher chloride concentrations (i.e. greater ionic strength), less acid is required to produce a given degree of saturation of the fibre. This explains the experimental observation, discussed in Section 2.2.2, that the titration curves for wool are shifted to higher pH values in the presence of electrolytes. An advantage of the Gilbert–Rideal theory [23] was that it could be extended to polybasic acids, such as sulphuric acid, with a good fit to experimental data [27].

2.2.3.2   Donnan Theory

The Donnan membrane effect relates to the distribution, or partition, of ionic species between two different phases separated by a membrane. Donnan showed, using a solution containing sodium chloride and a dye, that while the sodium and chloride ions were able to diffuse through a paper membrane into an outer solution, larger dye anions of Congo Red (C.I. Direct Red 28) were not able to penetrate the membrane and colour the outer solution. The membrane was described as 'semipermeable' because penetration depended on the size of the penetrating molecules. While Donnan demonstrated the principle using a semipermeable membrane, the presence of a 'membrane' is not essential.

Peters [10a] has described the application of the Donnan theory to wool. In this theory, anions are assumed to have no specific affinity for the protein, and the various groups in the protein are considered to come to equilibrium with the ions in the external phase. Equation 2.3, derived from this approach, can also be used to provide an explanation for the data of Steinhardt and Harris [25] regarding the adsorption of hydrochloric acid by wool in the presence of salt:

$$pH_i = pH_S - \log[Cl_S] + \log[H_a]/v \tag{2.3}$$

In the absence of salt, this becomes:

$$pH_i = 2pH_s + \log[H_a]/v \tag{2.4}$$

where subscripts i and s refer to the solution inside the fibre and the external solution, respectively, $[H_a]$ is the quantity of hydrogen ions absorbed, $[Cl_s]$ is the quantity of chloride ions absorbed and v is the internal aqueous volume of the wool fibre (assumed to be $0.3\,L\,kg^{-1}$).

It was arbitrarily assumed that the activity coefficients of all the ions in solution were unity.

Using these equations, Peters [28] explained the effect of salt on the titration of wool with hydrochloric acid and calculated the internal pH of the fibre. The pK_a value of the carboxylic acid groups was then calculated to be 4.3, which is very close to those for aspartic acid (4.2) and glutamic acid (3.9).

In spite of this good agreement, the Donnan equations do not provide a full explanation for the uptake of acids and salts by wool. Writing in terms of affinity (see Section 2.3) gives the following [10a]:

$$-\Delta\mu_H^0 = RT \, \ln \left[\frac{\theta_H^2}{1 - \theta_H} \right] - RT \, \ln \, [H_s] \, [Cl_s] + RT \, \ln \, S/\nu \qquad (2.5)$$

where $\Delta\mu_H^0$ is the chemical potential when $\theta_H = $ zero and S is the saturation value (see Figure 2.1).

On the basis of this equation, a plot of $\ln \, [\theta_H^2/(1 - \theta_H)]$ versus $\ln \, [H_s][Cl_s]$ should give a straight line of slope 1.0. Vickerstaff [9] has shown that such a plot yields a slope of 0.70 of the theoretical, whereas a similar plot based on the Gilbert–Rideal theory gives a slope of 0.88 of the theoretical. Thus, it appears that the Gilbert–Rideal model gives a better fit to the experimental data than does the Donnan theory. In any event, it is clear from both these models that affinity due to electrostatic attraction is not the sole force binding ions (e.g. dyes) to wool.

Sumner [29] developed a generalised equation based on the Donnan approach to determine the affinity of anionic dyes on wool and polyamide fibres. He claimed that, in contrast to models based on 'dye sites', the equations based on the Donnan approach predict that overdyeing can occur – without the need for extra sites in the fibre – and that their application to multivalent dyes is straightforward.

2.2.4 Modern Theories of Wool Dyeing

Later workers showed that the simple models based on ionic interactions, such as the Gilbert–Rideal and Donnan theories, are an oversimplification of the mechanism of wool dyeing. Thus, in contrast to Elöd's findings for the relatively hydrophilic, disulphonated dye Crystal Ponceau, Meybeck and Galafassi [30] demonstrated that the observed release of chloride ions during the course of dyeing was not always equivalent to the amount of dye adsorbed. For slightly hydrophobic dyes, only about 10% of the bound chloride ions were released; and for the more hydrophobic dyes studied, the release of chloride ions was not measurable. These results confirmed earlier work [31–36] which indicated that Coulombic interactions between cationic sites in the wool and dye anions play only a very small part in bonding the dye to the fibre at equilibrium (i.e. the affinity of the dye).

It is now known that the following combination of noncovalent interactions is important in wool dyeing [6,11a,30,31]:

- ionic interactions;
- van der Waals forces;
- hydrogen bonds;
- hydrophobic (solvophobic) interactions.

The ionic (Coulombic) interactions between anionic dye molecules and cationic sites on the wool were discussed in Section 2.2.2. They are important in the initial sorption of the dye and in attracting dye molecules to sites in the fibre, where they become strongly fixed.

Van der Waals forces are a group of intermolecular interactions that include dispersive and repulsive forces, such as dipole–dipole (Keesom forces), dipole–induced dipole (Debye forces) and London dispersion forces. Although these are relatively weak interactions, with a strength of less than $8\,\mathrm{kJ\,mol^{-1}}$ [37], because they occur extensively between molecules they are very important.

Meybeck and Galafasi [30] considered hydrogen bonding between dyes and wool fibres to be relatively unimportant among these interactions. They thought that dyes with a high affinity for wool (and therefore better wet fastness properties) must have a hydrophobic character. Hydrophobic interactions arise from the interaction of nonpolar parts of a water-soluble dye molecule with hydrophobic regions in wool, which results in a decrease in the overall surface area exposed to water. Thus, hydrophobic interactions are entropy-driven processes that decrease the free energy of a system by minimising the interfacial surface between water and the hydrophobic parts of dye molecules. The hydrophobic character of dyes is increased by placing the hydrophobic substituents as far as possible from the polar groups in the molecule. Iyer and Srinivasan [38], from thermodynamic studies on three anthraquinone-type dyes, also concluded that hydrophobic interactions are important in the dye–wool adsorption process and in dye binding, especially at low temperatures.

Lewis [6] has reexamined the role of nonpolar interactions in the wool-dyeing system, placing particular emphasis on the importance of π–π interactions. These are often used to explain interactions between aromatic molecules [37], but their overall contribution is very small compared with that of Coulombic interactions [39]. The planar configuration of dye molecules might be expected to favour a flat, stacked arrangement via van der Waals interactions. Furthermore, hydrophobic interactions resulting from the nonpolar, flat π-electron molecular surfaces will also favour stacking [40]. The overlap of the molecules is, however, not usually perfect, with molecules being offset to some extent. This has been explained by the action of electrostatic forces, which repel the moieties containing the π-electrons and push them away from a situation of maximum overlap [41]. Hunter [37] concluded that in a face-to-face alignment, aromatic molecules repel each other, but that other arrangements, such as face-to-edge and offset, produce molecular attraction.

The types of interactions that occur between wool and dyes are also responsible for the aggregation of dyes in aqueous solution [10b,11a], as discussed in Section 2.5.

2.3   Standard Affinity and Heat of Dyeing

Two terms are used to describe the interaction of a dye with a fibre: 'substantivity' and 'affinity' [11b]. These are defined by the Society of Dyers and Colourists (SDC) [42]. Substantivity is defined as:

> The attraction between a substrate and a dye or other substance under the precise conditions of test whereby the latter is selectively extracted from the application medium by the substrate.

In the context of the present chapter, 'selective extraction' refers to exhaustion from a long liquor.

Substantivity is often confused with affinity and the two terms are sometimes (incorrectly) used interchangeably.

Affinity is defined as:

> The quantitative expression of *substantivity* (q.v.). It is the difference between the chemical potential of a dye in its standard state in the fibre and the corresponding chemical potential in the dyebath. *Note:* Affinity is usually expressed in units of calories (joules) per mole. Use of this term in a qualitative sense, synonymous with substantivity, is deprecated.

L. Peters [11b] has pointed out that:

> Although the affinity of a dye for a substrate is related to the chemical potential of the dye in solution, it refers strictly only to the potential at equilibrium, so that to equate it to a driving force can lead to the error of assuming that dyes of high affinity should combine at a higher rate than those of low affinity. This is not necessarily so, because affinity is an expression of the ratio of the rate of absorption to that of desorption. Even if the former is rapid, the latter may be equally rapid, thus giving a zero value for the affinity. In practice, dyes that have high affinity tend to be of large molecular size and are absorbed at a relatively lower rate than dyes of lower affinity; their high affinity is caused by their even lower rate of desorption.

From this definition, the affinity of a dye for a particular fibre can be determined from the extent to which the dye concentration is lowered by the fibre when the dye–fibre system has reached equilibrium. Peters [11b] has suggested that a model for the dyeing process is necessary before an absolute numerical value can be calculated for affinity. Obviously, the magnitude of this affinity value will depend on the model chosen. Rys and Zollinger [43a] agree that the use of absolute values of standard affinities for dyeing processes is rather dubious and that they can, at best, only be used for meaningful comparisons of different dyes in the same dyebath–substrate system.

The standard affinity for the distribution of a dye between the fibre and the dyebath is proportional to the logarithm of the ratio of the absolute activities of the dye in the fibre and dyebath. Since the activity of the dye is assumed to be directly related to its concentration, one can write:

$$-\Delta\mu^{\theta} = RT \ \ln(C_f/C_s) \tag{2.6}$$

where $\Delta\mu^{\theta}$ is the standard affinity (J mol^{-1}), R is the gas constant (8.317 J K^{-1} mol^{-1}), T is the absolute temperature (K), C_f is the concentration of dye on the fibre at equilibrium (g L^{-1}) and C_s is the concentration of dye remaining in solution at equilibrium (g L^{-1}).

The affinity of dyes for fibres may also be determined by comparing the affinity of dye anions with that of chloride or sulphate anions using the 'displacement method' of Gilbert [44].

The affinity term, $\Delta\mu^{\theta}$, is composed of two parts: the heat of dyeing (ΔH^{θ}) and the entropy of dyeing (ΔS^{θ}) [11b]:

$$\Delta\mu^{\theta} = \Delta H^{\theta} - T\Delta S^{\theta} \tag{2.7}$$

For most dyeing systems, including the dyeing of wool, ΔH^θ is negative; that is, dyeing is an exothermic process.

The standard heat of dyeing is related to the standard affinity and absolute temperature by [11b]:

$$\Delta H^\theta = \frac{\delta \left(\dfrac{\Delta\mu^\theta}{T} \right)}{\delta \left(\dfrac{1}{T} \right)} \tag{2.8}$$

Thus, an estimate of ΔH^θ can be obtained at any temperature from the slope of a plot of $\Delta\mu^\theta/T$ versus $1/T$.

2.4   Classification of Dyes Used for Wool

The dyes commonly used on wool can be divided into the following groups: acid dyes, chrome dyes, premetallised (metal-complex) dyes and reactive dyes [1–12,45]. Considerations for optimising the dyeing of wool, both in theory and in practice, have been discussed by Egli [46].

The various types of dye used on wool are summarised in this section; a detailed description of each dyestuff class and its methods of application can be found in later chapters.

Although, strictly speaking, all wool dyes containing sulphonic acid groups can be described as 'acid dyes', this term is now restricted to levelling acid dyes, half-milling dyes, milling dyes and supermilling dyes. These subclassifications of acid dyes result from the methods used for their application, their levelling properties and their wash fastness. Levelling acid dyes (also called level-dyeing or equalising dyes) are of relatively low r.m.m. (300–600 Da) and are the simplest type of dye used on wool (see Chapter 7). In an acid levelling dye containing an azo bond, such as C.I. Acid Red 1, (**1**), water solubility is conferred by the sulphonic acid groups. Another structure commonly found in acid dyes is based on the anthraquinone group, for example C.I. Acid Blue 45 (**2**).

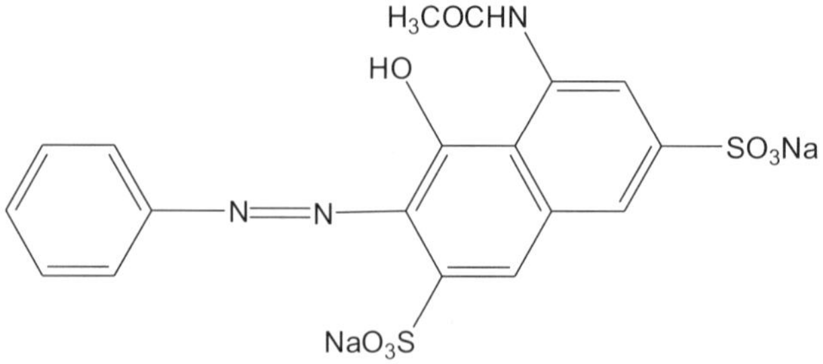

1

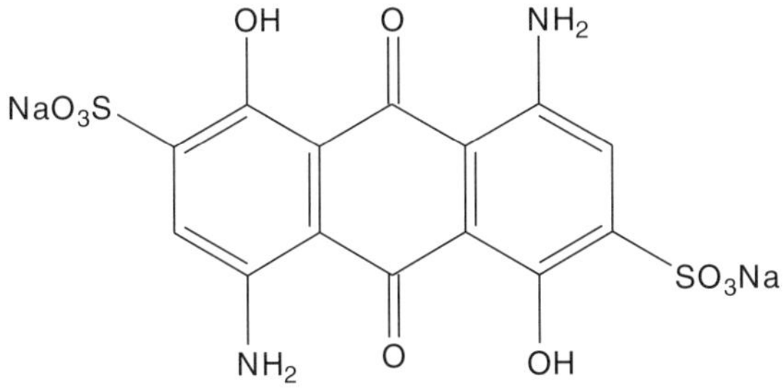

OH   O   NH$_2$

NaO$_3$S

SO$_3$Na

NH$_2$ O   OH

2

Levelling acid dyes are usually applied at the boil from dyebaths set to around pH 3 with sulphuric acid and sodium sulphate. Under these conditions, the high cationic charge carried by wool results in very rapid, but uneven, sorption of dye. As their name implies, levelling acid dyes have good migration properties, and redistribution of dye molecules occurs during the time the dyebath is held at the boil, giving an overall even result. Levelling of the dyes on the wool is assisted by the presence of sodium sulphate in the dye bath, because the anionic sulphate ions compete with dye anions for the positively charged amino groups in the fibre. In general, because of their strong reliance on ionic interactions, these dyes have poor resistance to wet treatments, but they are used when level dyeing is very important.

Milling acid dyes are so named because they are more resistant to extraction from wool during the milling process than are acid levelling dyes (i.e. they have a higher affinity). They are of higher r.m.m. (600–900 Da), are more hydrophobic and have a lower dependence on ionic attraction than levelling acid dyes. A consequence of the higher affinity of milling dyes, however, is that their migration properties – and hence their ability to level during application – are not as good as those of levelling acid dyes. In order to obtain level results, it is important to ensure that the rate of exhaustion is uniform and not too rapid. This is achieved by setting the dyebath at a higher pH ($\sim$4.5–6.0) than that used for levelling dyes, using a weak acid such as acetic acid, and by adding a levelling agent that promotes even uptake on wool. Such surfactant-based auxiliaries are discussed in Chapter 3. Half-milling dyes fall between levelling acid and milling dyes in r.m.m. (500–600 Da), migration properties and wet fastness. They are applied at a pH in the range 4.0–4.5.

Supermilling acid dyes are similar in r.m.m. to milling dyes but contain long alkyl groups, or other nonpolar groups, which make them very hydrophobic in character. Thus, the dye C.I. Acid Red 138, shown in Structure **3**, has the same structure as the acid levelling dye, C.I. Acid Red 1, shown in Structure **1**, except for the hydrophobic dodecyl group on the benzene ring. The presence of this group, and its location on the molecule away from the hydrophilic sulphonic acid groups, increases the affinity of C.I. Acid Red 138 for wool compared with C.I. Acid Red 1. The high affinity of these dyes is virtually independent of ionic interactions with the fibre; consequently, they have very good wet fastness.

3

Supermilling dyes show good exhaustion under almost-neutral (pH 5.5–7.0) dyeing conditions with the addition of ammonium acetate or ammonium sulphate. Their high affinity means that they have poor migration and relatively poor levelling properties, however. As discussed further in Section 2.5, aggregation of these dyes is a major factor in their poor migration/poor levelling properties. Despite the use of auxiliary products to assist level application, some supermilling dyes are only suitable for dyeing loose fibre and top, where any unevenness can be overcome by subsequent blending.

Chrome dyes (see Chapter 7) are acid dyes (r.m.m. 300–600 Da) that contain groups capable of forming complexes on reaction with a metal salt, usually sodium or potassium dichromate. The chrome–dye complex has lower solubility, and hence better wet fastness, than the parent dyestuff. Reaction between the dye molecule and chromium salt can be carried out before, during or after application of the dye to the wool. Modern practice is to carry out the chroming step after dyeing. Chrome dyes are relatively cheap and have good migration and level dyeing behaviour and excellent wet fastness properties. Their popularity has declined in recent years, except for black and navy blue shades. This is because of the need for prolonged dyeing cycles, fibre damage associated with oxidation of the fibre in the chroming step and environmental concerns about the use of chromium salts in the textile industry.

In many applications, chrome dyes have been replaced by metal-complex dyes (see Chapter 8), which have a very high affinity for wool. In these dyes, the metal complex is preformed during manufacture by reaction of one metal atom with either one (1 : 1 metal-complex dyes) or two (1 : 2 metal-complex dyes) dyestuff molecules containing groups capable of coordinating – usually with chromium, or occasionally with cobalt atoms. In general, metal-complex dyes produce duller shades than acid or milling dyes. The 1 : 1 metal-complex dyes are applied to wool from strongly acidic dyebaths at around pH 2. They are almost all monosulphonates (r.m.m. 400–500 Da) and have good levelling behaviours and wet fastness properties. A disadvantage of these dyes is that the low dyeing pH can cause some degradation of the fibre. The earliest type of 1 : 2 metal-complex dye was unsulphonated, with solubility being provided by nonionic polar groups, such as sulphonamide or methylsulphone. More recently, monosulphonated, disulphonated and some carboxyl-containing types have become available. These dyes range in r.m.m. from 700 to around 1000 Da. They are applied at pH values ranging from 4.5 to 7.0, depending on the degree of sulphonation and the molecular size of the dye.

Reactive dyes have r.m.m.s in the range 500–900 Da and usually contain two or three sulphonic acid groups (see Chapter 9). These dyes also contain groups that react covalently with wool, which gives them outstanding wet fastness properties. Shore [47] has stated that the main reactive groups in protein fibres which form covalent bonds with reactive dyes are the amino groups of lysine and the amino and imidazole groups of histidine, followed by cysteine thiols and the phenolic groups in tyrosine. Asquith and Chan [48] have reported that reaction also occurs with the hydroxyl groups of serine.

Reactive dyes are characterised by bright colours and moderate migration and levelling properties, provided careful attention is paid to ensuring that the exhaustion rate is not too rapid. This is achieved through careful control of the dyebath pH and by using special amphoteric levelling agents. Reaction with the fibre means that uneven uptake is very difficult to rectify. Reactive dyes are relatively expensive. Currently, their most important application is on wool that has been treated to withstand shrinkage in machine washing; they are, however, becoming increasingly important as alternatives to chrome dyes for the dyeing of all types of wool, particularly in situations where there are concerns over chromium in dyehouse effluent.

2.5   Dye Aggregation

A full understanding of the dyeing process requires a knowledge of the molecular state of the dye in the dyebath, because many of the problems associated with the dyeing of wool can be explained by the aggregation or colloidal state of the dye [10b,11a]. It is well known that most dyes are aggregated to some extent in aqueous solution and that the degree of aggregation depends on factors such as concentration, temperature and the presence of electrolytes. Many studies have been carried out that examine the role of dye aggregation in the dyeing of wool. The methods used have been reviewed by Duff and Giles [49]; the most productive of these are techniques such as light-scattering and diffusion-based measurements [9,10b,11a].

Much of the early work involved measuring the diffusion of dyes in solution across a membrane. In spite of the problems associated with electrostatic forces 'pulling' dye anions across the membrane, thereby requiring the addition of electrolyte to the system, valuable data were obtained. In 1935, Robinson [50] reported that Benzopurpurine 4B (C.I. Direct Red 2) and 'meta' Benzopurpurine have aggregation numbers of about 10 in 0.5% solution, and that in both cases about 25% of the sodium is included in the dye aggregate. Valko [51] studied the effect of both salt concentration and temperature on the aggregation of Chlorazol Sky Blue FF (C.I. Direct Blue 1). He found that increasing the amount of salt increased the degree of aggregation, while increasing the temperature reduced it.

Speakman and Clegg [52] investigated the influence of the chemical composition of some acid dyes on their colloidal behaviour. They found that, in general, the greater the degree of aggregation of a dye in solution, the more easily the dye will be salted out by an electrolyte and the less level the dyeings will be. They also found that, while the tendency of dyes to aggregate increases with molecular size, increasing the number of sulphonic acid groups on the molecule decreases aggregation.

From this early work on diffusion, it was believed that, while some dyes may be aggregated at low temperatures, acid dyes would not be aggregated to any appreciable extent in the presence of the amounts of salt and acid and at the high temperatures normally employed in dyeing [45]. However, it is now believed that many wool dyes remain aggregated even at the boil.

Many dyes of low solubility are difficult to apply evenly to wool; these difficulties might be associated with either:

(1) The formation of dye aggregates [10a,10b] that cannot easily enter the fibre due to their large size; or
(2) A high affinity of the dye for the fibre; this would make it difficult to achieve a uniform distribution on the wool by a mechanism involving desorption and readsorption of the dye.

Of the methods used to measure dye aggregation [49], diffusion-based and light-scattering techniques should allow measurement under near-practical dyeing conditions, especially dyeing temperature. Both methods require measurements to be made in salt solutions, which reflects real dyeing conditions.

The tetrasulphonated dye, C.I. Acid Red 41 (**4**), has been shown to be monodisperse in aqueous solution by both diffusion-based [53] and light-scattering [54] techniques. This result is to be expected, because the four sulphonate groups make the molecule very hydrophilic. Decreasing the number of sulphonate groups to one, as in C.I. Acid Red 88 (**5**), is expected to increase the degree of aggregation.

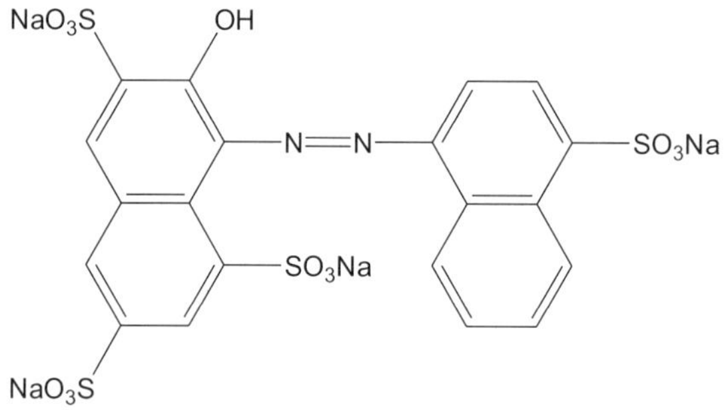

4

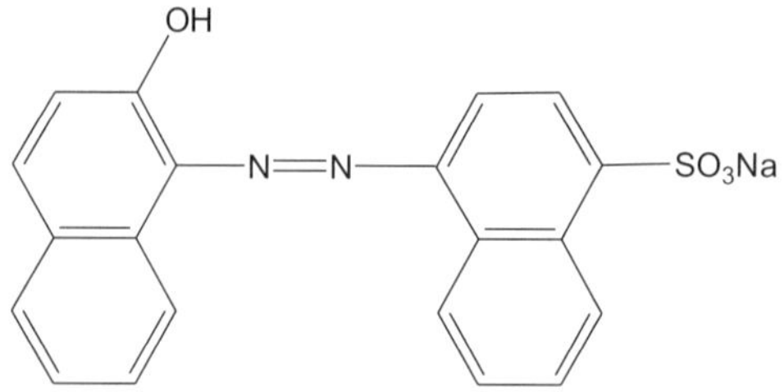

5

The aggregation of C.I. Acid Red 88, determined by diffusion methods [55] and by polarography [56] at 25 °C, was found to range between 2 and 5. El Mariah *et al.* [57] reported that C.I. Acid Red 88 dissociated at higher temperatures and was monomeric at 50 °C. Datyner *et al.* [54] reported that light-scattering measurements gave an aggregation number of 370 ± 50 at 25 °C. Diffusion-based methods measure the number-average r.m.m. of the aggregates, while light-scattering techniques determine their weight-average r.m.m. Results from the two methods can be combined with a suitable distribution function to estimate the distribution of aggregate sizes. The Schulz–Zimm distribution was used, in conjunction with combined diffusion and light-scattering data, to determine the distribution of particle sizes for C.I. Acid Red 88 at 25 °C. [54]. It was concluded that most of the dye exists as particles with a radius of less than 2 nm.

The aggregation of Orange II (C.I. Acid Orange 7; **6**) was examined by Frank [58], who found that the aggregation number varied with temperature and concentration and reached a value of 110 at 5 °C. Orange II and C.I. Acid Red 88 have the same structure, except that in Orange II the sulphonate group is on a benzene rather than a naphthalene ring. Thus, Orange II is more hydrophilic than C.I. Acid Red 88 and can be expected to be more soluble in water and less aggregated. A rate equation, based on the Langmuir sorption theory and Arrhenius plots, has been used to show that the activation energy for Orange II is higher below 60 °C than it is above this temperature [59]. It was suggested that this could be explained on the basis that Orange II was aggregated at temperatures below 60 °C.

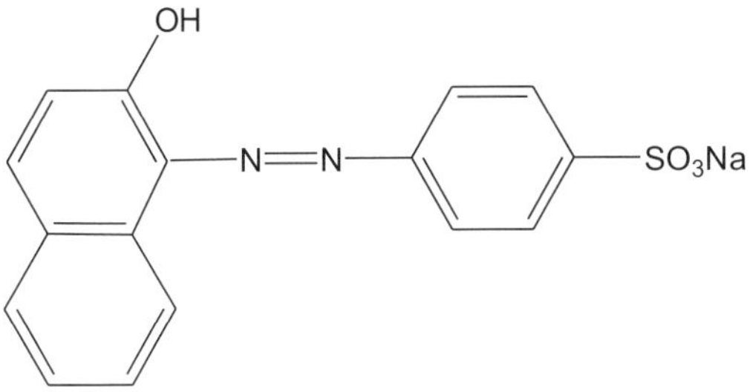

6

The effect of increasing the hydrophobic character on the aggregation of wool dyes has been demonstrated by many workers, some of whom have examined a series of model dyes [54,56,60–62]. These studies show that many wool dyes are highly aggregated and that the more hydrophobic the dye, the more likely it is to aggregate. As dyes become more hydrophobic in character, both the number-average and the weight-average aggregation numbers dramatically increase. This has been demonstrated for two model dyes [54]. Dye (**7**) contains a disulphonated naphthalene ring and an n-butyl-substituted benzene ring. Dye (**8**), however, has an n-octyl group on the benzene ring and would therefore be expected to be more hydrophobic than Dye 7. The study shows this to be the case, because compared with Dye 7, Dye 8 is found to have a much greater propensity for aggregation, which is strongly concentration-dependent. Thus, the aggregation number for Dye 7 is found to be

52 ± 10 at 55 °C, whereas the aggregation number for Dye 8 is 2200 [54]. It is clear therefore that what might seem to be a relatively small change in the structure of a dye can have a very large effect on its tendency for aggregation

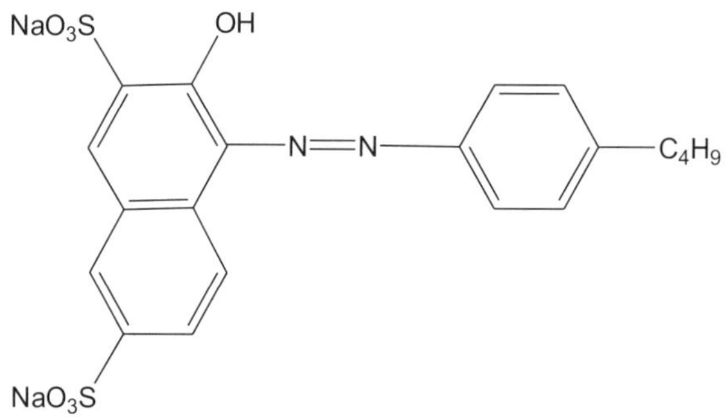

7

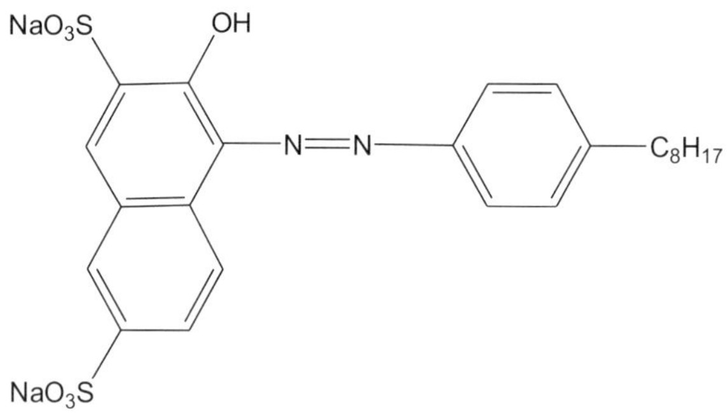

8

The general effect of dye aggregation on practical wool dyeing is that highly aggregated dyes, such as supermilling dyes, can present difficulties in the level dyeing of a wool. In some cases, the use of certain dyes may be restricted to loose stock or top dyeing, because these methods enable a level result to be obtained through blending of the dyed fibre to randomise any unevenness.

Datyner and Pailthorpe measured the effect of a range of dyeing auxiliaries (levelling agents) and urea on the aggregation of some highly aggregating dyes [62]. The levelling agents examined were AM20 (an ethylene oxide adduct of a fatty acid amide), Albegal A and B (Huntsman) and Antarox CO-880 (Solvay). The latter product, which is an adduct of a branched nonylphenol and ethylene oxide, containing an average of 30 ethylene oxide moles for each mole of nonylphenol, is also known as NP30.

NP30 was found to significantly decrease the weight-average aggregation numbers of the 1 : 2 metal-complex dyes to values of the order of 200 at both 55 and 95 °C, presumably by increasing the net solubility of the dyes in water as a result of disaggregation. The increase in dye solubility explains the earlier observation of Hine and McPhee that NP30 decreased the rate of sorption of metal-complex dyes [63]. In contrast, Nemoto and Funahashi [64] found that, in the presence of a polyoxyethylene nonylphenyl ether, the rate of uptake of acid dyes increased with the increasing hydrophobic character of the micelles formed in the dyebath. Three types of micelle were observed, and their formations depended on both temperature and salt concentration. AM20, Albegal A and Albegal B were found to disaggregate dyes, but their effects were highly specific to particular combinations of dye and levelling agent. For example, Albegal B was very effective in disaggregating some levelling acid and milling dyes, but was ineffective with 1 : 2 metal-complex dyes [62].

Of the reagents examined by Datyner *et al.* [62], urea was the only one that significantly disaggregated all the dyes studied at both 55 and 95 °C. Weight-average aggregation numbers were reduced to below 100 in all cases, meaning that the largest aggregate would be less than 2.5 nm in diameter, and hence should be able to penetrate the structure of wool.

The role of urea in the dyeing of wool has been a matter of considerable contention, because, in addition to disaggregating the dye in the dyebath, urea may also interact with wool [62]. Asquith *et al.* [65] suggested that urea does not penetrate wool fibres. Other explanations, however, have ranged from effects such as swelling of the fibre [66,67] to removal of part of the epicuticle [68]. Blagrove [69] has claimed that urea rapidly penetrates the wool fibre in an unhindered fashion as the fibre wets out, but this has been disputed by Gardner [70].

The interaction of urea with wool was clarified when it was shown that swelling does not occur with whole fibres, but that cut segments swell due to penetration of urea into the cut ends [71]. It was also concluded that the epicuticle of wool must be impervious to urea. Burdett and Galek [72] have shown that urea does not cause any significant swelling of wool yarns. It would appear, therefore, that no correlation exists between fibre swelling and accelerated dye adsorption. These authors [72] also investigated the effect of urea, thiourea and related compounds on the rate of dyeing of wool by C.I. Acid Red 18 at low temperature (25 °C). They concluded that urea is not unique in its ability to accelerate the rate at which this dye is absorbed by wool. They postulated three reasons for the accelerated rate of dyeing in the presence of these compounds:

(1) The long-range order in the structure of liquid water is disrupted, causing disaggregation of the dye. In addition, the binding forces at the interface are weakened, allowing easier penetration of the dye.
(2) Preferential absorption of the compounds by the fibre displaces water molecules.
(3) Weak complexes are formed between the compounds and the dye. The complexes bind to the wool at sites not normally available to the dye alone.

Thiourea has also been shown to increase the dyeing rate of wool with chrome dyes [73].

2.6   The Physical Basis for Wool Dyeing: The Role of Fibre Structure

When a textile substrate is dyed by an exhaustion method, the dyeing process occurs in four stages:

(1) Diffusion of dye through the aqueous dyebath to the fibre surface.
(2) Sorption of dye by the fibre.
(3) Transfer of dye across the fibre surface.
(4) Diffusion of dye from the surface throughout the whole fibre.

The rate at which dye is supplied to the fibre surface is largely determined by the circulation rate of the dye liquor. In a well stirred dyebath, diffusion of dye to the fibre surface is unlikely to be a critical factor in determining the overall dyeing rate [9]. The initial sorption of dye by wool is affected by the characteristics of the particular dye, the pH of the dyebath and the presence of inorganic salts and surfactants.

In order to obtain satisfactory shade development and fastness properties, complete penetration of dye into the fibre interior is essential. Incomplete dye penetration (i.e. 'ring dyeing') results in a lower depth of shade compared with that produced by the same concentration of dye uniformly distributed throughout the fibre [74,75]. Furthermore, ring dyeing usually results in dyed substrates with poorer fastness properties.

2.6.1   Diffusion of Dyes

The diffusion of dyes is usually described in terms of a diffusion coefficient (D). This parameter is determined by measuring the rate of decrease in the concentration of dye from a dyebath, or by measuring dye migration in a substrate [9,76]. Diffusion coefficients measured by indirect methods, such as change in dyebath concentration, are called 'bulk' diffusion coefficients; these may not always be related to the rate of dye diffusion within the fibres. The kinetics of the dyeing of wool have been described in detail by Jones [11c] and Peters [10c], while Rys and Zollinger [43b] described a solution for the 'non-steady-state' diffusion case. Fick defined two laws to describe diffusion [77]. The first states that under steady-state conditions, the amount of a substance that diffuses through a unit area of a plane in a fixed time is proportional to the concentration gradient and the diffusion coefficient. The second law expresses the rate of change in dye concentration at any point with time. The derivation of diffusion coefficients from these equations is very complex, because of the large number of variables involved. It has been usual, therefore, to simplify the calculations by making several assumptions, such as that fibres have infinite length, that they have circular cross-sections, that the diffusion coefficient varies only with temperature, that equilibrium between dye in solution and in the fibre occurs instantaneously, that dye uptake is a function of diffusion alone and that the dye absorption isotherm is linear [78]. Fick's original equations have been simplified by Hill [79]:

$$C_t/C_m = 4(Dt/\pi r^2)^{1/2} \qquad (2.9)$$

where C_t is the amount of dye in the fibre at time t, C_m is the amount of dye absorbed at equilibrium, D is the diffusion coefficient of the dye and r is the fibre radius.

Other, more complicated solutions to Fick's laws have been developed by Wilson [80], Crank [81,82] and Medley and Andrews [10c,83].

2.6.2   Pathways of Dye Diffusion into Wool

Early workers studying the uptake of dyes by wool were mainly interested in the thermodynamics of the dyeing process. This led to the development of theories based on models such as those of Gilbert and Rideal or Donnan, as discussed in Section 2.2.3 These approaches, which treated the wool fibre as a cylinder of uniform composition, were largely concerned with the situation applying when dyebath equilibrium had been attained. They provided little information on the mechanism of the dyeing process itself.

It is now recognised that the diverse morphological structure of wool is instrumental in determining its dyeing behaviour.

2.6.2.1   The Fibre Surface as a Barrier to Dyeing

The diffusion equations, based on Fick's laws of diffusion (Section 2.6.1), dictate that a plot of dye uptake versus the square root of time should be a straight line over most of the dyeing curve. Although this is the case for many fibres, for wool the initial part of the plot of dye uptake is concave, with the curve only becoming linear after some time [84]. This observation led to the assumption that a 'barrier', with a small capacity for dye, exists at the fibre surface [83]. The barrier was believed to be responsible for the non-Fickian dyeing isotherms obtained with wool [83,85,86]. The concept of a barrier on or near the fibre surface was supported by the observation that surface modification [87,88], including removal of the cuticle [89] and chlorination [90], results in a large increase in the rate of dye uptake.

Early workers identified the epicuticle with the barrier to dye penetration [91–93] This was based on the chemical resistance of the epicuticle (see Section 1.4.1) and the incorrect opinion that this component constituted a continuous membrane around the whole fibre [91,93]. The barrier was also ascribed to the whole cuticle [94] and to the highly crosslinked A-layer of the exocuticle [88]. All these suggestions regarding the nature of the barrier were associated with a common belief that dyes *must* diffuse through the cuticle cells in order to reach the fibre cortex (i.e. the transcellular route shown in Figure 2.3).

The epicuticle is not a continuous membrane, however, but surrounds each individual cuticle cell (see Section 1.4.1). As shown schematically in Figure 2.3, gaps exist between

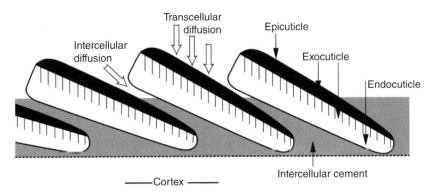

Figure 2.3   *Diffusion pathways of dyes into wool [95].*

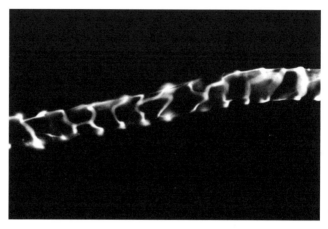

Figure 2.4   Light micrograph showing diffusion of a fluorescent dye at scale junctions.

the scales where the intercellular material extends to the exterior of the fibre; in fact, the intercellular material constitutes approximately 0.05% of the fibre surface [95]. The gaps between the scales make it possible for dyes to penetrate wool without diffusing through the cuticle (i.e. by the intercellular route in Figure 2.3) [95]. A light micrograph showing the initial stages of dye diffusion between cuticle cells is given in Figure 2.4.

This effect was first observed in 1908 by Bowman [96] and confirmed in 1937 by Hall [97], who stated that 'dyes gain access to the interior of the fibre via the junctions between the scales'. Millson and Turl [93] also observed uptake of dye at the edges of cuticle cells and found that the rate of uptake could be increased by distorting the fibre. However, these workers mistakenly believed that this was because rupture of the (supposedly continuous) epicuticle had occurred and that this facilitated penetration of dye into the cuticle [86,93]. This observation can also be explained in terms of an increase in dye accessibility caused by separation of the cuticle cells [98]. In support of this, an increase in dyeing rate has been reported when wool fibres are extended [99]. In this study, scanning electron micrographs of the extended wool fibres showed that the cusps of the cuticle cells were curved upwards and that the gaps between the cuticle cells were enlarged.

It has been suggested that lipids present at the cuticular junctions may hinder entry of dye into the fibre [95,100]; for example, treatment of wool with potassium tert-butoxide in anhydrous tert-butanol, as described in Section 1.4.1, markedly improves the dyeing rate [95]. This observation appears to be inconsistent with the fact that the anhydrous treatment is confined to the fibre surface, where it removes the F-layer lipids from the epicuticle [101]. The anhydrous alkali treatment would, however, be expected to remove lipids from the cell membrane complex at the point at which this component extends to the fibre surface (i.e. at the junctions where the cuticle cells overlap), thereby increasing accessibility of dyes between the cuticle cells [95]. In addition to increasing the overall dyeing rate, tert-butoxide treatment also improves the uniformity of uptake of anionic dyes and fluorescent brightening agents. This results in dyeings that are less skittery than those on untreated wool [95]. Extraction of normally scoured wool with lipid solvents also increases the dyeing rate [87,100,102,103]. This observation again supports the concept of a lipid barrier to wool dyeing located at or near the fibre surface. A significant finding is that surface lipids appear to be concentrated mainly at the edges of the cuticle cells [104].

Leeder *et al.* [98] used specially synthesised dyes to study the mechanism of wool dyeing. These metal-complex dyes contained platinum, palladium or uranium atoms, but in other respects were similar to conventional anionic wool dyes. The nuclear-dense, heavy-metal atoms in the model dyes had a high electron-scattering power, which enabled their location in the fibre at different stages of the dyeing process to be determined with the transmission electron microscope (TEM). This investigation provided the first unequivocal evidence that dye does in fact enter the wool fibre between cuticle cells (Figure 2.5), and showed that dye diffuses along the nonkeratinous endocuticle and cell membrane complex early in the dyeing cycle.

This finding supports the view that the cuticle [94] – probably the highly crosslinked A-layer of the exocuticle [88,105] – is a barrier to dye penetration, in that dyes are directed to the gaps *between* the scales in order to reach the cortex. It appears, however, that lipids present at the intercellular junctions impede the diffusion of dyes into the nonkeratinous regions of the cell membrane complex [95].

The intercellular mode of dye penetration applies to unmodified wool. Different dyeing behaviour (including transcellular diffusion) may be shown by fibres that have been substantially chemically or physically altered, for example by chemical reduction of the A-layer of the exocuticle [88], severe surface abrasion [88,92] or complete removal of the cuticle [89].

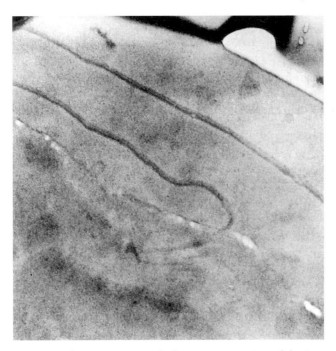

Figure 2.5 *Transmission electron micrograph showing penetration of dye into wool along the cell membrane complex [98].*

2.6.2.2   *Diffusion of Dye in the Cortex*

After dyes have penetrated into wool, they must diffuse throughout the entire fibre cross-section in order to obtain optimum colour yield and fastness properties. Several workers have suggested that the continuous network of the cell membrane complex provides a pathway for the diffusion of reagents into the cortex of wool. The vapours of organic solvents [106], the salts of zirconium and titanium [107], chromium [108] and phosphotungstic acid [109] all appear to penetrate the fibre by this route. Leeder and Rippon [110] have shown that the cell membrane complex swells in formic acid to a much greater extent than does the whole fibre (Table 2.2). They suggested that this disproportionately high degree of swelling is the reason why dye is taken up very rapidly from concentrated formic acid [111].

Further evidence for the importance of the cell membrane complex in the mechanism of wool dyeing was provided by observations on the effect of specific modifications of the fibre on dyeing rate. For example, treatment with the proteolytic enzyme pronase preferentially digests the nonkeratinous proteins of the cell membrane complex [107, 108,112–114], and extraction with formic acid or an n-propanol/water mixture removes lipid and nonkeratinous proteins [106,114–116] from this region. Both treatments produce a significant improvement in dyeing rate [103,117]. Other workers have also suggested that the nonkeratinous proteins [105,107] and/or lipids [100] of the cell membrane complex may impede diffusion of dyes throughout the fibre cortex.

The situation regarding the pathway for dye diffusion into the cortex of wool remained unresolved until the study by Leeder *et al.* [98] involving the TEM described in Section 2.6.2.1. This investigation demonstrated – unequivocally – the importance of all the nonkeratinous components of the fibre in wool dyeing. After dye has entered the fibre between the cuticle cells, diffusion occurs along the cell membrane complex. Dye also transfers progressively from the cell membrane complex into the other nonkeratinous regions, including the endocuticle and the intermacrofibrillar material (Figure 2.5). It is interesting that dye also appears in the nuclear remnants very early in the dyeing cycle, before it can be seen in the surrounding cortical cells. Early appearance of dye in nuclear remnants has also been observed by light microscopy [86]. The mechanism by which this occurs is not clear,

Table 2.2   *Comparison of swelling factors (as defined by Zahn [108,112]) of whole wool and intercellular cement [110].*

Amino Acid	Content/Mol%	
	Whole Wool [116]	Intercellular Cement [114,116]
Arginine	6.9	6.4
Lysine	2.9	3.9
Histidine	0.9	1.6
Aspartic acid	6.5	7.1
Glutamic acid	11.9	8.9
Total polar amino acids (A)	29.1	27.9
1/2-cystine (B)	10.3	1.3
Swelling factor (A/B)	2.8	21.5

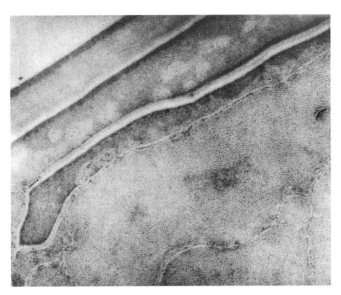

Figure 2.6 *Transmission electron micrograph showing dye located in the sulphur-rich regions of the fibre at equilibrium [98].*

but dyes may diffuse along 'membrane pores'. Kassenbeck [108] has suggested that these pores, which may have been important in the transfer of nutrients during fibre growth prior to keratinisation, connect the nuclear remnants with the endocuticle and the cell membrane complex. Although such pores may be swollen in a hot, aqueous dyebath, they may not be easily seen under the anhydrous conditions used in electron microscopy.

As the dyeing cycle proceeds, dye progressively transfers from the nonkeratinous regions into the sulphur-rich proteins of the matrix surrounding the microfibrils within each cortical cell (Figure 2.6). Dye also transfers from the endocuticle into the exocuticle, particularly the A-layer; this can be clearly seen at the top of Figure 2.6. It appears that, as would be expected, the hydrophobic proteins located in these regions have a higher affinity for wool dyes than do the more hydrophilic proteins that constitute the nonkeratinous regions. At the end of the dyeing process, the nonkeratinous regions, which were important in the early stages of the dyeing cycle, are virtually devoid of dye (Figure 2.7).

These findings on the importance of the nonkeratinous regions as pathways for diffusion of dyes into wool have been confirmed by fluorescence microscopy [85,86,118,119]. However, the lower resolution of the light microscope compared with the TEM restricts the amount of information that can be obtained by this technique. Light microscopy has also been used to show that the nonkeratinous intercellular regions participate in the initial stages of the diffusion of surfactants [120,121] and simple nonionic compounds [122] into wool. A study using the TEM has demonstrated that an anionic polymer (Synthappret BAP) diffuses into wool along the cell membrane complex and intermacrofibrillar regions [123]. At equilibrium, even this high-r.m.m. polymer (3000–10 000 Da) diffuses into the matrix proteins surrounding the microfibrils. It has also been shown that zinc, silver and polymer-based nanoparticles diffuse into wool via the cell membrane complex [124]. Recently, other

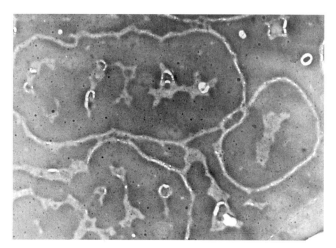

Figure 2.7   *Transmission electron micrograph showing undyed nonkeratinous regions at equilibrium [98].*

workers have confirmed the importance of the nonkeratinous regions in the mechanism of wool dyeing, showing in particular that removal of lipids from these regions facilitates the uptake of dyes by wool [125].

From the preceeding discussion it is clear that, for nonreactive dyes, thermodynamic equilibrium with wool is not established until the process of dye transfer into the keratinous regions is complete. This stage, which is not usually achieved until some time after the dyebath is exhausted, is the reason why a prolonged time at an elevated temperature is required in order to produce satisfactorily dyed wool [86,89,98]. If dye remains largely in the nonkeratinous regions, rapid diffusion out of the fibre can occur, and hence poor wet fastness properties are obtained. Reactive dyes, however, may show a somewhat different equilibrium distribution between the nonkeratinous and keratinous regions of wool [126]. Reactive dyes are capable of forming covalent bonds with the proteins of the nonkeratinous regions [127], and therefore at equilibrium these dyes may be present in the cell membrane complex and endocuticle to a greater extent than are their nonreactive analogues [126].

2.7   Effect of Chemical Modifications on Dyeing

The reactivity of wool with a wide range of chemicals [128] has enabled many industrial treatments to be developed for the modification of fibre and fabric properties. One treatment, involving arylation, is described in Chapters 4 and 10. This increases the affinity of wool for disperse dyes by making the fibre more hydrophobic. Some other treatments, such as those described in this section, change the uptake of wool dyes.

2.7.1   Chlorination

The effect of chlorination on wool printing is discussed in Chapter 12. Chlorination also affects the uptake of dyes in exhaust dyeing. Chlorination oxidises cystine in the A-layer of the exocuticle of wool to cysteic acid [7,128,129] and removes around 60% of the surface lipids from the epicuticle (F-layer) [130]. These changes, which make the fibre surface more hydrophilic, increase the dyeing rate but do not significantly affect the equilibrium dye uptake [131,132]. As discussed in Section 2.6.2, the surface-specific treatment carried out under nonswelling conditions, with potassium tert-butoxide in tert-butanol, also increases both the rate and the evenness of dye uptake [95]. This occurs not through initiation of transcellular diffusion but by increasing the rate of intercellular diffusion (Figure 2.3). It has been suggested that this happens as a result of the removal of lipids, which impede dye penetration, from the junctions between the overlapping cuticle cells [95]. It seems likely, therefore, that chlorination may also increase the rate of dye uptake, by promoting intercellular diffusion following removal of lipids at cell junctions.

A chlorination pretreatment is often used to increase the apparent depths of deep shades dyed with chrome black and navy blue. The mechanism by which this occurs is not understood. It is possible that with high dye concentrations, through increasing the rate of dye uptake, chlorination enables dyeing equilibrium within the fibre cortex to be reached in a commercially acceptable time. As was discussed earlier, for the same amount of dye, ring-dyed fibres have a lower depth of shade than do fibres in which the dye has fully penetrated the cross-section [74,75]. In support of this, the present author found no significant difference between the colour yield on chlorinated versus untreated wool of a very deep chrome black shade that had been dyed for a much longer time is than normally employed in commercial practice. As already discussed, chlorination removes a high proportion of the covalently bound surface lipids of the F-layer. It is possible that this might affect the colour depth of chlorinated wool by changing the specular and diffuse reflectance of the substrate.

2.7.2 Plasma Treatment

Plasma treatment has been studied over many years for shrinkproofing and as a pretreatment for improving the dyeing properties of wool, including increased dyebath exhaustion and shorter dyeing cycles [133,134]. However, other studies, using a range of dyes, found plasma-treated wool to have similar dyeing properties to untreated wool with respect to rate of dye uptake, equilibrium exhaustion and depth of shade [135]. The effect of plasma treatment on the rate of dye uptake appears to be dependent on the hydrophilic/hydrophobic properties of the dye. This is consistent with the plasma treatment increasing the wettability of the wool surface by removing bound lipids, thereby exposing the underlying hydrophilic protein material, and also by oxidising cystine to cysteic acid residues [136]. In the absence of a levelling agent, plasma treatment has been found to have little effect on the rate of adsorption of relatively hydrophobic, sulphonated dyes [136]. This is consistent with the greater importance of hydrophobic dye–fibre interactions than of ionic interactions for these dyes. For relatively hydrophilic dyes, however, the rate of dye uptake was increased by plasma treatment, which would be expected due to the greater importance of ionic interactions between the more strongly anionic, hydrophilic dye molecules and a more hydrophilic fibre surface. It has also been shown that when plasma-treated wool is dyed in the presence of a levelling agent, even hydrophilic dyes show similar rates of uptake on both the plasma-treated and untreated wool [137]. It should be noted that, because the plasma treatments used by most workers affect only the surface of the cuticle cells, these effects on dye uptake relate solely to surface sorption and not to diffusion in the cortex. Severe plasma treatments have been found to produce considerable oxidation of the cystine in the A-layer of the exocuticle, and in some cases cracking of the cuticle cells; these effects can result in transcellular diffusion of dye [138]. Under the milder conditions used by most workers, however, although plasma treatment enhances the uniformity of dye adsorption, the intercellular diffusion pathway shown by untreated wool still occurs [136].

Despite the large number of investigations into the effects of plasma treatment of wool, the technology has not been widely adopted by the industry for the improvement of wool properties. The reasons for this probably include the cost of the equipment and the high energy requirements. Regarding plasma treatment's use as a pretreatment for low-temperature wool dyeing, cheaper alternatives using special auxiliaries are available, as discussed in Chapter 3.

2.7.3   Differential Dyeing

The dyeing properties of wool can be modified by pretreatment with compounds that either resist or assist dye uptake [139]. Blending the treated wool with untreated substrate enables tone-on-tone and multicolour effects to be produced by fabric, yarn or garment dyeing methods [140]. Of the compounds investigated, two have been used commercially: Sandospace R and Sandospace DPE. Sandospace R is a 2,4-dichloro-s-triazine derivative of a sulphonated aromatic amine [141]. It reacts with nucleophilic sites in wool, in particular basic amino groups, thereby decreasing the cationic character of the fibre. Sandospace DPE is a fibre-reactive quaternary ammonium compound that increases the substantivity of acid dyes on wool [142].

2.8   Conclusion

Wool is a heterogeneous fibre of complex chemical and morphological structure. Theories of dyeing that treat wool fibres as homogeneous cylinders are now known to provide an oversimplified picture of the chemical and physical processes involved in dyeing wool.

The older theories on the dyeing of wool, which were based on ionic attraction between the dye and 'dye sites' in the fibre, are now known to be only partly valid. It appears that the ionic forces of attraction act predominantly to control the rate of dyeing, while hydrophobic interactions are largely responsible for the affinity of dyes for wool, and therefore for their wet fastness properties on the substrate.

The diffusion of dyes into virgin wool is a continuum of processes, not a series of discrete steps. The relatively highly crosslinked surface of the A-layer of the cuticle acts as a barrier to dye penetration. In the case of untreated wool, dyes diffuse into the fibre where the cuticle cells overlap. They then move into the nonkeratinous regions of the endocuticle, the intercellular cement and the intermacrofibrillar material. As the dyes diffuse through the nonkeratinous components, progressive transfer occurs into regions containing hydrophobic proteins; these have a higher affinity for wool dyes than the proteins of the more hydrophilic nonkeratinous regions. At equilibrium, for nonreactive dyes, the nonkeratinous regions are devoid of dye. This is necessary in order to achieve maximum wet fastness, because dye remaining in the cell membrane complex can easily diffuse out of the fibre.

In the case of reactive dyes, however, significant amounts of dye may be located in the nonkeratinous components. In this case, because the dye is covalently bound, wet fastness is not affected.

Wool dyes can be highly aggregated in the aqueous environment of the dyebath, even at the boil. In some cases, the size of the aggregates can impede diffusion into the fibre and auxiliaries are necessary to disaggregate the dyes in order to increase both the rate and the levelness of the dye uptake.

References

[1] D M Lewis, *Journal of the Society of Dyers and Colourists*, **98** (1982) 165.
[2] I B Angliss, *Textile Progress*, **12**(3) (1982) 6.
[3] D M Lewis, *Proceedings of the 8th International Wool Textile Research Conference, Christchurch*, **4** (1990) 1.
[4] D M Lewis, *Review of Progress in Coloration and Related Topics*, **8** (1977) 10.
[5] J R Christoe, R J Denning, D J Evans, M G Huson, L N Jones, P R Lamb, K R Millington, D G Phillips, A P Pierlot, J A Rippon and I M Russell, *Encyclopedia of Polymer Science and Technology* online edition http://www.mrw .interscience.wiley.com/epst/index.html (Chichester, UK: John Wiley & Sons Ltd, 2003), last accessed 1 February 2013.
[6] D M Lewis, in *Handbook of Textile and Industrial Dyeing, Part I: Textile Applications*, ed. M Clark (Cambridge, UK: Woodhead, 2011), Chapter 1, 3.
[7] J A Rippon and D J Evans, in *Handbook of Natural Fibres: Vol. 2: Processing and Applications*, ed. R. Kozlowski (Cambridge, UK: Woodhead, 2012), Chapter 3, 63.
[8] H Zollinger, *Proceedings of the 5th International Wool Textile Research Conference, Aachen*, **1** (1975) 167.
[9] T Vickerstaff, *The Physical Chemistry of Dyeing* (London, UK: Oliver and Boyd, 1954), 344.
[10] R H Peters, *Textile Chemistry, Vol III: The Physical Chemistry of Dyeing* (New York, USA: Elsevier Scientific Publishing Company, 1975), (a) 203; (b) 851; (c) 275.
[11] C L Bird and W S Boston (eds), *The Theory of Coloration of Textiles* (Bradford, UK: Dyers' Company Publication Trust, 1975), (a) C H Giles, 41; (b) L Peters, 163; (c) F Jones, 237.
[12] J A Bone, J Shore and J Park, *Journal of the Society of Dyers and Colourists*, **104** (1988) 12.
[13] M Feughelman, in *Encyclopedia of Polymer Science and Engineering*, 2nd Edn, **Vol. 8** (New York, USA: John Wiley & Sons Ltd, 1987), 566.
[14] J D Leeder and I C Watt, *Journal of Colloid and Interface Science*, **48**(2) (1974) 339.
[15] J B Speakman, *Transactions of the Faraday Society*, **40** (1944) 6.
[16] J W Rowen and R L Blaine, *Journal of Research of the National Bureau of Standards*, **39** (1947) 479.
[17] P Alexander and R F Hudson, in *Wool: Its Chemistry and Physics*, ed. C Earland (London, UK: Chapman and Hall, 1963), 180.
[18] H Beyer and U Schenk, *Journal of Chromatography*, **39** (1969) 491.

[19] W G Crewther, R D B Fraser, F G Lennox and H Lindley, in *Advances in Protein Chemistry*, **Vol. 20**, ed. C B Anfinsen Jr, M L Anson, J T Edsall and F M Richards (New York, USA: Academic Press, 1965), 191.

[20] J Steinhardt and M Harris, *Journal of Research of the National Bureau of Standards*, **24** (1940) 335.

[21] I D Rattee, *The Physical Chemistry of Dye Absorption* (London, UK and New York, USA: Academic Press, 1974).

[22] E Elöd, *Transactions of the Faraday Society*, **29** (1933) 327.

[23] G A Gilbert and E K Rideal, *Proceedings of the Royal Society*, **A182** (1944) 335.

[24] R H Fowler and E A Guggenheim, *Statistical Thermodynamics* (Cambridge, UK: Cambridge University Press, 1939), 426.

[25] J Steinhardt and M Harris, *Journal of Research of the National Bureau of Standards*, **24** (1940) 335.

[26] J Steinhardt, *Journal of Research of the National Bureau of Standards*, **28** (1942) 191.

[27] J B Speakman and E Stott, *Transactions of the Faraday Society*, **31** (1935) 1425.

[28] L Peters, *Symposium on Fibrous Proteins* (Bradford, UK: Society of Dyers and Colourists, 1946), 138.

[29] H H Sumner, *Journal of the Society of Dyers and Colourists*, **102** (1986) 341.

[30] J Meybeck and P Galafassi, *Proceedings of the 4th International Wool Textile Research Conference, San Francisco*, Part 1 (1970) 463.

[31] H Zollinger, *Journal of the Society of Dyers and Colourists*, **81** (1965) 345.

[32] A N Derbyshire and R H Peters, *Journal of the Society of Dyers and Colourists*, **71** (1955) 530.

[33] A N Derbyshire, *Hexagon Digest*, **21** (1955) 12.

[34] A N Derbyshire, *Transactions of the Faraday Society*, **51** (1955) 909.

[35] N S Allen, *Colour Chemistry* (Bath, UK: Pitman Press, 1971).

[36] E R Trotman, *The Dyeing and Chemical Technology of Textile Fibres*, 4th Edn (London, UK: Charles Griffin, 1970), 378.

[37] C A Hunter, *Chemical Society Reviews*, **23** (1994) 101.

[38] S R S Iyer and D Srinivasan, *Journal of the Society of Dyers and Colourists*, **100** (1984) 63.

[39] F M Cozzi, M Cinquini, R Annuziata and J S Siegel, *Journal of the American Chemical Society*, **115** (1993) 5330.

[40] W L Forgenson and D L Severence, *Journal of the American Chemical Society*, **112** (1990) 4768.

[41] C A Hunter and J A Sanders, *Journal of the American Chemical Society*, **112** (1990), 5525.

[42] Colour Terms and Definitions, 1988 edition (Bradford, UK: Society of Dyers and Colourists).

[43] P Rys and H Zollinger, *Fundamentals of the Chemistry and Application of Dyes* (New York, USA: Wiley Interscience, 1972), (a) 172; (b) 177.

[44] G A Gilbert, *Proceedings of the Royal Society*, **A183** (1944) 167.

[45] C L Bird, *The Theory and Practice of Wool Dyeing* (Bradford, UK: 1951), 26.

[46] H Egli, *Textilveredlung*, **18** (1983) 319.

[47] J Shore, *Journal of the Society of Dyers and Colourists*, **84** (1968) 408, 413, 545; **85** (1969) 14.

[48] R S Asquith and D K Chan, *Journal of the Society of Dyers and Colourists*, **87** (1971) 181.

[49] D G Duff and C H Giles, *Water*, **Vol. 4**, ed. F Franks (New York, USA: Plenum, 1975), 169.

[50] C Robinson, *Journal of the Society of Dyers and Colourists*, **50** (1934) 161.

[51] E Valko, *Journal of the Society of Dyers and Colourists*, **55** (1939) 173.

[52] J B Speakman and H Clegg, *Journal of the Society of Dyers and Colourists*, **50** (1934) 348.

[53] B R Craven and A Datyner, *Proceedings of the 4th International Conference on Surface Active Substances*, **3** (1964) 545.

[54] A Datyner, A G Flowers and M T Pailthorpe, *Journal of Colloid and Interfacial Science*, **74** (1980) 71.

[55] B R Craven, A Datyner and J F Kennedy, *Australian Journal of Chemistry*, **24** (1971) 723.

[56] D G Duff, D J Kirkwood and D M Stevenson, *Journal of the Society of Dyers and Colourists*, **93** (1977) 303.

[57] A A R El-Mariah, I E El-Sabbagh and A Labib, *Doklady Akademii nauk SSSR*, **234** (1977) 72.

[58] H P Frank, *Journal of Colloid Science*, **12** (1957) 480.

[59] M S Yen and H Y Lee, *J. Chin. Inst. Eng.*, **9**(6) (1986) 587.

[60] A Datyner and M T Pailthorpe, *Journal of Colloid and Interfacial Science*, **76** (1980) 557.

[61] A Datyner and M T Pailthorpe, *Proceedings of the 6th International Wool Textile Research Conference, Pretoria*, **5** (1980) 585.

[62] A Datyner and M T Pailthorpe, *Dyes and Pigments*, **8** (1987) 253.

[63] R J Hine and J McPhee, *Proceedings of the 3rd International Wool Textile Research Conference, Cirtel, Paris*, **3** (1965) 261.

[64] Y Nemoto and H Funahashi, *Proceedings of the 7th International Wool Textile Research Conference, Tokyo*, **5** (1985) 231.

[65] R S Asquith, A K Booth, K R F Cockett, I D Rattee and C B Stevens, *Journal of the Society of Dyers and Colourists*, **88** (1972) 62.

[66] K R F Cockett, I D Rattee and C B Stevens, *Journal of the Society of Dyers and Colourists*, **85** (1969) 461.

[67] K R F Cockett, D J Kilpatrick, I D Rattee and C B Stevens, *Applied Polymer Symposium*, **18** (1971) 409.

[68] R S Asquith and A K Booth, *Textile Research Journal*, **40** (1970) 410.

[69] R J Blagrove, *Journal of the Society of Dyers and Colourists*, **89** (1973) 212.

[70] K L Gardner and R J Blagrove, *Journal of the Society of Dyers and Colourists*, **90** (1974) 331.

[71] D J Kilpatrick and I D Rattee, *Journal of the Society of Dyers and Colourists*, **93** (1977) 424.

[72] B C Burdett and J A Galek, *Journal of the Society of Dyers and Colourists*, **98** (1982) 374.

[73] I A Ledneva and L V Anisimova, *Izv Vyssh Uchebn Zaved, Tekhnol Tekst Prom-sti*, **5** (1986) 58.

[74] D A Garrett and R H Peters, *Journal of the Textile Institute*, **47** (1956) T166.

[75] T H Morton, *Journal of the Society of Dyers and Colourists*, **92** (1976) 149.

[76] P R Brady, *Review of Progress in Coloration and Related Topics*, **22** (1992) 58.

[77] A Fick, *Annalen der Physik (Leipzig)*, **170** (1855) 59.

[78] J N Etters, *Textile Chemist and Colorist*, **12** (1980) 140.

[79] A V Hill, *Proceedings of the Royal Society*, **B104** (1928) 39.

[80] A H Wilson, *Philosophical Magazine*, **39** (1948) 48.

[81] J Crank, *Philosophical Magazine*, **39** (1948) 362.

[82] J Crank, *The Mathematics of Diffusion* (Oxford, UK: Oxford University Press, 1956).

[83] J A Medley and M W Andrews, *Textile Research Journal*, **29** (1959) 398.

[84] W Zhao and M T Pailthorpe, *Textile Research Journal*, **57** (1987) 579.

[85] J D Leeder and J A Rippon, *Proceedings of the International Symposium on Fiber Science and Technology, Hakone, Japan*, **II-9** (1985) 203.

[86] J D Leeder, L A Holt, J A Rippon and I W Stapleton, *Proceedings of the 8th International Wool Textile Research Conference, Christchurch*, **4** (1990) 227.

[87] J A Medley and M W Andrews, *Textile Research Journal*, **29** (1960) 855.

[88] G M Hampton and I D Rattee, *Journal of the Society of Dyers and Colourists*, **95** (1979) 396.

[89] H Zollinger, *Melliand Textilber*, English Edn, **9** (1987) E294.

[90] E R Trotman, *Dyeing and Chemical Technology of Textile Fibres*, 4th Edn (London, UK: Charles Griffin, 1970) 266.

[91] J Lindberg, E H Mercer, B Philip and N Gralén, *Textile Research Journal*, **19** (1949) 673.

[92] V Kopke and B Nilssen, *Journal of the Textile Institute*, **51** (1960) T1398.

[93] H E Millson and L H Turl, *American Dyestuff Reporter*, **39** (1950) 647.

[94] K R Makinson, *Textile Research Journal*, **38** (1968) 831.

[95] J D Leeder, J A Rippon and D E Rivett, *Proceedings of the 7th International Wool Textile Research Conference, Tokyo*, **4** (1985) 312.

[96] F H Bowman, *The Structure of the Wool Fibre* (London, UK: Macmillan and Co., 1908), 435.

[97] R O Hall, *Journal of the Society of Dyers and Colourists*, **53** (1937) 341.

[98] J D Leeder, J A Rippon, F E Rothery and I W Stapleton, *Proceedings of the 7th International Wool Textile Research Conference, Tokyo*, **5** (1985) 99.

[99] J Koga, K Joko and N Kuroki, *Proceedings of the 7th International Wool Textile Research Conference, Tokyo*, **5** (1985) 14.

[100] K Joko, J Koga and N Kuroki, *Proceedings of the 7th International Wool Textile Research Conference, Tokyo*, **5** (1985) 23.

[101] J D Leeder and J A Rippon, *Journal of the Society of Dyers and Colourists*, **101** (1985) 11.

[102] J Lindberg, *Textile Research Journal*, **23** (1953) 573.

[103] F J Harrigan and J A Rippon, *Proceedings of the Textile Institute World Conference, Sydney* (1988) 412.

[104] L E Aicolina and I H Leaver, *Proceedings of the 8th International Wool Textile Research Conference, Christchurch*, **4** (1990) 297.

[105] H Baumann and L D Setiawan, *Proceedings of the 7th International Wool Textile Research Conference, Tokyo*, **5** (1985) 1.

[106] J H Bradbury, J D Leeder and I C Watt, *Applied Polymer Symposium*, **18** (1971) 227.

[107] H Baumann, in *Fibrous Proteins: Scientific, Industrial and Medical Aspects*, **Vol. 1**, ed. D A D Parry and L K Creamer (London, UK: Academic Press, 1979) 299.

[108] H Zahn, *Plenary Lecture, 6th International Wool Textile Research Conference, Pretoria*, **1** (1980).

[109] J A Swift and A W Holmes, *Textile Research Journal*, **35** (1965) 1014.

[110] J D Leeder and J A Rippon, *Journal of the Textile Institute*, **73** (1982) 149.

[111] J D Leeder and J A Rippon, *Journal of the Society of Dyers and Colourists*, **99** (1983) 64.

[112] H Zahn, *Lenzinger Berichte*, **4** (1977) 19.

[113] J H Bradbury, in *Advances in Protein Chemistry*, **Vol. 27**, ed. C B Anfinsen Jr, J T Edsall and F M Richards (New York, USA: Academic Press, 1973), 111.

[114] J D Leeder, *Wool Science Review*, **63** (1986) 3.

[115] J D Leeder, D G Bishop and L N Jones, *Textile Research Journal*, **53** (1983) 402.

[116] J D Leeder and R C Marshall, *Textile Research Journal*, **52** (1982) 245.

[117] H D Feldtman, J D Leeder and J A Rippon, in *Objective Evaluation of Apparel Fabrics*, ed. R Postle, S Kawabata and M Niwa (Osaka, Japan: Text. Mach. Soc. Japan, 1983) 125.

[118] P R Brady, *Proceedings of the 7th International Wool Textile Research Conference, Tokyo*, **5** (1985) 171.

[119] P R Brady, *Proceedings of the 8th International Wool Textile Research Conference, Christchurch*, **4** (1990) 217.

[120] L A Holt and I W Stapleton, *Proceedings of the Textile Institute World Conference, Sydney* (1988) 420.

[121] L A Holt and I W Stapleton, *Journal of the Society of Dyers and Colourists*, **104** (1988) 387.

[122] L A Holt and J A Saunders, *Textile Research Journal*, **56** (1986) 415.

[123] J R Cook, B E Fleischfresser, J D Leeder and J A Rippon, *Textile Research Journal*, **59** (1989) 754.

[124] D G King and A P Pierlot, *Coloration Technology*, **125** (2009) 111.

[125] M Marti, R Ramirez, C Barba, L Coderch and J Parra, *Textile Research Journal*, **80** (2010) 365.

[126] D M Lewis and S M Smith, *Journal of the Society of Dyers and Colourists*, **107** (1991) 351.

[127] H Baumann, *Applied Polymer Symposium*, **18** (1971) 307.

[128] J A Maclaren and B Milligan, *Wool Science: The Chemical Reactivity of the Wool Fibre* (New South Wales, Australia: Science Press, 1981).

[129] J A Rippon, in *Friction in Textile Materials*, ed. B S Gupta (Cambridge, UK: Woodhead, 2008), Chapter 7, 253.

[130] A P Negri, H J Cornell and D E Rivett, *Textile Research Journal*, **62** (1992) 381.

[131] F Townend, *Journal of the Society of Dyers and Colourists*, **61** (1945) 144.

[132] J. Barritt and F F Elsworthy, *Journal of the Society of Dyers and Colourists*, **64** (1948) 23.

[133] K Chi-Wai and M Y Chun-Wah, *Textile Progress*, **39**(3) (2007) 121.

[134] H Thomas, in *Plasma Technology for Textiles*, ed. R Shishoo (Cambridge, UK: Woodhead, 2007), 228.

[135] K M Byrne, W Rakowski, A Ryder and S Havis, *Proceedings of the 9th International Wool Textile Research Conference, Biella*, **4** (1995) 422.

[136] M Naebe, P G Cookson, J A Rippon, P R Brady, X Wang, N Brack and G van Riessen, *Textile Research Journal*, **80** (2010) 312.

[137] M Naebe, P G Cookson and J A Rippon, X Wang, *Textile Research Journal*, **80** (2010) 611.

[138] H Höcker, H Thomas, A Küsters and J Herrling, *Melliand Textilber.*, **75** (1994) 506.

[139] V A Bell, D M Lewis and M T Pailthorpe, *Journal of the Society of Dyers and Colourists*, **100** (1984) 223.

[140] R R D Holt, *Proceedings of the 8th International Wool Textile Research Conference, Christchurch*, **4** (1990) 107.

[141] Sandoz, *Multicolour Effects on Wool*, No. 1570 (Basle, Germany: 1971).

[142] Sandoz Technical Information, *Sandospace DPE Liquid*, **8** (1987).

3

The Role of Auxiliaries in the Dyeing of Wool and other Keratin Fibres

Arthur C. Welham
The Dyehouse Doctor Ltd, UK

3.1 Introduction

The role of dyeing auxiliaries is generally not well understood, for various reasons. First, it is difficult to assess the performance or strength of auxiliaries – this is unlike dyes, which have the easily discernible and differentiated property of colour. Second, there is no authoritative guide to textile chemicals, such as a comparable publication to the *Colour Index*. Such a guide would be very difficult to produce, because of the widespread use of complex mixtures and the uncertainty as to the active content of commercial products. The dyer, therefore, cannot be certain as to the chemical nature of individual products or which products are similar in chemical structure or action. This situation has become even more confused following significant restructuring of the dye and chemical industry in recent years and an increase in the secrecy engendered by that thinking which is currently fashionable in industrial marketing. Third, comparatively little practically orientated work on the action of auxiliaries has been published; perhaps as a result, dyebath chemicals and their properties often do not figure largely in the formal education of dyers and textile chemists.

In fact, the action of many auxiliaries used in the dyeing of wool and other keratin fibres can be explained in fairly simple terms that, however inadequate they may be in relation to absolute truth or to modern dyeing theory, can be used to predict the action of a given product in overcoming a given problem.

Because of the large number of applications of some of the surface-active products, in this chapter we will first cover the general properties of each of the ionic groups of auxiliaries and then discuss each application sector in more detail.

The Coloration of Wool and other Keratin Fibres, First Edition. Edited by David M. Lewis and John A. Rippon.
© 2013 SDC (Society of Dyers and Colourists). Published 2013 by John Wiley & Sons, Ltd.

3.2   Surface Activity of Wool-Dyeing Auxiliaries

Many of the auxiliaries used in wool dyeing are surface-active, and it is essential that the general chemistry of these products is considered in order that their role in dyeing can be clearly understood. While some appreciation of the theory of wetting and detergency is necessary, this is not covered here; there are already many excellent texts on this subject [1,2]. Surfactants are usually divided into groups according to the charge carried by the larger ion in aqueous solution.

3.2.1   Anionic Auxiliaries

In these products, the larger part of the molecule takes on a negative charge in the dyebath, as do most of the dye classes used on wool. Anionic compounds include soaps, such as sodium stearate, and detergents, such as sodium oleylsulphate (**1**). The properties of these products can be changed by modifying the structure. For example, if the sulphate group in Structure **1** is replaced by a sulphonate group, the product becomes more polar and more resistant to hard water. If the alkyl chain is shortened, the compound loses its power of detergency but retains its properties as a wetting agent. Soaps, which were widely used in textile processing before the introduction of synthetic detergents, are not resistant to hard water or acids. Both calcium and magnesium salts are insoluble, as is the free acid, and scumming therefore occurs. The slight alkalinity of soap solutions can also lead to problems. These drawbacks are avoided in synthetic detergents, in which sulphate or sulphonate groups replace the carboxylate groups in soaps (i.e. detergents are derivatives of stronger acids). These products are resistant to hard water and do not precipitate at low pH. Furthermore, they produce neutral solutions, although they may be more effective as detergents when used in conjunction with alkali. Early commercial examples of anionic detergents included Lissapol D (ICI) (sodium oleylsulphate) and the corresponding sulphonate Igepon T (former IG Farben name). Modern household detergents and washing-up liquids are usually based on alkylbenzenesulphonates; sodium dodecylbenzenesulphonate (**2**), for example, is used in washing powders.

$$C_{17}H_{33}OSO_3^- \ Na^+ \qquad\qquad C_{12}H_{25}\!-\!\langle\!\langle\ \rangle\!\rangle\!-\!SO_3^- \ Na^+$$

$$\textbf{1} \qquad\qquad\qquad\qquad\qquad \textbf{2}$$

Anionic products of the simple detergent type are used in the dyeing of wool/polyamide blends, where they behave as colourless acid dyes and are able to compete for sites with the lower relative molecular mass (r.m.m.) acid levelling dyes. This behaviour leads to a more effective restraining action on polyamide fibres, which have considerably fewer dye sites than has wool; these anionic compounds can therefore be used to control the proportion of dye on the polyamide fibres, which would normally be dyed more deeply than wool in pale and medium depths. Commercial products include Thiotan LM (Clariant), Thiotan RMF (Clariant) and the former Matexil WA-HS (ICI), previously known as Calsolene Oil HS.

Anionic auxiliary products can also be produced by condensing naphthalenesulphonic acid with formaldehyde (**3**). Products similar to this, such as the former Matexil DA-AC (ICI), Irgasol DAM (Huntsman), Avolan IS (Tanatex) and Lyocol O (Clariant), can be used to exert a restraining effect and are more effective than the alkyl sulphates or sulphonates at neutral pH. They do diffuse into the wool fibres, however, and since the relatively large molecules are not removed by rinsing, their tendency to yellow on exposure to light can impair light fastness in all but very full depths of shade. These products are most widely used as dispersing agents in polyester dyeing.

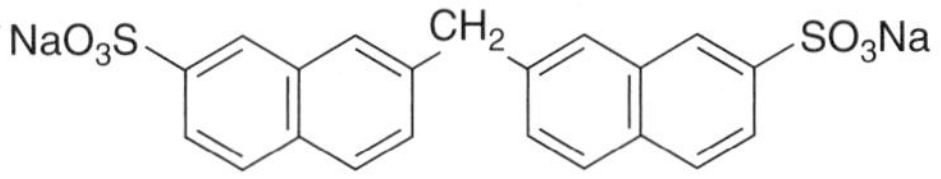

3

Condensation products from high-r.m.m. aromatic sulphonic acids and formaldehyde, similar to the syntans used in the leather industry, are widely used to aftertreat dyeings on polyamide fibres in order to improve wet fastness properties. Examples are Mesitol NBS (Tanatex), Nylofixan P (Clariant), Nylofixan PST (Clariant) and Erional RF (Huntsman). These products can also be used to control the partition of acid milling and 1:2 metal-complex dyes between wool and polyamide fibres in blends. Generally, the balance of the condensate is modified for the particular application; products specifically designed for wool/polyamide blend dyeing include Erional RF (CGY), Thiotan WPN (Clariant), Rewin KNR (CHT) and Mesitol HWS (Tanatex).

3.2.2 Cationic Auxiliaries

In these products, part of the molecule carries a positive charge, as do the basic dyes used widely in acrylic dyeing, but very rarely in wool dyeing. Examples include quaternary ammonium compounds, such as cetyltrimethylammonium bromide (**4**), and tertiary ammonium derivatives, which are more weakly cationic.

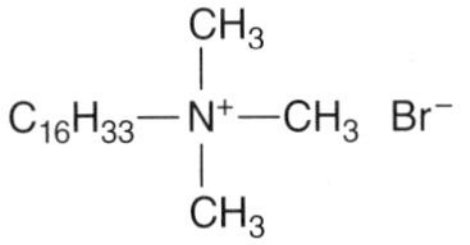

4

Cationic products are widely used to improve the wet fastness properties of dyeings, particularly when dyeing with direct dyes on cellulosic fibres, but also when using metal-complex and milling dyes on wool.

The most important group of cationic auxiliaries used in wool dyeing is the polyethoxylated cationic products (e.g. tertiary amines), which are used as levelling agents. These are discussed more fully in the next section.

3.2.3   Ethoxylated Nonionic and Cationic Auxiliaries

The term 'nonionic auxiliaries' refers almost universally to polyoxyethylene and poly-oxypropylene derivatives. The history of these products dates back to 1930, when C. Schöller, working in the Ludwigshafen laboratories of IG Farbenindustrie, discovered polyethoxylated compounds [3]. The main industrial use of such products today is as detergents, but their first use was in the textile industry as levelling agents in dyeing. The polyethoxylated oleic and stearic acids developed by Schöller will retard the adsorption of all ionic dyes.

There is a tremendous versatility in ethoxylation, in that by varying the hydrophobic group – the initiating group containing the necessary reactive hydrogen and the degree of polymerisation – an infinite range of products with an extensive array of specialised properties can be produced. Further to their widespread use as detergents, polyethoxylated nonionic and cationic products are still the most dominant class of level-dyeing assistant.

Nonionic and weakly cationic surfactants are produced from the addition of ethylene oxide to compounds containing one or more reactive hydrogen atom, such as fatty acids, fatty alcohols, alkylphenols, fatty mercaptans (thiols), fatty amines, fatty amides, polyols and amine oxides (Scheme 3.1). Of course, like all polymers, such products are not just one species, and commercial products may contain a wide range of chain lengths, from the totally unethoxylated, relatively hydrophobic starting material upwards. Nevertheless, most of the product from this type of reaction will be polyethoxylated to a chain length within a fairly narrow range. This is important with respect to the specific properties required.

Nonionic surfactants do not ionise in aqueous solution, but due to hydration at the ether oxygen atoms the polyoxyethylene chain does take on a small positive charge; the compounds thus have advantages as detergents and emulsifiers. They possess several unusual properties as a result of their solubility being dependent on the hydration of the hydrophilic polyether chain. Their solubility is higher at lower temperatures, and the solutions of many of them become turbid above a certain temperature. This turbidity is termed the 'cloud point'.

In order to have any degree of solubility, a minimum of from four to six ethylene oxide units per molecule is required, depending on the nature of the hydrophobic end group. Above the cloud point, solutions begin to separate into two phases: in one, the surfactant concentration is high, and in the other it is depleted. While the detergent properties of nonionic surfactants are comparatively poor above the cloud point, in other applications, such as low-temperature dyeing, this property of phase separation is useful.

To understand the way in which surfactants work, it is necessary to have some knowledge of the theories relating to detergency, wetting, emulsification and solubilisation. These are covered well in the existing literature [1,2,4]. Briefly, however, surfactants work by forming micelles in aqueous solution; by incorporating other molecules within the micelle, they obtain a detergent or emulsifying/solubilising effect. The critical micelle concentration (CMC) is an important property of all surfactants: it is defined as the lowest concentration at

$$ROCH_2CH_2OH + n\ H_2C{-}CH_2 \longrightarrow ROCH_2CH_2(OCH_2CH_2)nOH$$
$$\underset{O}{\vee}$$

Scheme 3.1   *The ethoxylation of aliphatic alcohols to produce nonionic surfactants.*

which micelles will form. The CMC of anionic surfactants increases with temperature; with nonionics, however, the converse is true: their CMC is at a minimum just below the cloud point, and consequently the detergency power at a given concentration is at a maximum at that temperature. Longer-chain polyoxyethylene products with more than about 20–25 ethylene oxide units per molecule do not cloud below the boiling point of water and can therefore find use either at or above 100 °C.

In recent years, one group of nonionic surfactants has received particular attention from environmentalists, due to the suspicion that its members are endocrine disruptors and may interfere with the hormone systems in animal life. This group is the polyethoxylates of alkyl phenols. Ethoxylated nonyl and octyl phenol have been widely used as detergents in many industries, including textiles, and as a component of domestic laundry detergents. In 1984 the formation of 4-nonyl phenol in wastewater treatment of effluent containing nonyl phenol ethoxylates was first reported [5]. Nonyl phenol is persistent in the environment [6], and the quantities of nonyl phenol ethoxylates used are very high. Therefore, despite being thought of as only a weak endocrine disruptor, due to its suspected ability to mimic oestrogen it is considered a major environmental risk. The use of such products is now restricted in the European Union and North America, but they continue to be used in other parts of the world, causing considerable concern [7–9]. Most major producers have replaced alkyl phenol ethoxylates with the more expensive alcohol ethoxylates, which have similar surface-active properties.

Examples of nonionic auxiliaries used in wool dyebaths include Matexil DN-VL (ICI), Ekaline F (Clariant) and Avolan IW (Tanatex), which are all ethoxylated aliphatic alcohols (e.g. **5**; $n = 20$–25). Others are polyethoxylated aromatic alcohols, such as the former Lissapol N (ICI) and Avolan SC (Tanatex) (**6**; $n = 9$–10). The two types of product have quite different properties. The aliphatic products can be used as levelling agents in neutral or weakly acid conditions, but cause some foaming problems. The aromatic products have too strong a retarding action under neutral conditions, but can be used as levelling agents at around pH 4. Nonionics of both types are used as levelling agents with 1 : 1 metal-complex dyes to enable the use of smaller amounts of sulphuric acid. Originally, as much as 12% on mass fabric (o.m.f.) sulphuric acid was recommended, but since wool is damaged by this concentration of acid, it is now preferred to use dyebaths set with 4–8% o.m.f. sulphuric acid, plus a nonionic auxiliary to give satisfactory levelness.

$$C_{18}H_{35}-O-(CH_2CH_2O)_nH \qquad\qquad C_9H_{19}-\!\!\!\langle\bigcirc\rangle\!\!\!-O(CH_2CH_2O)_nH$$

$$\textbf{5} \qquad\qquad\qquad\qquad\qquad \textbf{6}$$

Nowadays, the more important levelling agents are polyethoxylated amines or quaternary ammonium compounds, which have both nonionic and cationic properties. These products retard anionic dyes more effectively than do wholly nonionic auxiliaries, due to the formation of an ionic complex between the auxiliary and the dye. Ideally, this complex should break down slowly as the dyeing temperature rises, which permits a gradual and level uptake of dye. Some of these products, such as ethoxylated tertiary amines, can reduce the pH sensitivity of acid dyes as they are increasingly protonated at lower pH, which counterbalances the cationic character of the wool keratin (see Chapter 1). Polyethoxylated quaternary ammonium products are completely ionised at all pH values, and many are too strongly cationic to be truly effective levelling agents.

Commercial examples of polyethoxylated amines include Lyogen NH (Clariant), Lyogen MS (Clariant) and Albegal SW (Huntsman). These products have very long ethylene oxide chains (possibly as many as 90–120 units in two chains). Less highly ethoxylated products, which are therefore more cationic, include Levegal ED (Tanatex), Keriolan A2N (CHT) and the former Matexil LC-CWL (ICI). Even more strongly cationic and probably quaternised products include Seragal W-KW (DyStar), Lyogen WD (Clariant) and Albegal W (Huntsman). Structure **7** shows a typical example of a cationic levelling agent; it is important that the products used should have the right balance of nonionic and cationic properties.

$$C_{17}H_{35}N \begin{cases} (CH_2CH_2O)_4CH_2CH_2OH \\ (CH_2CH_2O)_4CH_2CH_2OH \end{cases}$$

7

The addition of sodium sulphate has occasionally been recommended in order to promote micelle formation with polyethoxylated levelling agents, but this is not really necessary. When dyeings of milling or metal-complex dyes are carried out at above the isoelectric point of wool, Glauber's salt can have the undesirable effect of increasing the rate of dyeing.

3.2.4   Amphoteric Auxiliaries

These products are usually polyethoxylates that possess both cationic and anionic groups. A typical structure is shown in Structure **8**, in which one of the terminal hydroxyl groups of an ethoxylated amine has been sulphated. Amphoteric auxiliaries are particularly effective in the coverage of tippy wool (i.e. the production of nonskittery dyeings), even with dyes that are very prone to this fault, such as fibre-reactive or disulphonated 1 : 2 metal-complex dyes. Commercial products include Albegal A (Huntsman), Albegal B (Huntsman), Uniperol SE (BASF), Lyogen FN (Clariant), Sarabid PAW (CHT), Albegal SET (Huntsman) and Lyogen ULN (Clariant). In general, amphoteric auxiliaries increase the rate of dye uptake by wool, although they decrease the rate of dyeing and improve dye levelness on chlorinated or chlorine–Hercosett-treated wool.

$$R-\overset{+}{N}H \begin{cases} (CH_2CH_2O)_n\,CH_2CH_2OH \\ (CH_2CH_2O)_n\,CH_2CH_2OSO_3^- \end{cases}$$

8

N-alkylbetaines have been investigated in a series of studies by Cegarra and Riva [10,11]. These products also form complexes with dye molecules. They generally increase the rate of uptake and cover dyeability variations in wool. Compared with ethoxylated amphoterics [12], betaines have a larger capacity to form complexes with dyes. They do not, however, have the solubilising characteristics associated with the ethylene oxide chain, which can have a positive effect on dye migration and on decreasing the rate of dye uptake in certain circumstances. This is particularly true of products containing many ethylene oxide units per molecule.

Other commercial products, such as Tipsol LR (HWL) and the former Lyogen TP (S), are pseudo-amphoteric in that they are mixtures of cationic and mildly anionic (e.g. carboxymethylated ethoxylates) surfactants. Many of these products are very effective and are much cheaper to produce than amphoteric compounds.

3.2.5 Other Auxiliaries

Many wool-dyeing auxiliaries are not surface-active. These include sequestering agents, brightening agents, wool protective agents, products used to improve dyebath effluent and products used to set a pH value or induce a shift in dyebath pH during dyeing. These are covered in the appropriate application sections.

3.3 Brightening Agents

The term 'brightening agents' is used here for products that can be added to the dyebath in order to eliminate the natural yellow colour of wool.

The light fastness of dyeings on wool, particularly in bright or pale shades, is often quite poor due to the inherent yellow pigmentation of the wool itself, which is very fugitive to light. This yellowness increases during wet processing but can be reduced by incorporating hydroxylamine salts [13] in the dyebath. One such salt is Lanalbin B (Clariant), which is believed to have a specific effect on the tryptophan residues in wool, thought to be associated with the production of the yellow colour in the wool fibres. Lanalbin B can also be used as an aftertreatment, and if for example goods are treated, after dyeing, for 30 minutes at the boil with 2% Lanalbin B, the effect on the whiteness of the ground is almost as beneficial as when they are dyed with Lanalbin B in the bath.

The improvement in brightness is obtained without appreciable wool damage compared with that which occurs during peroxide bleaching, for example. Table 3.1 shows the cystine content and alkali solubility of raw wool, wool dyed only, wool dyed with 2% Lanalbin B and wool which has been given a peroxide bleach. There is only a small increase in alkali solubility resulting from the use of Lanalbin B, compared with the significant increase caused by peroxide bleaching of wool to improve the base shade prior to dyeing. Even on a peroxide-bleached ground, the use of a brightening agent in dyeing may be advisable in order to prevent yellowing during dyeing.

Table 3.1   *Effect of dyeing in the presence of Lanalbin B and of peroxide bleaching on wool.*

	Cystine Content/%	Alkali Solubility/%
Raw wool:		
dyed with no addition at 90 °C, 30 minutes	11.5	11.5
dyed with 2% Lanalbin B at 90 °C, 30 minutes	11.5	10.0
alkaline peroxide bleached at 50 °C, 6 hours	10.5	19.5

Table 3.2   *Effect of dyeing in the presence of Lanalbin B and of sodium metabisulphite on the Martindale and Stoll abrasion resistance of wool fabric.*

	Dyed at 85 °C for 1 hour with Acetic Acid	Dyed at the Boil for 1 hour with Acetic Acid
Martindale abrasion test (light serge):		
1.5% Lanalbin B	10 500 rubs	8750 rubs
1.6% sodium bisulphite	7200 rubs	5700 rubs
Stoll flex abrasion test (blazer material):		
2.0% Lanalbin B	6563 flexes	6330 flexes
0.8% sodium bisulphite	–	4065 flexes
1.6% sodium bisulphite	4854 flexes	3860 flexes

Bisulphite-based products, such as Erioclarite B (Huntsman), can also be used as brightening agents; while these may give similar performance with respect to brightening, they are unsuitable for use with certain dyes with which they cause a substantial shade change. They also cause some wool damage. Table 3.2 shows the inferior abrasion resistance of wool dyed in the presence of sodium metabisulphite compared with that dyed in the presence of Lanalbin B. The Martindale (flat) abrasion test simulates gentle abrasion, such as occurs during garment wear. The Stoll Quartermaster flex abrasion test is a much more severe test, in which a strip of material is bent around the edge of a metal bar and then oscillated in both directions over the edge until the fabric breaks. A study of Table 3.2 demonstrates that the quantities of Lanalbin B and metabisulphite used on the light wool serge give equivalent brightnesses of shade; similarly, 2% Lanalbin B and 1.6% metabisulphite give equivalent brightnesses on the blazer material.

Lanalbin BE (Clariant), a similar product to Lanalbin B (Clariant), is particularly suitable for use where metal contamination is present in the water. It is also more suitable for use with dyes requiring a dyebath at neutral pH, due to the less acidic reaction it gives in aqueous solution.

A possible problem with hydroxylamine derivatives is that when dyeing wool with reactive dyes, interaction with the reactive group can cause reduced wet fastness. This does not apply to chlorodifluoropyrimidine reactive dyes, but reactive dyes containing other reactive groups may be inactivated. In particular, the dyes most frequently used on wool, based on bromoacrylamide or vinylsulphone (VS) chemistry, are affected. Since it is often uncertain nowadays which reactive group is present in the individual members of commercial reactive dye ranges, it is best to always add Lanalbin B towards the end of a reactive dyeing cycle, after the dye has exhausted on to the fibre and reacted with it.

3.4   Levelling Agents

The common faults encountered in the dyeing industry are off-shade dyeings, inadequate fastness and unlevelness. Of these, unlevelness can represent the biggest problem. All the faults may be correctable, but if all effort at correction has failed, material which is off shade or of poor fastness is still saleable, albeit at a reduced price, whereas a grossly unlevel dyeing is usually completely worthless. A good levelling agent can therefore pay for itself several times over by reducing the number of rejected dyeings.

A dyeing is said to be 'unlevel' when the material does not exhibit the same depth of shade over the whole of its area (and when the differences are undesirable). Unlevelness takes different forms: piece goods may be stripy in the warp or weft direction, or even randomly; the ends or edges of the fabric may differ in colour from the bulk; or there may be light and dark places, giving a cloudy or – in bad cases – patchy appearance. When the dark and light places are numerous and close together, the patchiness is called 'skitteriness'; it is commonly experienced in wool dyeing due to variations in dyeability between the roots and tips of individual fibres.

In yarn and fibre dyeing, small shade differences from different parts of the batch, or even poor reproducibility from batch to batch, can show up as marked unlevelness in the fabric. Poor dye penetration can also result in unlevelness at later stages of processing, or in use, when undyed portions of the material are rendered visible by, for example, surface abrasion. The causes of unlevel dyeing can be subdivided into two groups: material faults, and dyeing and processing faults.

3.4.1   Material Faults

Foreign matter on the material, such as soaps, fats, waxes, finishes or sizes, can impede dye penetration and cause unlevel dyeings. The answer to this problem lies with correct preparation. Variations in the substrate – both in blends of different fibres and with natural fibres, such as wool, which are of inherently heterogeneous composition – can lead to skittery dyeings. Uneven actions of chemicals in preceding processes, such as carbonising or chlorination, and physical influences, such as light, heat and mechanical damage, can also result in gross unlevelness. With synthetic fibres, of course, other influences are also important, such as the degree of polymerisation, stretching and differences in end-group content.

3.4.2   Dyeing and Processing Faults

The dye itself plays a significant role in levelling. Important factors are solubility, sensitivity to hard water, affinity, diffusion and migration properties. Some properties change with pH, temperature and electrolyte concentration; all of these therefore require control.

Unlevelness is often associated with the inherent nature of machinery (such as ending on a jig or rope marks from winches), certain machine variables (such as drum speed in a rotary garment dyeing machine or pump speed and direction of flow in a package machine), foaming or overloading. Even with optimum machinery and loading conditions, however, it is often impossible to obtain satisfactory levelness without using a level-dyeing assistant.

In order to obtain level dyeings, it is necessary either to control dye uptake so that exhaustion occurs gradually over an extended period or to promote migration of the dye after its initial adsorption on to the fibre, from areas of high dye concentration to areas of low dye concentration. Acid levelling colours perform adequately in terms of migration to give perfect levelness, whatever the levelness of the initial uptake might be. Increasingly, however, as standards improve, the wet fastness achieved with this type of dye is becoming inadequate. Where strict fastness standards require the use of dyes with better fastness properties it is necessary to control dye uptake in order to obtain levelness. There are several ways of controlling the rate and levelness of dyeing.

3.4.2.1   Method 1

The rate of temperature rise may be reduced, but this is often not enough on its own to give satisfactory levelness and it does not improve dyeability variations in tippy wool. Productivity considerations can also militate against the use of longer dyeing times.

3.4.2.2   Method 2

If there is a gradual change in pH from conditions favouring dye in solution to conditions favouring dye on fibre, the rate of exhaustion can be controlled. Ammonium salts can be used to give small changes in pH during the dyeing of wool. These occur following decomposition of the salt, which releases volatile ammonia, leaving acid in the dyebath. For this concept to be utilised completely with dyes applied from neutral and weakly acid baths, it is necessary to start dyeing under alkaline conditions using more sophisticated acid donors, such as Optacid V or VS (Clariant), Meropan EF (CHT) or Eulysin N-WP (BASF), which do not depend on the often unpredictable evolution of ammonia for a progressive pH change. Such products are often esters, which hydrolyse in the presence of hydroxide ions to produce the free acid. This technique is used for nylon, but the initial alkaline conditions, which are essential to preventing rapid exhaustion of dyes of high neutral affinity, would cause severe fibre damage in wool dyeing. Such pH shift techniques have however been used in wool dyeing with monosulphonated dyes of molecular sizes between those of acid levelling and those of acid milling dyes. Optilan MF (Clariant) dyes are frequently applied by such a technique, particularly on carpet yarn, where tightly twisted yarns and relatively poor level-dyeing Hussong-type machines present special problems. These dyes have low substantivity at pH values of around 7.5, but exhaust well at pH values of 6.2 or below. These characteristics make these dyes suitable for use with acid donors such as Optacid V, Optacid VS (Clariant), Meropan EF (CHT) and Eulysin N-WP (BASF).

3.4.2.3   Method 3

It is possible to use auxiliaries which have substantivity for the fibre and compete with dyes for dye sites in the fibre, thereby reducing the rate of dye uptake. This is a simple mechanistic interpretation but is surprisingly useful to the practical dyer. Anionic products could in fact be used successfully to control the rate of uptake of acid levelling dyes, but in such cases retardation is not usually necessary as the dyes possess excellent migration properties. Furthermore, because Glauber's salt can be used effectively to enhance the migration of acid levelling dyes by this mechanism, the cost of a more sophisticated chemical is unjustified. With dyes applied at a pH closer to neutral, the products that work well on nylon are found to be ineffective on wool, because of the much greater number of cationic sites available on this substrate. If an anionic product with sufficiently high substantivity under neutral conditions were used in quantities sufficient to give good levelling, this would inevitably lead to dye site blocking and poor exhaustion, as well as being expensive due to the quantity of chemical required.

3.4.2.4 *Method 4*

Nonionic or cationic auxiliaries interact with anionic dyes to form a weak complex in solution with a reduced mobility. In the case of cationic products, the electrostatic attraction of the anionic dyes for wool is also partially neutralised. These products, particularly the ethoxylated amines, which possess both cationic and nonionic properties, are the most important auxiliaries for control of the rate of dye uptake on wool.

3.4.3 Testing the Action of Levelling Agents

In order to test the effectiveness of levelling agents, some years ago Sandoz AG (now part of Clariant) developed a method in which dye liquor is circulated in one direction through a compact column of fabric discs. The liquor is heated indirectly by a poly(ethylene glycol) bath to the dyeing temperature, and the column of fabric discs is compressed in a metal cylinder by a threaded insert to give a constant density and resistance to flow. Dyebath conditions such as liquor ratio, temperature and speed of circulation can be varied, and the effect of different additives can be assessed from the degree of dye penetration within the column.

Another important test method is a migration test in which dyed patterns are treated in a blank dyebath at the top dyeing temperature with an equal mass of undyed material. The dyebath auxiliaries, time and temperature should all be identical to those usually employed in the proposed dyeing system. The degree of transfer from the dyed to the undyed pattern is a measure of the migrating properties of the dyeing system.

The amount of dye in the liquor is also important, and can be assessed directly by colorimetry or indirectly by dyeing on to a piece of white fabric. This can indicate possible problems with excessive retardation and hence poor bath exhaustion.

The most direct test method for rate of dyeing is a series of identical dyeings, each halted at a different stage. As each dyeing is stopped, a sample is removed and fresh material is entered, and the dyeing is then continued until equilibrium. This gives an enhanced picture of the degree to which the buildup of shade is on tone. In a perfectly compatible system, the rate of dyeing of the patterns will be a mirror image of the samples dyed from the residual liquors. The test is of special significance in assessing the combinability of different dyes used in mixture shades.

With the current widespread availability of reflectance spectrophotometers linked to colour-matching computers, numerical values can easily be given to the results of any of these tests, which have hitherto been used mostly in a qualitative fashion. Other test methods have been used to give numerical results in theoretical studies. Vickerstaff, for example, used what he called a 'strike test' [14]. In this, two identical pieces of fabric are entered into a dyebath, one a little later than the other. The time at which they both have the same colour is measured. In practical dyeing systems, however, it is probably impossible to obtain reproducible, meaningful results by this method, particularly for the higher-substantivity dyes. In practice, colour photographs are often used to illustrate the action of levelling agents.

3.4.4   Product Selection

There are several possible problems related to the use of levelling agents. The use of more, rather than less, of a cationic product gives a more powerful retardation effect, and therefore, it might be thought, a more level dyeing. In extreme cases, however, the complex formation may be too strong. Lyogen WD (Clariant), for example, which was developed as a levelling agent for acid milling dyes, can under certain conditions precipitate some 1 : 2 metal-complex dyes, such as C.I. Acid Black 170 or C.I. Acid Violet 66. This fault is clearly demonstrated in disc dyeing tests, in which the top discs are more deeply coloured than those in which no levelling agent has been used. More highly ethoxylated products, such as Lyogen MS (Clariant), cause no such problems; with its very long polyethoxy chain, this agent enhances the solubility of anionic dyes.

Due to the marked retarding action of more strongly cationic products and the lower manufacturing costs of products with ethoxylated chains, relatively cheap and apparently powerful levelling agents have been marketed. These are mixtures of short-ethoxy-chain cationic and nonionic compounds, with perhaps 20–25 ethoxy units, which prevent precipitation. These products give rise to other problems, however, such as poor dyebath exhaustion and exhaustion of the dye–auxiliary complex on to the fibre. This latter phenomenon leads to low fastness to wet rubbing and in wet contact tests, such as alkaline perspiration. It is not clear whether this problem arises because the complex formed is so strong that it can be taken up by the fibre intact or because residual auxiliary retained on the fibre surface exerts a localised stripping effect. The phenomenon is often perceived as the 'drainage effect', in which, although bath exhaustion is complete, coloured liquor is squeezed from the dyed wool on cooling; this loose colour can migrate and cause unlevelness in wet goods awaiting drying. Post-treatment with anionic surfactants, such as Irgasol DAM (Huntsman) or Sarabid DLO (CHT), has been suggested for the removal of residual cationic product and the breaking of any dye–auxiliary complex.

The more highly ethoxylated products (Lyogen MS (Clariant) and Albegal SW (Huntsman)) seem to exhibit fewer of these problems. They also give the best performance with respect to migration, presumably due to their very high solubilising action and lower tendency to form complexes compared to more cationic products. These therefore seem to be the best type of product with respect to controlling surface levelness when dyeing with most acid milling or 1 : 2 metal-complex dyes. This is not necessarily the case with dyes requiring more acid conditions, such as chrome dyes, or with very incompatible combinations of acid milling dyes. With these, products such as Lyogen WD (Clariant) or Albegal W (Huntsman) (see Section 3.2.3) may be more suitable.

3.4.5   Coverage of Skittery or Tippy-Dyeing Wool

Where coverage of tippy wool is the main problem, the type of auxiliary selected may be different from those described so far. Although cationic surfactants all improve coverage of dyeability variations, it is often necessary to use specialised products.

The cause of skittery dyeing of wool lies in the action of light and weathering on the tip of the wool fibre while it is on the sheep's back. The cuticle, or outer sheath, of the fibre is hydrophobic (Chapter 1) and usually impedes the penetration of hydrophilic dyes into the fibre. Photodegradation partially removes the epicuticle and oxidises cystine in the underlying exocuticle to anionic cysteic acid, making the fibre surface hydrophilic. These changes occur to a greater extent at the more exposed fibre tips, which results in preferential adsorption of dye at the tip compared with the rest of the fibre. Damage to the epicuticle can also be caused by mechanical processing, and skittery dyeings caused by such damage are indistinguishable from natural tippiness. Acid levelling dyes migrate at the boil and therefore, although the initial dyeing may be skittery, they do finally produce a nontippy result. Hydrophilic (polysulphonated) acid milling dyes do not migrate, however, and the initial unequal penetration of the fibre results in a final skittery dyeing, even after the full dyeing time at the boil. The main effect of skitteriness with a straight shade of a single dye is a reduction in the apparent colour yield, but dyeing with a hydrophilic dye in combination with, for example, a more hydrophobic dye can give two colours; this is often called 'positive dichroism'.

Wool-reactive dyes represent an extreme example when dyeing tippy wool, since they are usually di- or tri-sulphonated (and, therefore, very hydrophilic) and generally have poor migration properties. The earlier levelling agents were inadequate for such severely skittery-dyeing dyes. The highly sulphonated copper phthalocyanine dyes previously presented a similar problem, which was overcome by using a strongly cationic product such as Lyogen BPN (Clariant), which forms a salt with the sulphonate groups of the dye, thus making it more hydrophobic and therefore less skittery-dyeing. The dye consequently suffers a severe loss in solubility, however, and a powerful dispersing agent such as Ekaline F (Clariant) (an ethoxylated alcohol) must be added to prevent precipitation.

With the introduction of reactive dyes, new amphoteric levelling agents were introduced [15]. These products have a strongly cationic head and form a complex with the dye in a similar manner to Lyogen BPN (Clariant). The complex formed with these amphoteric compounds is more surface-active than the Lyogen BPN–dye complex and is therefore self-dispersing. Because of the presence of a sulphonic acid group in the auxiliary, the resultant complex also has substantivity for wool, which prevents an undue decrease in the final dyebath exhaustion.

The effect of amphoteric products in the dyeing of chlorine–Hercosett-treated wool is entirely different. In this case, not only has the hydrophobic fibre surface been altered by chlorination, but a very hydrophilic cationic polymer (Hercosett 125: Hercules Chem. Co.) has also been applied to the fibres. With untreated wool, amphoteric auxiliaries actually increase the rate of dyeing by effectively increasing the hydrophobic nature of the dye, but with the very hydrophilic Hercosett wool the effect is to reduce the rate of dyeing; that is, to exert a straightforward retarding action.

Before wool-reactive dyes were introduced, the existing levelling agents were adequate to deal with skitteriness. Lyogen WD (Clariant), for example, has a strong effect with milling dyes. Conventional, 1 : 2 metal-complex dyes containing nonionic solubilising groups are hydrophobic in nature and therefore inherently non-skittery-dyeing. In recent years, however, sulphonated metal-complex dyes have presented new problems. Amphoteric auxiliaries similar to those used with reactive dyes, such as Uniperol SE (BASF), have been recommended for disulphonated metal-complex dyes. An anionic product with some nonionic properties, Lyogen SU (Clariant), has been recommended for the prevention of skitteriness with monosulphonated metal-complex dyes. Lyogen SU presumably acts partly as a nonionic levelling agent and partly like the anionic auxiliaries used for the coverage of barriness in nylon.

Increasingly, it appears that the complexity of the dye ranges available and of the combinations in which they may be used is leading to the investigation and development of very specialised products, based on amphoteric or pseudo-amphoteric surfactants, for the covering of tippy wools. Modern, highly compatible and therefore combinable dye ranges such as Lanasan CF (Clariant) and Lanaset (Huntsman) require the use of amphoteric auxiliaries such as Albegal SET (Huntsman) and Lyogen UL (Clariant) for optimum coverage of dyeability variations. These ranges often use dye mixtures to produce combination elements of almost identical dyeing properties. Inspiration for this has been drawn from the work of J. Carbonel [16], in which the effect of substrate, machinery, dyes and chemical auxiliaries on the thermodynamics of dyeing is considered. These ranges of mixtures of dyes offer great advantages in practical dyeing.

'Negative dichroism' is a possibility, and in some cases the use of too much auxiliary can cause the undamaged parts of the fibre to be dyed more deeply than the damaged areas. Thermodynamically (as opposed to kinetically), wool dyes have a higher affinity for the root than the tip of the fibre, and this sometimes gives rise to reverse tippiness with the Optilan MF range of dyes, which exhibit high migrating power but are larger and therefore more hydrophobic than acid levelling dyes.

A further level-dyeing problem, which can occur in the dyeing of fully fashioned garments or socks, is that of seam penetration. Often, this problem is exaggerated by poor chemical penetration of the seams during prechlorination of the garments: a treatment that is carried out to enable garments to be dyed without undue felting and also to confer subsequent washability. As such, the problem of poor chemical penetration is analogous to tippiness, and auxiliaries should be selected with this in mind.

Penetration of thick wool felts, such as those used in pianos, or of hard-twist carpet yarns is a purely physical problem. It is usually necessary to select acid levelling dyes which migrate strongly and use extended boiling times. Dyes of the Optilan MF type are used where higher wet fastness standards are required. To improve matters further, disaggregating products such as the pyridine- and anionic surfactant-based Lyocol FDW (S) (formerly Tetracarnit) were formerly used. Unpleasant odour and possible toxicological problems led to the replacement of this product by powerful wetting agents, however, of which Lyogen WPA (Clariant) is an example.

3.5   Restraining and Reserving Agents in Wool Blend Dyeing

Section 3.2.1 described products similar in nature to the syntans used to improve the wet fastness of dyeings on nylon. These can be used to reduce the uptake of dye on the nylon portion of a wool/nylon blend, and are widely used to restrain milling and 1 : 2 metal-complex dyes. They act by retarding uptake on the nylon fibres in the early stages of a dyeing. With nonmigrating dyes, this initial partition between the fibres is retained, but if syntan-type products are used with acid levelling dyes, migration from the wool to the nylon occurs at the boil. This is difficult to control, and therefore with acid levelling dyes the partition is poorly reproducible, particularly with extended dyeing times. With migrating dyes, simple anionic surfactants are used, which compete with the dye for cationic groups in the polyamide fibre and form an equilibrium which does not change even on prolonged boiling. These products, which are typically alkylbenzenesulphonates, alkylnaphthalene-sulphonates, alkanol sulphates or sulphonated castor oils, are of insufficient substantivity to compete with, and therefore be effective with, high-substantivity dyes when dyeing at neutral pH. Thiotan RMF (Clariant), which foams less and has higher affinity, has been introduced specifically for use with Optilan MF dyes. This product is also used with acid levelling dyes, where its low-foaming behaviour is advantageous.

The required quantity of such products is related to depth of shade, decreasing with increasing depth. With the syntan-type products, for a given milling or metal-complex dye combination on a given blend, a standard quantity of restrainer can be used for all depths.

Where syntan-type anionic products are used together with cationic levelling agents, the dyer may live in the forlorn hope that 'restrainer + levelling agent = solidity + levelness'. In fact, the effect of adding a cationic levelling agent is to negate the effect of the restrainer. Furthermore, the effect of the levelling agent is also negated, because in practice excess restrainer is used in order to obtain satisfactory partition. If syntan-type restrainers are used in wool/nylon dyeing then levelness must depend on temperature control and the retarding action of the syntan itself.

Restraining agents are also used in wool/cotton blend or union dyeing. Syntan-type products are used, as well as specialised anionic surfactants such as Thiotan HW (Clariant), in order to inhibit cotton direct dyes staining the wool fibres.

3.6   Antiprecipitants

Several methods are used in the dyeing of wool/acrylic blends: one bath, one bath/two stage, two bath and so on. If both cationic and anionic dyes are employed, it is usually necessary to take some action to avoid precipitation. In general, nonionic products with good solubilising properties prevent precipitation in all but extreme cases. For maximum effect, and to minimise cross-staining, it is usual to incorporate a small amount of either cationic or anionic product. Commercial products include Lyogen AB (Clariant).

3.7   Wool Protective Agents

Wool is easily damaged by aqueous treatments, particularly at high temperature or high pH. Wool protective agents are thus often used, especially in reprocessing or in high-temperature dyeing of wool/polyester blends. These agents fall into three categories:

(1) Products based on water-soluble proteins.
(2) Products based on fatty sulphonic acid esters or alkylsulphonic acids.
(3) Products based on formaldehyde or other chemicals capable of crosslinking wool.

The first group of products was developed following the observation that wool is less damaged if dyed in previously used dyebaths, since the dissolved protein residues in the standing dyebath decrease the rate of hydrolysis of the wool protein. This situation was reproduced in fresh dyebaths by introducing protein residues in the form of commercial

Table 3.3   *The protective effect of Lanasan PW in wool dyeing (for experimental details, see text).*

	Alkali Solubility/%
Undyed wool	15.9
Dyed, no Lanasan PW	18.7
Dyed, 3% Lanasan PW	13.4
Dyed, 6% Lanasan PW	10.7

products such as Egalisal CS (Grunau). Such products are often used in the stripping and reprocessing of wool or when extended dyeing times are necessitated by shading additions. Recent work has shown that the use of protein derivatives gives only a small degree of fibre protection but can lead to fastness problems [17].

The second group is claimed to be effective due to the bonding of large molecules at the surface of the wool, leading to the formation of a thin hydrophobic coating that resists penetration by the liquor.

Products in the last group act by crosslinking the wool and thus replacing broken cystine disulphide crosslinks ($-CH_2-S-S-CH_2-$) with more stable lanthionine-type groups ($-S-CH_2-S-$). Formaldehyde has been used for many years in this context [18], but it suffers from the disadvantages of being unpleasant, toxic and allergenic, and also giving a harsh handle to the wool.

Recently, formaldehyde derivatives, such as Irgasol HTW (Huntsman) [19], Meropan SK (CHT) and Lanasan PW (Clariant), have been introduced. These contain negligible quantities of free formaldehyde and, since they introduce a longer and more flexible crosslink, do not have the same adverse effect on handle. Table 3.3 shows alkali solubility figures for wool gabardine fabric dyed with 1.7% Sandolan Milling Blue F-BL 180% at 120 °C for 45 minutes at a liquor ratio of 30:1 and at pH 5.5. In this table, the protective effect of Lanasan PW is clearly demonstrated. The most common application for this type of product is in the dyeing at 120 °C of wool/polyester blends [20], which cannot normally be dyed at temperatures above 105 °C, due to excessive wool damage. At temperatures as high as 120 °C, however, higher-r.m.m. disperse dyes of higher wet fastness can be used for the polyester fibre. Disperse dye staining of the wool is also minimised at these high temperatures, contributing to enhanced wet fastness. Protein crosslinking products have also been recommended for use with 1:1 metal-complex dyes, in order to protect the fibre from severe damage usually caused by the very low dyebath pH used in their application.

3.8   Low-Temperature Dyeing

Attempts to dye wool at low temperature go back many years [21–23]. Initially, a prime motive was energy conservation, but more recently there has been an increasing awareness of the advantages to be gained from protecting the wool fibre by limiting the temperature, or the time at top temperature, during dyeing.

It is relatively easy to obtain good dyebath exhaustion at temperatures down to 80–85 °C. Indeed, with lower-substantivity dyes, such as acid levelling dyes and some chrome dyes, perceptibly higher exhaustions are obtained at lower temperatures even in normal dyeing times. Particularly with dyes of higher molecular size, however, a major problem is ring dyeing. Micrographs of fibre cross-sections show clearly that these molecules do not readily penetrate into the fibre at lower temperatures. This can give rise to problems such as reduced fastness in subsequent processing, reduced wet contact fastness, shade change during steaming or finishing and poor rubbing (crocking) fastness.

Several techniques have been suggested for improving the diffusion of dye into the fibre, thus making lower-temperature dyeing viable:

(1) Solvent-assisted dyeing.
(2) The use of alkyl phosphates.
(3) The use of ethoxylated alcohols with short polyethoxy chains.

It seems likely that all these methods depend upon the formation of an auxiliary-rich phase of high dye concentration adjacent to the fibre surface. All the chemicals that have been successfully used operate most effectively at or close to their solubility limit (parallels can be drawn to the effectiveness of urea at concentrations of $200–300\,g\,l^{-1}$, as used in pad–batch dyeing of wool). The auxiliaries used all have very high dissolving power for ionic dyes, and the consequent high dye concentration at the fibre surface increases the rate of dyeing without impairing levelness. The aggregate size of the dye in the auxiliary-rich phase is at a minimum, and this permits faster diffusion into the wool fibre.

The first proposal for low-temperature dyeing of wool appears to have been made by Schöller as early as 1934. He used cationic surfactants together with dispersing agents. Later, the use of organic solvents was proposed [21]. Water-soluble solvents such as butanol were also suggested, but the concentrations needed were very high. Lister [24] suggested the use of chlorinated hydrocarbons, which have more limited solubility, but in the light of present-day concerns these would be unsuitable on toxicological grounds. Beal [22] developed the benzyl alcohol [21] concept further, allowing the industry to dye wool at 60–70 °C. This again required very high concentrations (up to $30\,g\,l^{-1}$) and was thus rather expensive. The Ciba-Geigy company then proposed the Irga-solvent processes [25] for the low-temperature dyeing of several textile substrates, including wool. Alkyl phosphates were also examined for use in low-temperature dyeing and Dilatin VE (S), a commercial form of tributyl phosphate, received some attention.

During the mid-1960s, several low-temperature dyeing methods using nonionic surfactants were published following work at the Commonwealth Scientific and Research Organisation (CSIRO) [23]. Optimum results were obtained using a polyethoxylated nonylphenol having 6–10 ethylene oxide units per molecule. Although the practical application of these techniques was complicated – due to the need for different recommendations for use with different dyes, together with problems of dissolution of nonionic surfactants with a very low cloud point – the physical advantages to the wool that came from dyeing at 85 °C and pH values between 3 and 5 were clearly demonstrated. It was suggested that the use of these short-chain ethoxylated products might be made easier if they were incorporated with a longer-chain dispersing agent.

Ultimately, the two determining factors as to the approach to be adopted in commercial practice are, first, toxicological considerations and, second, cost. On both these counts, the use of nonionic products is favoured. Commercially available products that incorporate both short-chain ethoxylated alkylnonylphenols and longer-chain dispersing agents include Lanasan LT (Clariant) and Keriolan W (CHT). These products appear to be more effective than those suggested initially by CSIRO, and it may be that the use of premixed products permits the use of chemicals with polyethoxy chains as short as four or five ethylene oxide units. More recently, CSIRO has carried out more work on low-temperature dyeing, which is discussed later [26,27].

Concerns about the possibility of ethoxylated nonyl and octyl phenols breaking down in the environment to produce chemicals suspected of being endocrine disruptors have resulted in the reformulation of several commercial nonionic products, through the use of ethoxylated alcohols instead of ethoxylated phenols.

Problems have been experienced with all low-temperature dyeing methods regarding certain dyes: in particular, 1 : 2 metal-complex dyes have presented difficulties, while the incompatibility between dyes in combination is exaggerated by dyeing at lower temperatures. Readily combinable ranges of 1 : 2 metal-complex and acid milling dyes are now available in the Lanaset (Huntsman) and Lanasan CF (Clariant) dye ranges. Sandoz (now Clariant) [28] have strongly promoted the application of the Lanasan CF dyes at 85 °C with the addition of their nonionic auxiliary (Lanasan LT). Equivalent performances of reproducibility of shade and colour fastness compared with dyeings produced at the boil have been claimed. Coincidentally, nonionic surfactants suitable for use as auxiliaries for low-temperature dyeing are also effective in covering fibre dyeability variations (tippiness).

CSIRO and ICI (Australia) developed the Sirolan LTD process during the 1990s, which uses a novel chemical, Valsol LTA-N (later manufactured by APS Chemicals [26,27]), an amphoteric ethoxylated hydroxy sulphobetaine, which works via a different mechanism to the other chemicals: it removes labile lipid material from within the fibre and speeds up the rate of dye diffusion through the cell membrane complex. Unlike other amphoteric auxiliaries, it can be used with all types of wool dye, with no retardation of exhaustion or draining effects – because it does not form strong complexes with the dyes. In fact, dyebath exhaustion is usually much better than that which is obtained with conventional methods. An added bonus is that the exhaustion of mothproofing agents is enhanced, with less remaining in the exhausted dyebath. It has been used for this reason alone, where dyehouses had a problem with insect-resist agents in the dyebath [29].

An important additional feature of Valsol LTA-N is that it can also be used to dye wool for a short time at the boil. Both the low-temperature and the short-boiling-time methods produce significant decreases in fibre damage, without any loss of fastness, compared with conventional methods.

3.9   Correction of Faulty Dyeings

Wool dyeings which require reprocessing may be chemically stripped using, for example, Arostit ZET (S), which is zinc formaldehyde sulphoxylate. It is necessary to use a reducing agent that is reasonably stable at acid pH, as obviously the hydros/soda ash stripping methods commonly used with other fibres would cause severe damage. Other stabilised dithionite- or hydrosulphite-based stripping agents may also be used, but they are less effective than zinc formaldehyde sulphoxylate.

It is often possible to level up a dyeing, or to reduce the depth in order to allow a shade correction, by using partial stripping treatments with cationic surfactants. Polyethoxylated quaternary ammonium compounds are particularly effective and are often used in conjunction with mildly alkaline conditions (e.g. with ammonium hydroxide). Quaternised products are fully ionised even at pH 8, whereas polyethoxylated tertiary amines are not.

Unexpectedly poor fastness is another fault. Poor wet fastness can often be corrected with the specialised cationic products covered in Section 3.10. Poor rubbing fastness can be improved by post-scouring with nonionic dispersing agents or other specialised soil dispersants. Poor light fastness, resulting from poor dye selection for example, would require a stripping treatment followed by redyeing with more suitable dyes. In pastel shades, the poor light fastness of the natural or dyebath-induced yellow pigmentation of wool may be a problem, but this can be rectified by an aftertreatment with a dyebath brightening agent, such as hydroxylamine sulphate.

3.10   Aftertreatments to Improve Wet Fastness

The first cationic aftertreatment designed specifically for wool was launched in late 1979 [30]. This product, Sandopur SW (S), is recommended for the aftertreatment of 1 : 2 metal-complex dyes and acid milling dyes. In many cases, the washing and alkaline perspiration fastness can be improved to a level that meets the Woolmark Company requirements for machine-washable wool. Like other similar products, however, Sandopur SW is quite dye-selective and is only really effective when applied to wool that has been given an oxidative shrink-resist treatment. Nevertheless, it can produce improvements in fastness that are quite dramatic, particularly with sulphonated metal-complex and certain acid milling dyes.

Prior to the introduction of cationic aftertreatments, reactive dyes were recommended in most shade areas for the dyeing of machine-washable or Superwash wool. The high cost of reactive dyes, however, together with problems of skittery dyeing and relatively poor levelling, led several dye companies to investigate the use of nonreactive dyes for this purpose. Unfortunately, the shrink-resist processes most frequently used considerably diminish the wet fastness of nonreactive dyes. Post-scouring treatments with specialised nonionic surfactants were often recommended and gave positive but small improvements. Cationic aftertreatments of the type used for direct cotton dyes were known to improve wet fastness on wool to a limited degree, but at the expense of impaired rubbing fastness and in some cases reduced light fastness.

Following the introduction in the 1970s of several ranges of reasonably priced mono- and disulphonated metal-complex dyes with inherently good wet fastness, the advantages to be gained from a successful aftertreatment seemed very attractive. Sandopur SW contains a selected high-molecular-size cationic product which exhausts well on to wool, along with a combination of nonionic surfactants that prevent the buildup of a dye–cation complex on the fibre surface, which would impair rubbing fastness. There also appears to be a synergistic action between the nonionic surfactants and the cationic species with respect to increasing the wet fastness of the dyeings.

Other products have since been developed. Initially, products containing cationic – instead of nonionic – surfactants were introduced, such as Sandofix L (Clariant), Fixogene MW (ICI) and Croscolor PMF (Crosfields). These were especially effective on chlorine/Hercosett-treated wool, where the removal of dye held in the cationic resin layer was necessary, although subsequent reductions in the amount of applied resin by using the Kroy chlorination–Hercosett process have minimised the amount of dye held in this area. More importantly, products have been introduced which are even more effective in improving wet fastness, including Lanasan MW (Clariant) and Basolan F (BASF) [31]. Lanasan MW is a reactive product which, under the conditions of application, reacts with wool and with itself to form a large resin-like structure. These reactions are induced by mildly alkaline conditions at 40 °C or by treatment at elevated temperature. When applied to chlorinated wool, the products also have a shrink-resist effect similar to that of Hercosett 125 resin.

These more recent cationic aftertreatments do, however, have certain disadvantages with respect to reprocessing difficulties and soiling with loose anionic dye in domestic laundering. Their use has therefore been restricted to very full shades.

3.11   Effluent Control in Chrome Dyeing

Because of their economy, their high fastness and particularly their intensity of shade in blacks and navies, chrome dyes continue to be widely used (see Chapter 7). They still represent the greatest mass of any dye class used on wool. Effluent problems and control of the dyeing system have favoured the afterchrome method and this is now the only technique that is commercially important.

Even in afterchrome dyeing, however, effluent problems have had to be addressed. Every major chrome dye manufacturer has published data relating to chrome additions and dyeing techniques. It is now widely accepted that the lowest residual chromium is achieved by chroming at pH 3.5 in the absence of sulphate ions and using the dye manufacturer's recommended amount of sodium or potassium dichromate (up to a maximum of 1.5% o.m.f.). In order to achieve the low chromium concentrations now required in discharged dyebath effluent (typically a maximum of $2\,mg\,l^{-1}$), further chemical additions are necessary. Although chromium(VI) is more toxic than chromium(III), uncertainty as to subsequent oxidation during incineration of sewage sludge has led to the authorities setting limits for total chrome rather than separate limits for different valency states.

The addition of sodium thiosulphate towards the end of chroming has been shown to reduce residual chromium to negligible amounts [32]. A proprietary product, Lyocol CR (Clariant) [33], is also effective in removing residual chromium or reducing it to very low levels. Lyocol CR seems to be more reliable in bulk practice than thiosulphate and also eliminates the possibility of sulphur precipitation. Lyocol CR is a two-component product, containing a mild reducing agent which reduces residual chromium(VI) to chromium(III) and a very specific chelating agent of high r.m.m. which complexes with chromium(III) and then exhausts on to the wool. There is some evidence that, by reducing the crosslinking of the wool with chromium(III), an improvement in wool quality (with respect to brittleness) is obtained.

Both the thiosulphate and Lyocol CR techniques permit low-temperature dyeing of wool at 80–90 °C, and the thiosulphate method permits chroming at 85–90 °C. The Lyocol CR technique demands a chroming temperature of 95–98 °C. These low-temperature techniques are very important in maintaining wool quality.

3.12 Antifrosting Agents

Several continuous and semicontinuous dyeing methods have been developed for wool:

- pad–steam [34];
- International wool secretariat (IWS) cold-pad–batch [35];
- Vicontin CR process (Carp-O-Roll);
- Lanasan pad–store;
- Fastran EDF.

The latter two involve the use of radio-frequency energy.

In continuous dyeing on a textile material, a surface frostiness (lighter-coloured fibre at the surface) can develop, resulting from dye migration to the inside of the material during fixation due to reduction in the water content at the surface of the material. This problem is overcome using mixtures of powerful film-forming wetting agents (usually anionic) and short-chain ethoxylated products, which establish a second liquid phase of higher dye concentration close to the fibre surfaces (known as a coacervate system). Commercial products include Irgapadol P (Huntsman). Products such as Lyogen V and Lyogen CW (Clariant) are also recommended for separate addition (i.e. not premixing). Part of the action of the anionic part of these products in pad–steam dyeing is often related to the production of an unstable foam during steaming, which induces dye migration outwards. Other products, such as Sandogen WAF (Clariant), are mixtures of amphoteric and nonionic surfactants.

3.13 Antisetting Agents

Keratin fibres can be permanently set into a different conformation when steamed or heated in water, such as in a dyebath at the boil [36]. This can cause significant problems in wool textiles, including reduced tensile strength, reduced bulk and resilience in package-dyed yarns. Permanent setting is also the main cause of hygral expansion in piece-dyed fabrics, which is the source of subsequent irreversible puckering of garments along sewn seams. Setting is due to a simultaneous rearrangement of hydrogen bonds and disulphide bonds. The so-called 'thiol disulphide interchange' reaction, which is a key element of this, is initiated by the presence of cystine thiol groups and involves a rearrangement of cystine disulphide crosslinks and the formation of new crosslinks from cystine, in particular lanthionine. The actual chemistry is more complex, but since it is clear that in every case the presence of the thiol groups of cystine are essential to these reactions, many of the successful attempts to control this phenomenon have been based on blocking the thiol groups. Since they are very reactive, this is not difficult, but in many cases there are side effects such as yellowing. A study of various alternatives has been published in *Coloration Technology* [37]. An early successful commercial product was Lanasan AS [38], which is used to assist in maintaining the bulk and resilience of package-dyed carpet yarns. This is a product based on sodium bromate, together with an activator such as sodium metavanadate and a corrosion inhibitor (e.g. sodium nitrite). Another commercially successful system, developed jointly by CSIRO, IWS and BASF [39], is based on the use of hydrogen peroxide and a special

$$Na_2(Na_4P_6O_{18}) + 2CaCl_2 \rightarrow Na_2(CaP_6O_{18}) + 4NaCl$$

Scheme 3.2   *Sequestration of calcium ions by Calgon T.*

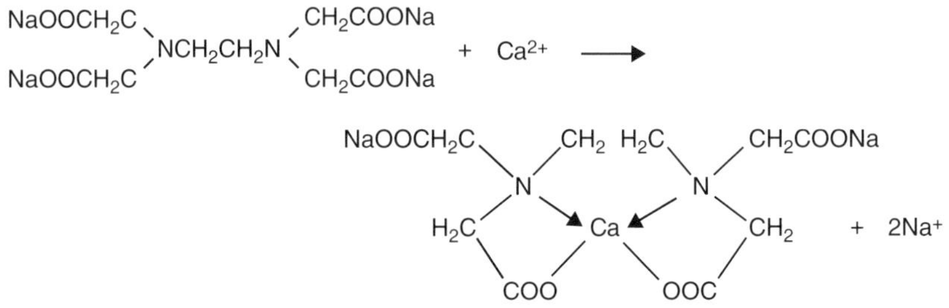

Scheme 3.3   *Sequestration of calcium ions by ethylenediaminetetraacetic acid.*

auxiliary, Basolan AS-A. This reagent inhibits the degradative effect of hydrogen peroxide on many dyes at an atmospheric-pressure boil and extends the effective life of the oxidant.

3.14   Sequestering Agents

It is often necessary in the dyeing of any fibre to take action to restrict the impact of dissolved metal salts, through the addition of sequestering agents. In wool dyeing, it is important to avoid the use of sequestering agents with 1 : 1 metal-complex dyes, where the metal may be removed from the dyes, and in chrome dyeing, where a sequestrant may impede the chroming reaction.

Sodium hexametaphosphate, reacts in solution with calcium compounds (Scheme 3.2). This allows the calcium to be held in a form that will not interfere with dyes or soaps. At very high calcium concentrations, the insoluble calcium hexametaphosphate is precipitated; sodium hexametaphosphate is nevertheless very useful, and is widely used to soften water of low hardness.

EDTA (ethylenediaminetetraacetic acid) is also widely used in wool dyeing. EDTA forms stable coordination complexes with most metals and is particularly useful for calcium, copper and iron (Scheme 3.3). EDTA is a much more powerful sequestrant than Calgon T, and unlike sodium hexametaphosphate withstands prolonged boiling in alkaline conditions.

3.15   Conclusions

From the large number and range of types of auxiliary products used in wool dyeing, it is clear that the modern dyer finds the use of auxiliaries essential in the pursuit of their trade. Increasingly, the subject of wool protection is being brought to the fore by the need to make maximum use of an expensive raw material. Fibre losses in processing and productivity losses due to such faults as yarn breakages must be minimised. Furthermore, the consumer must be presented with articles of sufficient quality to justify a high price tag. In the future, it is likely that many new auxiliary developments will be made in the area of protecting the wool fibre.

Development is also likely to be driven by the increased awareness throughout industry of both ecological and toxicological issues, and by force of legislation. New auxiliaries will have to be developed which are more biodegradable or which present fewer toxicological problems than those in use today. It may be necessary to further improve dyebath exhaustion or to introduce auxiliaries that will minimise the toxicological impact of dyebath effluents.

Whatever happens in the future, it will remain essential to understand the considerations that must be taken when selecting suitable auxiliaries, and to know which aspects of a product's performance have to be appraised.

References

[1] M Schick, *Nonionic Surfactants* (New York, USA: Marcel Dekker, 1967).
[2] N Schönfeldt, *Surface Active Ethylene Oxide Addition Products* (New York, USA and London, UK: Pergamon Press, 1971).
[3] C Schöller, *Melliand Textilber*, **48** (1967) 1212.
[4] E I Valko, *Review of Progress in Coloration and Related Topics*, **3** (1972) 50.
[5] W Giger, P H Brunner and C Schaffner, *Science*, **225** (1984) 623.
[6] D Y Shang, R W Macdonald and M G Ikonomou, *Environmental Science and Technology*, **33** (1999) 1366.
[7] Greenpeace International, *Dirty Laundry* (July 2011).
[8] Greenpeace International, *Dirty Laundry 2* (August 2011).
[9] Greenpeace International, *Dirty Laundry: Reloaded* (March 2012).
[10] J Cegarra and A Riva, *Journal of the Society of Dyers and Colourists*, **102** (1986) 59; **103** (1987) 32.
[11] A Riva and J Cegarra, *Journal of the Society of Dyers and Colourists*, **105** (1989) 399.
[12] J Cegarra and A Riva, *Journal of the Society of Dyers and Colourists*, **104** (1988) 227.
[13] D K Clough, private communications and Sandoz internal reports.
[14] T Vickerstaff, *The Physical Chemistry of Dyeing*, 2nd Edn (London, UK: Oliver and Boyd, 1954).
[15] W Mosimann, *Textile Chemist and Colorist*, **1** (1969) 182.
[16] J Carbonell, R Haslet, R Walliser and W Knöbel, *Melliand Textilber*, **54** (1973) 68.
[17] E Finnimore and R Jerke, *Proceeding of DWI Arbeitstagung* (1987).
[18] Wygand, *Textile Chemist and Colorist*, **1** (1969) 446.
[19] P Liechti, *Journal of the Society of Dyers and Colourists*, **98** (1982) 284.
[20] H Baumann and H Müller, *Textilveredlung*, **12** (1977) 163.
[21] L Peters and P B Stevens, *Dyer*, **115** (1956) 327.
[22] W Beal, K Dickinson and E Bellhouse, *Journal of the Society of Dyers and Colourists*, **70** (1960) 333.
[23] J Hine and J R McPhee, *Dyer*, **132**(7) (1963) 523H.
[24] G Lister, *Textil Rundschau*, **II** (1956) 463.
[25] W Beal and G S A Corbishley, *Journal of the Society of Dyers and Colourists*, **87** (1971) 329.

[26] J A Rippon, F J Harrigan and A R Tilson, *Proceedings of the 9th International Wool Textile Research Conference, Biella, Italy*, **3** (1995) 122.

[27] CSIRO, J A Rippon and F J Harrigan, US patent 5496379.

[28] A C Welham, *Wool Record* (April 1988), 73.

[29] J Barton, *International Dyer*, **185**(9) (2000) 14.

[30] Sandoz Products Ltd and A C Welham, BP application 32130/79.

[31] K Reincke, *Melliand Textilber*, **67** (1986) 191.

[32] P Spinacci and N C Gaccio, *Proceeding of the 12th IFATCC Congress* (1981).

[33] A C Welham, *Journal of the Society of Dyers and Colourists*, **102** (1986) 126.

[34] I B Angliss, P R Brady and J Delmenico, *Journal of the Society of Dyers and Colourists*, **84** (1968) 262.

[35] D M Lewis and I Seltzer, *Journal of the Society of Dyers and Colourists*, **84** (1968) 501.

[36] A J Farnworth and J Delmenico, *Permanent Setting of Wool* (Watford, UK: Merrow Publishing, 1971).

[37] J Kim and D M Lewis, *Coloration Technology*, **118** (2002) 121, 181; **119** (2003) 108, 119.

[38] A C Welham, Sandoz Products (UK) Ltd, *Internal Reports* (1988–1989).

[39] P G Cookson, P R Brady, K W Fincher, P A Duffield, S M Smith, K Reincke and J Schreiber, *Journal of the Society of Dyers and Colourists*, **111** (1995) 228.

4

Ancillary Processes in Wool Dyeing

David M. Lewis

Department of Colour Science, University of Leeds, UK

4.1 Introduction

This chapter discusses the chemical processes that may be carried out before, during or after the dyeing process, and emphasises their influence on dye selection, dye application and the subsequent performance of the dyed material. Such processes include scouring, bleaching, photostabilisation, moth-proofing, carbonising, flame-proofing, water- and oil-proofing, setting and shrink-proofing, which were all detailed in the earlier work, Wool Dyeing (SDC 1992). This chapter contains additional sections on antisetting and fibre arylation agents in wool dyeing. The bleaching section has been transposed as a separate chapter, 'Bleaching and Whitening of Wool: Photostability of Whites', authored by Dr Keith R. Millington (Chapter 5).

4.2 Wool Scouring

Wool fleeces usually contain less than 50% clean fibre, being heavily contaminated by wool wax, skin flakes, suint, sand, dirt and vegetable matter. (Wool wax is secreted by the sebaceous glands of the sheep and suint is the dry residue of the secretions from sudoriferous glands. Bleached and purified wool wax is sold to the cosmetic industry as lanolin.) To achieve satisfactory dyeings, these contaminants need to be efficiently removed by scouring with sodium carbonate and nonionic surfactants. The pollution load from a wool scouring mill can be equivalent to the normal discharge from a small town, and steps must therefore be taken to recover at least some of the contaminants before discharge [1,2]. Comprehensive scouring systems have been developed by the Wool Research Organisation of New Zealand

The Coloration of Wool and other Keratin Fibres, First Edition. Edited by David M. Lewis and John A. Rippon.
© 2013 SDC (Society of Dyers and Colourists). Published 2013 by John Wiley & Sons, Ltd.

(WRONZ) in their minibowl technology and by the Commonwealth Scientific and Research Organisation (CSIRO) in their Siroscour process. These technologies employ minimum volumes of water without sacrificing final wool cleanliness.

Solvent scouring offers clear pollution control advantages; a system based on hexane (de Smet process) and another using 1,1,1-trichloroethane (Toa/Asohi process) were in commercial operation in Taiwan and Japan [3].

Both neutral and alkaline aqueous scouring systems based on nonionic surfactants are currently used; earlier systems used nonylphenolethoxylates but the adverse environmental effects of these agents, in particular their oestrogen mimicry, have resulted in their complete removal from wool scouring. In alkaline aqueous scouring systems, which are the more usual, sodium carbonate is employed as a builder to improve grease removal and prevent redeposition. Saponification of the fatty glyceride esters which make up much of the wool wax is not important in wool scouring. Saponification of wool grease takes several hours at the boil in caustic alkali.

Pesticides in wool grease are an important ecological issue, as reviewed by Shaw [4]. These agents are applied by farmers during sheep dipping and can be detected, even at very low levels, in dyehouse effluents. At one time, even lindane was detected in Yorkshire rivers, which was attributed to the activities of wool processors; however, further analysis indicated that 80–90% of the residues came from sources other than raw wool scouring. Strict controls of pesticides in sheep dips in many countries as Australia, New Zealand and South Africa are now practised.

Yarn and piece scouring are extremely important to the dyer, these processes usually being carried out in the dyehouse immediately prior to dyeing. To facilitate spinning, wool is lubricated with acidic emulsions of olein-based oils and emulsification aids such as polyglycols; proprietary mixtures are nowadays carefully formulated for easy removal in scouring. The scouring process is thus fairly mild, being typically a treatment in nonionic surfactant at approximately 50–60 °C for 15–30 minutes. Inadequate removal of oil may lead to problems of uneven dyeing, and it is important that dyehouse quality controls are in place to check the efficiency and levelness of oil removal. For a comprehensive overview of yarn and piece scouring, the review by McPhee and Shaw is recommended [5].

4.3   Wool Carbonising

Raw wool contains seeds, burrs and other pieces of vegetable matter. Much of this may be removed in scouring and in the worsted process during carding and combing. Combing is not used in the woollen system, and vegetable matter is only partially removed in carding. In some worsted processes, small amounts of residual vegetable material get through to the fabric; in such cases, carbonising is essential to remove the residues.

The carbonising process requires that scoured wool fabric is padded, either in rope form or in open width, with a liquor containing dilute sulphuric acid (5–7% by mass; approximately 65% wet pickup), and dried at 65–90 °C in order to concentrate the acid. Baking at 125 °C for 1 minute chars the cellulosic material. The charred vegetable material is brittle and easily crushed on passing through rollers; it can then be removed as dust during subsequent mechanical working. Following carbonising, the wool should be rinsed

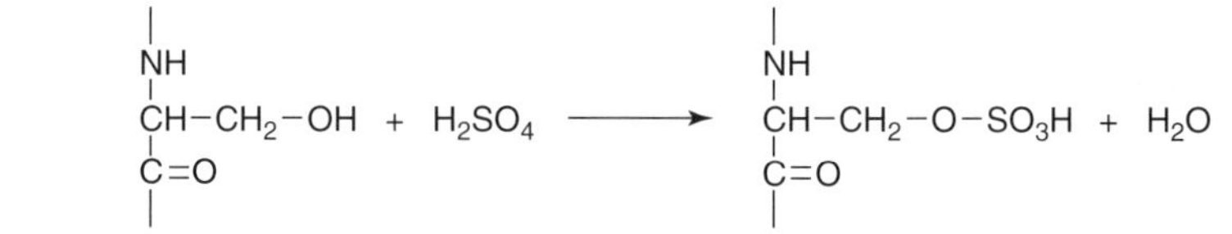

$$
\begin{array}{c}
| \\
\mathrm{NH} \\
| \\
\mathrm{CH{-}CH_2{-}OH} \\
| \\
\mathrm{C{=}O} \\
|
\end{array}
\;+\; \mathrm{H_2SO_4} \longrightarrow
\begin{array}{c}
| \\
\mathrm{NH} \\
| \\
\mathrm{CH{-}CH_2{-}O{-}SO_3H} \\
| \\
\mathrm{C{=}O} \\
|
\end{array}
\;+\; \mathrm{H_2O}
$$

Scheme 4.1 *Sulphation of wool serine by sulphuric acid in carbonising.*

and neutralised by a wet process. Such neutralisation should be carried out immediately after baking, or else fibre damage will occur during storage of the wool in such an acidic state. It is convenient to neutralise prior to dyeing, but uneven neutralisation leads to unlevel dyeings. Many dyers of carbonised piece goods thus choose to dye with acid levelling or 1 : 1 metal-complex dyes – the latter at pH 1.8 with 6–8% on mass fabric (o.m.f.) sulphuric acid.

If faster dyes such as milling or 1 : 2 metal-complex dyes are required, very careful neutralisation with ammonia and ammonium acetate gives the best results [6].

The substantivity of acid dyes for carbonised wools is generally reduced, presumably by the formation of sulphate esters at serine residues (Scheme 4.1) [7]. The dyer may see white resist spots, which clearly arise from localised high concentrations of sulphuric acid giving rise to large amounts of serine sulphate residues.

It is generally accepted that the loss in strength during carbonising (or, more importantly, during storage of carbonised goods) is due to the N→O migration (peptidyl acyl shift) reaction at serine and threonine residues (Scheme 4.2). Chain cleavage occurs at the ester linkage in Scheme 4.2, resulting in breakage of the peptide chain and consequently reduced strength.

Given the drawbacks of acid carbonising, such as excessive fibre damage and environmental pollution from the relatively large amount of acid used, it is not surprising that researchers are looking at the potential role of enzymes; three papers were presented at the Aachen Wool Research Conference [8–10], and a further one at the Leeds Wool Research Conference [11]. Bereza *et al.* [8] described the Polish Pectopol PT 100 process, in which the raw, vegetable-containing wool fabrics are first reacted with a mixture of cellulolytic and proteolytic enzymes and the fabrics are then padded with 1.0–1.5% by weight sulphuric acid. In normal carbonising, 4% sulphuric acid is used, and therefore the reduction in acid strength leads to better physical properties. Ruers *et al.* [9] used a mixture of cellulase, pectinase and xylanase enzymes to attack the vegetable matter, whereas Bucur *et al.* [10] used cellulases plus sequestrants such as Heptol ESW (CHT/Bezema) to substantially reduce cellulosic impurities and allow their complete removal by subsequent mechanical treatment. Metal ions such as iron and copper have a deleterious effect on the enzyme

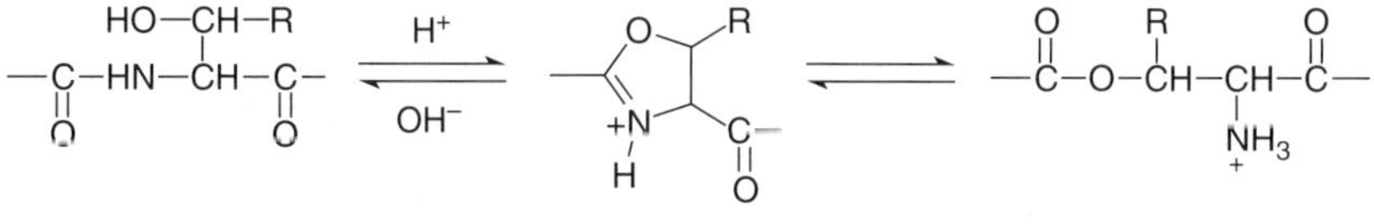

Scheme 4.2 *N→O shift in wool carbonising.*

systems, and thus careful selection of chelating agent is important. A useful review of enzyme-based finishes for wool and cotton was prepared by Heine and Höcker [12].

4.4   Shrink-Resist Treatments

Machine-washable wool garments are now well established and are especially important to the wool knitwear industry. The wool fibres in such garments must be given some sort of chemical treatment (either degradative or additive) in order to prevent felting. Felting during aqueous washing (or in aqueous processing) is attributed to the differential friction effect of the surface cuticular scales. An excellent review of wool felting mechanisms and the wide variety of chemical treatments designed to prevent felting is given by Makinson [13]. The chemical treatments employed commercially to produce washable wool often adversely influence the wet-fastness properties of the dyes employed. This section considers the various shrink-resist processes available and describes changes produced in the dyeing properties of the treated wools.

4.4.1   Top Shrink-Resist Processes

Mercer's observation that chlorination treatments prevent felting is still the basis of the most popular industrial processes. Acid chlorination processes are preferred to alkaline systems as they have a minimal effect on handle, improve lustre and do not overtly yellow the wool. An important disadvantage of chlorination procedures is the reduction in wet-fastness properties of dyeings produced with anionic (acid) dyes; in comparison to their wet fastness on untreated wool, a drop of one or two points is often recorded. This drop is even more severe if the prechlorinated wool is subsequently aftertreated with reactive polyamide–epichlorohydrin resins such as Hercosett 125 (Hercules Chemical Co). Alternative nonchlorine, oxidative shrink-resist processes – in particular those based on ozone or permomosulphuric acid – produce a similar drop in acid dye wet fastness.

To date, the most successful process for producing truly machine-washable wool is the chlorine–Hercosett top process [14,15]. The Kroy Deepim chlorinator has tended to replace the acid–hypochlorite bowl. Fleissner and Woolcombers also developed improved chlorination equipment. Precision Processes (Textiles) Ltd promoted a similar chlorination–resin system using a special polymer, Dylan GRC [16]. Many polycationic products apparently reinforce the level of shrink resistance achievable with oxidative processes. Some of these products, such as Hercosett, are reactive and capable of further crosslinking by reaction with both thiol groups in the fibre and secondary amino nucleophilic residues in the resin itself, whereas others such as Basolan F are clearly not capable of further crosslinking or of covalent reaction with the fibre. Since both types give the same improvement in shrink resistance to meet machine-washability specifications, it is necessary to look for a common explanation. It appears that the initial surface oxidation procedure, whether with chlorine- or peroxy-based chemistry, is mainly responsible for imparting shrink resistance, but the effect at that stage will not withstand repeated washing and needs to be reinforced by aftertreatment with polycations.

Surface oxidation converts many of the cuticular disulphides to sulphonic acids (cysteic acid residues). These residues are highly water-attracting, thus increasing the aqueous swelling potential of the surface proteins, altering both the physical nature (softening) and the shape of the scales. Undoubtedly, changing of the fibre surface from hydrophobic to hydrophilic is largely responsible for the change in felting properties. This change, however, is brought about by the production of degraded proteins containing hydratable, strongly anionic residues, which may be progressively dissolved from the fibre during repeated washing; thus the overall antifelt protective effect is gradually lost. By precipitating these anionic proteins and anchoring them more firmly to the fibre following treatment with a polycation, the antifelting effect is retained even after repeated launderings.

Chlorination–resin treatments are commonly carried out continuously on wool in top form, although there is also significant production in garment form. The commonest resin employed is Hercosett 125 (Hercules), a water-soluble, cationic polyamide–epichlorohydrin polymer that has substantivity for anionic surfaces such as those produced on chlorinating wool. This resin contains the azetidinium cation, which is reactive to a variety of nucleophiles; such reactions lead to resin insolubilisation, probably by the formation of relatively few crosslinks between the azetidinium group and secondary amino residues in the polymer itself [17]. Additionally, covalent bonding to the wool surface through reaction with cysteinyl thiol groups is likely [18].

The cationic character of the 'cured' polymer depends on two residues:

(1) unreacted azetidinium cations;
(2) protonated tertiary amino groups.

Since the pK_a of protonated aliphatic tertiary amines lies in the region of 9–10, this contribution to the 'cured' resin's cationic character is probably small under home laundering conditions (pH >10) but large under neutral or acidic conditions. Cationic character due to residual quaternised, unreacted azetidinium groups is probably removed by alkaline hydrolysis of this group in the washing process, giving a substituted 1,2-dihydroxypropane residue. Thus under neutral to acidic conditions, such polymer finishes retain their cationicity and interact strongly with anionic dyes; under washing conditions, however, their cationic character is weak, explaining the observed reduced wet fastness of anionic dyes.

BASF have proposed [19] the use of a fully quaternised resin, Basolan F, for both shrink-proofing of wool and improving the wet fastness of anionic dyes, especially milling and 1:2 metal-complex dyes. The effect of this cationic, water-soluble polymer on the wet fastness of dyeings is quite remarkable; even black shades dyed on prechlorinated grounds with 5% o.m.f. Acidol Black M-SRL show good fastness to washing, potting and light following aftertreatment with Basolan F.

In general, the wet fastness of 1:2 disulphonated metal-complex dyeings generally follows the order: Basolan F aftertreated dyeings on chlorinated wool > nonaftertreated dyeings on chlorinated wool > Basolan F aftertreated dyeings on chlorine–Hercosett wool > nonaftertreated dyeings on chlorine–Hercosett wool. An interesting consequence of using Basolan F to improve the wet fastness of acid dyes on chlorinated wool is that the shrink resistance of the treated wool is enhanced.

In addition to improving the wet-fastness properties of disulphonated 1:2 metal-complex dyes, Basolan F has a beneficial effect on the wet fastness of monosulphonated, asymmetric 1:2 complexes (Lanasyn S (Clariant), Isolan S (DyStar), Lanacron S (Huntsman)), nonsulphonated, sulphonamide- or sulphone-methyl-solubilised, 1:2 complexes (Ortolan

(Dohmen), Irgalan (Huntsman)) and 1 : 1 metal-complex dyes (Palatin Fast (Dohmen) or Neolan (Huntsman)).

For mill operation, BASF recommends the following procedures in top finishing:

- chlorinate with acid hypochlorite;
- antichlor, rinse;
- dry, store;
- dye;
- apply Basolan F.

Basolan F may be applied after dyeing or printing in any type of dyeing machine, in a winch or by padding. For application in a dyeing machine, the following recipe is suggested: dye, rinse, aftertreat with 2–6% o.m.f. Basolan F from a bath set at pH 7.0–8.5 (ammonium hydroxide or other suitable alkali), raising the bath temperature to 40–50 °C and running for 10 minutes.

It is clear that the processing sequence of oxidation (to produce a surface containing many anionic cysteic acid residues) followed by a cationic polymer treatment is decisive in producing wool articles which possess excellent resistance to shrinking/felting in washing.

Schumacher-Hamedat *et al.* [20] have pioneered the use of FTIR (Fourier transform infrared) analysis to study the sulphur oxidation products of various shrink-resist processes. Attenuated total reflectance (ATR) measurements with the KRS 5 crystal, which measures to a surface depth of 3 μm, gave the results shown in Table 4.1. This sort of analysis clearly demonstrates why optimum shrink resistance is achieved in the peroxymonosulphuric acid process only after bisulphite treatment; the concentration of anionic groups necessary to change the hydration of the fibre surface is achieved following the reaction of cystine monoxide residues with bisulphite to give the cystine-*S*-sulphonate.

X-ray photoelectron spectroscopy (XPS) has also been employed [22–24] to study the surface chemistry of sulphur in commercial shrink-resist procedures. The results generally confirm the formation of cystine monoxide, cysteic acid and cystine-*S*-sulphonate in oxidative wool shrink-resist processes.

One particular problem of oxidatively shrink-resist-treated wools is that when dyeing above a critical packing density, fibres become so closely 'glued' together by soluble protein

Table 4.1 *Relative amounts of sulphur oxidation products formed during shrink-resist processing [21].*

Oxidation Product	Frequency/cm^{-1}	Treatment	Quantity
RSO$_3^-$	1042	chlorine–Hercosett	+++
	1042	KHSO$_5$	++
	1042	KHSO$_5$ + bisulphite	+
RSOSR	1076	chlorine–Hercosett	+
	1076	KHSO$_5$	+++
	1076	KHSO$_5$ + bisulphite	+
RSSO$_3^-$	1024	chlorine–Hercosett	++
	1024	KHSO$_5$	+
	1024	KHSO$_5$ + bisulphite	+++

residues that an additional expensive backwashing step is required [25]. The solution to this problem is to use a special silicon-based softener [26], which alters the surface characteristic of the fibre without interfering with the uptake of dyes.

Environmental pressures, coupled with increased awareness of the production of organochlorine compounds in chlorination processes, make it likely that the next generation of top shrink-resist processes will be either peroxy acid- or enzyme-based. Haefely [27] has described a process based on a padding treatment of wool tops with enzymes that gives a high level of shrink resistance. Specially designed enzymes have been produced by Biochemie A-Kundl in Switzerland in collaboration with Schoeller-Hardturm AG, but only a few have met the minimum criteria for knitwear of less than 10% shrinkage in a washing machine test coupled with less than 10% loss in fibre strength.

4.4.2 Garment Shrink-Resist Treatments

Knitted garments, usually in botany, Shetland or lambswool styles, are often shrink-resist-treated in paddle or rotating drum machines. Several processes are employed to give full machine washability [28].

4.4.2.1 *Chlorine–Resin Processes*

In this case, the garments are chlorinated, usually with dichloroisocyanuric acid (DCCA), antichlorinated with bisulphite and then finished by a substantive treatment with a cationic resin such as Hercosett, Basolan F or Dylan GRB [29]. One of the drawbacks of this procedure is the yellowing associated with the chlorination. Bereck and Reincke [30] have developed a nonyellowing procedure which uses hydrogen peroxide as the antichlor agent, rather than bisulphite (Scheme 4.3). Peroxide as the antichlor agent acts synergistically with chlorine to improve the shrink-resist effectiveness of the system; thus the amount of chlorine used can be significantly reduced, alleviating the problem of absorbable organohalogens (AOX) in the effluent, reducing wool damage and giving a softer handle. Finally, the soft polymer Basolan SW (BASF) is applied at 2.5% o.m.f. in the presence of bisulphite to give a fully machine-washable garment.

In many such procedures, dyeing is carried out after chlorination, followed by application of the cationic resin. The polycation then performs the dual function of improving both shrink resistance and dye wet fastness. Thus BASF recommended the following procedure for both garment and yarn treatments:

- chlorinate with Basolan DC (DCCA);
- antichlor, rinse;
- dye;
- apply Basolan F.

The clear advantage of these methods is that they give the dyer the opportunity to produce fully machine-washable articles in standard dyeing machinery. Disadvantages include the

$$OCl^- + H_2O_2 \rightarrow Cl^- + H_2O + O_2$$

Scheme 4.3 *Reaction of residual chlorite anion with peroxide.*

impossibility of overdyeing satisfactorily once Basolan F has been applied, the increased tendency to soil (more evident in pale shades) and the increased likelihood that goods finished with this strongly cationic system will pick up loose colour during home laundering.

Sandoz (now Clariant) have also been active in the development of cationic polymers to improve the wet-fastness properties of wool dyeings [31], with a process for aftertreating wool dyeings with a liquor containing the cationic fixing agent formed by the condensation of formaldehyde, dicyandiamide and ammonium chloride; thus Sandopur SW liquid was sold for the aftertreatment of dyeings produced with milling and 1 : 2 metal-complex dyes on either prechlorinated wool, chlorine–Hercosett wool or untreated wool. The following fresh-bath application conditions were recommended [21]:

- *Chlorinated wool:* Set the bath at 50 °C; add 8–10% o.m.f. Sandopur SW liquid; run for 20 minutes.
- *Untreated wool:* Set the bath at 70 °C and pH 7.5–8.0 (ammonium hydroxide); add 6–8% o.m.f. Sandopur SW liquid; run for 20 minutes.
- *Hercosett wool:* Set the bath at 50 °C and pH 7.5; add 8% o.m.f. Sandopur SW liquid and 2% o.m.f. Lyogen WD liquid; run for 20 minutes.

A possible disadvantage of many of these cationic aftertreatments lies in the susceptibility of the dye–cationic polymer complex to break down in hot steam pressing or pressure decatising. Despite this potential shortcoming, these agents became popular because their use allowed dyers to dye deep shades on machine-washable wool more economically than with the use of reactive dyes.

4.4.2.2   Polymer-Only Systems

Although the process is, as yet, not as successful as the chlorine–resin garment systems, much industrial experience has been gained with polymer-only shrink-resist finishing of woollen knitwear. The initial work, carried out in the International Wool Secretariat (IWS) laboratories [32], showed that polyether-based polymer solubilised by reactive Bunte salt or carbamoyl sulphonate head groups had substantivity for wool substrates at an artificially induced cloud-point temperature of about 50 °C. Following absorption of the polymer on to the wool, surface curing or resin crosslinking was achieved by the addition of ammonium hydroxide [33,34]. The inventive step in this procedure was the observation that low concentrations of magnesium chloride hexa-hydrate (about $4\,g\,l^{-1}$) caused aqueous solutions of these polyether polymers to acquire turbidity at 40–50 °C, and only these turbid solutions gave the physical form of the reactive polymer capable of being absorbed by the wool surface. The polyether polymers most widely employed in this technique were the Bunte salt derivative Securlana (Cognis) and the carbamoyl sulphonate Synthappret BAP (Dohmen). Organic solvent processes for the application of reactive polymers are still employed by some producers; of special interest are the reactive silicones.

4.4.3   Fabric Shrink-Resist Treatments

It is possible to apply shrink-resist treatments to fabrics using long-liquor oxidative treatments on winches, but these processes are difficult to control. Thus, padding processes have acquired some degree of success. The Kroy fabric system is capable of giving significantly high levels of machine washability, as is the DCCA pad–batch (5 minutes) system [35,36].

The use of reactive crosslinkable polymers to impart shrink resistance through fibre immobilisation has been widely adopted. Usually the goods are padded through an aqueous solution or emulsion of the polymer and baked to ensure crosslinking of the polymer at the fabric surface. Electron micrographs of such finishes show extensive fibre–fibre bonding, with the cured polymer acting as a durable adhesive medium [37,38].

Pad–bake processes of the Sirolan BAP type are based on mixtures of the reactive polyether Synthappret BAP (Dohmen) and a polyurethane dispersion. Sodium carbonate is used as the catalyst, and following drying the polymer is cured by baking at 150 °C [39–41].

Other proposed pad–dry–bake polymer shrink-resist cure processes include the application of thiol-terminated polyethers [42], thiomaleate esters of polytetrahydrofuran [43], Bunte salt acetate esters of polyethers [44] and aziridine-terminated polyethers [45].

All the pad–dry polymer shrink-resist processes, with the exception of silicon-based polymers, impart a harsher or stiffer handle to the fabrics, which has seriously reduced the popularity of such finishes. Lewis [46] has therefore investigated processes that obviate the need for such a curing step by developing pad–batch–wash-off processes. Using pad liquors containing the Bunte salt-terminated polyether Nopcolan SHR3 (now Securlana (Cognis)), sodium sulphite and sodium carbonate, an excellent level of shrink resistance can be achieved without interfering with the original fabric handle. The advantages of this system include reduced energy costs, elimination of thermal yellowing and the production of a full flat-set fabric; the substrate may be directly dyed, or may be printed without any further pretreatment. In principle, such pretreatments should be of interest in the stabilisation of wool piece goods against felting in low-liquor-ratio jet-dyeing machines.

Lewis [47] has shown that the pad–batch process can be used as a combined dye/shrink-resist procedure with thickened pad liquors containing anionic dye, Bunte salt polymer, sodium di-iso-octylsulphosuccinate ($10\,g\,l^{-1}$), sodium metabisulphite ($10\,g\,l^{-1}$) and urea ($300\,g\,l^{-1}$). Reactive dyes are preferred because the dyeings exhibit the excellent wet-fastness properties required for machine-washable fabrics. The process is economical as it saves energy, water and processing times while simultaneously producing high colour fastness and machine washability.

4.4.4   Miscellaneous Developments

Rakowski [48] has shown that wool tops and fabrics can be very effectively shrink-resist-treated using a plasma treatment; the equipment and mode of operation are fully described. Such a system would offer significant economic and ecological advantages as compared with aqueous processes.

Leeder and Rippon [49] have described a shrink-resist process for wool based on a treatment with potassium t-butoxide. It appears that this procedure is much less disruptive to the wool cuticle surface than is chlorination; it is therefore postulated that chemical alteration of the outermost hydrophobic layers around the epicuticle is sufficient to obtain shrink resistance.

Several workers have shown that pretreatment of wool with amines such as tetraethylene-pentamine greatly enhances the shrink-resist effectiveness of reactive polymers such as Synthappret BAP (BAY) [50]. An interesting study by Erra *et al.* [51] has shown that wool may be shrink-resist-treated from a bath containing cationic surfactant and sulphite.

Table 4.2 *Minimum fastness requirements for machine-washable wool.*

Test	Shade Change	Stain on Hercosett Wool	Stain on Cotton
TM 193	3–4	4	3–4
TM 174	3–4	4	3–4
TM 165	Dry or wet, −3		
TM 5	Under 1/12 st. depth, −3; above 1/12 st. depth, −4		

4.4.5   Colour-Fastness Requirements for Machine-Washable Wool

The Australian Wool Innovations (AWI) company [52] has established stringent colour-fastness specifications to meet the demands of the home laundering of chlorine–Hercosett-treated wool. These include a light-fastness test (TM 5), a washing test in perborate containing heavy-duty detergent (TM 193), a close-contact alkaline perspiration test (TM 174) and a rub-fastness test (TM 165). In the case of the wet tests, a special requirement is that one of the adjacent fabrics be knitted from chlorine–Hercosett wool. Table 4.2 summarises these requirements; suitable dyes are selected from ranges of reactive dyes, chrome dyes, 1 : 2 metal-complex dyes and milling dyes.

As mentioned in Section 4.4.1, the most common shrink-resist substrate available for dyeing in top, hank package and piece form is chlorine–Hercosett wool. The cationic character of the surface-deposited Hercosett resin affects the dyeing process in many ways, including:

- Increased strike of anionic dyes.
- Reduced wet fastness of nonreactive and chrome dyes.

Thus, dyeing is usually commenced at a slightly higher pH than normal, and at a lower temperature. In the early days of dyeing these substrates, reactive and chrome dyes were particularly preferred, both being given an ammonium hydroxide aftertreatment in medium to full depths in order to achieve maximum wet fastness; more recently, economics have encouraged the use of milling and 1 : 2 metal-complex dyes up to medium depths, provided a suitable cationic aftertreatment is given.

4.5   Insect-Resist Treatments

4.5.1   Insect Pests

Unlike any other textile fibres, wool and other animal fibres are subject to attack by the larvae of certain moths (Lepidoptera) and beetles (Coleoptera). The important wool-digesting insect pests are [53]:

- *Tineola bissiella* (Hummel) (common or webbing clothes moth). This pest is distributed worldwide as a single species. It is therefore widely employed as a laboratory test insect.
- *Tinea* spp. (case bearing clothes moth). This pest exists in several forms in subtropical and temperate regions.

- *Hofmannophila pseudosprettella* (Stainton) (brown house moth). It is common in the moist temperate climates of New Zealand and western European coastal areas [54, 55]. Special developments in moth-proofing agents and procedures have subsequently controlled this voracious pest.
- *Anthrenus* spp. (varied carpet beetle). This pest is prevalent in subtropical and temperate regions. The species widely used in laboratory testing is *Anthrenus flavipes* (Le Conte).
- *Attagenus* spp. (fur beetle). There are two important species of *Attagenus*: *Attagenus piceus* (Oliver), the black carpet beetle (also known as *Attagenus megatoma*), a native of Japan; and *Attagenus pellio* (L), the furrier's beetle, which is indigenous to North America. Both species are now widely distributed.

4.5.2   Insect-Resist Agents

Various chemicals have been applied to wool to control larval attack, but considerable environmental restrictions have been placed on the type of agent that may be employed. Many useful reviews of the field are available [56–60].

Insect-resist agents (IR agents) need to be applicable in many different processes and to be resistant to hydrolysis in boiling dyebaths. On the treated wool, they should exhibit adequate stability to washing and light. For carpets – the main product area – these fastness requirements are modest, but for those product areas where stringent fastness requirements exist (such as machine-washable wool), an excess of IR agent must be applied, in order to compensate for subsequent losses.

A further important requirement is that an IR agent should be safe in use, both during mill application and in the consumer situation, and should pose no environmental hazards. Since these agents are biologically active, they attract the attention of authorities and organisations responsible for worker and consumer safety and environmental protection. For example, in the UK there is strict control of IR agent discharge into sewers; regional water companies set consent limits for the maximum permitted concentration of IR agents in the sewage discharge of mills, and these are reviewed at frequent intervals.

Modern IR agents fall into two categories: those that have been developed specifically as IR agents for wool and have no uses in other fields, and those that are formulations of agricultural pesticides specially designed for wool textile application. Most of the members of the former group presently available are anionic polychlorinated aromatic compounds (Products 1–7 in Table 4.3). Members of the latter group are based on synthetic pyrethroid insecticides (Products 8–15 in Table 4.3). It is probable that both classes of IR agent enter an insect's system only through the digestive tract, since moth-proofed wool has no insecticidal activity towards species that do not ingest the fibre.

Given the relatively small market [61], it is now unlikely that a novel IR agent specific to wool could be economically commercialised, bearing in mind additional factors such as the costs of registration and ecotoxicological testing. The available market is simply not big enough to support the cost of such testing. As a consequence, new developments in IR agents for wool are likely to follow developments in pesticide chemistry in the agrochemical industry.

Lipson and Hope [61] were the first to demonstrate that the highly hydrophobic pesticides developed by the agrochemical industry could be applied to wool in dyebaths, provided they were formulated as emulsifiable concentrates. In hot aqueous dyebaths, the wool

Table 4.3   *Commercially available IR agents.*

Product	Manufacturer	Systematic or Trivial Name of Active Ingredient	Active Ingredient in Product (%)	Year of Introduction (Where Known)
1 Mitin FF high conc.	CGY	sulcofenuron	80	1939
2 Eulan U33	BAY[a]	chlorphenylid	33	1958
3 Eulan WA new	BAY[a]	chlorphenylid	20	1961
4 Mitin LP	CGY[a]	chlorphenylid/ flucofenuron	12 / 8	1972
5 Molantin P	Chemapol	chlorphenylid	32	
6 Eulan BLS	BAY[a]	trichlorobenzene-*N*-chloromethyl-sulphonamide	15	
7 Mitin 4108	CGY[a]	chlorphenylid/ sulcofenuron	16.5 / 14.5	
8 Perigen	Wellcome	permethrin	10	1980
9 SMA-V	Vickers	permethrin	20	1980
10 Antitarma NTC	Dalton	permethrin	7	1982
11 Mitin BC	CGY	permethrin	10	1982
12 Mitin AL	CGY	permethrin/ hexahydropyrimidine derivative	5 / 5	1983
13 Eulan SP	BAY	cyfluthrin	3	1982
14 Cirrasol MPW	ICI (Aus)	cyhalothrin	5.6	1985
15 Eulan SPN	BAY	permethrin	10	1987

[a]Withdrawn in 1989.

is swollen and the pesticide emulsion destabilised; exhaustion of the pesticide from the dyebath and deposition within the fibre can thus occur, resulting in acceptable fastness properties.

The first such pesticide emulsion to be applied commercially in this way contained dieldrin (**1**). Although dieldrin was widely used in industry during the 1960s and 1970s, its high mammalian and fish toxicity, coupled with its extraordinary persistence in the environment, has led to an almost universal ban on this product. Since 1983, dieldrin has not been allowed for the protection of Woolmark products against insect pests.

1

4.5.2.1   *Mitin FF High Conc. (CGY)*

This product contains the active ingredient sulcofenuron (**2**) (80% by mass) and has been used as an IR agent since 1939; although relatively expensive, it is employed where very good fastness to washing and light is demanded (for example, in uniform fabrics).

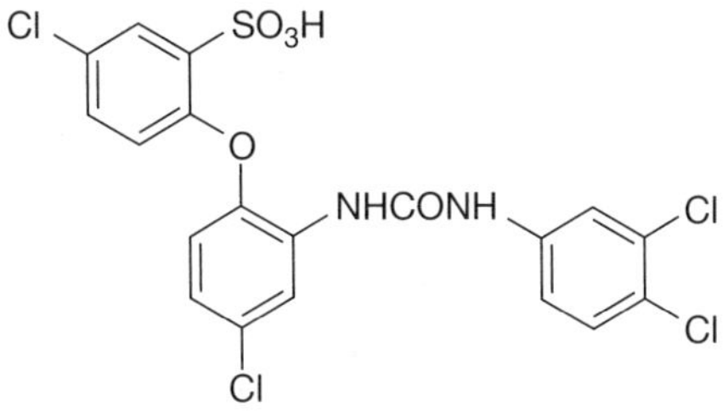

2

Of all the products available, Mitin FF is the only one that possesses substantivity for the fibre in a manner analogous to that of anionic wool dyes; this property is conferred by the sulphonate residue, which imparts water solubility.

4.5.2.2   *Eulan U33, Eulan WA New (BAY)*

The active ingredient in these products is chlorphenylid (**3**). Eulan U33 contains 33% by mass and Eulan WA new 20% by mass of this active ingredient. The chloromethyl-sulphonamido side chain of chlorphenylid imparts water solubility under alkaline conditions.

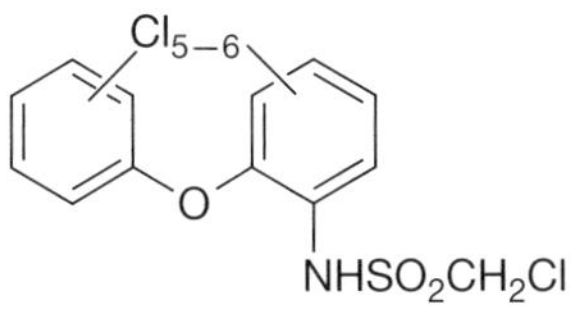

3

These products are formulated as solutions of the sodium salt at about pH 10; on acid-ification for application to wool, a milky dispersion of the free sulphonamide is formed (Scheme 4.4).

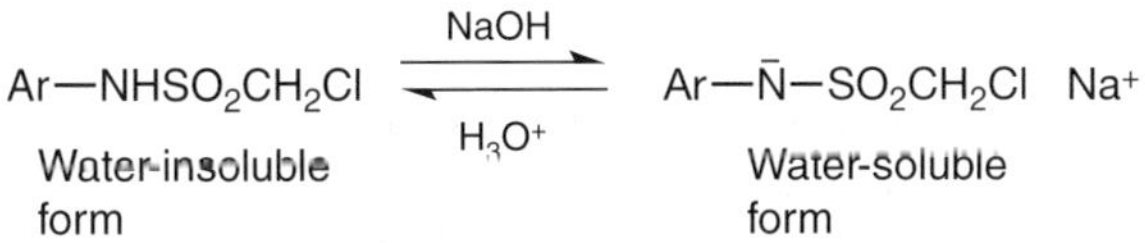

Scheme 4.4   *Formation of water-soluble chlormethylsulphonamide under alkaline conditions.*

In the past, it was generally accepted [62] that chlorphenylid behaves as a colourless acid dye, being initially absorbed by Coulombic forces from aqueous solutions. Work by Wolf and Zahn [63], however, has shown that the predominant mechanism of uptake may be through weak nonpolar interactions and hydrogen-bonding forces, through the un-ionised sulphonamide residue.

During manufacture of chlorphenylid, chlorination is carried out as a last stage. Since this process gives rise to environmentally problematic chlorinated compounds, Bayer, the main manufacturer, has ceased its production, even though the toxic impurities can be removed from the product before sale.

Chemapol supplied Molantin P, a chlorphenylid that has a reduced amount of chlorine in the molecule. Higher applications of Molantin P were thus required to give the same level of insect resistance as Eulan WA new. Pyrethroid-based insect-resist agents have been widely used, and these include: Perigen (Wellcome), SMA-V (Vickers), Mitin BC (CGy), Antimara (Dalton) and Eulan SP/SPA/WBP (Bay). The active ingredient in these products is permethrin (**4**), a synthetic pyrethroid pesticide. Permethrin is toxic to fish and aquatic invertebrates, and water authorities therefore enforce very stringent requirements on the amount that can be discharged. These restrictions mean that special methods of application are now required [64,65], or that very fish-safe agents should be brought into use [64,66].

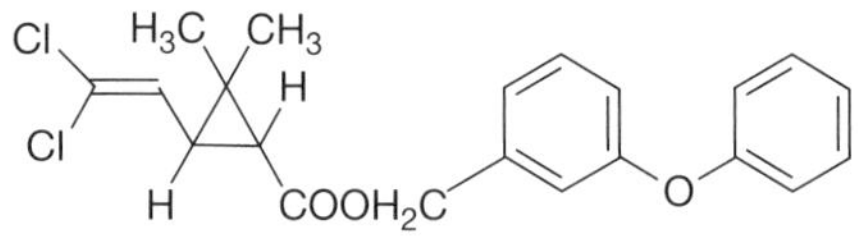

4

4.5.2.3   Mitin AL (CGY)

This product contains two active ingredients: permethrin and a specially developed hexahydropyrimidine derivative (HHP) (**5**).

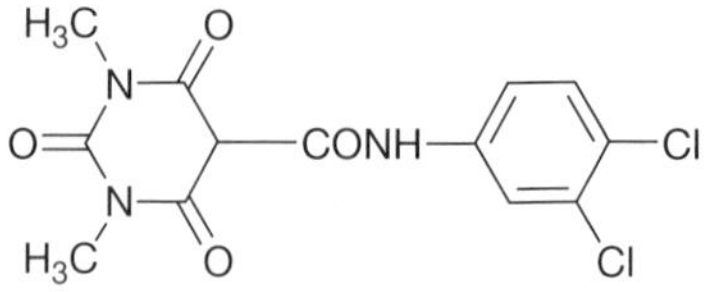

5

Pyrethroid-based IR agents [67,68] have two possible disadvantages: potential for resistance to buildup in the target species, and the so-called 'beetle-gap'; that is, the borderline effectiveness of pyrethroids against a particular *Anthrenus* beetle: *Anthrenus flavipes* var. *seminiveus* (Casey). Mitin AL was claimed to have combined the excellent moth protection of permethrin with improved *Anthrenus* beetle protection from the HHP component.

4.5.3 Application Methods for IR Agents

4.5.3.1 Application During Dyeing

The application of IR agents during dyeing is still popular, even though environmental questions will probably result in this method being replaced with cleaner procedures. All that is necessary is to add the appropriate quantity of IR agent to the dyebath along with the dyes and auxiliaries. All dyeing methods that give satisfactory dyeing will normally give satisfactory insect resistance, although exhaustion of IR agents may be inhibited by certain dyeing auxiliaries and some agents may be destroyed if prolonged boiling of the dyebath is carried out for shade-matching purposes.

Blends of wool and polyamide are commonly used in the carpet sector. In the UK, for example, a wool-rich polyamide blend predominates (80 : 20, wool/nylon). Often these blends are union dyed, mainly in yarn form, and the IR agent is applied at this stage. Mayfield [62] has clearly demonstrated that the chlorphenylid derivatives (Eulan WA new/U33, Molantin P) partition in favour of the polyamide component in the 80 : 20 wool/nylon blend in the approximate ratio of 4 : 1. The corresponding ratio for the pyrethroid-based agent Perigen was 4 : 3, indicating a clear advantage for this type of product in the treatment of wool/polyamide blends. Mayfield [62] also showed that the chlorphenylid agent absorbed by the polyamide portion was ineffective against insect pests.

A significant proportion of wool tufted carpet production outside the UK is manufactured by the so-called dry-spun route. There are undoubted technical difficulties both in applying IR agents and in achieving adequate fastness properties on Berber blends containing only a small proportion of stock-dyed wool. In these cases, it is common to apply the IR agent using the 'spinning lubricant' method, which gives unsatisfactory fastness properties.

The observation that the best overall fastness performance can be achieved only by dyebath application of IR agent has led to a method involving overtreating a portion of a dry-spun wool blend in the dyebath and blending with untreated wool to give the 'normal' overall treatment level [69]. If a sufficient proportion of the blend is of dyed wool, this method presents no technical problems. If only a very small fraction of the blend is dyed – the remainder being undyed wool – then some of the undyed wool must be treated with IR agent in a blank dyebath. In laboratory trials, it was found possible to blend untreated wool and overtreated wools up to a maximum blend ratio (untreated : overtreated) of 40 : 1 before serious insect damage occurred, provided the amount of IR agent on the total wool was sufficient to control the insect pests. The effectiveness of this method of insect-proofing has been demonstrated both for intimate fibre blends and for blends produced by folding overtreated and untreated yarns.

In mill conditions, it may be that such high blend ratios as 40 : 1 cannot be used with safety, but ratios as high as 10 : 1 have been used successfully in fullscale practice.

4.5.3.2 Application During Scouring

Insect-proofing during scouring is the second most important technique used in industry. Carpet wools may be proofed in continuous processes during raw-wool scouring, or more commonly when the yarns are scoured in hank form. Blankets, upholstery fabrics and uniform fabrics may be proofed in batchwise treatments during the scouring of piece goods.

These treatments are carried out at lower temperatures and for shorter times than dyebath proofing processes. The IR agent therefore does not penetrate so deeply into the wool fibres; as a consequence, its fastness is lower, and higher levels of agent should be applied to compensate. For carpet yarns, proofing during yarn scouring is important when Berber-style carpets are fashionable. These yarns are composed, at least in part, of undyed wool, and treatment of the yarns with IR agent during the scouring process to remove spinning lubricants is a convenient method of proofing. Scouring is normally done in hank form, the hanks being carried through the machine by tapes or brattices. The machine normally has four bowls, into which the yarns are immersed in turn; they are then passed through a mangle. The first two bowls contain alkali (sodium carbonate or hydrogen carbonate) and detergent respectively, the third is used for rinsing with water and the fourth contains acid and IR agent. Although the time of immersion of the yarns in the last bowl is very short (10–90 seconds, depending on the type of machine and the speed at which it is operated) and the temperature is relatively low (usually below 50 °C), significant exhaustion of the IR agent on to the yarn takes place. IR agent and acid are added to the bowl continuously, preferably by metering devices, to keep concentrations in the last bowl and uptake by the wool at a constant level.

4.5.3.3   *Application by Addition to Spinning Lubricants*

In terms of the volume of wool treated, this is the third most important method used for IR treatment of carpet yarns, but it is confined to so-called dry-spun yarns, which are carded and spun with such low additions of lubricant that scouring is regarded as unnecessary. The IR agent is mixed with the spinning oil and water, and the mixture is applied to the loose wool in the normal way. The fastness of IR agents applied in this way is lower than when any other method is used, because the treatment is largely superficial. It is usual to apply higher amounts of IR agent to compensate for the lower fastness.

Application of IR agent from spinning lubricant cannot be recommended for high-quality products because of the lack of fastness of the finish. The method also has other potential disadvantages in that mill operatives and consumers may be exposed to direct contact with IR agent. Even though modern IR agents are not regarded as harmful according to accepted definitions, such exposure should be avoided if at all possible.

4.5.3.4   *The Current and Future Status of Wool Moth-Proofing*

The history of wool moth-proofing is a great success story [5,60], following the introduction of such products as Eulan New and Eulan CN Extra in the 1930s. It is perhaps a worry that less and less moth-proofing is being carried out on wool garments, as retailers have taken steps to eliminate pesticide treatments of wool articles received from their suppliers. The threat against pyrethroids being able to be used as moth-proofers also comes from the European REACH process; suppliers are thus withdrawing from the market [70]. Unfortunately, over the past 5 years the consumer may have become increasingly frustrated to find that relatively expensive all-wool and cashmere garments are increasingly prone to attack by moth larvae; this problem is widespread and requires urgent attention, or else consumers will be driven to buy cotton and acrylic sweaters. It appears that the problem of avoidance of moth-proofing was created by powerful retailers and mixed up with their environmental policies.

One of the most successful modern moth-proofers was based on the pyrethroid insecticide permethrin; it was reasoned that since most moth-proofing was carried out in dyebaths, the discharge created a problem for fish, and especially for their food sources, such as daphne and gammerus insects. In fact, the use of pyrethroids in agriculture far outweighs their use as moth-proofing agents; an estimate would be by many million times. Notwithstanding this, wool scientists at AgResearch in New Zealand have invented the Lanaguard method of applying pyrethroids to wool, in a polymer carrier system, which does not result in any aqueous discharge [71].

A paper by Lehmann *et al.* [70] did offer an alternative, by replacing pyrethroids with potassium hexafluorozirconate, which forms the basis of the Zirpro flame-retardant finish for wool (see Section 4.6). However, it is likely that this would not represent a long-term solution since the water authorities would not view favourably the news that potassium hexafluorozirconate is an insecticide. It is clearly critical, in order to maintain the virtues of wool as a luxury fibre, that research be carried out to render the amino acids in wool keratin indigestible to moth larvae; simple, cheap chemical modifications, carried out in the dyebath, are not impossible.

4.6   Flame-Retardant Treatments

The wool fibre under most conditions may be regarded as reasonably resistant to burning, because of its high nitrogen content. However, with the introduction of stringent test methods for aircraft carpets and upholstery, and latterly for contract furnishings, it became apparent that wool required further chemical treatment to enhance its flame retardance.

One of the most successful methods has been to employ fluoro complexes of titanium and zirconium, especially hexafluorotitanate and hexafluorozirconate. As potassium salts, these complex anions are readily adsorbed by wool fibre from aqueous solutions [72,73]. This is conveniently carried out by the dyer as an aftertreatment following dyeing. Usually the pH is reduced to 3.0, potassium hexafluorozirconate (about 8% o.m.f.) is added and the bath is run for 30 minutes at 60 °C. Although the titanium complex is slightly more effective than its zirconium counterpart, it tends to give slight yellowing. The hexafluorozirconate is thus usually preferred, except in the case of dark shades.

The measurement of smoke emission is becoming of increasing concern and a modified treatment leading to reduced smoke emission has been developed. This incorporates zirconium acetate in addition to potassium hexafluorozirconate and uses formic and citric acids rather than hydrochloric acid. These treatments are currently applied principally to aircraft textiles.

A further variant of the Zirpro treatment incorporates tetrabromophthalic acid to ensure that treated products have very short or zero afterflaming times. This treatment can also be beneficial if a fabric contains a synthetic fibre, has a shrink-resist finish or has to be neutralised.

An alternative method of flame-proofing wool was suggested by Lewin [74]. This process, which never became commercial, depended on the introduction of a large number of sulphate ester and sulphamate residues in the fibre through padding a mixture of sulphamic acid and urea on the fibre. Covalent attachment of sulphonic acid residues at hydroxyl and amino groups was achieved by baking for 1 minute at 150 °C. The biggest drawback of this process was the disastrous effect it had on the wet fastness of anionic dyeings, with the possible exception of wool dyed with reactive dyes.

Perachem Ltd recently launched Peralana ZR, which is a durable, environmentally friendly alternative to Zirpro. The product does not contain halogen, zirconium, antimony or formaldehyde. It is conveniently applied to wool in all its forms in conventional dyeing machines at 60 °C for 30 minutes.

4.7   Antisetting Agents

Permanent setting is a contributory factor leading to loss in wool fibre strength during dyeing. It is also the main cause of increased hygral expansion of wool fabrics following piece dyeing, the reason for surface marks such as 'crows' feet' in piece dyeing and the source of reduced bulk or yarn leanness following package dyeing [75–77]. Aside from these wholly negative effects, permanent setting can sometimes be seen as beneficial; examples include the setting of wool yarn in hank form when using hank dyeing machines: this gives extra bulk and resilience to yarns and explains why this dyeing route remains popular for the production of wool carpets. It is necessary to summarise the various chemistries involved in the production of permanent set in wool dyeing.

Wool is a heterogeneous material made up of keratin proteins, a small amount of nonkeratinous proteins and even smaller amounts of lipid and fatty acid materials. The keratinous or cystine disulphide crosslinked proteins are responsible for most of wool's physical properties and thus the reactivity of the cystine disulphide residue is of paramount importance. The cystine content of wool varies, but an appropriate average value is 450 μmol g^{-1}, and for its reduced form, cystine, 30 μmol g^{-1}. The chemistry of these residues bound to the protein chains through the amide linkage is illustrated in Structures **6** and **7**.

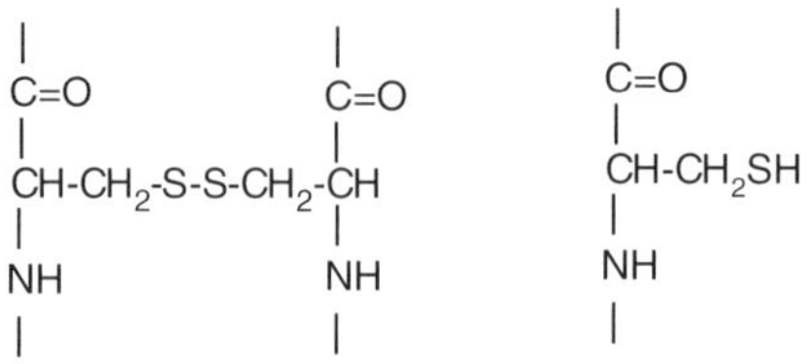

Both of these amino acid residues can readily undergo a *trans* 1,2-β-elimination reaction to form the dehydroalanine residue, which contains an activated double bond capable of subsequent Michael addition with suitable nucleophiles. These reactions occur in water; their extent and nature depend particularly on pH, temperature and time [78–80]. They are activated by the electron-withdrawing character of the adjacent amide carbonyl residue. In general, the thiol form, cystine, more readily undergoes β-elimination than does the disulphide form, cystine. Even at pH 3 in boiling aqueous dyebaths, there is some elimination of hydrosulphide ion from the cystine residue, whereas the cystine

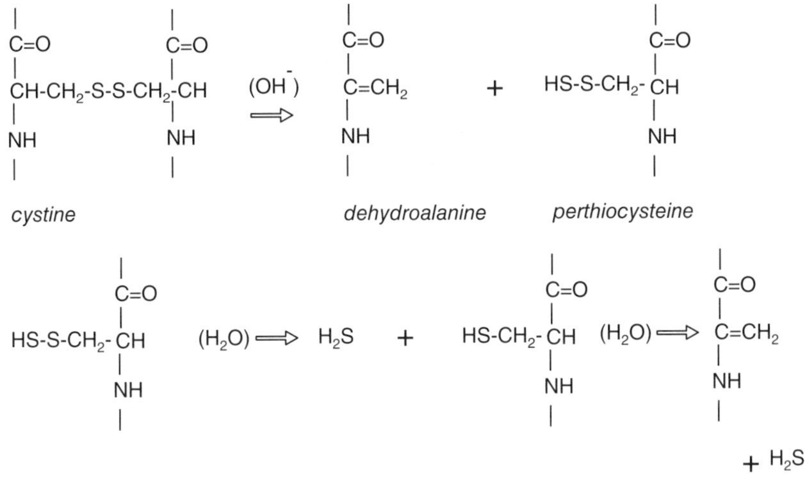

Scheme 4.5 *Elimination reactions of protein-bound cystine and cystine residues.*

disulphide only undergoes such reactions above pH 7. Scheme 4.5 summarises these reactions.

Hydrogen sulphide, or hydrosulphide anion, produced in these reactions, is capable of ready reaction with cystine disulphide residues to produce further cystine thiol residues. In boiling dyebaths, these will undergo rapid β-elimination to dehydroalanine and hydrogen sulphide, which is clearly the start of a runaway degradation reaction. The reactive entity, dehydroalanine, will undergo Michael addition with amino nucleophiles present in histidine and lysine residues and with thiol nucleophiles present in cystine; in this way, more stable crosslinks such as histidino-alanine, lysino-alanine and lanthionine are formed [81,82]. The extent and exact composition of these new crosslinking amino acids vary greatly with the pH and temperature of treatment, since the nucleophilicities of amino and thiol residues increase with increasing pH and temperature. The chemistry of these crosslinking reactions is exemplified, for lysine and cystine residues, in Scheme 4.6.

These new crosslinks will not undergo degradation or elimination reactions under conditions normally encountered in wool dyeing and are thus likely to be of great importance in explaining the phenomena of permanent setting during wool dyeing. Also of some importance in setting is the so-called thiol-disulphide interchange reaction [83,84].

It is clear from this discussion that control of setting in dyeing can be achieved by the addition of chemicals which scavenge hydrosulphide anions as they are liberated, or which rapidly modify free cystine thiol residues to prevent the elimination reaction; in practice, this can be achieved in two ways:

(1) Inclusion of oxidants in the dyebath.
(2) Inclusion of fibre substantive electrophiles in the dyebath.

It is thus important to measure set following dyeing, and most of the published research in this area uses Køpke's crease angle method to achieve this [85]. Typically, blank dyeings of wool fabric in pH 5 buffer for 1 hour at the boil, without antisetting agent present, give

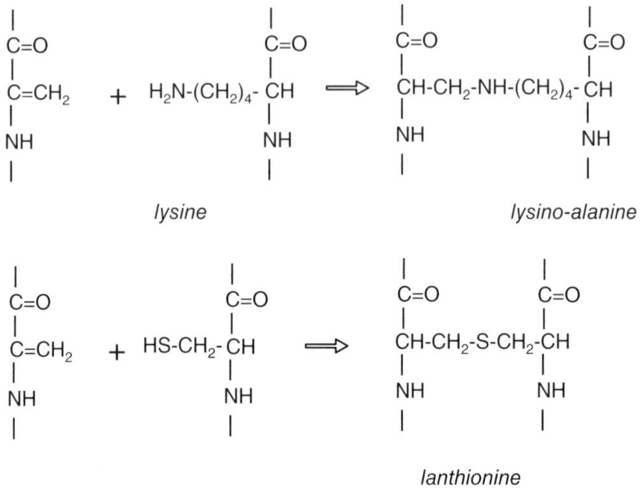

Scheme 4.6   *New crosslinks formed from the elimination reactions of cystine and cystine.*

set values of ~70%, whereas including an effective antisetting agent gives a set value of ~30%.

4.7.1   The Role of Oxidants in Preventing Setting in Dyeing

One of the most useful antisetting systems for use in wool dyeing was developed by workers at CSIRO, IWS and BASF [86]. This system, offering improved fibre physical properties, was based on a mixture of hydrogen peroxide and a special auxiliary Basolan AS; the latter auxiliary performed two functions:

(1) Inhibition of the degradative effect of hydrogen peroxide on some wool dyes.
(2) Enhancement of the stability of the oxidant in the boiling dyebath.

The BASF/IWS/CSIRO process [86] recommended dyeing wool in the presence of hydrogen peroxide (35%) at a level of $1\,cm^3dm^{-3}$ (minimum 2% o.m.f.) and Basolan AS at $0.5\,g\,l^{-1}$ (minimum 1% o.m.f.).

Kim and Lewis [87] studied the effect of hydrogen peroxide concentration on set after boiling wool fabric for 1 hour at pH 3, 5, and 7 in the presence of citric acid/phosphate buffers (McIlvaine buffers); their results are reproduced in Figure 4.1. Hydrogen peroxide (30%) concentrations as low as $2\,cm^3dm^{-3}$ gave set values of 21% at pH 3 (control without oxidant gave 56% set), 27% at pH 5 (control 59%) and 43% at pH 7 (control 67%). The effect of pH on set parallels the effect of pH on cystine degradation: if the mechanism is β-elimination of cystine, as shown in Scheme 4.6, then a higher concentration of oxidant should be more effective at controlling set at pH 7 where elimination is more rapid. This hypothesis is confirmed in Figure 4.1, which shows that a concentration of $4\,cm^3\,dm^{-3}$ hydrogen peroxide gives the best control of set. The oxidant clearly functions by rapidly oxidising cystine back to cystine or cysteic acid; of possible equal importance is the

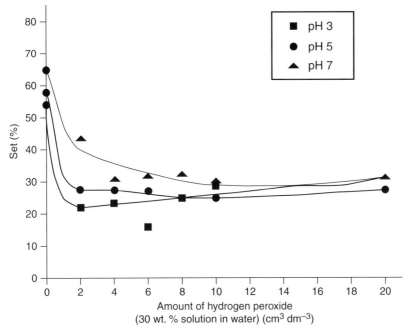

Figure 4.1   *Percentage set after boiling wool, at pH 3, 5 and 7, with H_2O_2.*

oxidation of hydrogen sulphide to bisulphate anions, thus removing this reactive reducing nucleophile from the system.

Kim and Lewis [87] used FTIR second-derivative spectroscopy to follow the production of cysteic acid when treating wool at the boil with hydrogen peroxide ($0–10\,\mathrm{cm}^3\,\mathrm{dm}^{-3}$). By measuring the sulphonate band intensity, attributed to cysteic acid, at $1040–1044\,\mathrm{cm}^{-1}$, it was shown that overoxidation to cysteic acid was most significant in those treatments carried out at pH 7.

Other oxidants shown to have antisetting properties include sodium bromate and sodium tetrathionate [88].

It is not possible to use oxidants such as hydrogen peroxide when dyeing with reactive dyes as the perhydroxy anion is a potent nucleophile which reacts at the electrophilic site in the reactive dye, rapidly producing the hydrolysed, inactive dye.

4.7.2   The Role of Electrophilic Reagents in Controlling Setting in Dyeing

At the same time as the Basaolan AS-A system, based on hydrogen peroxide, was launched, a system based on application of an electrophile Basolan AS-B (a formulation containing maleic acid) was also marketed.

Kim and Lewis [89] showed that sodium maleate (MAS) was a very effective antisetting agent when included in boiling pH 3 wool dyebaths; their results are reproduced in Figure 4.2, which clearly demonstrates that the maleate anion is most effective at controlling set at pH 3; it shows only modest effect on set at pH 5 and has no effect at pH 7. The

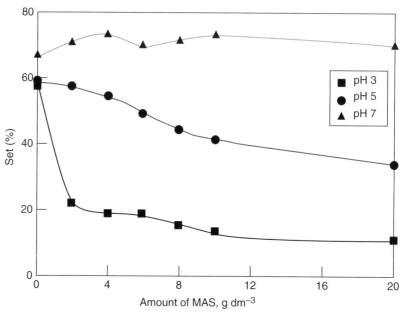

Figure 4.2 *Percentage set after boiling wool, at pH 3, 5 and 7, with sodium maleate (MAS).*

reason for this strong pH dependence lies in the modest substantivity of the maleate ion for wool under acidic conditions and its almost total lack of substantivity at pH 7 and above. It was estimated [89], from capillary electrophoretic analysis of the treatment baths, that when using $2\,g\,dm^{-3}$ MAS for the 1-hour boiling treatment, maleate anion uptake was 22% at pH 3, 15% at pH 5 and 4% at pH 7. The mechanism by which maleate anions reduce the extent of wool setting in boiling aqueous treatments is described in Scheme 4.7.

Liao *et al.* [90] synthesised N-naphthylmaleimide and showed that it inhibited wool setting and protected the wool component of a wool/polyester blend from serious damage during the high-temperature dyeing step required to adequately dye the polyester component with a disperse dye. These authors and others [86] attribute the main cause of setting (and hence wool damage) in hot aqueous wool treatments to the thiol-disulphide interchange reaction, without implicating cystine residue β-elimination reactions.

4.8   Fibre Arylating Agents (FAA)

The covalent incorporation of bulky aryl residues in wool fibres, at a level $\sim$13% o.m.f., results in the following interesting properties: disperse dyeable wool [91], shrink-resist wool [91] and heat-settable wool [92]. Undoubtedly these effects are due to an increased number of aromatic interactions [93], which arise from the negatively charged π-electrons in aromatic systems interacting with the less negative σ-electrons; these aromatic

Reaction with wool cysteinyl residues:

WOOL-SH + Na$^+$$^-$OOC-CH=CH-COO$^-Na^+$

Cysteine residue

⇓

WOOL-S-CH(COO$^-$Na$^+$)-CH$_2$-COO$^-$Na$^+$

Reaction of nonabsorbed maleate with H$_2$S in solution:

H$_2$S + Na$^+$$^-$OOC-CH=CH-COO$^-Na^+$

⇓

Na$^+$$^-$OOC-CH$_2$-CH(-COO$^-Na^+$)-SH

⇓

Na$^+$$^-$OOC-CH$_2$-CH(-COO$^-Na^+$)-S-CH(-COO$^-Na^+$)-CH$_2$-COO$^-Na^+$

Scheme 4.7   *Reaction of maleate anion with cysteinyl thiol.*

interactions are geometric in action and are seen as being very important in determining crystal orientation [94].

Most of the successful early laboratory-scale work used benzoic anhydride, applied from dimethylformamide, as the electrophile in the modification of predried wool fabric. After a typical treatment for 2–3 hours at 70 °C, the modified wool fabric was of good colour and strength and was resistant to felting under machine-washing conditions. In addition, the wool could be dyed or printed with disperse dyes to a high standard of brilliance, thus offering the opportunity to develop the wool/polyester market by simplifying the coloration process. It would also be of interest to investigate the possible advantages of using novel bright chromophores available only in disperse dye ranges, in order to reproduce such shades on wool. Clearly, the disadvantages of this procedure, hindering its acceptance by the wool processing industry, are the necessity of working from organic solvents and the relatively high degree of fibre modification required to get the optimum effects.

It is interesting to reflect that a similar procedure was developed in Japan for the modification of cellulosic fibres so as to render them transfer-printable with existing disperse dye printed papers; this was the so-called Shikibo-Uni process and involved a pad pretreatment of the cotton fabric with sodium hydroxide solution, passing the alkaline wet fabric through neat benzoyl chloride and finally washing-off with sodium hydroxide [95]. Such a process using the acyl chloride did not work well on wool fabric as it produced unacceptable yellowness.

Due to these restrictions, there has been a concerted effort to develop water-soluble acylating/arylating agents capable of incorporating covalently bound bulky aromatic residues in either wool or cotton fibres. The leaving group in the reaction with fibre nucleophiles has to be the water-solubilising group. The following are some examples of the compounds evaluated:

- 2-chloro-4,6-di(aminophenyl-4-β-sulphatoethylsuphone)-1,3,5-triazine (XLC) (**8**) [96];

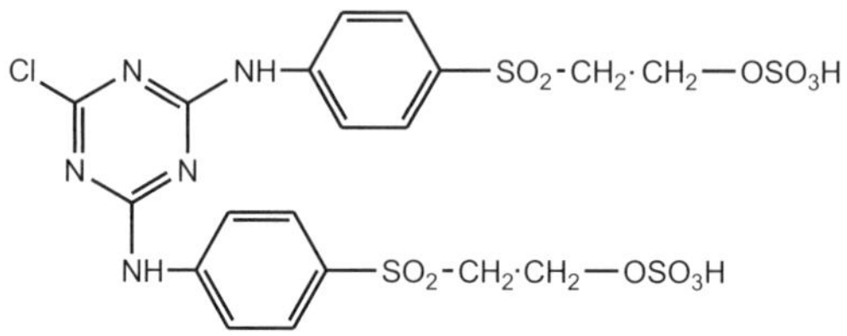

$$\text{Cl} \quad \text{N} \quad \text{NH} - \text{C}_6\text{H}_4 - \text{SO}_2\text{-}\text{CH}_2\text{·}\text{CH}_2 - \text{OSO}_3\text{H}$$

$$\text{NH} - \text{C}_6\text{H}_4 - \text{SO}_2\text{-}\text{CH}_2\text{·}\text{CH}_2 - \text{OSO}_3\text{H}$$

8

- benzoyl-thioglycollate (BTG) (**9**) [97];

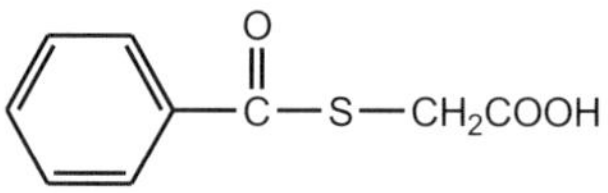

$$\text{C}_6\text{H}_5 - \overset{\overset{\text{O}}{\|}}{\text{C}} - \text{S} - \text{CH}_2\text{COOH}$$

9

- (4-phenylsulpho)benzoate (**10**) [98];

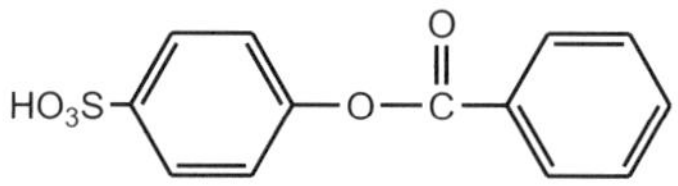

$$\text{HO}_3\text{S} - \text{C}_6\text{H}_4 - \text{O} - \overset{\overset{\text{O}}{\|}}{\text{C}} - \text{C}_6\text{H}_5$$

10

- AC24 (**11**) [99].

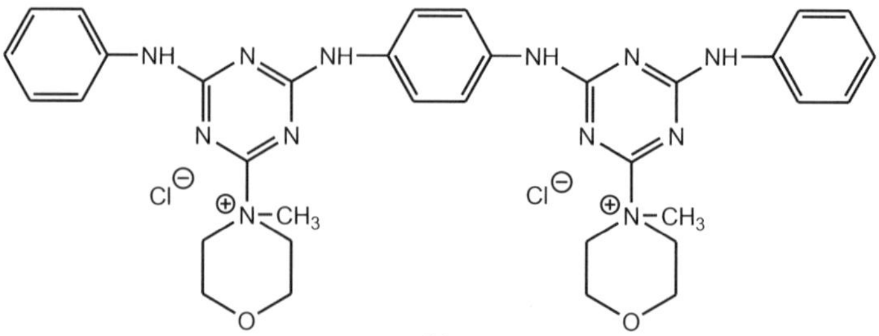

11

These compounds all contain water-solubilising groups that act as leaving groups in nucleophilic substitution reactions; apart from AC24, all are anionic and hence potentially co-applicable with commercially available disperse dyes that contain anionic dispersing agents. In the case of cotton fabrics, most promise was shown by the BTG system. This was applied by padding from an aqueous pad-liquor containing Na_2CO_3, followed by steaming to fix. However, this system cannot be used on wool as the leaving group, thioglycollate, would damage the fibre at the high temperatures involved in dyeing. In addition, if this agent were to be used for dyeing wool/polyester blends, the thioglycollate leaving group would reduce many disperse dye chromophores.

In order to achieve a one-bath, one-dye system for the dyeing of wool/polyester blends, it was decided to develop a water-soluble acylating agent for wool that was applicable from long liquor, had good solubility, was compatible with commercial disperse dye dispersions and formed covalent bonds with the fibre in the region of pH 5–6. In addition, the agent was designed to significantly protect the fibre from damage even when the dyebath temperature was raised to 120 °C in order to fix the disperse dye on the polyester component. A number of compounds were thus synthesised (Scheme 4.8) based on the *s*-triazine/*p*-base system (*p*-base is anilino-4-*β*-sulphatoethylsulphone). One compound was selected [100] which gave promising disperse dyeings when included in a dyebath containing commercial disperse dye (virtually 100% dyebath exhaustion even at 2% disperse dye o.m.f.); this compound was coded as FAA 200 (**12**).

FAA 200 (10% o.m.f.) was applied to wool for 1 hour at the boil from dyebaths set at pH 2–7. The amount of FAA 200 absorbed and covalently fixed to wool was determined by extracting the treated fabric four times with sodium sulphite solution (10% w/w) at 80 °C. By combining the extracts and measuring the absorbance at λ max, the amount of non-fixed FAA 200 was determined. This method of extraction functions in two ways: first, the pH is set at 8.5, which effectively desorbs the unreacted SES form of FAA 200, and second, the sulphite anion adds on to the activated vinylsulphone (VS) form to give the sulphonate-ethylsulphone analogue. The VS form, being nonionic and water insoluble, is difficult to desorb if it remains in a noncovalently bonded form on the wool, whereas the sulphonate-ethylsulphone form, being nonreactive, anionic and water soluble, is easily desorbed at pH 8.5. Figure 4.3 shows the plot of FAA 200 uptake versus bath pH. It can

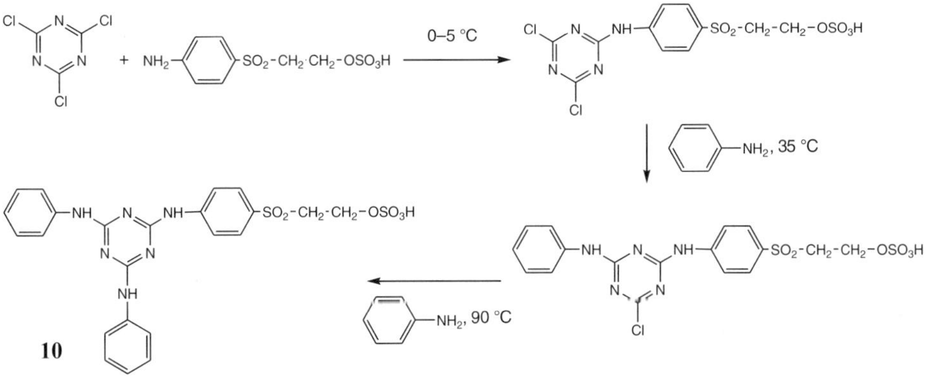

Scheme 4.8 *Synthesis of the fibre-reactive arylating agent, FAA 200.*

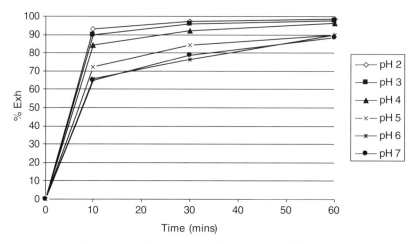

Figure 4.3   *Effect of bath pH on uptake of FAA 200.*

be seen that although the compound generally behaves like a mono-sulphonated reactive dye, showing good substantivity in the pH region 2–6, the high exhaustion values at pH 7 are not typical of reactive dyes; these results may be explained by the formation of the VS form, which will behave like a reactive disperse dye, its uptake being independent of pH. Disperse dyes were applied to all-wool fabric in the presence of 0, 5 and 10% o.m.f. FAA 200, dyeing for 1 hour at the boil; control dyeings (pH 6) were carried out on polyester fabric at 130 °C for 45 minutes. In all cases, dyebath exhaustion was greater than 95%. Table 4.4 shows colour yield (f_k) results for the final dyeings from a selection of disperse dyes; it demonstrates that even at a level of 5% o.m.f. FAA 200, good substantivity for disperse dyes may be achieved.

Following the BS EN ISO 105 C02 (soap flakes 5g per litre; 50°C, 15 mins) wash test, all the dyeings produced in the presence of 5% o.m.f. FAA 200 gave a shade change

Table 4.4   *Colour yield values from dyeings produced with selected disperse dyes applied to all-wool fabric and polyester fabric (PES).*

| | Colour Yield (f_k) Values | | | |
| | Wool | | | |
Disperse Dye	No FAA	5% FAA	10% FAA	PES
C.I. Disperse Red 50	50	120	170	160
C.I. Disperse Orange 25	23	42	83	93
C.I. Disperse Violet 26	14	21	28	39
C.I. Disperse Red 167 : 1	48	106	128	127
C.I. Disperse Blue 79 : 1	37	81	117	136
C.I. Disperse Red 60	15	21	32	72
C.I. Disperse Yellow 119	12	51	60	73

rating of 4–5, whereas dyeings produced in the absence of FAA 200 gave a colour change rating of 1–2. It was noted that some disperse dyes were very prone to reduction, C.I. Disperse Blue 79 giving dull bluish-brown dyeings instead of navy blue, for example. The prevention of these reduction processes was studied and promising effects were observed when dyeing in the presence of hydrogen peroxide or sodium thiocyanate; the latter agent controlled dye reduction more effectively than peroxide, especially at temperatures above 110 °C. Equally, the addition of thiocyanate reinforced the fibre-protective effect of the FAA 200, allowing dyeing to be carried out even at 130 °C, if required. The reason for these phenomena is that extensive β-elimination of cystine at high temperatures results in the copious production of H_2S or NaSH, which damages the wool by promoting further disulphide bond breakdown [101,102] and reduces some azo dyes very rapidly. Thiocyanate reacts with cystine through its iso-thiocyanic acid tautomer, the formation of which is favoured by acidic conditions; the dithiocarbamate has a much-reduced tendency to β-eliminate [103]. Scheme 4.9 summarises this reaction.

The excellent colour yields of the disperse dyeings on wool and their promising wet-fastness properties are worthy of comment. Lewis and Pailthorpe [91] demonstrated that wool acylated with benzoic anhydride showed high substantivity for disperse dyes and good wash-fastness (ISO2) properties. However, similar acylation procedures with propionic anhydride, while giving modified wool exhibiting good substantivity for disperse dyes, gave dyeings of poor wash fastness. Even more interesting perhaps was the observation that the benzoylated wool did not felt at all when tested to the Superwash standard (TM193), whereas the propionylated wool showed some felting (5% Area Felting Shrinkage, AFS); the acetylated wool gave 35% AFS and the untreated 65% AFS. Milligan and Wolfram [92], in an evaluation of the heat-setting properties (Hoffman press for 15 seconds) of acylated wools, showed a remarkable difference between aromatic acylations and aliphatic acylations: benzoylated wool gave a crease angle of 63°, whereas propionylated wool gave a value of 137° and nonacylated wool >170°. These observations indicate that the incorporation of aromatic residues is highly desirable, increasing significantly the forces of interaction between adjacent protein chains and between the modified proteins and the disperse dye. These interactions are clearly being played out in the FAA 200 modified wool. Lewis [93] points out that multiple aromatic interactions have their basis in the relatively 'positive' σ-electron system interacting with the strongly negatively charged π-system on the benzene ring or in the extended π-electron system in the chromophore of the disperse dye. Hunter [94] has reviewed the nature of such aromatic interactions and their geometry-determining properties, which influence molecular assembly and shape. The fibre-protective effect of FAA 200 when dyeing wool at high temperatures was assessed by carrying out the following treatments on the 100% wool fabric: 0 and 5% o.m.f. FAA

NaSCN + HX → NaX + HN=C=S

isothiocyanic acid

WOOL-CH$_2$-SH + HN=C=S → WOOL-CH2-S-(C=S)-NH2

wool-cysteine-S-thiocarbamate

Scheme 4.9 *Reaction of isothiocyanic acid with cysteinyl thiol residues.*

Table 4.5   *Wet burst strength (WBS) of untreated and treated fabrics.*

Fabric	WBS ($kg\,cm^{-2}$)	% Strength Loss
Untreated	5.63	0.0
No FAA 200, 100 °C, 60 minutes	4.75	15.6
5% FAA 200, 100 °C, 60 minutes	5.82	0.0
No FAA 200, 120 °C, 30 minutes	3.97	29.5
5% FAA 200, 120 °C, 30 minutes	5.15	8.5
10% FAA 200, 120 °C, 30 minutes	5.41	3.9

200, pH 5.5, 100 °C for 60 minutes; 0, 5 and 10% o.m.f. FAA 200, pH 5.5, 120 °C for 30 minutes. The wet burst strength (WBS) properties of the fabrics were measured according to IWS TM 29; the results are shown in Table 4.5, which demonstrates the remarkable efficiency of the VS-reactive group in FAA 200 in blocking free thiolate anions as they are formed, thus preventing setting in the high-temperature treatment processes used; these results are similar to those achieved with VS-reactive dyes [104]. When dyeing at 120 °C without FAA 200, after 30 minutes' treatment the wool was brittle and yellow, in contrast to its excellent condition when dyed in the presence of FAA 200. Therefore, FAA 200 achieves two important effects when dyeing wool/polyester blends above the boil with a single dye class: the covalently incorporated aryl residues act as sites for the disperse dye and the VS-reactive group blocks thiol residues, ensuring that fibre damage is reduced to less than that produced during conventional wool dyeing at the boil.

References

[1] G F Wood, *Textile Progress*, **12** (1982) 9.

[2] R G Stewart, *Wool Scouring and Allied Technology* (Christchurch, New Zealand: Caxton Press, 1983).

[3] D S Taylor, *Proceedings of the 7th International Wool Textile Research Conference, Tokyo*, **1** (1985) 33.

[4] T Shaw, *Proceedings of the 8th International Wool Textile Research Conference, Christchurch*, **4** (1990) 533.

[5] J R McPhee and T Shaw, *Review of Progress in Coloration and Related Topics*, **14** (1984) 59.

[6] G Blankenburg and M Breuers, *Schriftenreihe des Deutschen Wollforschungsinstitutes*, **87** (1982) 402.

[7] M Cole, J C Fletcher, K L Gardner and M C Corfield, *Applied Polymer Symposium*, **18** (1971) 147.

[8] M Bereza and Sedelnik, *Proceedings of the International Wool Textile Research Conference, Aachen*, CD of Papers EN-P2 (2000).

[9] A Ruers, E Heine and H Höcker, *Proceedings of the International Wool Textile Research Conference, Aachen*, CD of Papers, EN-P3, (2000).

[10] M S Bucur, C Poescu and M D Stanescu, *Proceedings of the International Wool Textile Research Conference, Aachen*, CD of Papers EN-P4 (2000).

[11] I C Gouveia, J Queiroz and J M Fiadeiro, *Proceedings of the International Wool Textile Research Conference, Leeds*, CD of Papers 34-CCF (2005).

[12] H Heine and H Höcker, *Review of Progress in Coloration and Related Topics*, **25** (1995) 57.

[13] K R Makinson, *Wool Shrinkproofing* (New York, USA: Marcel Dekker, 1979).

[14] D Feldman, J R McPhee and W V Morgan, *Textile Manufacturer*, **93** (1967) 122.

[15] J Lewis, *Wool Science Review*, **55** (1978) 23.

[16] Anon, *Hosiery Trade Journal*, **79** (1972) 82.

[17] N A Bates, *Tappi*, **52** (1969) 1157.

[18] A Swanepoel, *Textilveredlung*, **5** (1970) 200.

[19] K Reincke, *Melliand Textilber*, **67** (1986) 191.

[20] U Schumacher-Hamedat and H Höcker, *Textilveredlung*, **21** (1986) 294.

[21] Sandoz, Technical Information Leaflet T.Da.13.

[22] U Schumacher-Hamedat, J Föhles and H Zahn, *Proceedings of the 7th International Wool Textile Research Conference, Tokyo*, **4** (1985) 120.

[23] C M Carr, S F Ho, D M Lewis, E D Owen and M W Roberts, *Journal of the Textile Institute*, **76** (1985) 419.

[24] H Baumann and L Setiawan, *Proceedings of the 7th International Wool Textile Research Conference, Tokyo*, **4** (1985) 108.

[25] P A Duffield and R R D Holt, *Proceedings of the 33rd Arbeitstagung des Deutschen Wollforschungsinstitutes*, (1989).

[26] D L Connell and K M Huddlestone, *Textile Technology International*, (1989) 257.

[27] H R Haefely, *Textilveredlung*, **24** (1989) 271.

[28] IWS, *Textile Tech. Manual, Dyeing and Finishing of Machine-washable Knitwear*, **11** (1986).

[29] K R F Cockett, *Wool Science Review*, **56** (1980) 2.

[30] A Bereck and K Reincke, *Melliand Textilber*, **70** (1989) 452.

[31] Sandoz, BP 2059477A.

[32] D Allanach, K R F Cockett and D M Lewis, BP 1571188.

[33] K R F Cockett, D M Lewis and P Smith, *Journal of the Society of Dyers and Colourists*, **96** (1980) 214.

[34] D Allanach, M J Palin, T Shaw and B Craven, *Proceedings of the 6th International Wool Textile Research Conference, Pretoria*, **5** (1980) 61.

[35] Technical Information Bulletin, *Guidelines for the Non-felt Finishing of Wool with Basolan DC and Basolan SW* (Ludwigshafen, Germany: BASF, 1983).

[36] P G Cookson, *Wool Science Review*, **62** (1986) 3.

[37] V A Bell, K M Byrne, P G Cookson, D M Lewis and M A Rushforth, *Proceedings of the 7th International Wool Textile Research Conference, Tokyo*, **4** (1985) 292.

[38] J R Cook, *Journal of the Textile Institute*, **70** (1979) 157.

[39] G B Guise and M B Jackson, *Journal of the Textile Institute*, **64** (1973) 665.

[40] G B Guise and M A Rushforth, *Journal of the Society of Dyers and Colourists*, **92** (1976) 265.

[41] F Reich, *Textilveredlung*, **13** (1978) 454.

[42] T Shaw, *Wool Science Review*, **46** (1973) 44.

[43] D J Kilpatrick and T Shaw, *Proceedings of the 5th International Wool Textile Research Conference, Aachen*, **5** (1975) 19.

[44] V A Bell and D M Lewis, *Proceedings of the 5th International Wool Textile Research Conference, Aachen*, **3** (1975) 595.

[45] S B Sello and G C Tesoro, *Applied Polymer Symposium*, **18** (1971) 627.

[46] D M Lewis, *Textile Research Journal*, **52** (1982) 580.

[47] D M Lewis, *Journal of the Society of Dyers and Colourists*, **93** (1977) 105.

[48] W Rakowski, *Melliand Textilber*, **70** (1989) 780.

[49] J D Leeder and J A Rippon, *Journal of the Society of Dyers and Colourists*, **101** (1985) 11.

[50] J R Cook and D E Rivett, *Textile Research Journal*, **51** (1981) 596.

[51] P Erra, M R Julia, P Burgués, M R Infante and J Garcia, *Proceedings of the 7th International Wool Textile Research Conference, Tokyo*, **4** (1985) 332.

[52] P J Smith, *American Dyestuff Reporter*, **62** (1973) 35.

[53] J R McPhee, *The Mothproofing of Wool* (Watford, UK: Merrow Publishing, 1971).

[54] C T Page and A J Fergusson, *WRONZ Communication*, **71** (1981).

[55] T Shaw and J D Shepley, *Journal of the Textile Institute*, **72** (1981) 92.

[56] J R McPhee, *Wool Science Review*, **27** (1965) 1; **28** (1965) 33.

[57] R J Mayfield, *Textile Progress*, **11**(4) (1982).

[58] T Shaw and M A White, *Handbook of Fibre Science and Technology, Vol. 2, Chemical Processing of Fibres and Fabrics*, ed. M Lewin and S B Sello (New York, USA and Basel, Switzerland: Marcel Dekker, 1984), 380.

[59] J Haas, *Bayer Farben Review*, **35** (1983) 27.

[60] D M Lewis and T Shaw, *Review of Progress in Coloration and Related Topics*, **17** (1987) 86.

[61] M Lipson and R J Hope, *Proceedings of the 1st International Wool Textile Research Conference, Australia*, **E** (1955) 523.

[62] R J Mayfield, *Journal of the Society of Dyers and Colourists*, **101** (1985) 17.

[63] K Wolf and H Zahn, *Melliand Textilber*, **66** (1985) 817.

[64] T Shaw, *Proceedings of the 8th International Wool Textile Research Conference, Christchurch*, **4** (1990) 533.

[65] D Allanach, *Proceedings of the 8th International Wool Textile Research Conference, Christchurch*, **4** (1990) 568.

[66] J Haas, *Proceedings of the 8th International Wool Textile Research Conference, Christchurch*, **4** (1990) 558.

[67] D. Reinehr, J P Feron, A Rauechle and W Schmidt, *Textilveredlung*, **21** (1986) 137.

[68] B De Sousa, W Schmidt, H Hefti and D Bellus, *Journal of the Society of Dyers and Colourists*, **98** (1982) 79.

[69] F W Jones, *Journal of the Society of Dyers and Colourists*, **101** (1985) 137.

[70] K H Lehmann, E Schuh, O Robertus, M Cludts, P Mesnage, J Knott and J Bias, *IWTO CTF 04, Edinburgh* (2007).

[71] J. Barton, *International Dyer*, **185** (2000) 14.

[72] L Benisek, *Journal of the Society of Dyers and Colourists*, **87** (1971) 277.

[73] L Benisek, *Melliand Textilber*, **53** (1972) 931.

[74] M Lewin, P K Isaacs and B Shafer, *Proceedings of the International Wool Textile Research Conference, Aachen*, **5** (1975) 73.

[75] A J Farnworth and J Delmenico, *Permanent Setting of Wool* (Watford, UK: Merrow Publishing, 1971).

[76] J R Cook and J Delmenico, *Journal of the Textile Institute*, **62** (1971) 27.

[77] P G Cookson, K W Fincher and P R Brady, *Journal of the Society of Dyers and Colourists*, **107** (1991) 135.

[78] I Steenken and H Zahn, *Journal of the Society of Dyers and Colourists*, **102** (1986) 269.

[79] B Milligan and J A Maclaren, in *Wool Science: The Chemical Reactivity of Wool* (New South Wales, Australia: Science Press, 1981).

[80] H Zahn, *Plenary Lecture, 9th International Wool Textile Research Conference, Biella*, **I** (1995) 1.

[81] L R Mizell and M Harris, *Journal of Research of the National Bureau of Standards*, **30** (1943) 47.

[82] J B Speakman, *Journal of the Society of Dyers and Colourists*, **52** (1936) 335.

[83] F Sanger, A P Ryle, L F Smith and R Kitai, *Proceedings of the International Wool Textile Research Conference, Melbourne*, **C** (1955) 49.

[84] A P Ryle, F Sanger, L F Smith and R Kitai, *Biochemical Journal*, **60** (1955) 541.

[85] V Køpke, *Textile Research Journal*, **61** (1970) 361.

[86] P G Cookson, P R Brady, K W Fincher, P A Duffield, S M Smith, K Reincke and J Schreiber, *Journal of the Society of Dyers and Colourists*, **111** (1995) 228.

[87] J Kim and D M Lewis, *Coloration Technology*, **118** (2002) 121.

[88] J Kim and D M Lewis, *Coloration Technology*, **119** (2003) 112.

[89] J Kim and D M Lewis, *Coloration Technology*, **118** (2002) 181.

[90] Q Liao and P R Brady, *Proceedings of the 10th International Wool Textile Research Conference, Aachen*, CD paper code F1-15 (2000).

[91] D M Lewis and M T Pailthorpe, *Journal of the Society of Dyers and Colourists*, **99** (1983) 354.

[92] B Milligan and L J Wolfram, *Journal of the Textile Institute*, **63** (1972) 515.

[93] D M Lewis, *International Journal of Cosmetic Science*, **18** (1996) 123.

[94] C A Hunter, *Chemical Society Reviews*, **23** (1994) 101.

[95] Shikibo, *Japanese Textile News*, **255** (1976) 75.

[96] Wool Dev. Intl., USP 4563189 and EP 118983.

[97] P J Broadbent and D M Lewis, *Dyes & Pigments*, **43** (1999) 51.

[98] V Giannakis, PhD Thesis, University of Leeds (2000).

[99] P J Broadbent and D M Lewis, WO 95/12585 (1995).

[100] P J Broadbent, M Clark, S N Croft and D M Lewis, *Coloration Technology*, **123** (2007) 29.

[101] D M Lewis, *Advances in Coloration Science and Technology*, **7** (2003) 1.

[102] P G Cookson, P R Brady and K W Fincher, *Journal of the Society of Dyers and Colourists*, **107** (1991) 135.

[103] J Kim and D M Lewis, *Coloration Technology*, **119** (2003) 108.

[104] D M Lewis and S M Smith, *Journal of the Society of Dyers and Colourists*, **107** (1991) 357.

5

Bleaching and Whitening of Wool: Photostability of Whites

Keith R. Millington

CSIRO Materials Science and Engineering, Geelong, Victoria, 3216, Australia
Co-operative Research Centre for Sheep Industry Innovation, University of New England,
NSW, 2800, Australia

5.1   Introduction

Until the discovery of dyeing in around 1000 BC, wool from Middle Eastern sheep was highly pigmented and the fleece was most probably brown in colour, similar to wild Mouflon sheep still living today in isolated parts of Europe such as Sardinia and Corsica [1]. Textile remains from the first millennium BC indicate that a gradual increase in the amount of unpigmented wool occurred, due to the selection of sheep with large areas of unpigmented wool and of all-unpigmented mutants by early shepherds. By the turn of the last millennium, almost all wool textiles found in areas of the Roman Empire bordering the Mediterranean were produced using dyed unpigmented wools, and a thriving wool-dyeing industry had been established.

The Romans also developed the first chemical bleaching process for unpigmented wool, which involved exposure of the wet fabric to burning sulphur in closed chambers, known as stoving. It is perhaps surprising that this remained the dominant commercial bleaching process for wool until the 1930s, together with the use of sodium bisulphite in acid conditions, which is based on the same chemistry. In his book on dyeing from 1888, Sansone makes the comment, 'if it were not so dear, peroxide of hydrogen...would be the best' [2]. Hydrogen peroxide manufactured by electrolysis of sulphuric acid became available in 1926, which reduced its cost. Stabilised hydrogen peroxide was then adopted for wool bleaching and has remained the dominant commercial process to the present day [3]. This

The Coloration of Wool and other Keratin Fibres, First Edition. Edited by David M. Lewis and John A. Rippon.
© 2013 SDC (Society of Dyers and Colourists). Published 2013 by John Wiley & Sons, Ltd.

certainly supports the view that 'the wool dyer and finisher was conservative by nature and traditionalist by inclination' [4].

Selection of unpigmented sheep from 1000 BC onwards has resulted in the cream-coloured wools that are familiar today. However, in terms of its achievable whiteness, wool compares poorly to cotton and synthetic fibres. Wool remains cream in colour even after bleaching, whereas cotton and synthetic fibres are bright and white. Current commercial bleaching of wool still relies exclusively on the use of hydrogen peroxide, which is relatively inefficient on wool compared to other fibres, and reduces wool's photostability. Although fluorescent whitening agents can be applied to bleached wool to produce a much whiter base fabric, they are seldom used due to the extreme rate of photoyellowing of the treated wool, particularly when wet. This can render a garment unwearable after a single laundering cycle. Improving the whiteness and photostability to a point where wool is able to achieve the same range of photostable brilliant whites and bright pastel shades as competing fibres, while maintaining its superior performance in other areas, remains a significant challenge for textile and fibre research.

5.2 Wool Colour

Ideally, all undyed textile fibres should be perfectly white. This is because the whiter the clean untreated fibre, the greater the range of shades that can be obtained during coloration, including brilliant whites and bright pastels. The cone cells in the retina of the human eye responsible for colour vision at high ambient light levels, known as photopic vision, are sensitive to wavelengths from 750 to 380 nm, which define the limits of the visible spectrum. In physical terms, white materials reflect light strongly across the whole of this spectral region. The higher and more uniform this spectral reflectance, the whiter the material appears.

Common synthetic fibres, such as polyester, nylon and acrylic, contain no visible chromophores and hence exhibit high reflectance across the visible range. They may also contain a mineral delustrant such as titanium dioxide, which has excellent reflectance properties above 400 nm. Cellulosic fibres, such as cotton and linen, are composed of simple chains of 10 000–15 000 glucose units that contain no light-absorbing chromophores, although there may be traces of lignin (a dark-coloured dendritic network polymer of phenyl propene units found in plant cell walls), which is removed during bleaching. Wool, silk and other unpigmented animal fibres are complex proteins composed of 20 different amino acid residues, with traces of visible yellow chromophores of unknown origin and structure that are introduced during growth. These chromophores give rise to the familiar creamy off-white appearance of natural wool and silk. Commercial bleaching processes for wool and silk are inefficient as they are unable to fully remove these chromophores. This makes it difficult for protein fibres to compete when brilliant whites and pastel shades are required [5]. Further research is required to resolve this serious issue.

5.2.1 Measuring Wool Colour

The colour of fibrous materials can be objectively measured using a reflectance spectrophotometer fitted with a standard illuminant (D65) that is equivalent to outdoor daylight. A

10° field of view is normally used to analyse reflected light from the sample, and the measured tristimulus values are quoted as D65/10° data. The colour of unprocessed wools is most often expressed in terms of the three tristimulus values X, Y and Z obtained from reflectance measurements. They represent the amount of the red/orange (X), yellow/green (Y) and blue/indigo/violet (Z) components of the spectrum of white light reflected from the sample and are related to the three sets of cone cells present in the human retina, which are sensitive to these three ranges. In 1970 King showed that the simple relationship

$$YI = \frac{100\,(X - Z)}{Y} \tag{5.1}$$

(where YI represents yellowness index) could be used as a single coordinate measure of the yellowness of Merino wools at each stage of processing from greasy wool through to bleached wool [6], but this was not adopted by the wool industry. Instead, a standard test method was developed using two parameters to quantify wool colour: yellowness (Y–Z) and brightness (Y). This method requires a spectrophotometer with a special cell fitted with glass windows at both ends, which is packed with wool (using 5 g scoured samples) to constant density or pressure. The International Wool Textile Organisation approved this test method for scoured wool colour (IWTO-56-03) [7], along with a similar method for use on wool sliver and top [8]. An overview of wool colour measurement has been published by Wood [9].

For Merino wool, the mean yellowness is ∼8−8.5 and mean brightness is ∼75, as shown in the histograms in Figure 5.1. These show the range of colour data measured using the IWTO-56-03 standard test method for midside samples from 2553 Merino progeny from eight genetically-diverse Australian flocks in different regions of Australia totalling 4500 ewes during 2009, run by the Cooperative Research Centre (CRC) for Sheep Industry Innovation, Armidale, NSW [10]. Almost all of these sheep (99%) had wool with yellowness in the range 6−11, with only 1% having highly yellowed wool in the range 11.5−16.0.

In textile mills and dyehouses, measurements of whiteness or yellowness on fabric are derived from the same XYZ tristimulus values, but are normally expressed as yellowness

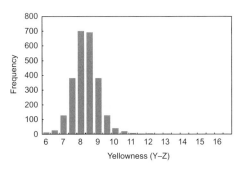 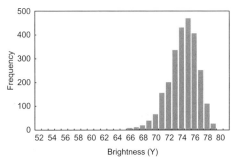

Figure 5.1 *Histograms showing the distribution in wool yellowness (Y–Z) and wool brightness (Y) for midside fleece samples taken from 2553 Merino sheep in 2009. These animals were the progeny of eight genetically well-characterised flocks (the information nucleus, IN) located in different regions of Australia and totalling ∼4500 ewes.*

(YI) or whiteness (WI) indices, of which there are many variants. The equations for deriving the various different YI and WI values from XYZ have been summarised [11].

5.2.2   Improving Wool Colour by Selection

As described in Section 5.1, the discovery of dyeing led to the selection of sheep with patches of unpigmented wool in their fleece and of unpigmented mutants. This selection process over 3 millennia has led to the natural cream-coloured wools we are familiar with today. An important question is whether further improvement in wool whiteness is possible by selective breeding; the distributions of yellowness and brightness shown in Figure 5.1 suggest that it is. Several studies have measured scoured wool colour and calculated the phenotypic and genetic correlations and heritability. Studies on Collinsville Merinos suggested that scoured wool colour is heritable [6,7], which was later confirmed by studies on both New Zealand [8] and Australian Merinos using New South Wales [9] and South Australian [10] flocks. The heritability estimates from these studies were in the range 0.25–0.54, showing that scoured wool colour is a moderately to highly heritable trait and suggesting that scoured colour would respond fairly rapidly to selection. Recent analysis of Merino data from the Sheep CRC information nucleus (IN) flocks in Australia [10] has shown that colour has a heritability of 0.70 ± 0.11, which is substantially higher than that reported in earlier work [12].

The major incentive for improving wool colour for bright whites and pastel shades by selection would be to avoid the need to carry out oxidative bleaching during processing, which adversely affects its photostability. Recent analysis has shown that genetic selection for whiter wool will produce wools with higher photostability [12]. To improve the yellowness of scoured wool to the level achieved by peroxide bleaching (Y–Z = 3.0–3.5 for fleece wool) by selection would require a reduction in mean Y–Z of about 4–5 units, which is outside the current range of yellowness values for scoured wool from Merino sheep. This would take many generations to achieve, and it is not yet known whether such a large improvement is feasible [13].

Selection for finer mean fibre diameter (MFD) results in whiter wool, due to a strong correlation between MFD and clean colour [14,15]. This effect is attributable mainly to a reduction in the ratio of optical scattering to light absorption with increasing MFD. Synthetic fibres of varying MFD have been used to create a model to correct for the effects of MFD on fibre colour, and the model has been successfully applied to data from the IN flocks [16].

There is little point in improving the colour of wool by selection if it is later yellowed during processing. The results of a commercial trial, where two bales of white wool were sourced and the colour was maintained through each processing stage, have recently been described [17]. Significant improvements in whiteness can be achieved by minimising the impact of each processing stage on wool colour, and in particular through the use of activated hydrogen peroxide as an alternative to chlorine/Hercosett to impart shrink resistance.

5.2.3   Improving Colour in the Scour

In commercial wool-scouring hydrogen peroxide is often added to the final bowl ($<10\,\mathrm{g\,l^{-1}}$ H_2O_2) before drying the wool. This leaves traces of peroxide present on the fibre, which slowly bleaches in the bale. This practice improves the colour of the wool and secures a higher price, but can result in unnecessary fibre damage, particularly for wools that are later dyed to dark shades.

Working on crossbred wools, New Zealand workers found that performing an acid extraction during scouring produces a much brighter fibre with a higher Y tristimulus value [18]. Acid extraction, performed at pH 2–3 using sulphuric acid and ethylenediaminetetraacetic acid (EDTA), removes adsorbed iron from the wool. This agrees with previous work by Simpson, who detected iron (10 mg/kg) and copper (1 mg/kg) ions after extraction of scoured wool fabric with molar sulphuric acid and with a phosphate buffer at pH 7 [19]. Although acid extraction improves the brightness of treated wool, its perceived yellowness (Y–Z) also increases because the Z tristimulus value is not improved significantly. The new scouring process, marketed as Glacial® wool, therefore includes a peroxide bleaching stage and an optional reductive bleach. Incorporating acid extraction and oxidative and reductive bleaching stages into the scouring process produces wool at least 7 Y tristimulus units brighter than conventionally scoured wool. The Z value is also improved by ~8 units, making the wool appear significantly less yellow.

5.2.4   Nonscourable Yellowing

Figure 5.1 showed that a small number of sheep produced yellow scoured wool with Y–Z in the range 11.5−16.0. This high level of scouring-resistant discoloration is due to staining of the wool and may occur via several mechanisms, including bacterial damage [20], protein oxidation and the increased pH of suint during growth [21]. The problem tends to be more prevalent in crossbred sheep than in Merino flocks.

A number of studies have extracted yellow compounds from nonscourable wools and carried out analyses of these. A study on crossbred wool found N-formylkynurenine (NFK), kynurenine and unidentified phenolic compounds covalently bound to the fibre [22]. NFK and kynurenine are oxidation products of tryptophan residues. Highly coloured phenazine compounds produced as secondary metabolites of *Pseudomonas* bacteria have been identified in the cuticular region of discoloured Merino wool [23]. The pH of aqueous suint solution has a wide range, from 5–6 for fine white wools to 10.0–10.5 for stained wools [24], which can result in alkali yellowing.

5.2.5   Wool Colour Compared with Cotton and Synthetics

Figure 5.2a compares the reflectance spectra of woven bleached cotton and polyester fabrics with wool, peroxide-bleached wool and double-bleached wool fabrics. The bleached cotton and polyester fabrics have high reflectance values across most of the visible range (from 750 to 450 nm), whereas wool, even after double bleaching, shows significantly lower reflectance below 500 nm. It is often more convenient to talk about absorption of light by materials rather than reflection, since absorption can be more easily related to the presence of coloured species (chromophores) distributed through the material, and to their concentration. Figure 5.2b shows the diffuse reflectance spectra of the same five fabrics, expressed as the remission function, $F(R)_\infty$, against wavelength. $F(R)_\infty$ is determined by the Kubelka–Munk equation:

$$F(R)_\infty = \frac{k}{s} = \frac{(1 - R_\infty)^2}{2R_\infty} \tag{5.2}$$

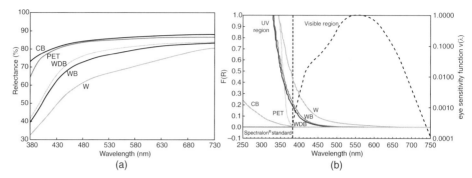

Figure 5.2   *(a) Reflectance spectra of bleached cotton (CB) and polyester (PET) fabrics compared with ecru (undyed) wool (W), peroxide-bleached wool (WB) and double-bleached wool (WDB) fabrics. (b) Diffuse reflectance spectra of the same fabrics plotted using the Kubelka Munk $F(R)_\infty$ function to show the influence of visible absorbing species in wool. Also shown is the eye sensitivity function $(V(\lambda)$, dashed line), which sets the wavelength limits and peak sensitivity for colour vision in the human eye.*

where k is the molar absorption coefficient, s is the scattering coefficient and R_∞ is the relative diffuse reflectance of the test material relative to a nonabsorbing white standard, such as magnesium oxide or Spectralon©, a pure form of polytetrafluorethylene. Transformation of reflectance spectra to the Kubelka–Munk form, $F(R)_\infty$, also allows qualitative comparison with conventional ultraviolet (UV)-visible absorption spectra.

Also shown in Figure 5.2b is a plot of the eye sensitivity function $V(\lambda)$, often referred to as the CIE 1978 $V(\lambda)$ function [25]. It is interesting that the two fabrics that appear bright and white, bleached cotton and polyester, both have zero $F(R)_\infty$ values above the 380 nm UV limit, where the eye sensitivity function $V(\lambda)$ begins to increase, and across the whole visible range. In contrast, wool, even after double bleaching, absorbs light significantly below 470 nm; this is why even double-bleached wool remains noticeably cream-coloured. It also shows that the current methods used to bleach wool are inefficient (this is discussed further in Section 5.3).

Table 5.1 compares the whiteness and yellowness indices of wool fabric before and after typical commercial bleaching processes with those of bleached cotton and polyethylene terephthalate fabrics.

Wool keratin contains significant amounts of the amino acids tryptophan, tyrosine, phenylalanine and cystine, which absorb in the UV region. Figure 5.3 shows the diffuse reflectance spectra in $F(R)_\infty$ format of these amino acids as finely ground powders, compared with the three wool fabrics shown in Figure 5.2b. All of these amino acids are white and have virtually no absorption above 380 nm, so they cannot influence the colour of wool. However, it is interesting that Figure 5.3 shows that solid cystine has the strongest UV absorbance of the four amino acids measured in the solid state using reflectance spectroscopy and the $F(R)_\infty$ function [26], whereas in conventional UV-visible studies in solution it is well known that tryptophan is the main absorbing species [27].

Wool keratin contains visible chromophores that absorb light from 660 nm down to the UV limit at 380 nm. Bleaching with hydrogen peroxide can completely remove species

Table 5.1 *Comparison of XYZ colour tristimulus values, yellowness and whiteness indices for wool, cotton and polyester fabrics.*

| Fibre | Tristimulus Values | | | | Yellowness Index E313 [11] | Whiteness Index CIE [11] | Whiteness Index Ganz [11] |
	X	Y	Z	Y–Z			
Polyester (PET)	80.6	85.2	87.7	−2.6	4.7	73.4	59.0
Bleached cotton	81.5	86.0	88.3	−2.3	5.2	73.4	57.9
Double-bleached wool	75.3	80.0	75.9	4.1	13.4	45.0	2.2
H_2O_2-bleached wool	73.3	77.6	72.4	5.2	15.7	38.0	−10.6
Ecru (natural undyed) wool	64.5	68.0	59.2	8.8	23.3	9.4	−62.5

that absorb above 470 nm, but significant amounts of blue light-absorbing chromophores remain, even after double bleaching, which make wool appear creamy compared with bleached cotton and polyester. Surprisingly little is known regarding the identity of these natural yellow chromophores in wool. It is likely that they are a complex mixture of compounds, which may include protein oxidation products. Fundamental knowledge of the origin, chemical nature and location within the fibre of these natural chromophores is well overdue; this would allow projects aimed at improving wool colour using both genetic and new bleaching approaches to be better targeted.

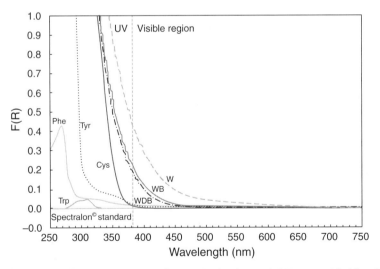

Figure 5.3 *Diffuse reflectance spectra of ecru (undyed) wool (W), peroxide-bleached wool (WB) and double-bleached wool (WDB) fabrics and the UV-absorbing amino acids cystine (Cys), tyrosine (Tyr), tryptophan (Trp) and phenylalanine (Phe) as finely ground powders plotted using the Kubelka–Munk $F(R)_\infty$ function.*

5.3   Wool Bleaching

Chemical bleaching agents attack and break down coloured molecules present in textile fibres, many of which have a series of conjugated double bonds and aromatic groups as chromophores.

5.3.1   Oxidative Bleaching

Hydrogen peroxide is exclusively used for the oxidative bleaching of wool, and is commercially supplied as 35 or 50% w/w acid-stabilised solutions. Wool is normally bleached at alkaline pH (8.5−9.0), where the active bleaching species is the perhydroxy anion ^-OOH [28].

$$H_2O_2 + OH^- \rightleftharpoons {}^-OOH + H_2O \tag{5.3}$$

Other active species, such as the superoxide radical anion $O_2^-\cdot$ and the hydroxyl radical $\cdot OH$, were suggested as alternative active species in peroxide bleaching of textiles [29] due to the observation of a decrease in the bleaching rate above pH 12. However, more recent studies on dye solutions [28] suggest that the decrease in bleaching rate at high pH is probably due either to greater repulsion between the negatively charged fabric and the ^-OOH ions, or to the lack of a metal sequestering agent in the bleaching solution. Further research in solution [30] using free radical and singlet oxygen trapping agents, and experiments on cotton fabric [31], strongly support the perhydroxy anion being the active bleaching agent at alkaline pH. Hydroxyl radicals have been shown to have little or no influence on the alkaline bleaching of wood pulp [32].

Typically, wool is bleached at pH 8.5−9.0 for at least 1 hour at 60 °C, with a solution of hydrogen peroxide (0.75% w/w) containing 3−4 g l^{-1} of a suitable stabiliser. An undesirable side effect is the rapid decomposition of hydrogen peroxide to water and oxygen, which is catalysed by transition-metal ions.

$$2H_2O_2 \xrightarrow{M^{n+}} 2H_2O + O_2 \tag{5.4}$$

Transition-metal ions can also lead to the generation of hydroxyl radicals, which can damage wool fibres.

$$H_2O_2 + M^{n+} \longrightarrow M^{(n+1)+} + OH^- + \cdot OH \tag{5.5}$$

Wool always contains trace amounts of transition-metal ions strongly complexed to the fibre, particularly iron and copper, which are not fully removed during scouring. Therefore, a stabiliser which sequesters metal ions is always added to the bleaching liquor in order to prevent both of these side reactions from occurring. Historically, the most common stabilisers for alkaline wool bleaching were phosphates, in particular tetrasodium pyrophosphate (TSPP). A comparison of three stabilisers used in alkaline peroxide bleaching – sodium nitrilotriacetate (NTA), sodium diethylenetriaminopentamethylphosphonate (DTPMP) and Stabiliser C (a mixture of tetrasodium pyrophosphate and ammonium oxalate) – found that the most effective in terms of whiteness was Stabiliser C [33]. However, concern over excessive phosphate release to the environment led to the development of alternative stabilisers based on silicates, and an alkaline peroxide bleaching process using a sodium metasilicate product, now Clarite WO (Huntsman), was developed [34].

Wool bleached under alkaline conditions should be given a final acid sour after rinsing with formic or acetic acid ($1 \, cm^3 \, l^{-1}$). This ensures that the wool is left slightly acidic, which minimises any thermal yellowing during drying.

Unless it is well controlled, alkaline bleaching with hydrogen peroxide can cause fibre damage due to oxidation of the cystine residues in the keratin protein, ultimately forming cysteic acid. Oxidation decreases the extent of disulphide bonding. This leads to loss of fibre mechanical strength and can be determined by an increase in alkali solubility. Urea-bisulphite solubility, which is a less sensitive test for assessing the degree of chemical damage to wool [35], does not increase to the same extent. Dyeing processes following bleaching further increase fibre damage, so that pastel shades are normally applied at low temperatures, often at 80 °C. Because of the low concentration of dye used for pastel shades, this temperature is usually sufficient to achieve full exhaustion and adequate fastness properties.

To reduce the risk of fibre damage, hydrogen peroxide can also be used to bleach wool under slightly acid conditions at pH 5, but bleaching is much slower and requires the presence of a suitable peracid activator. Prestogen W (BASF, a proprietary mixture of organic salts that generates percarboxylic acids) and citric acid are both reported to be effective activators at pH 5.5 [36]. The mechanism by which citric acid activates peroxide bleaching at low pH is not clear, since the perhydroxy anion is usually required to form peracids (see Section 5.3.8). Typical bleaching conditions are $0.75-1.00\%$ w/v H_2O_2 with $\sim 5 \, g \, l^{-1}$ Prestogen W for 1 hour at 80 °C. However, the level of whiteness achieved by peroxide bleaching under acid conditions is significantly less than is obtained by conventional alkaline peroxide [37].

While alkaline hydrogen peroxide remains the most effective means of bleaching unpigmented wools, Figure 5.2 shows that it does not remove all of the yellow chromophores that absorb light in the $470-380 \, nm$ region. Identification of these natural chromophores in wool would allow more effective bleaching processes to be developed.

5.3.2 Reductive Bleaching

Reductive bleaching of wool is less effective than hydrogen peroxide and is now seldom used without a preceding oxidative bleaching stage. Most reductive bleaching of wool is normally carried out using stabilised sodium dithionite (also known as sodium hydrosulphite), but other reducing agents, including thiourea dioxide, sodium formaldehyde sulphoxylate and zinc formaldehyde sulphoxylate, can also be used. A typical reductive bleaching process uses stabilised dithionite ($2-5 \, g \, l^{-1}$) at pH 5.5–6.0 and 45–65 °C for 1 hour. Thiourea dioxide is more expensive than sodium dithionite, and $1-3 \, g \, l^{-1}$ is applied at 80 °C and pH 7 for 1 hour.

A new reductive bleaching technology for wool, ColorClear™ WB, was developed by the Commonwealth Scientific and Research Organisation (CSIRO) and Rohm & Haas [38]. It uses the reaction between sodium borohydride and sodium bisulphite to produce the active bleaching species sodium dithionite *in situ*:

$$NaBH_4 + 8NaHSO_3 \longrightarrow 4Na_2S_2O_4 + NaBO_2 + 6H_2O \tag{5.6}$$

This combination of reagents is also used in the textile industry to strip dyestuffs from synthetic fibres. Commercial trials on wool and wool blends confirmed that the developed procedures are technically robust and give significant benefits over the use of conventional reductive bleaching in terms of improved whiteness.

5.3.3   Double (or Full) Bleaching

To bleach wool to the best achievable whiteness, a two-stage process is used. First the wool is bleached with hydrogen peroxide under alkaline conditions, as described in Section 5.3.1, and then a reductive bleaching is conducted, usually using either sodium hydrosulphite or thiourea dioxide, as described in Section 5.3.2.

5.3.4   Bleaching of Pigmented Wools

Heavily pigmented wool, such as Karakul, requires a more severe approach, known as mordant bleaching. The wool is first treated with a metal salt, usually iron (II) sulphate, in the presence of a reducing agent, and then with hydrogen peroxide. In the first stage, the melanin pigment granules in the wool preferentially adsorb the metal cations, which become strongly complexed to the melanin. It is important to carry out this reaction in the presence of a strong reducing agent, usually hypophosphorous acid, in order to avoid oxidation of Fe(II) to Fe(III). After mordanting, the excess uncomplexed metal ions are thoroughly rinsed from the wool. The rinse stage is usually carried out with hot (80 °C) water, containing $0.5 \, cm^3 \, dm^{-3}$ hypophosphorous acid. In the second stage, the Fe(II) cations bound to the melanin catalytically decompose hydrogen peroxide to produce highly aggressive hydroxyl free radicals. These selectively attack and bleach the melanin, while unpigmented regions experience a normal peroxide bleach [39].

5.3.5   Bleaching in the Dyebath

Wool has a tendency to yellow during dyeing; the extent of this is governed by the pH of the dyebath, the temperature and the treatment time. The effects of these factors on wool yellowness are shown in Figures 5.4, 5.5 and 5.6 [40].

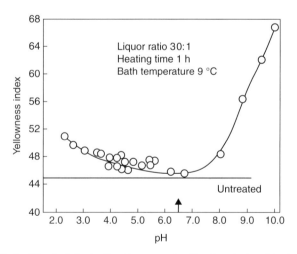

Figure 5.4   *Effect of dyebath pH on wool yellowing during blank dyeing.*

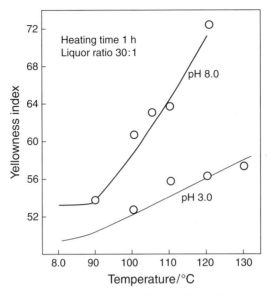

Figure 5.5 *Effect of temperature on wool yellowing during blank dyeing at pH 3 and 8.*

The yellow chromophores formed in wool during blank dyeing have not been formally identified, but it has been speculated that they may arise from carbonyl groups such as glyoxyl and pyruvyl residues, which are known to be produced during exposure to sunlight [41]. Wool pre-exposed to ultraviolet B (UVB) underwent further significant yellowing when heated in water at 80 °C [42], as shown in Figure 5.7. Unirradiated wool experienced little change in colour after heating in water at 80 °C for 2 hours, and the YI remained in the range 10–12.

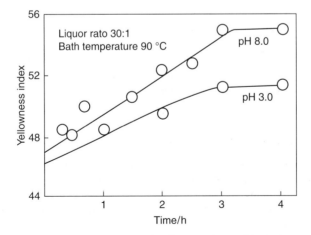

Figure 5.6 *Effect of treatment time on wool yellowing during blank dyeing at pH 3 and 8.*

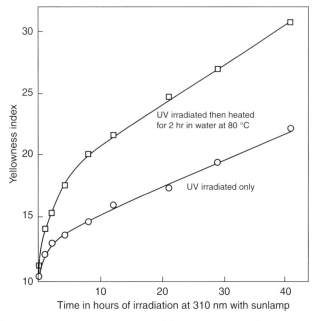

Figure 5.7   *Effect of pre-exposure of wet wool fabric to UVB radiation on yellowness index after heating in water for 2 hours at 80 °C. Adapted from F G Lennox, Studies in wool yellowing. XXV: Ultraviolet sensitisation to hydrothermal yellowing,* Text. Res. J., **39**, *700–702 (1969) with permission of Sage UK.*

The yellow chromophores produced in dyebath yellowing are very sensitive to subsequent photobleaching and can cause rapid changes in shade on exposure to sunlight. For the dyeing of bright pastel shades, a bleaching agent is normally added to the dyebath in order to minimise yellowing. Suitable additives are a mixture of sodium bisulphite with a chelating agent or hydroxylamine sulphate (HAS). HAS (0.1% w/v) reacts quantitatively with the carbonyl groups in wool under fairly mild conditions (80 °C, pH 4, 15 minutes) and substantially reduces hydrothermal yellowing [41].

A new additive for maintaining the whiteness of wool in the dyebath was recently described by Cai [43], using *tert*-butylamine borane. This additive is claimed to be superior to HAS, particularly under mildly acidic conditions (pH 4.5–6.5).

5.3.6   Biobleaching of Wool Using Enzymes

Levene reported that the whiteness achievable by both oxidative and reductive bleaching can be improved by adding a protease enzyme to the bath, which makes the wool fibre more susceptible to the whitening action of the bleach [44]. The whiteness improvement is due to a rapid initial surface-etching effect by the protease, which opens up the fibre surface for the bleaching agent. However, the improved whiteness comes at the cost of severely reduced fabric weight and strength, with at least a 3% weight loss.

5.3.7   Activated Peroxide Bleaching

It is possible to increase the rate of textile bleaching with hydrogen peroxide by using a suitable activator. In the early 1950s it was found that H_2O_2 was activated by a perhydrolysis reaction with an amide or ester to form the corresponding peracid:

$$RCL + \ ^-OOH \longrightarrow RCOOH + L^- \tag{5.7}$$

This reaction requires the perhydroxy anion, and therefore occurs more readily at alkaline pH. The first peracid that was exploited for textile bleaching was peracetic acid, due to its low molecular weight and known oxidation chemistry, and the variety of acetate esters and amides that were available as potential activators [45]. One of the first to be commercially exploited was tetraacetylethylenediamine (TAED). Although cost remains a barrier to its widespread adoption, it has been applied to the bleaching of nylon, cotton, viscose, cellulose acetate and linen [46]. Peracetic acid is not however an effective means of bleaching wool, although it can be used for wool/cotton blends [47].

A new method of improving the effectiveness of alkaline peroxide bleaching using dicyandiamide as a new activator has been proposed by Cardamone *et al.* [48]:

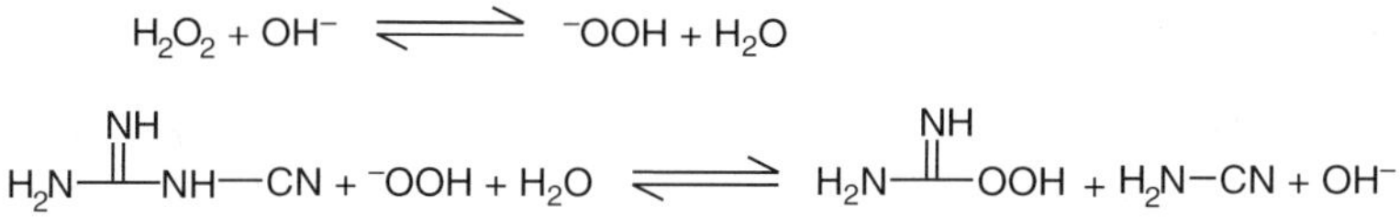

$$H_2O_2 + OH^- \rightleftharpoons \ ^-OOH + H_2O$$

$$H_2N{-}\overset{\overset{\displaystyle NH}{\|}}{C}{-}NH{-}CN + \ ^-OOH + H_2O \rightleftharpoons H_2N{-}\overset{\overset{\displaystyle NH}{\|}}{C}{-}OOH + H_2N{-}CN + OH^- \tag{5.8}$$

Dicyandiamide reacts with the perhydroxy anion to form peroxycarboxymic acid ($NH_2NHCOOH$). This is a powerful oxidant that can apparently bleach wool at 30 °C in 30 minutes at pH 11.5 to a high level of whiteness [48]. However, there are concerns about the high toxicity of cyanamides and dicyanamides. A subsequent publication provides whiteness data on several bleached fabrics using this method; however, it includes no comparison of the process against conventional alkaline bleaching, nor any data on the degree of fibre damage at the very high pH required [49]. The process is the oxidative stage of a new shrink-resist process for wool, aimed at producing wool uniforms for the US military.

Another powerful bleach activator that has been applied to wool bleaching and shrink-resist processes is percarbamic acid [50]:

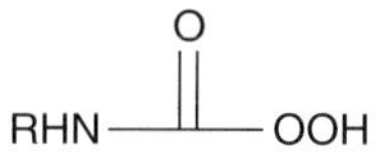

$$RHN{-}\overset{\overset{\displaystyle O}{\|}}{C}{-}OOH \tag{5.9}$$

Percarbamic acid can be conveniently generated by reacting a suitable carbamate compound ($RHNCOX$, where X is a leaving group) with hydrogen peroxide. A simple and effective activator can be formed by reacting sodium cyanate and sodium bisulphite. Percarbamic acid and its analogues are powerful bleaching agents but their nonionic character tends to damage the fibre and the process can be difficult to control.

5.3.8   Catalytic Peroxide Bleaching

In Section 5.3.4, the process of mordant bleaching of pigmented wools was described, which uses Fe(II) ions complexed to melanin granules to catalyse the generation of hydroxyl radicals from hydrogen peroxide. Similarly, catalytic peroxide bleaching uses small amounts of a metal salt as a catalyst to increase the rate of bleaching at low temperatures. One of the first catalytic bleaching agents commercialised by Unilever in 1994 was a manganese complex with the nitrogen heterocycle triazacyclononane (TACN), which was the active ingredient of a room-temperature laundry detergent called Persil Power [51]. This agent was highly effective at removing stains. Unfortunately, however, because it also caused a high level of damage to natural fibres and bleached out certain dyes, it was subsequently withdrawn from sale.

Despite this setback, research on bleach catalysts has continued, motivated in part by the extremely low dosage levels required for bleach catalysts compared with bleach activators [52]. However, to date, no catalytic bleaching process for use on unpigmented wool has been developed.

5.3.9   Novel Bleaching Methods for Wool

Photobleaching occurs when wool is selectively exposed to light in the 380–600 nm range, with blue light (400–450 nm) being the most effective [53]. Launer [54] used a high-pressure mercury arc fitted with a UV filter to photobleach dry wool fabric in around 15 seconds. Other workers [55] passed scoured wool through a conveyor system fitted with overhead blue-light fluorescent tubes. Simpson [56] found that photobleaching using similar low-intensity blue fluorescent tubes could be sped up significantly by irradiating the wool in the presence of alkaline hydrogen peroxide at pH 10.0–11.5. Using a parallel array of low-power blue fluorescent tubes (Philips TL03), irradiation times of 20–30 could obtain a good bleaching effect using peroxide-padded wool. Millington described a continuous photobleaching treatment that involved padding fabric with either an oxidative or a reductive bleach and exposing it to high-intensity visible light [57]. The light source was part of a conveyorised system for curing polymer coatings. The light source was a doped mercury arc, which produced significantly higher output in the blue light (400–450 nm) range, and a heat-resistant glass filter to attenuate UV wavelengths. However, to date, photobleaching of wool has not been used commercially.

Commercial use of ultrasonics to speed up heterogeneous processing in a liquid phase is widespread. The use of ultrasound as a pretreatment before wool bleaching has recently been reported [58]. Ultrasound caused a slight increase in yellowing of wool fabric before bleaching, possibly due to heating of the wool or the generation of hydroxyl radicals during ultrasonic cavitation of water. However, the ultrasound pretreated wool was slightly whiter than a control fabric after bleaching.

5.4   Fluorescent Whitening of Wool

Fluorescent whitening agents (FWAs) are colourless dyes that absorb ultraviolet A (UVA) light in the range 340–380 nm and emit lower-energy blue light in the 400–450 nm region.

Cream and yellow materials absorb light in the blue light region, so application of an FWA corrects for any blue light absorption and deceives the eye into perceiving a bright white surface, provided that the material is illuminated with light containing UVA wavelengths, such as sunlight. FWAs were discovered in 1929 by Krais [59] and rapidly grew into a large industry, since most domestic laundry detergents contained them in order to boost whiteness. FWAs for application to wool are essentially colourless acid dyes solubilised by sulphonic acid groups. They are usually applied to wool during reductive bleaching with stabilised hydrosulphite at pH 4–5 and at temperatures in the range 60–80 °C, in order to minimise thermal yellowing.

Application of FWAs to cotton and synthetics results in high levels of whiteness that are reasonably photostable. FWA application to wool and silk significantly boosts their whiteness but also severely reduces their photostability, to such an extent that very little pure wool is treated with FWAs, due to the high risk of consumer complaints. Nowadays, bis-stilbene derivatives such as Structure **1**, with light fastness ratings of only 2–3 on the SDC blue scale, are the most stable products available for wool. The photostability of white wools, including FWA-treated wool, is discussed in Section 5.5.

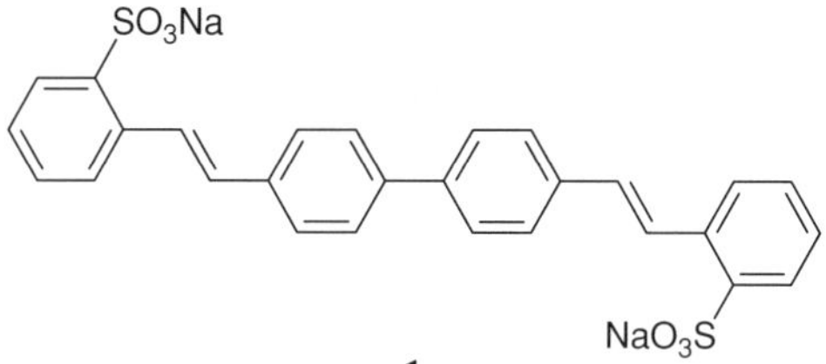

1

The ColorClear^{TM} WB reductive bleaching technology, described in Section 5.3.2, can also be used for the whitening of wool/polyester blends [38]. A disperse FWA is first applied to the polyester component, followed by a sequential double bleaching procedure using H_2O_2, then reductive bleaching with a mixture of sodium borohydride and sodium bisulphite. In this process, the ColorClear^{TM} WB carries out a dual role: it not only acts as a reductive bleach to whiten the wool component but also strips any disperse FWA from the wool. This is necessary because any FWA remaining on the wool would lead to rapid photoyellowing on exposure to sunlight.

5.5 Photostability of Wool

When wool is exposed to sunlight, it undergoes photoyellowing and photobleaching con-currently. The colour change observed depends on the initial colour of the wool and the relative intensities of the different wavelengths [60]. White wool is yellowed by the UV wavelengths (<380 nm), whereas visible light (400–600 nm) photobleaches the existing yellow chromophores. This effect is well illustrated by the action spectrum of wool, shown in Figure 5.8. UVB wavelengths (280–320 nm) yellow wool the most rapidly, and blue light (400–450 nm) is the most effective at photobleaching wool. Also shown in Figure 5.8 is a

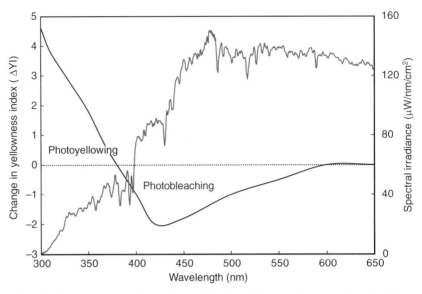

Figure 5.8 *Action spectrum of natural wool obtained for equal energy doses (redrawn from [53]) and spectral irradiance of natural sunlight measured in Geelong, Australia (latitude 38.18° S, longitude 144.34° E, 4 November 2008, 16:00).*

typical plot of the spectral irradiance of sunlight against wavelength (measured in Geelong, Australia). The intensity of UVB wavelengths in sunlight is dependent upon time of day, season, altitude and latitude. Seasonal variations in UVB intensity explain why wool may be yellowed in summer sunlight but photobleached during the winter [61].

Yellower wools are bleached by sunlight during the early stages of exposure, due to the higher intensity of blue light relative to UV, whereas whiter fleece wools are yellowed from the outset. The effect of irradiation by light sources producing different wavelengths on fleece wools having a wide range of initial yellowness was studied by Lennox and King [60]. They found that only irradiation with UVB produced yellowing in all cases, and that this reached a saturation point beyond which no further yellowing occurred. For this reason, a UVB source was chosen for a new test method aimed at measuring the photostability of large numbers of scoured fleece wool samples [62].

Figure 5.9 shows a plot of the change in yellowness $\Delta(Y\text{–}Z)$ after 4 hours of exposure to a UVB source against initial yellowness for 2557 fleece wool samples taken from eight Merino flocks in different regions of Australia. For the whiter wools (Y–Z <5), there is a range of $\Delta(Y\text{–}Z)$ values, which suggests that variance in fleece wool photostability may exist. Using these data, analysis [12] has shown moderate heritability of photostability (0.18), but environmental factors, including trace metal content, may also have some influence.

Wool photoyellows more rapidly under wet conditions, particularly after bleaching and fluorescent whitening [63], as shown in Figure 5.10. The chemistry of wool photoyellowing and the experimental techniques that have been applied to its study have been reviewed in detail [64].

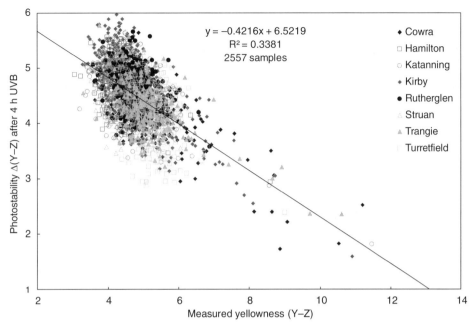

Figure 5.9 *Plot of yellowness (Y–Z) against photostability Δ(Y–Z) to UVB radiation for Merino midside samples taken from progeny of the Sheep CRC IN flocks [10] in 2009. Scoured fleece wool samples were irradiated for 4 hours in a UVB irradiator [62]. Whiter wools tend to photoyellow more than yellow wools, and for a given original yellowness, particularly in the Y–Z range 4–6, there is a range of photostability.*

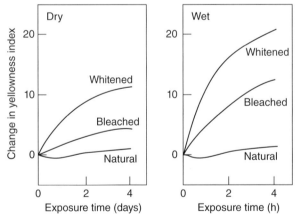

Figure 5.10 *Comparison of the photoyellowing rates of dry and wet wool fabrics exposed to simulated sunlight. Bleaching and whitening treatments significantly increase the rate of yellowing, especially when wet. Adapted from I H Leaver and G C Ramsay, Studies in wool yellowing. Part XXVII: the role of water in the photoyellowing of fluorescent whitened wool, Text. Res. J., **39**, 730–733 (1969) with permission from Sage UK.*

Prolonged sunlight exposure also has a highly detrimental effect on the mechanical properties of wool fabrics, resulting in loss of tensile strength, elasticity and abrasion resistance. This process is known as phototendering (or tendering), and its effects are of particular significance for wool upholstery, curtaining and carpets exposed to filtered sunlight through window glass, which absorbs the more energetic UV wavelengths below 350 nm. The lower-energy UVA wavelengths (350–400 nm), which have greater penetration into the fibre, are mainly responsible for the loss of tensile strength observed in fibres and fabrics exposed to sunlight, as fibre strength is determined by the cortical cells.

5.5.1   Mechanism of Wool Photoyellowing

Most polymers and biopolymers exposed to UV light or elevated temperatures in the presence of oxygen undergo degradation via the free radical autoxidation mechanism (Scheme 5.1) established by Bolland and Gee [65,66]. There are several primary light-absorbing species (Initiation I in Scheme 5.1) present in wool; these include the tryptophan (Trp) and tyrosine (Tyr) residues absorbing in the UV region, and the natural

$$I \xrightarrow{h\nu} I^* \tag{1}$$

Initiation

$$I^* \longrightarrow A\cdot + B\cdot \tag{2}$$

$$A\cdot + PH \longrightarrow P\cdot + AH \tag{3}$$

$$P\cdot + O_2 \longrightarrow POO\cdot \tag{4}$$

Propagation

$$POO\cdot + PH \longrightarrow POOH + P\cdot \tag{5}$$

$$POOH \xrightarrow{h\nu \text{ or } M^{n+}} PO\cdot + \cdot OH \tag{6}$$

$$PO\cdot + PH \longrightarrow POH + P\cdot \tag{7}$$

Branching

$$\cdot OH + PH \longrightarrow H_2O + P\cdot \tag{8}$$

$$POOH + PH \xrightarrow{h\nu \text{ or } M^{n+}} PO\cdot + P\cdot + H_2O \tag{9}$$

$$2\,POOH \xrightarrow{h\nu \text{ or } M^{n+}} POO\cdot + PO\cdot + H_2O \tag{10}$$

$$P\cdot + P\cdot \longrightarrow P\!-\!P \tag{11}$$

Termination

$$P\cdot + POO\cdot \longrightarrow POOP \tag{12}$$

$$POO\cdot + POO\cdot \longrightarrow POOP + O_2 \tag{13}$$

PH = polymer

I = primary photochemical absorber (radical initiator)

A· and B· = radical products from excited state of initiator

M^{n+} = catalytic metal ions

Scheme 5.1   *The mechanism of free radical photooxidation for polymers and biopolymers, including wool.*

$$POO^· + POO^· \longrightarrow [POOOOP] \tag{14}$$

$$\text{CL emission} \quad [POOOOP] \longrightarrow P_2'C{=}O^* + P''OH + O_2 \tag{15}$$

$$P_2'C{=}O^* \longrightarrow P_2'C{=}O + h\nu_{CL} \tag{16}$$

Scheme 5.2 *Russell mechanism for generation of chemiluminescence ($h\nu_{CL}$) by peroxy radical termination via a tetroxide intermediate [73].*

yellow chromophores and any yellow photooxidation products present that absorb in the visible region.

Electron spin resonance (ESR) spectroscopy has shown that, in the absence of atmospheric oxygen, both carbon- and sulphur-centred free radicals are formed directly in wool following UV irradiation [67]. In the presence of air or oxygen, the intensity of the ESR signals is reduced, indicating that the radicals react with oxygen. This is a key stage of the Bolland and Gee mechanism (Equation 4 in Scheme 5.1).

When wet wool keratin is irradiated with UVA or blue light in the presence of a highly specific fluorescent probe (the terephthalate anion), hydroxyl ($·OH$) radicals are produced [68]. Wool produces higher concentrations of hydroxyl radicals than other fibres, and the attack of these highly oxidising species on the aromatic amino acid residues Trp, Tyr and phenylalanine (Phe) present in the protein leads to yellowing. Trace metal ions that are strongly bound to wool, in particular iron and possibly copper, are involved in $·OH$ generation in photoirradiated wool, because the concentration of hydroxyl radicals is significantly reduced in the presence of the metal chelator deferoxamine [68].

Studies on the trypsin digests of heavily irradiated wool fabrics using high-performance liquid chromatography tandem mass spectrometry (HPLC/MS/MS) techniques, have confirmed the presence of 13 different yellow chromophores in 25 photomodified peptide sequences. The generation of all of these chromophores, which are oxidation products derived from Trp, Tyr and Phe residues in the keratin intermediate filaments (KIFs) and high glycine tyrosine proteins (HGTPs) present in the matrix, is consistent with a free radical oxidation mechanism [69,70]. The mechanisms of wool photoyellowing have been reviewed in detail [71].

Recent studies have shown that irradiated wool keratin and other fibrous proteins emit photoinduced chemiluminescence (PICL) [72], which arises from the bimolecular reaction of two macroperoxy (POO·) radicals via a tetroxide intermediate (Scheme 5.2) [73]. As expected, UV irradiation produces PICL with higher intensity than is produced by visible light. Chemiluminescence emission provides further evidence for a free radical mechanism for the photooxidation of wool.

5.5.2 Mechanism of Photoyellowing of Fluorescent Whitened Wool

Much research has been devoted to finding out why FWA-treated wool and silk yellow so rapidly, whereas other FWA-treated fibres, including cotton, nylon and polyester, have high light fastness. It has often been assumed that the presence of the FWA merely increases the rate of the photoyellowing reactions that occur in untreated wool, but this has been shown to be incorrect [74]. Photoyellowing experiments were conducted using FWA-treated collagen, a fibrous protein found in skin. Unlike keratin, collagen contains no Trp or Cys residues. While untreated collagen was far more photostable than wool or silk, FWA-treated collagen photoyellowed rapidly, and to a similar extent to FWA-treated wool. This clearly showed that Trp is not essential for the rapid photoyellowing of FWA-treated fibrous proteins.

Wet FWA-treated wool irradiated with simulated sunlight also produces significant amounts of H_2O_2 and superoxide radical anions ($O_2^-\cdot$), and this process also occurs in aqueous FWA solutions [75]. However, in solution, the rate of production of $O_2^-\cdot$ and H_2O_2 is far more rapid in the presence of an electron donor amino acid, such as Trp or arginine (Arg) [76]. This strongly suggests that a photoinduced electron transfer (PET) reaction occurs in irradiated FWA-treated wool, where electrons are transferred from the wool protein to the FWA. Electron-rich amino acid groups in the wool act as the electron donors. The PET reaction leads to high concentrations of $O_2^-\cdot$ and H_2O_2 at the fabric surface. Irradiation of H_2O_2 near the surface subsequently leads to the formation of hydroxyl radicals, which are highly reactive and cause rapid yellowing of both the wool and the FWA. A similar PET mechanism has previously been observed between *trans*-stilbene and tertiary and secondary amines [77], which suggests that amino acid residues with secondary or tertiary amino groups in wool (Arg, proline, possibly also histidine and Trp) may be the electron donor species.

Removal of electrons from wool is facile, and it is interesting that a wool cloth is often used as a good electron donor system in electrostatic experiments. Photoreduction of the FWA by wool is also in accord with a study carried out many years ago, which showed that for most dyed fibres irradiated in air, the dye is photooxidised and the fibre is photoreduced, whereas for the proteinaceous fibres wool and silk, the dye is photoreduced and the fibre is oxidised [78]. Since FWAs can be considered colourless fluorescent dyes, this theory explains very well the observed differences in the photostability of FWA-treated wool and silk and other FWA-treated fibres. It is also supported by a previous study showing the formation of reduced FWA photoproducts on irradiated FWA-treated wool, which would otherwise be highly unlikely in an oxidising environment [79]. A reaction mechanism for PET between FWAs and wool has been published [5,74].

A photocatalysis mechanism, similar to that which occurs for irradiated porphyrin and phthalocyanine dyes [80], can also explain the rapid photoyellowing of FWA-treated wool and the photoproduction of hydrogen peroxide. The FWA can be considered the photocatalyst, producing electrons and holes, and wool keratin acts as the electron donor. Aggregates of FWA molecules may be required to enable charge separation to occur. The holes can oxidise amino acid residues in wool and other FWA molecules to produce yellow products, and the electrons reduce dioxygen to hydrogen peroxide, as shown in Scheme 5.3.

Irradiation of aqueous FWA solutions doped with H_2O_2 produces yellow FWA oxidation products [74]. An aqueous distyrylbiphenyl (DSBP) FWA (Uvitex NFW) irradiated in the presence of H_2O_2 and analysed using HPLC/MS/MS has confirmed the presence of mono- and dihydroxylated DSBPs and a yellow stilbene quinone chromophore [81]. Stilbene quinones are highly coloured and have been implicated in the photoyellowing of cellulosic materials [82]. It is likely that the dihydroxylated DSBP initially arises via attack by hydroxyl radicals against the FWA, followed by oxidation and loss of H_2 (possibly by H atom abstraction by $\cdot$OH) to form the stilbene quinones, as shown in Scheme 5.4.

	FWA	$\xrightarrow{h\nu}$	FWA*	(17)
photocatalysis				
	FWA*	$\longrightarrow$	$h^+ + e^-$	(18)
	wool + 2h$^+$	$\longrightarrow$	oxidised wool (yellow)	(19)
oxidation				
	FWA + 2h$^+$	$\longrightarrow$	oxidised FWA (yellow)	(20)
reduction	O^2 + 2e$^-$ + 2H$^+$	$\longrightarrow$	H$_2$O$_2$	(21)

Scheme 5.3 *Photocatalytic generation of hydrogen peroxide by irradiated FWA-treated wool.*

It is not possible from mass spectrometry (MS) to determine whether the stilbene quinone adopts an *ortho-* or *para-* geometry or is a mixture of isomers, but both forms would be highly coloured. *Ortho*-quinones, in particular dopaquinone, are involved in the formation of melanin.

5.5.3 Methods for Improving Photostability

Few treatments have been described in the literature that can improve the photostability of wool, and only one has been found suitable for commercialisation, which is used

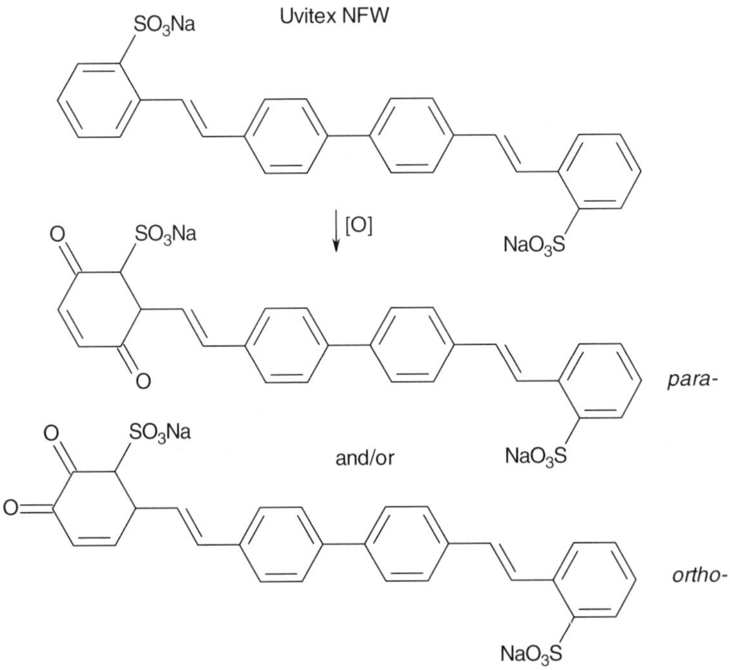

Scheme 5.4 *Formation of highly coloured ortho- and para-stilbene quinones from the distyrylbiphenyl (DSBP) fluorescent whitening agent Uvitex NFW.*

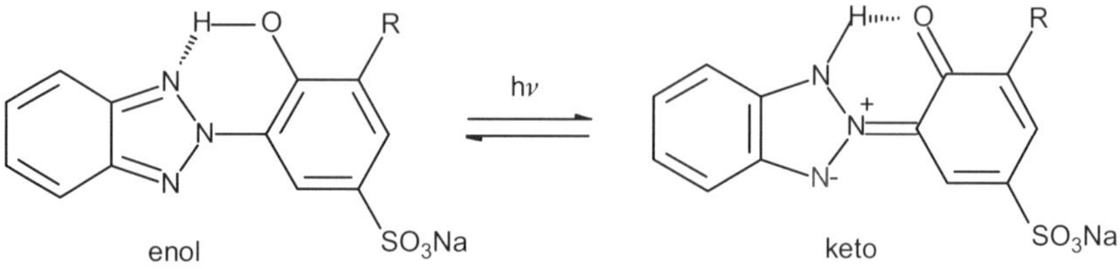

Scheme 5.5   *Tautomerisation of a 2-hydroxyphenylbenzotriazole UV absorber.*

mainly as a treatment against phototendering rather than photoyellowing. A UV absorber with a sulphonated 2-hydroxyphenylbenzotriazole structure was developed by CSIRO in Australia. This offers wool good protection against the reduced tear strength, abrasion resistance and dye fading usually experienced following prolonged sunlight exposure, with some reduction in photoyellowing for undyed wools [83]. Intramolecular hydrogen-bonded 2-hydroxyphenylbenzotriazoles are commonly used as UV absorbers in other polymers; they dissipate the absorbed UV energy as heat through conversion to a tautomeric form in the excited state (Scheme 5.5).

Hydroxyphenylbenzotriazoles used commercially for the stabilisation of other polymers are highly hydrophobic materials that have no substantivity for wool. Introduction of the sulphonate group imparts both water solubility and substantiveness for application to wool under slightly acid conditions, much like acid dyes. However, for wool the key to development of an effective additive is in the nature of the R group. A bulky hydrophobic substituent is required to shield the internal hydrogen bond from interaction with polar groups in wool [83]. This chemistry was first commercialised in 1990 and is now available as UVFast W (Huntsman), mainly for use on upholstery and curtaining fabrics. It cannot be used on FWA-treated wool as it absorbs the UV wavelengths necessary to excite the FWA.

Based on the metal-catalysed oxygen free radical yellowing reactions (Equations 9 and 10 in Scheme 5.1), a combination of a water-soluble antioxidant and a metal chelator was expected to be a more effective treatment for preventing the rapid photoyellowing of FWA-treated wool than an antioxidant alone. Rinsing FWA-treated wool with low concentrations of an antioxidant, such as N-acetylcysteine (NAC), is highly effective against photoyellowing when combined with a metal chelator, such as oxalic acid, and provides much better photoprotection than the antioxidant alone, particularly under wet conditions [84]. Unfortunately, the benefits of the rinse treatment are lost on laundering, and efforts to identify a substantive antioxidant–metal chelator combination that can be applied during wet finishing and provide effective protection against photoyellowing under both wet and dry conditions have so far been unsuccessful.

Improving the photostability of wool to photobleaching is also important for certain applications. When wool carpets are exposed to sunlight through window glass, they undergo photobleaching soon after the carpet is laid, during a period in which consumers are particularly sensitive to product performance. The effect is particularly apparent when the yellowness of the wool is fairly high and the carpet is dyed to a pale shade. New Zealand workers developed a dyebath additive that photoyellows at the same rate at which photobleaching occurs, so that the overall colour is unchanged throughout exposure. This is now marketed by Clariant as Lanalbin APB.

References

[1] M L Ryder, *Scientific American*, **256** (1987) 100.

[2] A Sansone, *Dyeing: Comprising the Dyeing and Bleaching of Wool, Silk, Cotton, Flax, Hemp, China Grass &c* (Manchester, UK: A Heywood & Son, 1888).

[3] R Levene, in *Handbook of Fibre Science and Technology Vol I*, ed. M Lewin and S B Sello (New York, USA: Marcel Dekker, 1984).

[4] I Holme, *Review of Progress in Coloration and Related Topics*, **33** (2003) 85.

[5] K R Millington, in *Advances in Wool Technology*, ed. N A G Johnson and I M Russell (Cambridge, UK: Woodhead, 2009).

[6] M G King, *Journal of the Textile Institute*, **61** (1970) 513.

[7] IWTO Standard Test Method 56-03, *Method for the Measurement of Colour of Raw Wool* (Brussels, Belgium: International Wool Textile Organisation, 2003).

[8] IWTO Standard Test Method 35-03, *Method for the Measurement of Colour of Sliver* (Brussels, Belgium: International Wool Textile Organisation, 2003).

[9] E Wood, *Wool Technology and Sheep Breeding*, **50** (2002) 121.

[10] J H J van der Werf, B P Kinghorn and R G Banks, *Animal Production Science*, **50** (2010) 998.

[11] ASTM Standard E313-00, *Standard Practice for Calculating Yellowness and Whiteness Indices from Instrumentally Measured Colour Coordinates* (West Conshohocken, PA, USA: ASTM International, 2000).

[12] S Hatcher, P I Hynd, A A Swan, K J Thornberry and S Gabb, *Animal Production Science*, **50** (2010) 1089.

[13] K R Millington, M del Giudice, S Hatcher and A L King, *Journal of the Textile Institute* (in press, doi: 10.1080/00405000.2012.744734).

[14] M R Fleet, K R Millington, D H Smith and R J Grimson, *Proceedings of the Association for the Advancement of Animal Breeding and Genetics*, **18** (2009) 556.

[15] J Smith and I W Purvis, *Proceedings of the Association for the Advancement of Animal Breeding and Genetics*, **18** (2009) 390.

[16] H Wang, T Mahar, X Liu, P Swan and X Wang, *Journal of the Textile Institute*, **102** (2011) 1031.

[17] K R Millington, A L King, S Hatcher and C Drum, *Coloration Technology*, **127** (2011) 297.

[18] J McKinnon, J R McLaughlin, M E Taylor, D A Rankin, P G Middlewood, P Le Pine, P G R Mesman and S B Manson, WO 99/16942 (1999).

[19] W Simpson, *Journal of the Textile Institute*, **78** (1987) 430.

[20] E B Fraser and A P Mulcock, *Nature*, **177** (1956) 628.

[21] R Sumner, S Young and M Upsdell, *Proceedings of the New Zealand Society of Animal Production*, **63** (2003) 155.

[22] J E Wood, R R Sherlock and M H G Munro, *International Journal of Sheep and Wool Science*, **53** (2005) 20.

[23] J M Dyer, S D Bringans, G D Aitken, N I Joyce and W G Bryson, *Coloration Technology*, **123** (2007) 54.

[24] J L Hoare and R G Stewart, *Journal of the Textile Institute*, **62** (1971) 455.

[25] J J Vos, *Color Research and Application*, **3** (1978) 125.

[26] K R Millington, *Amino Acids*, **43**(3) (2012) 1277.

[27] M I Anson, K Bailey and J T Edsall, *Advances in Protein Chemistry*, **7** (1952) 319.

[28] K M Thompson, W P Griffith and M Spiro, *Chemical Communications* (1992) 1600.

[29] J Dannacher and W Schlenker, *Textile Chemist and Colorist*, **28** (1996) 24.

[30] K M Thompson, W P Griffith and M Spiro, *Journal of the Chemical Society, Faraday Transactions*, **89** (1993) 4035.

[31] R E Brooks and S B Moore, *Cellulose*, **7** (2000) 263.

[32] G C Hobbs and J Abbot, *Journal of Wood Chemistry and Technology*, **14** (1994) 195.

[33] J Cegarra, J Gacen, D Cayuela and M C Riva, *Journal of the Society of Dyers and Colourists*, **110** (1994) 308.

[34] P A Duffield, *Review of Wool Bleaching Processes*, Technical Bulletin, The Woolmark Company, UK (1996).

[35] J Cegarra and J Gacen, *Wool Science Review*, **59** (1983) 2.

[36] W Karunditu, C M Carr, K Dodd, P Mallinson, I A Fleet and L W Tetler, *Textile Research Journal*, **64** (1994) 570.

[37] J Gacen and D Cayuela, *Journal of the Society of Dyers and Colourists*, **116** (2000) 13.

[38] Australian Wool Innovation, *Superwhite Australian Merino using ColorClear WB Bleaching Technology*, http://www.csiro.au/resources/ColorClear.html, last accessed 1 February 2013.

[39] A Bereck, *Review of Progress in Coloration and Related Topics*, **24** (1994) 17.

[40] H-J Meiswinkel, G Blankenburg and H Zahn, *Melliand Textilber*, **63** (1982) 160.

[41] W S Simpson, WRONZ Report R213 (1997).

[42] F G Lennox, *Textile Research Journal*, **39** (1969) 700.

[43] J Y Cai, *Fibres and Polymers*, **10** (2009) 502.

[44] R Levene, *Journal of the Society of Dyers and Colourists*, **113** (1997) 206.

[45] K Grime and A Clauss, *Chemistry and Industry* (1990) 647.

[46] W S Hickman, *Review of Progress in Coloration and Related Topics*, **32** (2002) 13.

[47] J Y Cai, D J Evans and S M Smith, *AATCC Review*, **1** (2001) 31.

[48] J M Cardamone, *Textile Research Journal*, **76** (2006) 99.

[49] J M Cardamone, *Proceedings of the 12th International Wool Textile Research Conference, Shanghai* (2010) 166.

[50] D M Lewis, J Yao, J S Knapp and J A Hawkes, US 7159333 B2 (2007).

[51] R Hage, J E Iburg, J Kerschner, J H Koek, E L M Lempers, R J Martens, U S Racherla, S W Russel, T Swarthoff, M R P van Vliet, J B Warnaar, L van der Wolf and B Krijnen, *Nature*, **369** (1994) 637.

[52] N J Milne, *Journal of Surfactants and Detergents*, **1** (1998) 253.

[53] M G King, *Journal of the Textile Institute*, **62** (1971) 251.

[54] H F Launer, *Textile Research Journal*, **41** (1971) 311.

[55] C Garrow, E P Lhuede and C M Roxburgh, *Textile Institute and Industry*, **9** (1971) 286.

[56] W S Simpson, WO 92/15744 (1992).

[57] K R Millington, *Proceedings of the 11th International Wool Textile Research Conference, Leeds, UK* (2005).

[58] S J McNeil and R A McCall, *Ultrasonics Sonochemistry*, **18** (2011) 401.

[59] R Anliker, in *Fluorescent Whitening Agents*, ed. R Anliker and G Muller (Stuttgart, Germany: Georg Thieme, 1975).

[60] F G Lennox and M G King, *Textile Research Journal*, **38** (1968) 754.

[61] J A Maclaren and B Milligan, *Wool Science: The Chemical Reactivity of the Fibre* (Marrickville, NSW, Australia: Science Press, 1981).

[62] K R Millington and A L King, *Animal Production Science*, **50** (2010) 589.

[63] H Leaver and G C Ramsay, *Textile Research Journal*, **39** (1969) 730.

[64] K R Millington, *Coloration Technology*, **122** (2006) 169.

[65] J L Bolland, *Quarterly Reviews of the Chemical Society*, **3** (1949) 1.

[66] J L Bolland and G Gee, *Transactions of the Faraday Society*, **42** (1946) 236.

[67] A Shatkay and I Michaeli, *Radiation Research*, **43** (1970) 485.

[68] K R Millington and L J Kirschenbaum, *Coloration Technology*, **118** (2002) 6.

[69] J M Dyer, S D Bringans and W G Bryson, *Photochemical and Photobiological Sciences*, **5** (2006) 698.

[70] J M Dyer, S D Bringans and W G Bryson, *Photochemical and Photobiological Sciences*, **82** (2006) 551.

[71] K R Millington, *Coloration Technology*, **122** (2006) 301.

[72] K R Millington, C Deledicque, M J Jones and G Maurdev, *Polymer Degradation and Stability*, **93** (2008) 640.

[73] G A Russell, *Journal of the American Chemical Society*, **79** (1957) 3871.

[74] K R Millington and G Maurdev, *Proceedings of the 11th International Wool Textile Research Conference, Leeds*, CD, Paper 102FWSA (2005).

[75] K R Millington and G Maurdev, *Journal of Photochemistry and Photobiology A*, **165** (2004) 177.

[76] G Maurdev and K R Millington, *Proceedings of the 11th International Wool Textile Research Conference, Leeds*, CD, Paper 99FWSA (2005).

[77] F D Lewis, *Accounts of Chemical Research*, **19** (1986) 401.

[78] J Cumming, C H Giles and A E McEachran, *Journal of the Society of Dyers and Colourists*, **72** (1956) 373.

[79] L A Holt and B Milligan, *Textile Research Journal*, **44** (1974) 181.

[80] J Premkumar and R Ramaraj, *Journal of Molecular Catalysis A: Chemical*, **142** (1999) 153.

[81] J M Dyer, C D Cornellison, S D Bringans, G Maurdev and K R Millington, *Photochemical and Photobiological Sciences*, **84** (2008) 145.

[82] B Ruffin and A Castellan, *Candian Journal of Chemistry*, **78** (2000) 73.

[83] W Mosimann, L Benisek, K Burdeska, I H Leaver, P C Myers, G Rienert and J F K Wilshire, *Proceedings of the 8th International Wool Textile Research Conference, Christchurch*, **IV** (1990) 239.

[84] K R Millington, *Coloration Technology*, **122** (2006) 49.

6

Wool-dyeing Machinery[1]

Jamie A. Hawkes[1] and Paul Hamilton[2]
[1]*Perachem Limited, UK*
[2]*Bulmer & Lumb Group Limited, UK*

6.1 Introduction

As reported by F.W. Marriott in Chapter 5 of 'Wool Dyeing' there is a positive trend away from fibre dyeing in loose stock and worsted top form towards late-stage coloration. The traditional breakdown of wool coloration is now estimated to be:

- top dyeing: 16%;
- loose stock: 16%;
- yarn–hank and package: 40%;
- piece and garment dyeing: 28%.

Quick response has become an important factor in the wool processing pipeline, and the industry has identified late-stage coloration as a means of delaying production commitments until the final possible stage. The current emphasis is on improving application techniques in package dyeing, piece dyeing and garment dyeing in order to reduce stock holding and shorten delivery times.

Robotic handling is well established in key application areas where high capital investment is linked directly to increased productivity, leading to a decrease in unit labour costs. Automation is now standard, giving increased efficiency and productivity and reduced operator dependency.

---

[1]Some information in this chapter is based on that published in Chapter 5 of *Wool Dyeing*, ed. David M. Lewis, published SDC, 1992 and is used with permission from the Society of Dyers and Colourists.

The Coloration of Wool and other Keratin Fibres, First Edition. Edited by David M. Lewis and John A. Rippon.
© 2013 SDC (Society of Dyers and Colourists). Published 2013 by John Wiley & Sons, Ltd.

Fibre damage in wool dyeing has been reduced by new machinery design, which incorporates controlled liquor flow pressure and optimised drying procedures. Drying has become of great importance, and a section of this chapter is devoted solely to this topic.

When dyeing wool in the form of tops, loose stock, yarn packages, muffs and warp beams, radial flow machines are generally used. While the general design of a machine remains the same, the design of the carrier differs depending on the type of substrate.

The most important factors in the dyeing of wool fibres are the ability to achieve even dye distribution, levelness of shade and targeted colour fastness, and to address operating costs and environmental impact. Modern machines are aimed at optimising dyeing conditions while minimising water and energy usage.

6.2   Top Dyeing

Dyeing wool in this form has a number of advantages:

- The degree of accuracy that can be achieved in large batches.
- The opportunity to blend the sliver after coloration, meaning that batches of 10 000 kg can be processed to a uniform shade.
- The ability to blend wool with other fibres that have been dyed under optimum conditions, ensuring solid results and optimum fastness for both types of fibre.

Historically, liquor ratios of around 12 : 1 were typically used to dye keratin fibres in top or package forms. This ratio has now been reduced to as low as 3 : 1 on some machines, and often 8 : 1 can be achieved through a system that involves strategic positioning of the heat exchanger in the circulation system and optimised loading by means of efficient material carrier designs (Figure 6.1). To avoid the need for an expansion tank, but to compensate for liquor expansion and shading corrections, the kier now uses an internal air pad.

This form of dyeing is ideal where colour continuity is essential and forward forecasting possible; the main disadvantages are the need for relatively large batches and the slow speed of turnaround due to the use of colouring at such an early stage of fabric manufacture.

Bump tops are made by coiling 10–20 kg of top sliver into a can between a central mandrel and the can wall, on top of a woven plastic fibre disc placed on a false bottom; they are then secured by strings tied around the top. Bump tops are loaded into a series of cans (as shown in Figure 6.2), located on a material carrier and locked into place. The carrier is then placed into a suitable dyeing machine, such as the Thies (Germany) eco-bloc vertical assembly shown in Figure 6.3. When a dyeing cycle is complete, the carrier is lifted by a hoist on to a transporting bogie and each individual perforated cage is unlocked from the base and lifted away by the crane. The base of the perforated cage is then opened and the bumps slid out into a suitable collection vessel.

The latest industry requirements are for the highest flexibility in machine loading capacity; historically, machines were often used at 50% capacity, which caused such problems as liquor ratio variations, increased energy usage and shade continuity. Interchangeable dyeing carriers now allow dyers to vary loading capacities, which, combined with analogue-level control and the use of an 'air pad', lowers the liquor level to achieve a constant liquor ratio.

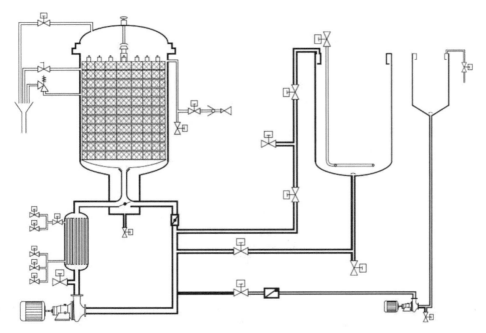

Figure 6.1   *Schematic layout of a Thies eco-bloc Machine (photographs kindly supplied by Thies GmbH & Co. KG, Germany).*

Figure 6.2   *Perforated bump cans attached to a bogie (photographs kindly supplied by Thies GmbH & Co. KG, Germany).*

Figure 6.3   *Thies eco-bloc in a vertical assembly (photographs kindly supplied by Thies GmbH & Co. KG, Germany). See colour plate section for a full-colour version of this image.*

Constant liquor ratio offers the advantages of:

- recipe correlation;
- increased reproducibility;
- constant process costs.

Quick and level dyeing is assured through the use of a high liquor turnover, with adjustments of the flow rate and a stepless increase in the differential pressure regulated by the machine controller. Multiple machines are now connected to a central control system, which serves to allow remote monitoring of the dyehouse, as well as fault finding, batch identification and report generation.

6.2.1   Longclose (UK) Large Bump Tops

Longclose (UK) developed a system for handling large bump tops. These are made by coiling 20–22 kg of top sliver into a can between a central mandrel and the can wall, on top of a woven plastic fibre disc placed on the false bottom of the can. A second woven disc is put on top of the coiled sliver, and the mass of sliver compressed in a bumping press. The bump top is then secured by strings tied around the top. Four of these tops are compressed into a dyeing cage, which consists of an outer removable perforated cage and a removable perforated spindle. Four cans can be located on a material carrier and locked into place, allowing 320 kg wool to be dyed in one batch.

For unloading after dyeing, the individual cages are lifted by crane on to a transporting bogie, and the outer perforated cage is unlocked from the base and lifted away by the crane. The central perforated spindle is then unlocked from the internal seatings on the material

Figure 6.4 *Longclose (UK) large bump top dyeing: unloading after dyeing.*

carrier base and lifted vertically from the column of dyed tops, again by means of the crane. Figure 6.4 illustrates the unloading operation, with the material carrier shown on the left of the picture. In the background, a material carrier for conventional ball tops or 10 kg bump tops can be seen.

6.2.2 Obem Big Form

The Obem (Italy) Big Form machine is capable of dyeing multiples of 100 kg of wool. Worsted top in 100 g m^{-1} sliver weight is coiled (as shown in Figure 6.5) into a perforated circular container with a large-diameter centre: 100 kg of wool is press-packed to a density of 0.3 g cm^{-3} into one container at the loading stage, and up to four containers can be dyed in one autoclave. After dyeing, the containers are removed from the dyeing vessel, centrifuged and then dried.

6.2.3 Vigoreux Printing

Although not strictly a dyeing process, Vigoreux printing of top sliver is a recognised method of producing mélange effects on wool. The original method was patented in 1863 by Jacques Stanislas Vigoreux of Reims, France. Conventional Vigoreux printing is carried out on a machine of very simple construction, consisting of the gillbox (combs or separating pins) and the printing compartment. Between 10 and 16 sliver bands of about 20 g m^{-1} are combined in the gillbox to produce a uniform fleece with a draft of 1 : 4 to 1 : 6; this is fed to the printing rollers, which carry a relief pattern of diagonal stripes. The dye paste is transported from the trough to a felt-covered bowl by means of a rubber-coated dipping roller. The raised bars of the printing roller (engraved roller), which is placed above the

Figure 6.5   Obem Big Form.

bowl, press the sliver against the felt-covered roller carrying the dye. The dye paste will impregnate the sliver only at the points of contact. The printed sliver is either coiled into perforated cans or plaited on to pallets, which are then placed in a steam autoclave for fixation.

In a system known as Siroprint, developed by CSIRO, Division of Wool Technology (Australia) and commercialised by OMP (Bodega, Italy), the printing head eliminates the need for pregilling of slivers and prints directly from a gravure roller, thereby avoiding the need for a felt-covered roller. This has led to improvements in production speeds, shade reproducibility and shade matching.

Use of the steamer in the new system – Sirosteam – led to a completely continuous process, where the printed and steamed tops are fed directly to a backwasher. The Sirosteam atmospheric steamer is suitable for metal-complex, milling or reactive dyes. For chrome dyes, batchwise autoclave steaming is required (Figure 6.6).

6.3   Loose Stock Dyeing

Various types of machine are used to dye wool in loose stock form. These include conical-pan, pear-shaped and radial-flow machines.

In the former two, the liquor is pumped through the pack of wool, which is packed relatively loosely in the container. In the conical-pan machine it usually flows from the bottom to the top of the pan, overflowing and returning to the pump to be recirculated. In the pear-shaped machine, the flow is usually from top to bottom. In both types it is possible to reverse the flow.

Because of the relatively loose packing of the wool in these two machines, relatively low flow pressures are required to give adequate penetration of the dye liquor through the

Figure 6.6   *Siroprint Vigoreux printing system.*

pack. This gentle action causes minimal damage to the wool fibre, but productivity is rather low. To increase productivity, conical-pan machines are frequently stamper-loaded, which causes fibre damage during dyeing.

In the radial-flow machines, a cage with a central perforated column accommodates the loose wool, which is usually press-packed into the cage by stamper loading to give a density of approximately $0.25\,\mathrm{g\,cm^{-3}}$. The packed cage is loaded into a kier, and dye liquor is circulated at a high flow rate to ensure level dyeing. Again, the wool fibres can be damaged under these conditions. There are three main contributory factors:

(1) *Effect of stamper loading:* Stamper loading techniques are widely used to increase package density and productivity in the dyehouse. The Wool Research Organisation of New Zealand (WRONZ) has confirmed that the physical compression exerted during stamper loading increases the number of bends and curves of the wool fibre in the pack, which are then set during dyeing. During carding and spinning, these bent fibres break, giving rise to reduced carding and spinning performance and increased wastage.

(2) *High liquor flow:* With increased package density due to stamper loading, conical-pan and radial-flow machines are used with high liquor flow to ensure level dyeing. The actual flow levels employed far exceed the minimum requirement for level dyeing. As the dyeing temperature increases, the wool pack becomes thermoplastic and softens; the velocity of the dye liquor then increases and the high flow rate through the wool pack at maximum temperature increases the degree of fibre damage.

(3) *Drying conditions:* Drying is a major factor in fibre damage, related to both overdry-ing (to avoid mildew formation) and uneven drying, which normally occurs under production-drum-dryer and brattice-dryer conditions in the absence of moisture control.

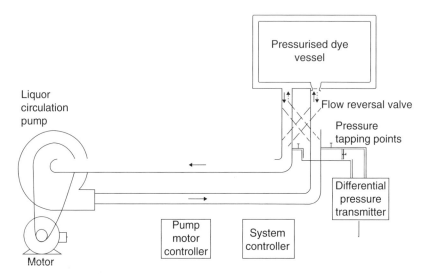

Figure 6.7   *Schematic diagram of the WRONZ Soft-Flo System.*

To reduce mechanical damage during loose stock dyeing, the WRONZ Soft-Flo System was developed. In this system, fibre damage is minimised by dyeing at a constant minimum flow pressure throughout the entire dyeing cycle, sufficient to give level dyeing but with reduced mechanical damage to the wool. A schematic diagram of the system is given in Figure 6.7. Two pressure sensors monitor the dye liquor pressure in the flow and return lines, respectively. The differential flow pressure is fed to a microprocessor controller, which opens or closes the valve in the flow line to maintain the required minimum flow pressure, predetermined in commissioning trials; alternatively, the pump speed can be controlled by a variable-speed drive operated by the microprocessor controller.

The temperature sensor, connected to the microprocessor controller, brings the Soft-Flo System into operation at a specific 'set-point' temperature, normally 60 °C. At temperatures above 60 °C, the selected reduced flow pressure is then maintained automatically by the controller, related to differential flow pressure sensor output. The microprocessor controller and temperature readout may be located either in the dyer's office or in a splash-proof cabinet adjacent to the dyeing machine. The system may also be linked to a process controller.

Typically, flow pressures are decreased from 1.05–1.12 bar to 0.07 bar, the control of the reduced pressure coming into operation as the dyebath reaches 60 °C. Figure 6.8 shows a comparison of a typical flow pressure profile in conventional loose stock dyeing of lambswool with that of Soft-Flo controlled dyeing. Although differences in the fibre properties are most significant between conventional and Soft-Flo dyeings, improvements in physical properties have also been observed in yarn spun from wool dyed using the Soft-Flo system. The system does not lower dye exhaustion or the efficiency of the aftertreatment process, and shade reproducibility and colour fastness performance are directly comparable to those obtainable in conventional loose stock dyeing. The technology was marketed by Longclose (UK) and was also installed on top-, package- and piece-dyeing machinery.

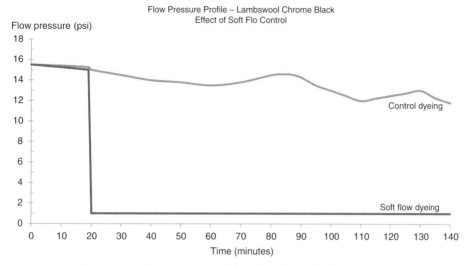

Figure 6.8 *Flow pressure profiles: the effect of Soft-Flo control.*

6.3.1 Continuous Dyeing of Loose Stock

A machine has been marketed by Fastran (UK) for the continuous dyeing of loose stock. A lap of loose stock is passed through a horizontal pad mangle, the dye liquor being contained within the radii of the pad bowls. The padded fibre is then passed into a dye-fixation unit by a hydraulic ram to ensure continuous processing under pressure, and radio frequency (RF) energy is applied. It is evenly heated to the required temperature in the time taken for the fibre to pass through the RF field (approximately 15 minutes), and a further 10–15 minutes in the fully insulated dwelling zone ensures full fixation of the dyes. Continuous backwashing and RF drying of the dyed loose stock completes the process, with the dyed, dried wool emerging at a controlled moisture content of $\pm 1\%$ (Figure 6.9).

A system was also designed for worsted top; this was capable of dyeing 50–500 kg h^{-1} in a fully continuous operation with minimum labour costs. It was fully computerised, and achieved high fixation levels for a wide range of dye types (except chrome dyes)

Figure 6.9 *Fastran system for the continuous dyeing of loose stock by RF fixation.*

with minimum fibre damage. The liquor ratio was low (4 : 1 including backwashing) and installation required only mains electricity and water supply.

6.4   Hank-Dyeing Yarn

Yarn for carpets, hand knitting and machine knitting is still predominantly dyed in hank form, although there are developments taking place which will allow it to be package-dyed.

6.4.1   Carpet Yarn

Carpet yarn may be dyed on single-stick Hussong machines (Figure 6.10). Heating is by open- or closed-coil steam pipes, situated below a perforated false bottom. The dye liquor is circulated over a weir and through the yarn by means of a reversible impeller. In modern machines, the yarn is suspended from V-shaped sticks, which have perforations to prevent stick marking.

Improvements have been made to Hussong-type machines by the introduction of central impeller compartments to reduce the distance through which liquor has to flow. It is also possible to link two machines in order to enable larger batches to be dyed, a development that is particularly relevant to the carpet industry.

Two-stick machines were developed from the one-stick machine. In one-stick machines of the Hussong type, the direction of flow is mainly up through the hanks, which causes the yarn to form a dense pack that impedes liquor flow. This, in turn, results in unlevel dyeing, particularly if dyes of higher wet fastness (with inferior levelling characteristics) must be used. The use of a second stick at the bottom of the hanks prevents the mass being lifted by the flow and allows a greater rate of flow to be used without severe tangling. The distance between each pair of sticks must be adjusted according to the hank length, so that the yarn is not stretched tight during dyeing – this can cause severe stick marking at both the top and the bottom of the hanks. It is customary to leave about 4 cm free space between the bottom

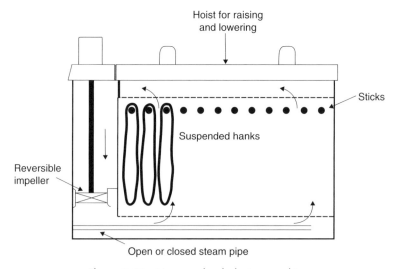

Figure 6.10   *Hussong hank-dyeing machine.*

Figure 6.11 *Thies HankMaster with six rods (photographs kindly supplied by Thies GmbH & Co. KG, Germany). See colour plate section for a full-colour version of this image.*

stick and the hank. This adjustment is especially critical when high-bulk yarns are being processed, since allowance must be made for the shrinkage that will occur in the hank as bulk is developed.

Modern machines, such as the Thies HankMaster, can use liquor ratios as low as 1 : 5, depending on the loading and yarn quality. These machines avoid high spots and lustre marks on the hanks by rotating the hanks on 960 mm-long rods. The machines can be supplied with from two to ten rods (Figure 6.11 shows a HankMaster with six rods), each capable of holding 10 kg of yarn. For bigger batch sizes, a number of machines can be coupled together and controlled by one control system.

6.4.2   Hand-Knitting and Machine-Knitting Yarn

For hand-knitting yarns, a cabinet hank-dyeing machine is often used. In these machines (Figure 6.12), the hank carrier, mounted on a trolley, is loaded outside the dyeing cabinet and then wheeled into the cabinet for dyeing. At least two hank carriers are required for each cabinet; downtime may thus be kept to a minimum, loading and unloading of the carriers taking place while alternate lots are being dyed. This type of machine also obviates the need for hoists to load and unload the hank carriers from the dyeing compartment.

Flainox (Italy) has modified its cabinet machine to minimise wasted space in the dyeing compartment, and has claimed that wool machine-knitting yarn can be dyed at an effective liquor-to-goods ratio of 15 : 1. Figure 6.13 illustrates a Flainox carrier with dyed hanks awaiting unloading.

Figure 6.12   *Cabinet hank-dyeing machine.*

A further development in hank-dyeing machinery is the use of a carrier that will fit into circular radial-flow machines, originally intended for the dyeing of tops, loose stock and yarn on package (Figure 6.14). Within such carriers, which are based on the two-stick principle, the hank sticks are situated in concentric circles around the frame; because each consecutive circle accommodates a different number of hanks, loading must be carried out

Figure 6.13   *Flainox hank carrier.*

Figure 6.14   *Typical circular hank carriers.*

with care. The distance between sticks is adjustable to allow for different hank lengths and, in general, the earlier comments regarding two-stick machines apply.

The dyeing of hanks on this type of machine should only be considered if the dyehouse is equipped with radial-flow machines that are not fully utilised for loose-stock or yarn-package dyeing; in such dyehouses, it provides a degree of versatility, however.

6.4.3   Robotic Handling

Robotic systems for handling hanks are available from Galvannin (Italy) and Minetti (Italy). The Galvannin robotic-handling equipment for hank drying is modular, but together the components provide a fully automated unloading and drying operation. Hydro-extraction is effected by squeezing the hanks in two different positions to achieve uniform moisture content. The hanks are then dried on rotating sticks and can be automatically packaged for despatching. Figure 6.15 demonstrates the loading/unloading procedure for undyed and dyed hanks.

The Minetti robotic hank-handling system (Figure 6.16) loads the hanks on to the sticks, loads the sticks in the carrier and loads and unloads the carrier in the dyeing machine fully automatically. The dyed hanks are then removed from the sticks in preparation for drying. The hank dryer operates by first squeezing the hanks and then passing them through a hot-air dryer, ventilated from all sides; this is claimed to improve drying uniformity. Operator handling is unnecessary until the hanks are wrapped and packaged for despatch.

6.4.4   Space Dyeing of Yarn

Space dyeing of yarn has been established for many years [1]; previously, the best-known method was the Texinox (UK) system, in which an impregnation head, comprising 40 nozzles, can be connected to any of the four colour distributors. The head is situated over a brattice, on which the yarn hanks are loaded traverse to the brattice movement. The brattice passes the hanks under the jet nozzles in the impregnation head, and a four- or five-coloured space-dyed effect is achieved.

Figure 6.15   *Galvannin robotic handling of hanks.*

The trend has now moved towards continuous yarn-printing machines, such as the ones offered by Biella Fancy Yarns of Italy. Here, the yarn is loaded on to cones on a feeding creel and then passed under pressure-driven rollers, each delivering a different colour fed by an independent dye tank complete with mixing device. The rollers are raised and lowered according to the required pattern, which is controlled via an integral programmable logic controller (PLC) with touchscreen display. Machines are available with four, six or eight colours, and up to 48 yarns may be processed simultaneously. The printed yarn is then

Figure 6.16   *Minetti robotic handling of hanks.*

Figure 6.17 *Superba continuous yarn space-dyeing machine.*

fixed with steam and the residual colour and chemicals removed by rinsing the yarn in hank form.

Fastness levels comparable with those of standard dyeing processes are achievable and the numbers of designs that can be created through the use of computer programs is endless. Current trends in hand knitting are making this form of coloration particularly fashionable.

Superba (France) make a continuous yarn space-dyeing machine, the SS/TVP (Figure 6.17). This is designed to space dye carpet yarns: a ground colour is sprayed on to the yarn sheet as it is fed through the machine, after which three space-dyeing heads project colour spots on to the yarn by means of a turbine fitted with up to 15 nozzles (depending on the result required); these are distributed randomly on the turbine perimeter. The length of the spots can be varied from 5 to 25 mm, depending on the speed of the turbines and the diameter of the nozzles.

After passing through the space-dyeing unit, the wool yarns are dried and then laid down in coils for steaming through the Superba steamer. From six to twelve ends of yarn can be processed simultaneously, depending on count; the production rate is $40–80 \, \text{kg} \, \text{h}^{-1}$.

A completely randomised continuous space-dyeing machine is made by SWA (Italy): the Spraychromatic 081. This was developed by IWS, and uses a computer-controlled method of randomised space dyeing fine-count singles yarn. From 10 to 12 ends of yarn are taken from a feed creel through four dyeing chambers, where colour is sprayed on to the yarn in either a selected, a predetermined or a randomised manner. The coloured yarn is then coiled on to the brattice of a continuous steaming unit, where the dye is fixed; this is followed by batchwise washing-off. The machine can run at up to $400 \, \text{m} \, \text{min}^{-1}$.

6.5 Yarn Package Dyeing

Package dyeing provides the textile industry with an opportunity to colour yarn at the latest possible stage prior to fabric manufacture. This is becoming increasingly important if the dyer is to respond rapidly to changes in fashion and consumer demands.

Due to the inability of further blending when dyeing in package form, it is preferable for each batch to be the exact required weight. In order to meet large weight demands, two or more machines may be linked. This enables a continuous interchange of liquor between the linked machines during the dyeing operation.

6.5.1   Package Preparation

Package dyeing is generally only as good as the package preparation. With the use of dyebath lubricants, biconical package centres unwind perfectly to the end of the cone. This virtually eliminates waste, thus reducing costs. They are ideal for use in direct warping and weaving.

Development of package centres has made it possible to improve level dyeing performance, to increase the size of the payload and to reduce fibre damage. For many years, perforated plastic cones have been widely used as support centres for package dyeing, but several disadvantages are associated with their use:

- Payloads in a given machine are limited.
- Spacing devices are required between the cones, making loading and unloading operations labour-intensive.
- Nonuniform column density requires high flow rates in order to achieve level dyeing.
- Cone slippage can occur.

A recent development that overcomes these disadvantages uses biconical package centres. The internal geometry of these centres allows packages to be prepared on conventional random winding or spinning machinery.

Grooves in the base of the formers correspond with spines on the top, so that interlocking and hence press packing can be achieved in order to form a parallel-sided dyepack (PSDP). The main manufacturers of these types of centre are K H Rost and Jos Zimmermann (both of West Germany), which produce the BIKO and Eisbar centres, respectively. A wide variety of centres is available to suit different winding machines, traverses and spindle diameters; completely parallel centres are also now available (Figure 6.18).

Press packing is generally carried out by 15–22%, to produce a uniform parallel-sided column. Package density is usually in the order of $350 \, \mathrm{g \, l^{-1}}$ prior to press packing, and may increase to $450 \, \mathrm{g \, l^{-1}}$ after pressing. A major advantage of using the press pack/PSDP

Figure 6.18   *Range of package centres.*

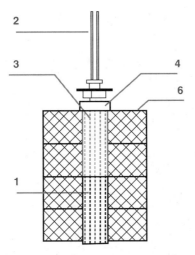

1 Perforated pipe spindle
2 Extended threaded spindle
3 Slip-over pipe
4 Locking device
5 Top plate

Figure 6.19 *Schematic of a column of package cones on a perforated spindle (photographs kindly supplied by Thies GmbH & Co. KG, Germany).*

method is that payloads in a given machine can be significantly increased and large packages can be dyed with complete levelness. Another advantage is in the loading and unloading operations, as the entire column can be unloaded and spacers are not required.

Self-adjusting caps are used to ensure a proper seal of the yarn column; as the packages are compacted, the self-adjusting locking cap and top plate automatically follow down the top package to maintain a proper seal (as shown in Figure 6.19).

Since the introduction of the parallel-sided system using BIKO dyepacks (Figure 6.20), flow rates can be effectively reduced from 30 to 12–15 l (kg per min), without any decrease in level-dyeing performance. Mechanical damage is reduced at the lower flow rate, with additional benefits in terms of a decrease in the amount of electrical energy consumed in running the pump and reduced wear and tear on the motor.

Biconical package centres are ideal for use in direct warping, weaving and possibly even knitting. The larger packages (up to 300 mm diameter) give many times the running length of the conventional package; this reduces the labour required by as much as 50% in assembling creels for warping and changing weft cones during weaving. Moreover, high-speed weaving machines, including air-jet weaving machines operating with very high weft insertion rates, can be fed directly by BIKO-style dyepacks without the need for rewinding.

Modern dyehouses may have automatic pressing machines, such as that shown in Figure 6.21. These are capable of simultaneously pressing all spindle stacks on a single carrier to a constant and monitored pressure. This ensures that all cones within a batch have been pressed equally, which in turn ensures that the liquor flow through each spindle is the same, resulting in highly reproducible dyeings.

Figure 6.20   Dyehouse using BIKO dyepacks.

Figure 6.21   Thies Automatic Press (photographs kindly supplied by Thies GmbH & Co. KG, Germany). See colour plate section for a full-colour version of this image.

6.5.2 Machinery

Historically, yarn-dyeing machines would fit into three categories:

(1) *Horizontal-spindle machines:* Examples include the Pegg GSH (UK), which was used to dye high-bulk, low-density packages, and the Thies horizontal-spindle eco-bloc machine (Figure 6.22). A known issue with this type of machine was its limited ability for press packing: sagging of the packages could occur, leading to unlevel dyeing.
(2) *Tube-type machines:* There are two kinds: one, typified by the Flainox Economy System F-1/AT-140, has vertical spindles on to which tubes are lowered; the other, of which the Obem API/O is an example, has horizontal tubes into which full spindles are loaded (Figure 6.23). Both types work at a low liquor-to-goods ratio of 4 : 1, enabling savings to be made in energy, water, effluent and chemicals.
(3) *Vertical-spindle machines:* These are the most commonly available and widely used machines for package dyeing. Press packing is possible, with the resultant advantage of higher payloads and the minimum liquor-to-goods ratio, leading to subsequent savings in resources and energy.

Historically, these machines would be installed vertically (as shown in Figure 6.24), with the majority of the vessel, pipework and pumps located in the basement, allowing operatives to work with the machines at ground level. This arrangement also allows for the maximum number of dyeing vessels to be accommodated in a limited space. Loading and unloading of the vessels is achieved via an overhead crane.

In recent years, systems have been developed by manufacturers such as Thies and Loris Bellini (Italy) in which a horizontal machine uses vertical spindles (Figure 6.25). These new horizontal versions have the advantage of performing like the traditional vertical-spindle machines, without the associated issues of unlevel dyeings observed with horizontal-spindle machines. Another benefit is the elimination of the underground civil works that used to be required, along with the removal of the need for the crane and relevant infrastructure. This enables new dyehouses to be constructed more economically on a single floor, rather than requiring a more complex structure, as shown in Figure 6.26.

This horizontal layout also lends itself to a system in which package carriers can be transported between processes (press packing to dyeing vessel to drying vessel etc.) at ground level by a rail-mounted trolley system, which minimises labour.

Figure 6.22 *Pegg GSH (left) and Thies horizontal-spindle eco-bloc.*

Figure 6.23   Obem horizontal-tube package-dyeing machine.

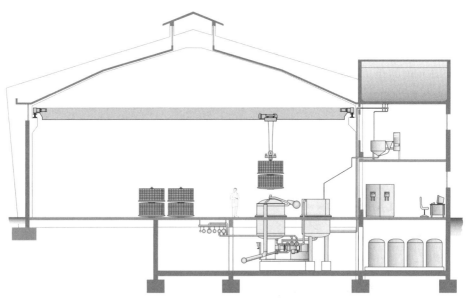

Figure 6.24   Traditional layout of a dyehouse with vertical machines (photographs kindly supplied by Thies GmbH & Co. KG, Germany).

Figure 6.25 *Thies Horizontal eco-bloc (with vertical spindles) (photographs kindly supplied by Thies GmbH & Co. KG, Germany). See colour plate section for a full-colour version of this image.*

Figure 6.26 *Packages being moved around a dyehouse on a rail-mounted carrier system (photographs kindly supplied by Thies GmbH & Co. KG, Germany). See colour plate section for a full-colour version of this image.*

6.6   Piece Dyeing

Traditionally, both worsted and woollen spun wool pieces were dyed in winches. These have been discussed in detail by Bird [2] and Bearpark, Marriott and Park [3].

A move from the traditional deep-draft wool winches to shallow-draft machines has improved the dyeing efficiency of winches, due to factors such as:

- Closed-coil steam heating under a false bottom, giving even temperature distribution throughout the winch.
- Lowering of the winch reel and driven jockey roller, reducing the drag on pieces being lifted out of the dye liquor.
- Variable machine speed, so that optimum speed for different fabrics can be selected.
- Aids to liquor circulation (pump or impeller).
- Even distribution of dyes and auxiliaries across the width of the winch, giving reproducibility of shade from piece to piece within one dye lot.

6.6.1   Jet and Overflow Dyeing

These were originally marketed as high-temperature piece-dyeing machines for textured polyester, designed in order to overcome the problems with carrier dyeing in atmospheric winches. The jet-dyeing principle was later extended to the dyeing of cotton knitted goods and polyester blends with cotton or wool.

It was logical that jet-dyeing machines should be considered for the dyeing of wool fabrics since replacements were needed for winches, and because a machine was required which would overcome the problems of winch dyeing, particularly the formation of running marks.

Holt and Harrigan [4] surveyed a variety of machines and identified five basic types:

(1) *Group 1:* Fully and partially flooded jet-dyeing machines with hydraulic fabric transport via a venturi nozzle. These machines, originally developed for textured polyester fabrics, generate a high-velocity flow of liquor that has a harsh action on wool. Partially flooded machines have a pronounced tendency to generate foam. Design modifications to reduce the intensity of direct liquor action while maintaining a high liquor interchange in the nozzle have improved the suitability of these machines for the dyeing of staple fibre fabrics, including wool.
(2) *Group 2:* Overflow machines with a driven winch reel: a combination of hydraulic transport and driven winch. These are usually partially flooded machines and are considered to be of gentle action.
(3) *Group 3:* Machines with a driven winch reel and a jet nozzle.
(4) *Group 4:* Combined overflow/jet-nozzle machines with a driven winch. These are normally partially flooded machines, regarded as offering considerable potential for wool piece dyeing.
(5) *Group 5:* Machines using a form of mechanical conveyance to assist fabric transport in addition to a winch or jet/overflow system.

Following this survey, practical dyeing trials were carried out on 11 jet machines from Groups 1, 3, 4 and 5. This work indicated that machines from Group 4 offered the most potential for wool fabrics. There appeared to be a limitation on the weight of the woollen fabrics that could be processed satisfactorily on this type of machine, but all the worsted-type fabrics processed were satisfactory. Since fabric transport can be very precisely controlled by means of the overflow cascade, the adjustable nozzle pressure and the driven winch reel, it is possible to eliminate the formation of running marks and overworking of the fabric surface. There are a number of the Group 4 machines available in the marketplace, from Thies, Then (Germany) and Scholl (Switzerland).

6.6.2   Beam Dyeing

Both woollen and worsted fabrics can be piece-dyed on beam-dyeing machines. The beam-dyeing process is a form of package dyeing in which the fabric is dyed in open width on a perforated cylinder. Atmospheric-pressure and high-temperature machines are available in which the direction of the liquor flow can be reversed and a high degree of automation can be achieved. The preparation of the beam of fabric must be carried out with care, since any creases formed during winding or shrinkage on the beam become permanent when processed at the boil; this is avoided by proper preparation. Dyes must be carefully selected, since particles of precipitated water-soluble dyes can be filtered out during dyeing.

One of the limitations of beam dyeing is the amount of fabric that can be dyed in a single lot. This is dependent upon the air porosity of the fabric: the more highly set the fabric, the lower its air porosity and the smaller the amount that can be dyed satisfactorily. An empirical relationship which gives a useful guide to the amount of fabric that can be dyed in one bath is:

$$\frac{N}{P} = K \tag{6.1}$$

where N is the number of turns of fabric on the beam, P is the air porosity of the fabric in $cm^3\,s^{-1}\,cm^{-2}$ per cm of water pressure and K is a constant, usually about 50, but varying according to the machine and type of fabric.

6.7   Garment Dyeing

Fully fashioned garments and body blanks for the cut-and-sew industry are increasingly dyed in garment form, as this allows the supplier to delay the choice of shade until the latest possible time before the garments appear on retail counters. This ensures that only the shades that are in popular demand are dyed; that is, fashion shades.

Side-paddle and overhead-paddle machines have traditionally been used for the dyeing and finishing of wool knitwear. Such machines are labour-intensive, as both loading and unloading are time-consuming. These problems can be overcome to some extent by the use of automatic controllers, with respect both to temperature and to chemical and dye additions to the dyebath.

Flainox developed a technique for unloading the side-paddle machine in which a gate at the end of the machine is opened to discharge the liquor and garments into a waiting truck (Figure 6.27).

Figure 6.27   *Flainox side-paddle machine.*

The type of machinery used in wool-garment dyehouses has changed, with the introduction of front-loading rotary-drum machines. A typical installation from Pellerin Milnor
(USA) is shown in Figure 6.28.

Because they operate at a low liquor-to-goods ratio (15 or 20 : 1, instead of 30 or 40 : 1 for
side-paddle machines), these machines offer cost savings through reduced water, energy and
manpower requirements. They are also equipped with process-control systems, enabling
better control of the dyeing procedure; this leads to better quality and better reproducibility
from batch to batch. The inner cage of the machine in which the garments are processed can
be either individual ('open pocket') or divided into two or three compartments, respectively
termed 'D-' and 'Y-pockets'. It is generally considered that either of the latter types enables
wool garments to be processed more gently than the open-pocket type.

Most of the rotary-drum machines incorporate a centrifuging cycle. This means that
the operator handles damp, rather than thoroughly soaked, garments at the end of the
dyeing cycle, and that dirty liquors are removed from the garments during the process (after
scouring, for example) much more efficiently than in side-paddle machines. The number of
rinsing cycles can therefore be reduced by 50%, with a consequent saving in water usage.

Barriquand (France) has developed a novel concept in garment dyeing with the Gyrobox
machine (Figure 6.29). This takes the form of a large wheel, divided into 12 compartments.
The compartment divisions, made from perforated stainless steel, are aligned nonradially,
which enables them to be opened more easily than conventional open- or Y-shaped compartments. A liquor ratio of 6 : 1 can be used for wool garments, at a speed of 1 rpm, as
there is constant circulation of liquor by pump. Loading and unloading of the Gyrobox are
easy, as on the loading side of the machine the floor of the compartment is level, and on the
opposite side the floor is tilted to allow the garments to slide out into waiting carts.

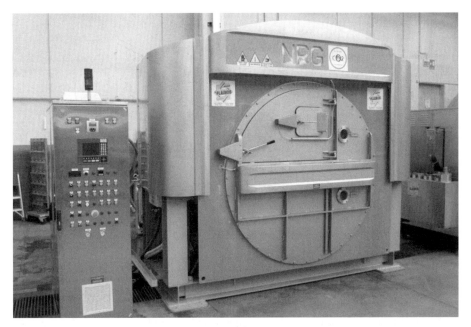

Figure 6.28   *Flainox rotary-drum machines.*

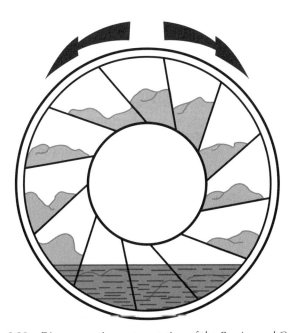

Figure 6.29   *Diagrammatic representation of the Barriquand Gyrobox.*

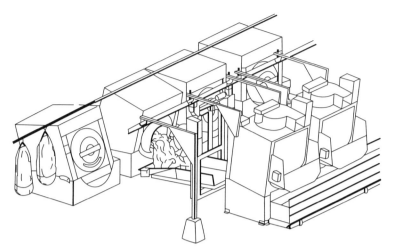

Figure 6.30   Schematic representation of the Milnor semiautomatic garment dyeing and drying system.

Pellerin Milnor rotary-drum garment-dyeing machines are located inline down one side of the dyehouse; opposite these is a line of tumble-drying machines, with conveyors linking the two lines (Figure 6.30). Once the garments are dyed, the dyeing machine is tilted and the garments are unloaded on to the lower end of the conveyor belt. They then travel on the conveyor into the open door of the tumble-drying machine. After drying, the tumble dryers are turned through 180°, the doors are opened and the goods are deposited into waiting carts.

Socks and half-hose have traditionally been dyed in side- or overhead-paddle machines, or in top-loading drum machines typified by the Smith's Drum. They are now being increasingly dyed in the front-loading rotary-drum machines, and for these garments the open-pocket type is preferred.

Modern garment-dyeing machines, such as the Cosmodye, manufactured by Cosmotex (Spain), use rotary-drum techniques. These have liquor ratios in the range 3 : 1 to 5 : 1, rather than the traditional 10 : 1 to 15 : 1, which reduces both the cost and the amount of effluent. Such machines can be provided with capacities ranging from 10 to 814 kg. Some of the larger machines can also be equipped with a hydraulic unloading system, which lifts the rear of the machine, effectively tilting it forward so that the contents can tumble out into a waiting container.

6.8   Carpet Piece Dyeing

There are five basic systems for the piece dyeing of carpets; most of these are represented by a range of machines from various manufacturers:

(1) *Pad–steam:* This system, of which there are several types, is based on the application of a thickened dye liquor to the carpet at open width, not necessarily by padding but more usually by a special applicator. This is followed by steaming (generally at 100–103 °C)

in saturated steam for 4–8 minutes and finally removal of auxiliaries and any unfixed dye by washing-off. All such machines are suitable for wool.

(2) *Pad–batch:* The only pad–batch system available for carpets is the Bruckner/Sandoz 'Carp-O-Roll' (Germany/Switzeland). In this system, an unthickened dye liquor at room temperature is applied to the open-width carpet. The carpet is then rolled, sealed with polythene film and rotated at four turns per minute for 12–48 hours, depending upon the depth of shade required. Any unfixed dye is removed by a mild afterwash in cold water. This system is suitable for wool, the carpet texture change being less marked than with any of the other systems due to the dyeing being carried out at room temperature; there is, however, a serious risk of streaks with unevenly set yarns. This unevenness can be alleviated by adding sodium bisulphite to the pad liquor [5] or by presetting the yarns prior to weaving with bisulphite (the WRONZ Chem-set process).

(3) *Continuous exhaustion:* This system, the 'Fluid-O-Therm', is based on the exhaustion of dye in a long shallow bath; the dye liquor is moved along the bath at the same rate as the carpet and is continuously replenished at the entry. The texture change is similar to that produced by pad–steam methods.

(4) *Winch/beck:* This is still commonly used, especially in the USA and for small batches. Carpets are normally dyed at open width in winches, and level dyeing across the width of the fabric is achieved using pumped liquor circulation. Because the carpet is constantly moving through boiling dye liquor for long periods of time (up to 12 hours if several shading additions are necessary), this method tends to change the texture of cut-pile wool products quite markedly. Only loop-pile textures and cut-pile carpets, with well-set yarn tufted in very low and very dense constructions, are considered suitable. The carpets must be well cropped after dyeing to remove surface fibres and restore the texture.

(5) *Foam dyeing and fluidyeing:* Very-low-liquor-ratio dyeing techniques such as foam dyeing and the Küsters Fluidye System (Germany) have been pioneered, but thus far have achieved limited success.

Various other continuous-dyeing systems have been developed (and still are being developed), particularly in the USA, based on the principle of the application of dye liquor to the carpet by means of sprays or jets, followed by steaming in the normal way. Modification of this system, using different colours in the jets, is used in jet-printing systems such as Chromotronic (Johannes Zimmer; USA) and Millitron (Milliken; USA).

6.9 Drying

The drying of keratin fibres can be subdivided into two forms: mechanical and thermal processes. The use of thermal energy in drying is expensive, and in general a prerequisite to any thermal drying process should be the efficient removal of excess moisture by mechanical methods. The most common of these are centrifuging, suction methods or the use of squeeze rollers.

6.9.1 Mechanical Moisture Removal

Hydro-extraction is the most commonly used mechanical method for removing moisture from loose stock, bumps, hanks and packages. The fibre is loaded into a circular vessel

Figure 6.31   *A typical hydro-extractor (photographs kindly supplied by Bulmer and Lumb Group, UK).*

(Figure 6.31), which is rotated at high speed, causing the water to spin out. This process can be labour intensive and the size and dimensions of some of the loads make an even distribution of weight in the vessel essential. A wide range of hydro-extraction machines are available, including ones from Broadbents (UK), Krantz (Germany) and Minetti & Rousselet-Robatel (Italy/France).

This system is not suitable for the hydro-extraction of packages because, due to their size and construction, they tend to get squashed and misshapen by the centrifugal forces during spinning of the hydro-extractor. However, Minetti and Rousselet-Robatel have designed systems in which inserts in the hydro-extractor enable packages to be held in place without distortion. This gives impressive loading capacity for mechanical moisture removal from package-dyed yarns.

The Dettin (Italy) centrifuge (Figure 6.32) consists of four separate units; 12 packages are loaded on to the separate spindles of two units, which are then lowered into two of the centrifuge units. Each spindle is spun individually. While these are in operation, the other two units are unloaded and reloaded with wet packages for centrifuging.

6.9.2   Thermal Moisture Removal

Traditionally, wool fibres were dried by hydro-extraction followed by stove or oven drying, or by some means of passage through a heated chamber. This system was problematic, with

Figure 6.32 *A Dettin hydro-extractor (photographs kindly supplied by Bulmer and Lumb Group, UK).*

constant overdrying of fibres on the outside and inadequate drying of those on the inside. RF drying of wool fibres is now an established industrial technology, with a wide range of companies offering a selection of RF units. The level of moisture left in the fibre after drying is of paramount importance for subsequent processing, and most customers now demand that the moisture content conforms to a pre-agreed tolerance.

Originally, conveyer-belt RF units were the most common, and they offered considerable savings in energy use compared to the heated systems previously used. These units offer the flexibility to dry wool fibres in package, top, hank or loose form using the same machine.

The new-generation RF machines (Figure 6.33), such as those manufactured by RF Systems (Italy), are equipped with modular polypropylene conveyer belts, multiposition upper electrode systems and an integral PC, usually connected via the Internet to the supplier company in order to assist in efficient running and fault analysis.

RF dryers can be sourced in a variety of powers; they were originally supplied in 40 and 60 kW models but now 80 kW RF dryers are common. A higher RF power allows the belt to be run at an increased speed, which increases the hourly drying capacity and throughput. Different power machines are recommended for specific fibres:

- cotton: 50/60 kW;
- acrylic: 30 kW;
- wool: 50 kW;
- polyester: 40 kW;
- viscose: 60/70 kW.

Figure 6.33   *An RF Systems (Italy) RF drier (photographs kindly supplied by Bulmer and Lumb Group, UK). See colour plate section for a full-colour version of this image.*

An alternative to RF drying is the use of a pressure dryer. Modern units are equipped with intelligent control units, high-performance heat exchangers and special blowers. They operate by allowing the carrier used for the dyeing process to be placed directly into the pressure-dryer vessel. Thies supplies both vertical and horizontal versions of its pressure dryer (Figures 6.34 and 6.35), in which the airflow during drying is carefully controlled to mimic the same liquor flow rates found in dyeing. This allows reproducible dyeing and drying cycles to be obtained from batch to batch.

6.10   Dyehouse Automation

6.10.1   Dyehouse Control Systems

Historically, dyehouses were controlled through manual systems that utilised planning boards, order sheets and individual cards controlled by technical dyers. While efficient and well-trained staff could accurately plan and carry out production to the required schedule, the advent of the digital era has led to considerable advancements in this field. The basic design and structure of dyeing machines has remained relatively static; however, the range of options available for planning and controlling these machines has changed vastly.

The main areas of control required in the operation of a dyehouse are:

- *Factory management systems:* Used for controlling substrate, dyewares, invoicing, batch identification, dye weighing and dispensing.

Figure 6.34   Thies eco-bloc vertical pressure drier (photographs kindly supplied by Thies GmbH & Co. KG, Germany).

Figure 6.35   Thies eco bloc horizontal pressure drier (photographs kindly supplied by Thies GmbH & Co. KG, Germany).

- *Process control:* Used for monitoring and controlling dyeing machines, fault finding, production reporting and diagnostics.
- *Effluent control systems:* Used for spot and composite sampling, effluent monitoring, effluent controlling and re-cycling.

6.10.2   Factory Management Systems

Most modern dyehouses now operate a computer-controlled total management system. The newer systems are usually Windows-based and tailored to the individual needs of the company.

The minimum requirements for such a system are to control the following areas:

- *Substrate control:* Incoming stock is entered into a database, dyeing orders are entered into the system and stock is allocated to each job. Often a visual flag of the required shade is attached to each order and a barcoded job sheet that contains the quality codes, dyeware recipe and required weight is printed.
- *Dyeware stock-control system:* A means of inputting stocks of dyes and chemicals is utilised, allowing simple and quick stock control, just-in-time ordering and stock-reconciliation facilities.
- *Recipe library:* The number of recipes required for some commission dyehouses can run into thousands, particularly where subtle alterations are required to cater for changes to blends and qualities on the same shade. The use of a database has largely removed the need for historical recipe files. The advent of the recipe database has made any adjustments required due to changing dye supplier/dye classes and so on relatively quick and simple.
- *Check-weigh systems:* Often the factory management system is linked to a means of issuing the correct amount of dyes and chemicals, through either a fully automatic dispensing system, such as that offered by Adaptive Controls (UK), or a partial system requiring operator interface,such as that developed by Bespoke (UK). In both cases, each dyeware dispensed is recorded in a history file; should a problem arise with the batch or finished product at a later date, the individual dye or chemical used can be identified. This requirement is considered a prerequisite for many of the documented quality-control systems now in place.
- *Batch identification:* The issuing of a barcode for the weighing and dispensing of the dyes and chemicals required readily lends itself to batch-wise identification of the lot as it travels through the factory to the customer. It is common for scanners to be located in the drying department, warehouse and dispatch facility. This enables the production planning department to quickly identify the location of every batch and accurately predict its availability to the customer.
- *Invoicing:* In the past, accounts departments would be responsible for the costing of each individual dye recipe and yarn sale. These are now generated automatically, allowing the dyehouse manager to accurately control costs and stocks.

6.10.3   Process Control

The last 30 years have seen rapid advances in the type and level of control offered in the actual controlling of the dyeing vessels.

Initially, the dyeing process was dependent on the skill and accuracy of the 'vesselman', who would be responsible for the operation of the steam valves, the pressure controls, the pump and the length of cycle. As knowledge of the dyeing process advanced, it was quickly understood that in order for a consistent product to be obtained, the dye cycle must be exactly reproduced.

Early process-control systems mimicked the procedures followed by the operative, by controlling the steam valves and pumps in an on/off manner. A pioneer of this control method was Beacon Controls (UK), which with the Beacon 101 controller offered a means of controlling the gradient of the heating cycle, holding the dye cycle at the required temperature and operating the pump. The pressure was still manually controlled in pressure-dyeing systems.

The next generation of controllers included proportional valve control, which gave a smoother gradient. Pump control, through the use of variable-speed motors, gave improvements in both energy usage and quality. A facility to link the individual controllers to a central system is usually located in the dyer's office.

The latest generation of controllers, such as the Plant Explorer, offered by Adaptive Controls (UK), offers a level of control never before available to the textile industry. Touch-screen technology and powerful computer software rapidly control and monitor machines around the factory, giving the dyehouse manager a real-time overview of ongoing and scheduled work, including lists of running steps, animated machine mimics and historical information (Figure 6.36).

With the latest technology comes the facility to run and coordinate multiple machines from one control system. The machines can be run either independently or in multiple combinations for maximum production flexibility. Figure 6.37 shows a screenshot from a controller running four machines simultaneously. This central system is based around the Plant Explorer facility, which collates and monitors all control data from around the

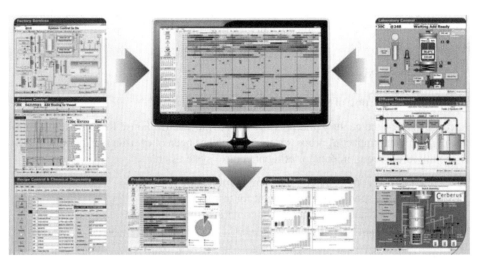

Figure 6.36 *Plant Explorer control system software (photographs kindly supplied by Adaptive Controls, UK). See colour plate section for a full-colour version of this image.*

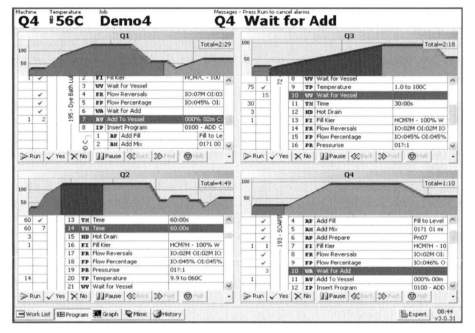

Figure 6.37   *Screenshot of a modern control system (photographs kindly supplied by Adaptive Controllers, UK). See colour plate section for a full-colour version of this image.*

factory, including equipment for effluent, plant and driers; information is stored in a standard Microsoft SQL database.

Adaptive Controls launched the smartphone Plant Explorer app (Figure 6.38) at ITMA 2011, which enables the dyehouse manager to monitor the plant via Wi-Fi or to make a VPN connection while on the move. This technology has made it possible to check the progress of a batch through the factory, or to monitor machines with just a few clicks on a mobile phone.

6.10.4   Effluent Control Systems

Since the turn of the century, the pressure on the modern dyehouse to accurately control the environmental impact of processes undertaken has been of ever-increasing importance.

The minimum requirements for effluent control systems are:

(1) Collection of a composite effluent sample for testing purposes.
(2) Collection of a spot sample of effluent for testing purposes.
(3) A system for balancing pH through automatic dispenser systems.
(4) Temperature control of effluent.

The costs associated with effluent disposal are generally dependent on the volume and content of the effluent streams. These costs, along with increasingly stringent consent limits for discharge, have led many modern dyehouses to install inhouse treatment of effluent.

Figure 6.38 *Adaptive Controls (UK) Plant Explorer app (photographs kindly supplied by Adaptive Controllers, UK). See colour plate section for a full-colour version of this image.*

One of the most advanced means of water recycling is through membrane technology. This was first developed in the 1960s and offers textile manufacturers the potential for cost savings through the following:

- Reduced water and energy costs.
- Recovery of chemicals.
- Reduced effluent volumes.
- Reduced disposal costs.
- Provision of a consistent supply of pure water, which is essential for reproducible dyeing.

One such system, supplied by Axium Process (UK) and shown in Figure 6.39, uses a combination of microfiltration, ultrafiltration, nanofiltration and reverse-osmosis technology to allow the separation of dissolved, colloidal or particulate constituents from a pressurised fluid.

Typical paybacks from such a system can be in the region of 2 or 3 years.

6.10.5 Colour Measurement

Most colour-measurement systems offer different degrees of sophistication to meet individual customer requirements. Features include:

- Recipe calculation for laboratory and production dyeing, to give minimum dye cost and metamerism.
- Recipe correction for laboratory and production dyeing.
- Recipe adjustment for multiple-batch dyeing of slubbing or loose fibre.
- Coloured fibre blend matching.

Figure 6.39   *Axium filtration system (photographs kindly supplied by Axium Process Limited, UK). See colour plate section for a full-colour version of this image.*

- Shade passing.
- Shade sorting.
- Fastness rating.
- Dye strength measurement.
- Shade library.

The introduction of colour measurement and control can reduce dyestuff costs and shading additions and increase production efficiency, with fewer customer rejections for off-shade products. Where these savings have been quantified, the payback time for equipment has been as little as 1 or 2 years.

6.11   Laboratory Dyeing

In order to facilitate the transfer of any new developments from laboratory to production, or to facilitate good quality control, modern dyehouses require accurate and dependable equipment which gives reproducible laboratory-to-production results. A number of factors must be controlled:

- quantity of dyestuffs (% on mass fabric (o.m.f.)) and other additives;
- liquor ratio;
- temperature gradients;
- flow rates and direction of flow;

- hydro-extraction prior to drying;
- Drying times/type of drying/temperature of drying.

The ideal scenario is to use laboratory-scale dyeing machines that can replicate the conditions used in larger production machinery, irrespective of machine capacity (e.g. a two-package machine needs to perform in the same way as a 200-package machine). Previous sections in this chapter have discussed production machinery for the dyeing of tops, loose stock, hanks, yarn packages, pieces and garments – all of which must also be dyed in the laboratory.

Mathis AG (Switzerland), Roaches (UK) and Ahiba (US) are the three most popular manufacturers of sample dyeing machinery (and are all featured in this section), although it should be pointed out that there are now a number of smaller companies based in China, India and other countries that are developing their own machinery and beginning to take a share of the market.

6.11.1   Tops, Loose Stock, Hanks and Package Yarn

Tops, loose stock, hanks and yarn can all be dyed in the laboratory using one of four methods.

6.11.1.1   *Moving Fabric and Moving Liquor*

This method has been used for a long time in the dyeing of piece and knit goods, yarns and loose materials, and fully fashioned articles such as socks, shorts, sweat shirts and T-shirts. It typically uses closed dye beakers, placed in a holder and rotated through a heated medium (water, glycol or ballotini) in such a way as to achieve liquor and fabric movement within each dyepot. The Zeltex Polycolor (USA) (Figure 6.40) rotates dyepots through either a water or a glycol bath. The number of dyepots in each machine varies, as does their size.

There are a number of disadvantages to this type of system:

- The machine can be messy; this is especially obvious when using a glycol bath.
- As the dyepots are removed, they are covered in glycol, which either needs to drain back into the bath or drips on to the bench/floor of the laboratory.
- The rubber seal within the dyepot lid has been known to fail occasionally, resulting in the sample becoming contaminated with glycol and the glycol bath being contaminated with dyestuffs.

A development in this type of machine came when infrared (IR) dyeing machines were introduced to the market. They work in the same way, but there is no liquid medium. The closed dye beakers are placed on to a revolving disc and heated by IR radiators while under constant rotation. One of the dye beakers has a temperature sensor, which allows constant monitoring of the temperatures inside the dyepot. The dye beakers are angled on the revolving disc to achieve optimum liquor circulation during rotation. There are a number of suppliers of this type of machine, including Mathis AG (Switzerland) (Figure 6.41), Coloursmith Ltd (UK) (Figure 6.42) and Datacolor Ahiba (US) (Figure 6.43).

- Liquor ratios can range between 1 : 6 and 1 : 50, and extra-short liquor ratios of 1 : 5 to 1 : 2 can be achieved with special liquor displacement bodies.

Figure 6.40 *Zeltex Polycolor (photographs kindly supplied by Perachem Limited, UK). See colour plate section for a full-colour version of this image.*

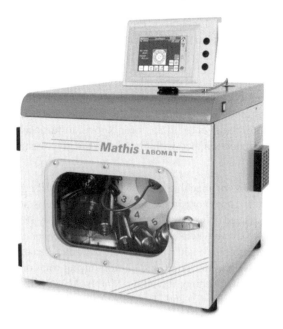

Figure 6.41 *Mathis AG Labomat BFA (photographs kindly supplied by Mathis AG, Switzerland). See colour plate section for a full-colour version of this image.*

Figure 6.42 *Coloursmith Ltd's Chroma Colour 5 (photograph kindly supplied by Coloursmith Ltd (UK). See colour plate section for a full-colour version of this image.*

- The disc speed is variable and the direction of rotation can be reversed in specified intervals.
- In some machines, the exteriors of the dye beakers are cooled by air, which is fed through a water-cooled heat exchanger.
- The number of dyepots used in each machine depends on the rotating disc and the dyepot size; this ranges from 50 ml to 5 l. Machines may hold up to 32 individual 50 ml dyepots or a single 5 l dyepot.
- Some manufacturers have introduced dosing nipples on the beakers, which allow additions to be made during the dyeing process without each individual dyepot having to be opened.

Figure 6.43 *Datacolor AHIBA IR (photographs kindly supplied by Datacolor Ahiba, USA). See colour plate section for a full-colour version of this image.*

6.11.1.2   *Stationary Beaker and Moving Liquor*

This type of machine works by pumping dye liquor through a stationary material sample at a constant pressure. The flow can be reversed (promoting levelness and replicating production machines) and the duration of each flow direction can be varied independently. These machines are particularly favoured for the dyeing of woven and knit goods, yarn (bobbin or hank), loose material, fibres and worsted tops. Examples include Datacolor AHIBA SE/1 (Switzerland) (Figure 6.43), Roaches Colortec (Figure 6.44), and the Mathis AG Colorstar CJ (Figure 6.45).

These machines all have:

- A control system for flow rate and differential pressure, including a special reversing device.
- pH measurement.
- Conductivity measurement.
- PC visualisation software for recording and registration of parameters ($^\circ$C, $l\,min^{-1}$, bar, pH).
- A system for spectrophotometric analysis of the dyebath.

6.11.1.3   *Stationary Beaker and Moving Liquor*

A further type of stationary-beaker dyeing machine uses a magnetic stirring system; such systems include the Turby Type T from Mathis AG (Figure 6.46).

Figure 6.44   *Roaches Colortec (photographs kindly supplied by Roaches, UK).*

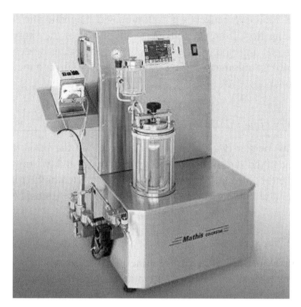

Figure 6.45 *Mathis AG Colorstar CJ (photographs kindly supplied by Mathis AG, Switzerland).*

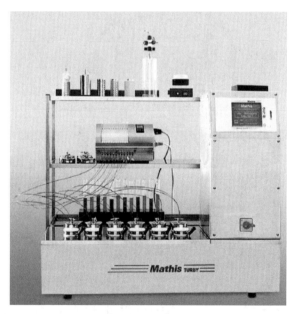

Figure 6.46 *Turby Type T from Mathis AG (photographs kindly supplied by Mathis AG, Switzerland). See colour plate section for a full-colour version of this image.*

At the bottom of each dye beaker is a Teflon-coated magnetic stirrer, which rotates at high speed, producing flow from the outside to the inside of the sample. Chemicals are normally added directly into the dye liquor; however, in certain models (such as the one shown in Figure 6.46) an automatic dosing pump is able to deliver a metered amount to all dyepots simultaneously.

6.11.1.4   Stationary Liquor and Moving Fabric

The 'moving textile' beaker dyeing apparatus operates on the basis that the dye liquor remains stationary, and the sample is agitated via a 'dipping and twisting' mechanism. The unit can be heated and cooled either by air circulation, as used in the Mathis Airboy (Figure 6.47), or by liquid circulation, as used in the Roaches DK Atmospheric (Figure 6.48).

6.11.2   Piece Dyeing

Piece dyeing in the laboratory is generally done using either a winch (Figure 6.49) or a jet machine (Figures 6.50 and 6.51); both are manufactured by Roaches and Mathis. The winch and the jet systems are suitable for the dyeing of piece goods (web or knitted) in rope

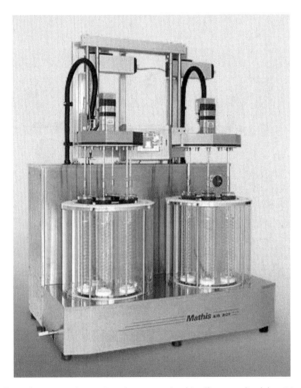

Figure 6.47   *Airboy from Mathis AG (photographs kindly supplied by Mathis AG, Switzerland).*

Figure 6.48 *DK Atmospheric by Roaches (photographs kindly supplied by Roaches, UK). See colour plate section for a full-colour version of this image.*

Figure 6.49 *Roaches Winch (photographs kindly supplied by Roaches, UK).*

Figure 6.50   *Internal mechanism of an overflow jet machine (photographs kindly supplied by Mathis AG, Switzerland).*

form or loose material. Laboratory-scale winches work using exactly the same principle as those used in full production, but on a much smaller scale.

Over the past few years, jet machines have become more popular. They can now be operated using the same modern control systems found on the larger-production machines.

There are two main types of jet:

(1) *Jet system:* The material, contained within a circular tube, is lifted by a small winch mechanism and passed through a jet. These machines are capable of running 3 or

Figure 6.51   *Mathis JFO (photographs kindly supplied by Mathis AG, Switzerland) and Roaches M10 (photographs kindly supplied by Roaches, UK) overflow jet machines.*

Figure 6.52 *Left to right: Mathis JFP Jet (photographs kindly supplied by Mathis AG, Switzerland), Roaches M25 Soft Flow Jet (photographs kindly supplied by Roaches, UK) and Thies miniMaster (photographs kindly supplied by Thies GmbH & Co. KG, Germany).*

4 kg of fabric, which makes them ideal for bridging the gap between laboratory and production. They are available from a number of suppliers, including Mathis, Roaches and Thies (Figure 6.52).

(2) *Overflow jet system:* The material, contained within a circular basket, is lifted by a small winch mechanism and passed through a large ring shape (the jet) (see Figure 6.51). The force of the liquor flow penetrates the fibres evenly and outgoing fabric causes the basket to rotate. These machines operate at a lower liquor ratio (1 : 5 and upwards) than standard jets (1 : 8 and upwards) and can run relatively small samples (e.g. 100 g), which is very useful for research-and-development (R&D) and quality-control (QC) applications.

Jet-overflow machines with this fabric arrangement are available from both Mathis and Roaches (Figure 6.50).

6.11.3 Garment Dyeing

The dyeing of garments in the laboratory is generally done using drum dyeing machines, such as those shown in Figure 6.53. These machines, which can be used to dye larger material samples or finished garments, are available in different sizes and can accommodate samples ranging from 1 to 30 kg. The drum housing, samples and dye liquor rotate in alternate directions for a period of time specified by the operator.

Some models, such as the Drum TWA machine by Mathis AG (Figure 6.53), are now available with dosing pumps (which enable additions to be made to the dye liquor during the dyeing process) and a hydro-extraction facility.

Figure 6.53   *Roaches Rotohose (photographs kindly supplied by Roaches, UK) and Mathis AG Drum Dyeing TWA machine (photographs kindly supplied by Mathis AG, Switzerland).*

6.11.4   Laboratory Machine Control Systems

Traditionally, control systems for laboratory dyeing machines have been relatively simple. Systems such as those shown in Figure 6.54 offered the user the ability to set simple programs to control the gradient of temperature rise, the final dyeing temperature and the duration of the dyeing process.

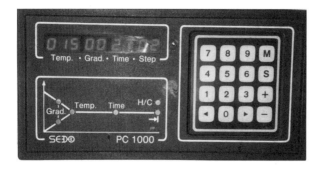

Figure 6.54   *Controllers from Zeltex PolyColour (left) and a Mathis Jig Machine (right) (photographs kindly supplied by Perachem Limited, UK). See colour plate section for a full-colour version of this image.*

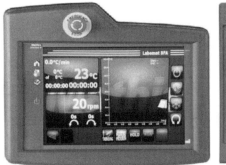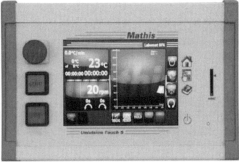

Figure 6.55 *Mathis AG Univision Controller (photographs kindly supplied by Mathis AG, Switzerland). See colour plate section for a full-colour version of this image.*

Modern laboratory machines may be installed with touchscreen controllers, which allow users to monitor and change more parameters, including the liquor flow rate, speed of fabric movement, direction of movement and. The Univision controller from Mathis AG (Figure 6.55) does all of this and can also record information to be downloaded on to a PC by the user.

References

[1] *The Bayer Farben Revue,* **Vol. 26** (1975).
[2] C L Bird, *The Theory and Practice of Wool Dyeing,* 4th Edn (Bradford, UK: Society of Dyers and Colourists, 1972).
[3] I Bearpark, F W Marriott and J Park, *A Practical Introduction to the Dyeing and Finishing of Wool Fabrics* (Bradford, UK: Society of Dyers and Colourists, 1986).
[4] I A Holt and F J Harrigan, *The Performance of Wool Piece Goods in Jet and Over-flow Dyeing Machines* (International Wool Secretariat/Commonwealth Scientific and Research Organisation, 1972).
[5] D M Lewis and I Seltzer, *Journal of the Society of Dyers and Colourists,* **84** (1968) 51.

7

Dyeing Wool with Acid and Mordant Dyes

Peter A. Duffield
Retired; Global Textile Associates Ltd, UK

7.1   Introduction

The term 'acid dye' covers, in Colour Index terms, the majority of dye classes that are conventionally applied to wool, except for mordant and reactive types. However, for the purpose of this chapter, the term will be restricted to those nonmetallised and nonreactive dyes that are applied to wool from an acidic dyebath. Even so, there is still some overlap, because a few dyes that fall within this description have the potential to react – albeit to a limited degree – with wool fibre. Similarly, some mordant dyes may also be used as acid dyes and applied to the fibre without mordanting.

Mordant dyes are those that are capable of forming complexes with metal ions or nonmetallic compounds in the wool fibre. This requires mordant dyes to have within their chemical structure groups that will form coordination complexes with metal ions or organic mordants. It must be stressed that commercially important mordant dyes are today invariably applied with chromium compounds: mainly dichromates, although some natural dyes use other metal salts. Nonmetallic mordants no longer have any commercial importance for wool but may be applied to a very limited extent by 'craft' dyers.

The history of the dyeing of wool with acid and mordant dyes, unlike that of other wool dye classes, is very long, stretching back to ancient times. Natural dyes from plant or animal sources were the only ones available to dyers prior to Perkin's development of mauveine in 1856. Most were applied to wool with a mordant in order to achieve acceptable fastness and the desired shade. Additionally, the range of mordants used also allowed dyers to achieve

The Coloration of Wool and other Keratin Fibres, First Edition. Edited by David M. Lewis and John A. Rippon.
© 2013 SDC (Society of Dyers and Colourists). Published 2013 by John Wiley & Sons, Ltd.

Table 7.1   *Chemical classes of acid and mordant dyes.*

Chemical Class	Acid	Mordant
Azo	✓	✓
Anthraquinone	✓	✓
Indigoid	✓	–
Quinophthalene	✓	–
Aminoketone	✓	✓
Phthalocyanine	✓	✓
Formazan	✓	–
Nitro, nitroso	✓	✓
Triarylmethane	✓	✓
Xanthene	✓	✓
Azine	✓	–

different shades from each natural dye, which, given the limited range of dyes available, was a valuable extra benefit. Some natural dyes, such as alkanet (*Alkanna tinctoria*), were applied without a mordant and could therefore be considered to be early forms of acid dyes, as they were applied from an acid dyebath.

At the time of writing, acid and mordant dyes are still commercially very important in wool dyeing, accounting for almost 50% of all wool dyes traded. Given that chrome mordant dyes are normally applied at high concentrations for dark shades, the proportion of wool dyed with these two dye classes is rather lower than 50%, but sill a significant amount. An important reason for the continued use of mordant dyes is the popularity of black and navy as apparel fashion shades, for which chrome dyes are particularly relevant in terms of cost and fastness performance. Mordant dyes now have commercial application only on wool and other animal fibres, and their use is declining as environmental pressure over the discharge of chromium grows.

A review of the Colour Index shows that acid and mordant dyes fall into a wide range of chemical classes, as shown in Table 7.1 [1].

It should be noted that although acid and mordant dyes contain a wide range of chromophoric groups, azo (including disazo) dyes account for around 70% of all commercial products. The azo reaction was discovered by Griess in 1858, after which date the Colour Index shows azo dyes were steadily developed and introduced to the industry. Typical examples of acid dyes in the most important chemical classes are shown in Structures **1**, **2**, **3** and **4**.

1

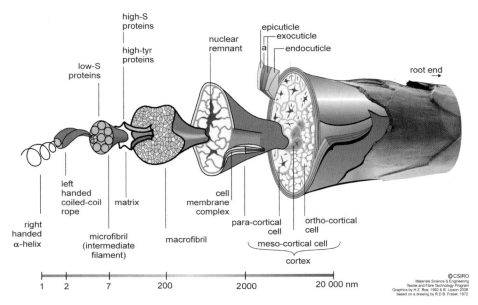

THE STRUCTURE OF A MERINO WOOL FIBRE

high-S proteins

high-tyr proteins

low-S proteins

nuclear remnant

epicuticle

exocuticle

a

endocuticle

root end

right handed α-helix

left handed coiled-coil rope

matrix

cell membrane complex

para-cortical cell

ortho-cortical cell

microfibril (intermediate filament)

macrofibril

meso-cortical cell

cortex

©CSIRO
Materials Science & Engineering
Textile and Fibre Technology Program
Graphics by H.Z. Roe, 1992 & B. Lipson 2008
based on a drawing by R.D.B. Fraser, 1972

1   2   7   200   2000   20 000 nm

Plate 1.6 *Diagram of the morphological components of a fine wool fibre (courtesy of CSIRO).*

Plate 6.3 *Thies eco-bloc in a vertical assembly (photographs kindly supplied by Thies GmbH & Co. KG, Germany).*

The Coloration of Wool and other Keratin Fibres, First Edition. Edited by David M. Lewis and John A. Rippon.
© 2013 SDC (Society of Dyers and Colourists). Published 2013 by John Wiley & Sons, Ltd.

Plate 6.11 Thies HankMaster with six rods (photographs kindly supplied by Thies GmbH & Co. KG, Germany).

Plate 6.21 Thies Automatic Press (photographs kindly supplied by Thies GmbH & Co. KG, Germany).

Plate 6.25 *Thies Horizontal eco-bloc (with vertical spindles) (photographs kindly supplied by Thies GmbH & Co. KG, Germany).*

Plate 6.26 *Packages being moved around a dyehouse on a rail-mounted carrier system (photographs kindly supplied by Thies GmbH & Co. KG, Germany).*

Plate 6.33   *An RF Systems (Italy) RF drier (photographs kindly supplied by Bulmer and Lumb Group, UK).*

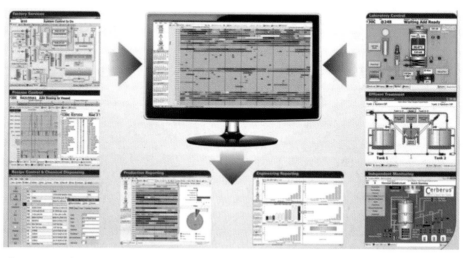

Plate 6.36   *Plant Explorer control system software (photographs kindly supplied by Adaptive Controllers, UK).*

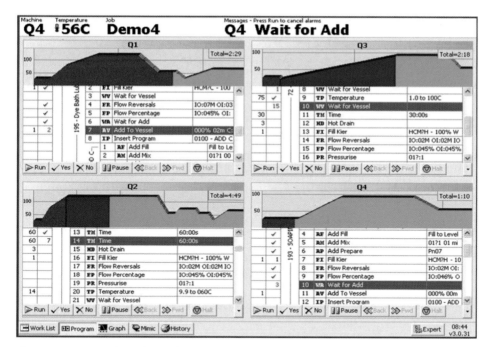

Plate 6.37 *Screenshot of a modern control system (photographs kindly supplied by Adaptive Controllers, UK).*

Plate 6.38 *Adaptive Controls (UK) Plant Explorer app (photographs kindly supplied by Adaptive Controllers, UK).*

Plate 6.39 *Axium filtration system (photographs kindly supplied by Axium Process Limited, UK).*

Plate 6.40 *Zeltex Polycolor (photographs kindly supplied by Perachem Limited, UK).*

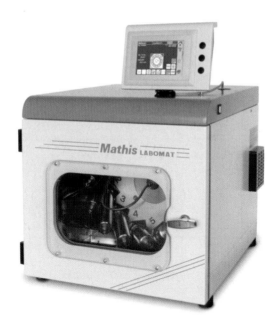

Plate 6.41 *Mathis AG Labomat BFA (photographs kindly supplied by Mathis AG, Switzerland).*

Plate 6.42 *Coloursmith Ltd's Chroma Colour 5 (photograph kindly supplied by Coloursmith Ltd (UK).*

Plate 6.43   *Datacolor AHIBA IR (photographs kindly supplied by Datacolor Ahiba, USA).*

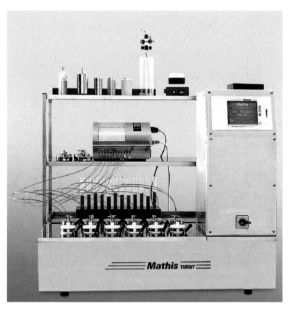

Plate 6.46   *Turby Type T from Mathis AG (photographs kindly supplied by Mathis AG, Switzerland).*

Plate 6.48 *DK Atmospheric by Roaches (photographs kindly supplied by Roaches, UK).*

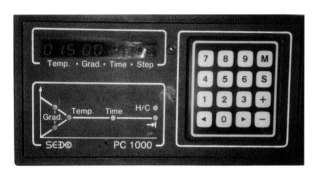 

Plate 6.54 *Controllers from Zeltex PolyColour (left) and a Mathis Jig Machine (right) (photographs kindly supplied by Perachem Limited, UK).*

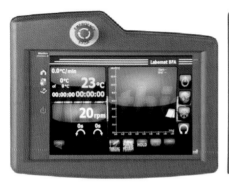 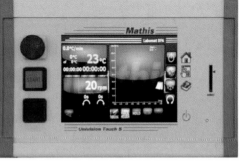

Plate 6.55 *Mathis AG Univision Controller (photographs kindly supplied by Mathis AG, Switzerland).*

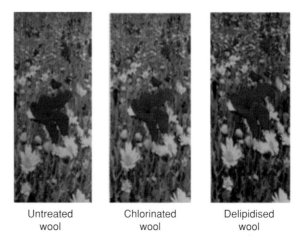

| Untreated wool | Chlorinated wool | Delipidised wool |

Plate 12.1 The effect of the wool pretreatment process on the colour yield of wool ink-jet printed with monochloro-s-triazine reactive dyes.

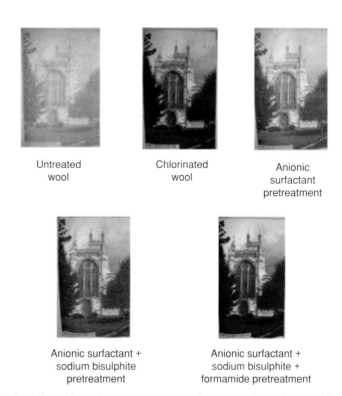

| Untreated wool | Chlorinated wool | Anionic surfactant pretreatment |

| Anionic surfactant + sodium bisulphite pretreatment | Anionic surfactant + sodium bisulphite + formamide pretreatment |

Plate 12.2 The effect of wool pretreatment conditions on the colour yield of wool ink-jet printed with difluoromonochloropyrimidine reactive dyes.

Plate 12.3 *Print-batch development of wool ink-jet printed with difluoromonochloropyrimidine reactive dyes.*

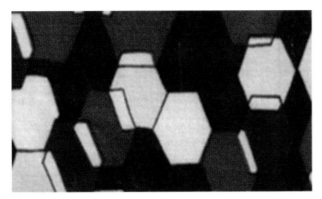

Plate 12.4 *Wool fabric pretreated with a chromium III salt and dry transfer printed with metallisable disperse dyes.*

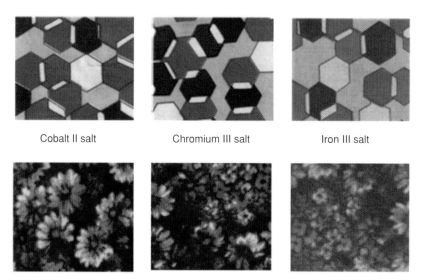

| Cobalt II salt | Chromium III salt | Iron III salt |

Plate 12.5 *The effect of metal salt type on the hue obtained for wool substrates dry transfer printed with metallisable disperse dyes.*

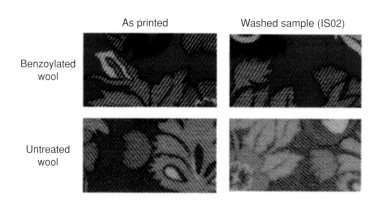

	As printed	Washed sample (IS02)
Benzoylated wool		
Untreated wool		

Plate 12.6 *Disperse dye transfer prints on benzoylated and untreated wool.*

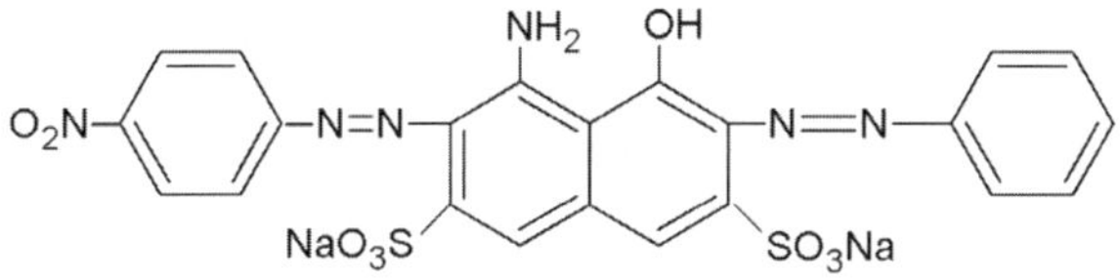

2

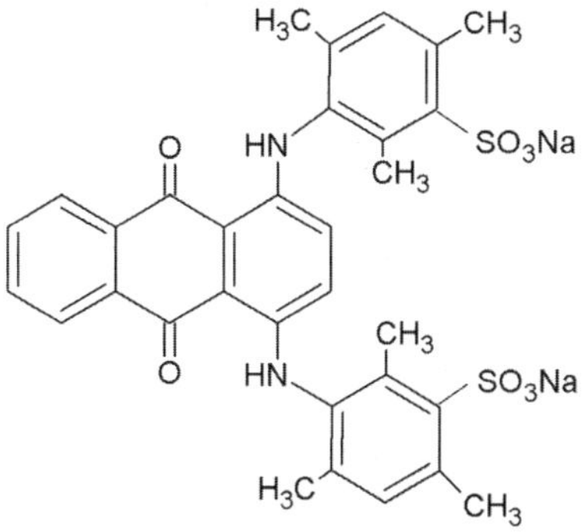

3

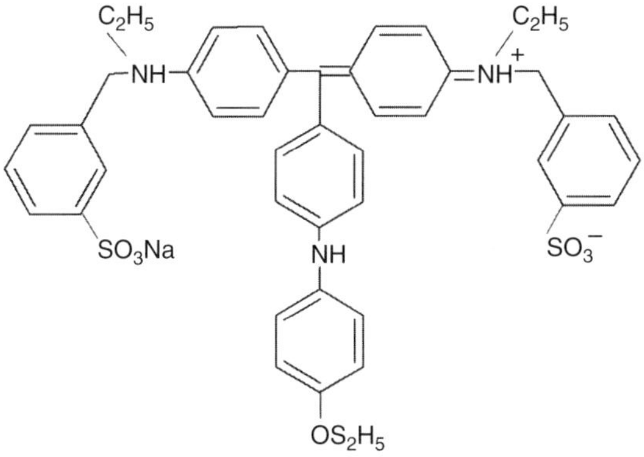

4

It should be noted that EU legislation [2] effectively prohibits the use of certain azo dyes that are capable of releasing to the environment amines which are carcinogenic or suspected carcinogens. When first implemented, this regulation was regarded by some as representing a ban on all azo dyes. This is patently not the case, and a vast majority of these dyes are still available for use. Members of the Ecological and Toxicological Association of Dyes and Organic Pigments Manufacturers (ETAD), to which most responsible dye manufacturers belong, have long since ceased the manufacture and supply of the prohibited dyes.

The fastness properties of the different acid dye chromophors depend mainly on the general chemical structure of the dye, particularly its molecular mass and solubilising and other substituent groups; details can be found in the Colour Index or in dye manufacturers' pattern cards. However, triarylmethane and xanthene dyes (not illustrated), although often bright in shade, generally exhibit lower light fastness than other types.

7.2   Acid Dyes

The Colour Index suggests that the term 'acid dye' derives from the presence in the dye molecule of one or more sulphonic acid or other acidic groups. However, because such groups are present in other dye classes, such as mordant, direct and reactive, the term is generally accepted as referring to an application class; that is, those water-soluble, anionic dyes that are applied from acidic or neutral dyebaths. The principle fibres to which they are applied are wool, other animal fibres, silk and polyamide.

7.2.1   Acid Dye Subclassification

Acid dyes are subclassified according to their dyeing properties and fastness performance. In practice, the choice of dye will be determined by the stage in the wool processing chain at which the fibre is dyed and the end use to which it will be put. Some typical processes and conditions that may be encountered are shown in Table 7.2 [3].

Quite clearly, dyes that are not fast to subsequent processing conditions cannot be used to dye wool that will be subjected to those processes. Dyeing behaviour, particularly migration – which is an important factor in determining level dyeing performance – is also critical in establishing the substrate form that may be dyed with an acid dye. The subclassifications are shown in Table 7.3, which also indicates the potential application areas for each of them.

The wet fastness of acid dyes for wool, as with metal-complex dyes, is inversely proportional to their migration properties. Figure 7.1 illustrates this fact and indicates the limitations on potential applications for acid dyes. The relative position of chrome mordant dyes has been included in this chart to illustrate the effect of mordanting on dyes that are in

Table 7.2   *Wet process treatment conditions for wool.*

Process	Typical Conditions
Scouring	Detergent, sodium carbonate (pH 8.5–10.0), 15–30 minutes at 40–50 °C
Alkaline milling	Detergent, sodium carbonate (pH 8.5–10.0), up to 2 hours at 40–50 °C
Acid milling	Detergent, sulphuric/formic acid (pH 2.0–4.0), up to 1 hour at 50–90 °C
Potting	Boiling water, 1–3 hours
Carbonising	Padding with 5% sulphuric acid, drying and baking for 3–15 minutes at 105–130 °C
Crabbing	Boiling water for 5 minutes, pressure steam 10–15 minutes at 110–130 °C
Decatising	Atmospheric or pressure steam for 10–15 minutes at 100–130 °C
Chemical setting	Padding with 1% sodium bisulphite, pressure steam for 5 minutes at 120 °C

Table 7.3 *Typical applications for acid dyes.*

Acid Dye Class	Potential Applications
Levelling or equalising	Carpet yarn, particularly high twist types for which penetration is an issue
	Woollen and worsted fabrics, particularly for bright shades on carbonised goods and for tightly constructed fabrics, such as gabardines
	Fashion and interior textile products for which washing and perspiration fastness are not required
Fast acid, half-milling or optimised half-milling	Piece goods, carpet yarn and knitting yarn that will not carry a washable care claim (pale-medium shades may meet hand-washable standards)
Milling	Medium-heavy bright shades on piece goods, weaving, carpet and knitting yarns
Super milling and optimised milling	Bright shades on loose stock, sliver, yarns that require high wet fastness to milling and similar processes. May even be used in pale-medium shades for machine-washable performance

many respects similar to acid levelling types. As previously indicated, mordant dyes exhibit very good wet fastness, while also possessing good migration properties.

7.2.1.1   Level-Dyeing or Equalising Acid Dyes

Commercial nomenclature for this class of dyes has traditionally used the letter 'E' (probably from the German 'egalisieren', rather than the English equivalent 'equalising'), although some manufacturers have now discarded this convention. There is a further subdivision within the level-dyeing acid dye group:

- disulphonated;
- monosulphonated.

The former have a relative molecular mass (r.m.m.) typically in the range 400–600 and often provide the cheapest and most level dyeing of wool dyes. Although they are very level

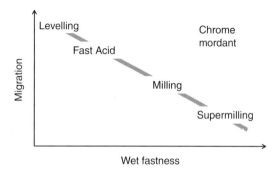

Figure 7.1 *Relative migration and fastness properties of acid dyes on wool.*

dyeing, these products are inferior to the monosulphonated types in terms of microlevelness; that is, coverage of inter- and intrafibre differences. The monosulphonated types are of lower r.m.m. (typically 300–500), slightly more level dyeing and have marginally better fastness than disulphonated products. Being of relatively low r.m.m., levelling dyes rapidly diffuse into and out of the wool fibre in a boiling dyebath. Therefore, any initial unlevelness may be overcome by dyeing at the boil until equilibrium dyebath exhaustion is achieved.

The level of equilibrium exhaustion is determined by the r.m.m., the level of anionic salt ions, dyebath pH and the number of solubilising groups on the dye molecule [4]. The principle mechanism by which level-dyeing acid dyes are attracted to the wool fibre is ionic, between the cationic protonated amine groups of the fibre and the anionic solubilising groups (normally sulphonic acid) on the dyes.

Both types of levelling dye are applied from an acid dyebath at pH 2.5–3.5. Under these conditions, a majority of the substituent amine groups (e.g. arginine, histidine and lysine) in the fibre will be protonated. Control of uptake of the dyes is achieved with sodium sulphate (Glauber's salt). Sodium sulphate is used as the cheapest way of providing this divalent anion, which competes with dye molecules for the cationic sites that are generated on wool in acid solution [5]. A typical time/temperature profile for the application of levelling dyes is shown in Figure 7.2.

If, after the allotted time at the boil, the dyeing is still unlevel, a further addition of sodium sulphate may be made to increase the concentration to 15–20%. Boiling should then be continued for a further 30 minutes or until uniformity of shade is achieved. A reduction in shade depth may be obtained by setting a fresh bath with acid and 15–20% sodium sulphate, raising it to the boil and running it for 30–60 minutes.

For off-shade dyeing, corrections may be made simply by turning off the steam supply, adding previously dissolved dyes, running for 5 minutes, returning to the boil and boiling for 30 minutes. However, it should be noted that disulphonated levelling dyes show an increased affinity for wool as the dyebath cools [3], which can lead to an increase in shade depth. Careful control of the dyeing process, including the cooling rate, is essential for batch to batch shade reproducibility.

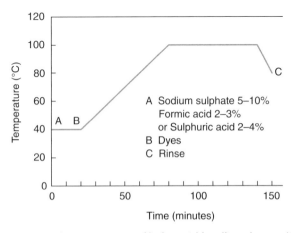

Figure 7.2 *Time/temperature profile for acid levelling dye application.*

Machinery   The excellent migration properties of acid levelling dyes mean that, as previously indicated, they can be used on substrates that are difficult to penetrate. They are also very useful for use in machines that have limited dye liquor–substrate interaction. Typical of such machines are the single stick hank dyeing types that are widely used for carpet yarns. In these, the liquor flow is relatively low and any unevenness in packing can lead to channelling or differences in liquor flow between the hanks. Similarly, in conventional winch machines, liquor–substrate interaction may be limited, and levelling dyes can be used to overcome this limitation.

7.2.1.2   Fast Acid or Half-Milling Dyes

The r.m.m. of this group of dyes is higher (typically 500–700) than that of acid levelling dyes and consequently they have better wet fastness properties. These dyes are therefore widely used for applications that require reasonably good migration/levelling and superior wet fastness to level-dyeing acid dyes. In particular, they find use on wool fabrics and yarns in medium to heavy shades on which fastness-to-wet-contact tests, such as water and shampooing, are needed. In pale shades, they may even meet handwash fastness on, for example, hand-knitting yarns or piece-dyed garments. The use of fast acid dyes has increased in recent decades due to the constant pressure from retailers and brand owners for improved fastness performance.

The reduced migration performance of fast acid dyes requires higher dyebath pH (4.5–5.0) for their application and the use of a levelling agent: typically a mildly cationic, dye-substantive product [6]. These conditions greatly improve migration characteristics; perhaps not to the same level as the previous class of wool dyes, but sufficient to allow application to a wide range of substrates for which shade levelness may be an issue. A typical application profile for fast acid dyes is given in Figure 7.3.

The profile is identical to that for level-dyeing acid dyes but the dyebath constituents are very different. Clariant has shown with their fast acid dye range, Optilan MF, that tippy dyeing wool (wool that has a significant difference in dye affinity between root and tip, due

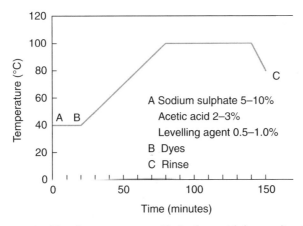

Figure 7.3   *Time/temperature profile for fast acid dye application.*

to weathering of the tip) can be satisfactorily dyed by adding 1.0–2.0% formic acid to the dyebath after 30 minutes at the boil and then boiling for a further 30 minutes.

Correction of faulty shades can be achieved by cooling the dyebath to 60–70 °C, adding predissolved dyes, running for 5 minutes, returning to the boil and boiling for 30 minutes.

7.2.1.3   Milling and Super-Milling Dyes

This range of dyes consists of products that are typically of r.m.m. 600–1000 and contain two or sometimes more solubilising groups. The super-milling dyes have an additional hydrophobic alkyl chain that increases dye–fibre attraction via van der Waals forces, and consequently also wet fastness. As the names indicate, the level of performance achieved is sufficient in most cases to enable dyed material to be subjected to milling processes. The price that must be paid for this increased fastness is greatly reduced migration, which means that these dyes cannot normally be levelled by continued dyeing at the boil.

A further issue with this class of dyes, and the super-milling products in particular, is that the dyeing performance of individual dyes can differ. They are therefore more suitable for the production of bright 'self' shades that are based primarily on just one dye. Dyeing neutral shades with, for example, a trichromatic yellow, red and blue combination will often lead to dichroic effects. Such shades are therefore most often dyed with metal-complex dyes or specifically developed compatible dye ranges.

The reduced migration performance of milling dyes requires the use of specific levelling agents and higher dyebath pH than for milling and half-milling dyes, to reduce the rate of dyeing and achieve uniformity of shade. Sodium sulphate is also essential to reducing the ionic attraction between fibre and dye, as previously noted for half-milling dyes, thereby allowing sufficient time for substrate–dye liquor contact to produce uniformity of shade. An appropriate levelling agent, normally weakly cationic, will complex with anionic solubilising groups on the dye, so reducing the exhaustion rate. Without these additions, the dyebath exhaustion rate of milling dyes would be so high as to produce unlevelness in most dyeing machines.

Given the very poor migration properties of milling dyes, it is essential to ensure uniformity of dye uptake, as subsequent boiling to achieve levelness will be ineffective. In addition to the use of auxiliaries, careful control of the heating phase of dyeing is an important factor in determining the rate of exhaustion.

A further step in controlling dye uptake and therefore level dyeing is the use of acid donor compounds. The simplest is an ammonium salt, such as ammonium sulphate, which at high temperature will break down to release ammonia. In an open dyeing vessel, this will lead to a steady fall in dyebath pH and a consequent increase in exhaustion. Such a process enables the dyer to start dyeing at relatively high pH values, at which dyebath exhaustion is low, but to achieve a lower final dyebath pH that will ultimately produce high exhaustion levels and consequently good shade reproducibility.

Ammonium salts have been used for this purpose in wool dyeing for many years, but their ability to reduce dyebath pH is limited. For this reason, chemical manufacturers have introduced other products, typically esters or lactones, that are able to liberate more acid in order to produce a greater drop in dyebath pH. Such products have been very useful in the application of milling, 1 : 2 metal-complex and reactive dyes to critical substrates, such as knitted garments. In this case, the use of an acid donor enables the slow exhaustion of dyes and penetration into the tight garment seams before the temperature is raised to that at which the acid donor begins to release acid. Such a process can produce good levelness without compromising shade reproducibility.

The environmental pressure to reduce or eliminate discharges of heavy metals from textile dyeing has led to the development of milling dyes with the normal high fastness, equivalent to monosulphonated 1 : 2 metal-complex types and with similar fibre coverage, which is obviously much better than traditional milling dyes. At the time of writing, the range of commercially available shades is still not complete, but it represents a potential solution to the increasing environmental pressure on heavy-metal discharges.

7.2.2   Optimised Dye Ranges

Commercial dyes are not pure products and contain diluents and other additives. They may also contain shading components: other dyes that may be added to ensure that the particular batch matches the standard shade for the dye, within specified tolerances. Other commercial dyes may be mixtures of individual products that have been produced to meet a specific shade, fastness or application requirement.

This approach was taken further in the latter decades of the twentieth century by the development of dyes that provided similar dyeing properties in a compact range covering a wide shade gamut. These ranges comprised products that were either individual dyes or specific mixtures and which had similar dyeing properties and fastnesses. A further benefit of these dye ranges was their ability to be applied, with specifically selected levelling agents, at the optimum pH for minimum wool fibre damage (around pH 4.5). The first range to be introduced that had similar fastness properties to fast acid dyes was Optilan MF (formerly Sandolan MF) from Clariant (formerly Sandoz). The advantage to the wool dyer was that a full range of shades could be produced with a compact range of dyes from the same manufacturer and with the same application process for all shades. The advantages in terms of reduced dye inventory and simplicity of process selection are obvious, and other manufacturers have now introduced dye ranges with similar properties.

For higher wet fastness that meets the performance of milling and 1 : 2 metal-complex dyes, a similarly compatible range was introduced by Huntsman (formerly Ciba), called Lanaset. Although systems for application of 1 : 2 metal-complex dyes at pH 4.5 had been proposed much earlier, this was the first compact range of dyes to offer compatibility across a wide gamut of shades that would normally be dyed with milling and 1 : 2 metal-complex dyes. Needless to say, the concept has been adopted by other manufacturers, and these optimised dye ranges are now widely used in the industry.

7.3   Natural Dyes

The term 'natural dye' is conventionally taken to apply to those products that are derived from plant sources, although dyes derived from insects, such as cochineal, kermes and lac (respectively *Coccus cacti*, *Coccus ilicis* and *Coccus lacca*), are also included. Given current global concerns over the environment and the clear benefits of using natural products, one must question why these dyes are not more dominant in the wool-dyeing market. Reviews of this subject [8,9] suggest several reasons:

- The supply of dyes does not meet the commercial demands of the wool-dyeing industry.
- The growing and extraction of natural dyes may be as environmentally damaging as the production of synthetic dyes, or more so.
- Application of many of the dyes requires the use of heavy-metal mordants, which, when discharged, have significant environmental impact.
- The consistency of shade of natural dyes is far lower than that of synthetic products.
- The fastness performance of many natural dyes does not meet commercial standards.

Even the Global Organic Textile Standards (GOTS), which, as the name suggests, sets standards for the manufacture of organic textiles globally, allow the use of some synthetic dyes. Despite these and other issues related to natural dyes, they continue to be used for specific limited applications, and by the 'craft' industry. There are many texts on traditional natural dyes and their application, and there has been a resurgence in publications relating to new dyes and/or processes. One source of details of modern optimised techniques for some traditional dyes is a publication by Dalby [10].

7.4   Mordant Dyes

It has already been noted that mordant dyes require substituent groups in their structure that will allow complex formation with metal ion mordants. Structure **5** is cochineal (carminic acid), an example of a dye that may be applied with a range of mordants, such as aluminium, chromium, copper, iron and tin. It shows the substituent hydroxyl and carbonyl groups, which provide ample sites for formation of coordination complexes.

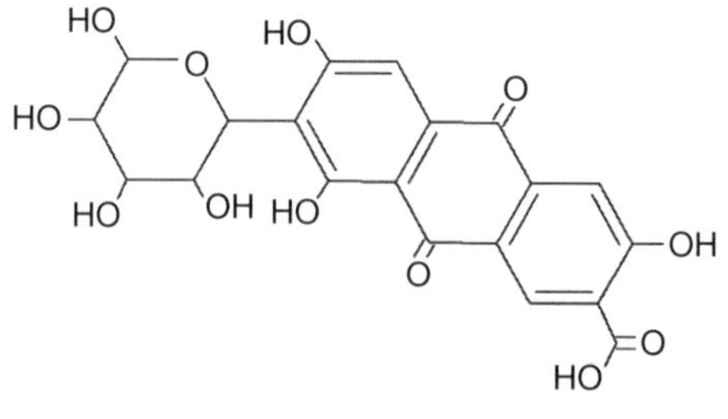

5

The different mordants that may be used with cochineal produce shades ranging from bright scarlet (tin) to red (aluminium) to purple (chromium or copper) to blue (iron) [10]. Historically, before the development of synthetic dyes, this polychromism was very useful, as it allowed the dyer to produce different shades from one dye. Indeed, many of the currently available synthetic mordant dyes exhibit similar characteristics when applied with different mordants. However, the modern wool dyer no longer needs to use this characteristic, because a full range of chrome dye shades is available without the need to resort to different mordants. Having said this, a technique was developed in which yarns that had been mordanted with different metal salts plus unmordanted fibre were incorporated in the same product, so that when it was dyed with selected chrome dyes a multicoloured product was generated from just one dyeing process.

Chromium is now the metal of choice for the commercial mordant dyeing of wool, and it has been so since the middle of the 20th century. It is for this reason that mordant dyes for wool are now invariably named 'chrome dyes'. The reason for this dominant position is the generally good wet and light fastness properties of dyeings with this mordant. Although other metals may provide good fastness with a few dyes, none provide as good a performance as chromium across the range of mordant dyes.

The substituent groups in azo chrome mordant dyes that confer the ability to chelate with metal salts are often hydroxyl groups that are ortho to the azo bond, although those based on salicylic acid are also produced. Some examples of azo mordant dyes are shown in Table 7.4 [11]. It can be seen from this table that the dyes illustrated, which are typical of

Table 7.4   *Structures of some azo mordant dyes.*

Azo Dye Class	Structure (Example)	Name
σ,σ′-dihydroxyazo dyes	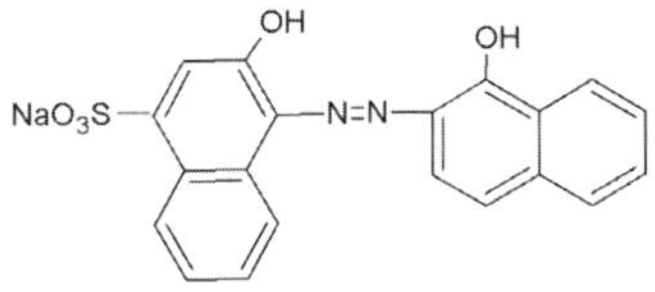 *[chemical structure: NaO₃S-substituted naphthol —N=N— naphthol, each bearing OH]*	C.I. Mordant Black 3
σ-amino-σ′-hydroxyazo dyes	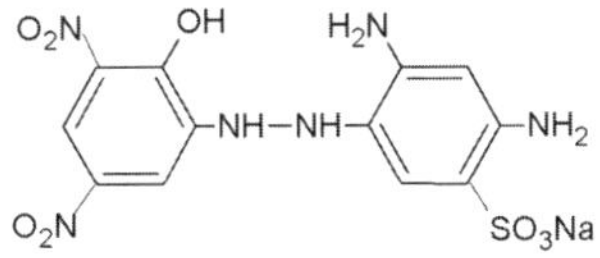 *[chemical structure: O₂N, OH, O₂N-substituted ring —NH—NH— ring with H₂N, NH₂ and SO₃Na]*	C.I. Mordant Brown 33
Salicylic acid dyes	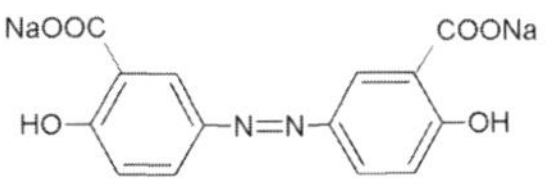 *[chemical structure: NaOOC, HO-substituted ring —N=N— ring with COONa, OH]*	C.I. Mordant Yellow 5
σ-hydroxy-σ′-carboxyazo dyes	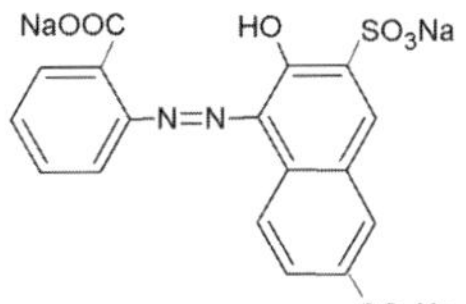 *[chemical structure: NaOOC-substituted ring —N=N— naphthalene with HO, SO₃Na and SO₃Na]*	C.I. Mordant Red 9
Azo dyes oxidised to quinone form when complexed	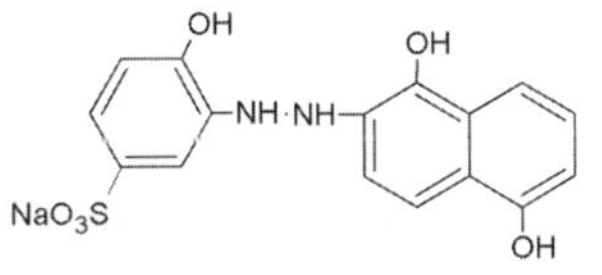 *[chemical structure: OH, NaO₃S-substituted ring —NH·NH— naphthalene with OH and OH]*	C.I. Mordant Black 9

mordant dyes, are simply level-dyeing acid dyes with substituent groups that are capable of complex formation. In fact, some dyes have a dual role and are used as both acid and mordant dyes; for example, C.I. Acid Red 14 is the same dye as C.I. Mordant Blue 79. This example of a dye with dual application also illustrates the very different shades produced without and with mordanting. In this particular case, the unmordanted dye is a bright crimson shade, but when chromed it is dark blue. This great shift in hue upon mordanting occurs with many, although not all, chrome dyes.

Because chrome dyes behave essentially as acid levelling dyes, they are relatively easy to apply to all wool substrate types, with good shade levelness. When chromed, their general fastness properties are very good, so these dyes provide a simple route to achieving high fastness performance without the risk of unlevelness. Prior to the introduction of reactive dyes for wool, chrome dyes provided the highest wet fastness performance of any range on wool. The dyes have an added advantage in that their simple structures mean that they can be cheaply manufactured, leading to very economical recipes. Many dyers also claim that the 'bloom' (a specific glow that arises when some chrome-dyed shades are viewed from an oblique angle) cannot be achieved with other dye classes.

Among the disadvantages of chrome dyes is their relative lack of brightness, although this is not universal, and C.I. Mordant Blue 1, a triphenylmethane structure, produces a bright blue shade when chromed. Because of their economy, good fastness and general dullness of shade, chrome dyes have typically found application in the production of navy, black, dark brown and bordeaux shades on wool. Other disadvantages such as long dyeing times, greater fibre damage than other dye ranges and the discharge of chromium salts in exhausted dyebaths have meant that these dyes are increasingly restricted to navy and black shades, for which their benefits are judged to outweigh the disadvantages [12]. Even for these shades, reactive dyes have been shown to offer the advantages of reduced fibre damage and improved processing efficiency with similar dyeing costs. Nevertheless, the significant use of chrome dyes continues, although it is still declining, and considerable research and development effort has been expended on minimising the potential detrimental aspects of the use of chrome dyes in wool dyeing. An area where they have found use is in dull medium to heavy shades for fibre dyeing in Asia and some other regions, to provide fastness to potting in fabric finishing. Potting is a treatment to stabilise woven wool fabrics by subjecting them to boiling water, while they are rolled on a beam. As manufacturers eliminate this process for many of their fabrics, the need for chrome dyes diminishes.

7.4.1   Chrome Dyeing Processes

Traditional methods of mordant dyeing relied in most cases on the application of the mordant prior to the dye. For chrome dyes, this was termed 'prechrome', 'on-chrome' or 'chrome mordant processing' and was the standard method of application until the afterchrome technique was developed by Nietski in the late 19th century with C.I. Mordant Yellow 1 [13]. Subsequently, in 1900, the metachrome process was introduced by Berlin Aniline Co. [14], in which chrome mordant and dye were applied simultaneously. The afterchrome process is now almost universally used, while the prechrome and metachrome methods have advantages for specific, limited applications. The different processes and their relative merits are described in this section.

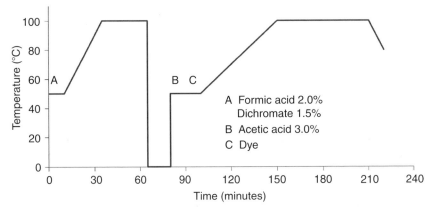

Figure 7.4 *Prechrome dyeing process.*

7.4.1.1 Prechrome Dyeing

In this process, chromium, in the form of potassium or sodium dichromate, is applied to the wool fibre and then a fresh bath is set for application of the dye, as illustrated in Figure 7.4. The levels of auxiliary additions are typical but may be adjusted according to the shade being dyed.

This process gives better coverage of any dye affinity differences within or between wool fibres and, because shading additions are possible, provides the opportunity for closer shade matching than the afterchrome technique (see Section 7.4.1.3). An additional benefit is the greater uptake of chromium on the fibre and hence lower discharge to effluent [15].

7.4.1.2 Metachrome Dyeing

The name of this process was based on the Berlin Aniline Company's name for their chrome dyes. It relies on the simultaneous application of dye and mordant. A relatively high dyebath pH is employed (6.0–7.0), which restricts the use of the process to those dyes that have good neutral substantivity for wool. Additionally, given the reduced dyebath exhaustion that is achieved at such high pH values, the process is generally unsuitable for very heavy shades, such as blacks. There is a risk of dye/chromium complex formation in the dyebath, rather than in the fibre, which can lead to reduced rubbing fastness.

Despite these negative aspects, metachrome dyeing provides a simple one-bath method of application and the higher dyebath pH provides good macro levelness. The process has therefore been widely adopted for piece dyeing with suitable dyes, such as C.I. Mordant Brown 48. A further advantage is that interference from contaminating metals, such as iron and copper, is minimised, as their salts are generally insoluble at pH 6.0–7.0. This is a valuable attribute for those dyehouses that do not have access to treated water and for which metal contamination of process water is an issue. In this case, the polychromatic nature of chrome dyes can be a disadvantage, because it can lead to off-shade dyeings.

It should also be noted that, because of the relatively high application bath pH, the exhaustion of dichromate is lower than for prechrome and afterchrome methods, leading to considerably higher residue discharge to effluent. Of all the issues related to metachrome dyeing, this is the one that has most led to its demise, as authorities around the world tighten restrictions on discharges of chromium.

7.4.1.3   Afterchrome Dyeing

It has already been indicated that, following the development of the afterchrome dyeing method, it became the most widely used technique for the application of chrome dyes to wool. In fact, until the last decade, chrome dyes applied by this method had the greatest sales of any wool dye class. The reasons for this situation are that afterchrome dyeing produces better fastness than either prechrome or metachrome methods and has no restriction on dye selection. Additionally, the application process can be carried out in one bath, which provides savings in time and energy when compared with the prechrome method (Figure 7.5).

The main disadvantage of afterchrome dyeing is that the final shade is not developed until the chroming stage is completed. It is not possible to add further chrome dye to adjust the shade, unless the addition is followed by yet another chroming process. As this would be impracticable, except for very major adjustments, shading additions are often made with milling or 1:2 metal-complex dyes. These products will have sufficiently good wet fastness properties at the low levels employed, but it must be ensured that the dyes used do not interact with dichromate or trivalent chromium ions present in the dyebath.

The dyeing profile illustrated is slightly modified from the traditional method in order to ensure shade levelness and reduce the concentration of residual chromium, although further optimised techniques will be describe later. The addition of dichromate at 70 °C before any formic acid provides an opportunity for the chroming agent to exhaust uniformly. At low pH, dichromate ions have a strong affinity for the wool fibre, and if added at pH 3.5–3.8, they might exhaust too rapidly, leading to unlevelness. By delaying the addition of formic acid, this issue is overcome and shade levelness is more readily achieved, which is particularly relevant in yarn and piece dyeing.

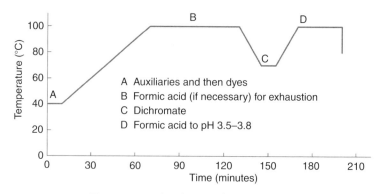

Figure 7.5   *Afterchrome dyeing process.*

The issue of levelness in chroming may also be overcome by chroming in a fresh bath, starting at 40–50 °C. This route has the additional benefit of producing greatly reduced chromium residues.

7.4.2 Theoretical Aspects

The series of papers by Hartley [16–20] and the review by Maasdorp [21] are particularly recommended to any reader seeking additional information on the chroming process applied to wool. Given that the afterchrome process dominates in the mordant dyeing of wool, it is on this method that the present author will concentrate, although much of the information is also relevant to prechrome and metachrome dyeing.

Potassium dichromate was introduced as a mordant for wool around 1840, about a decade after it was first manufactured on a commercial scale [22]. It was recognised that chrome dyes complexed with trivalent chromium, but the substantivity of these salts for wool was much lower than for dichromate and other hexavalent chromium anions [20]. For this reason, trivalent chromium cations were less suitable for conventional exhaust applications and therefore hexavalent dichromate products gained rapid acceptance as the chromium species of choice for chrome dyeing. Either potassium or sodium dichromate may be used, but the latter is hygroscopic and, even though it is less costly than the potassium salt, is therefore only rarely used. The exception is when it is supplied in liquid form, which is suitable for automatic dispensing systems and avoids the need for operatives to handle the product. In this case, hygroscopicity is not an issue. The toxicity of dichromate is well known, and in many countries the handling of dichromate salts is controlled, so automatic dispensing can provide an acceptable safeguard for operatives.

Hartley concluded that, at room temperature, it was dichromate ions that were involved in the binding of chromium to wool. Confirmation that anionic species were involved was given by observations that uptake was decreased by acetylation but increased by esterification of the fibre. The early suggestion that monobasic chromic acid was the principal species involved in chroming [24–27] has now given way to Hartley's theory. Although dichromate ions predominate in normal chrome mordanting baths, other species will also be present; these include chromate (CrO_4^{2-}), bichromate ($HCrO_4^-$) and polychromates; their relative concentrations are shown in Figure 7.6 [16]. Any of these may be involved in the reaction with wool, but Dobozy concluded that, whatever the species present in solution, it was chromate that was the oxidising agent in this reaction [23]. This is supported by the data in Figure 7.6, which show that at the pH values typically used for chroming (3.5–5.5), dichromate predominates in solution.

The uptake of hexavalent chromium by wool (room temperature at pH 4.5) has been shown to follow the Freundlich isotherm and to be inhibited by the presence of potassium chloride. Acetylation, which blocked 55% of amino groups, also reduced chromium uptake, whereas esterification of carboxyl groups increased uptake. This clearly suggests that ionic attraction is an important factor in the exhaustion of dichromate from the chroming bath, and, as with anionic dyes, a decrease in application bath pH will increase the exhaustion level by generating a greater number of positively charged amino groups and reducing the number of negatively charged carboxyl groups. However, the nature of the acidifying agent can also influence the degree of chromium uptake. Dibasic acids, such as sulphuric, have

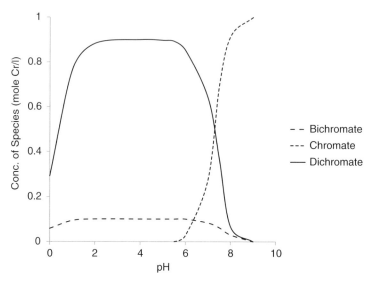

Figure 7.6 *Concentrations of different hexavalent chromium species in solution: effect of pH.*

been found not to be as effective as monobasic acids such as hydrochloric or nitric [28]. The inhibiting effect of sulphate ions, such as those from Glauber's salt, has also been indicated by Benise [29]. It may be concluded that the decreased uptake of chromium anions in the presence of Glauber's salt results from increased competition for protonated amine groups in wool by the divalent sulphate ion. Other polyvalent anions would, presumably, have a similar effect.

The kinetics of absorption of chromium(VI) by dyed and undyed fibres was further studied by Maasdorp [30], who found that, on undyed wool, increasing temperature led to a decrease in uptake of chromium, but the reverse was true for dyed wool. The reduced exhaustion of chromium from the application bath and diffusion into the undyed fibre at higher temperatures was explained by the formation of chromium(III), which crosslinked amino and carboxyl groups, or the absorption of hydrogen ions during the reduction of chromium(VI) (Scheme 7.1), thereby changing the internal pH of the fibre.

It was postulated that the behaviour of dyed wool could be explained in terms of steric hindrance due to the presence of dye on the fibre, higher liquor pH and repulsion of the chromium(VI) anion by the negatively charged dye molecule. More energy would thus be required to promote adsorption of chromium by the wool.

It has been noted that it was trivalent chromium that formed complexes with mordant dyes and that hexavalent chromium on the fibre must therefore first be reduced in order for complex formation to take place. That this reduction occurs may be readily seen if dichromate is applied to undyed wool, such as in the prechrome dyeing process. As

$$Cr_2O_7^{2-} + 14H^+ + 6e^- \longrightarrow 2Cr^{3+} + 7H_2O$$

Scheme 7.1 *Overall process for the reduction of dichromate.*

the dichromate exhausts from the application bath, the wool turns bright yellow. As the temperature increases, the colour slowly turns to grey/green, which is the colour of the chromium(III) ion.

The reduction of chromium(VI) to chromium(III) on wool has been shown to occur readily only at or above 60 °C [31]. This reduction was formerly attributed to the disulphide bonds of cystine via hydrogen sulphide [32,33], but Hartley showed [18] that, at low pH values, solutions of chromium(VI) oxidised disulphide bonds directly at boiling temperature. Other amino acids were shown to be oxidised – for example, tyrosine – but some, such as lysine, were only affected at pH 7 or above. Since these high pH values are not normally encountered in chrome dyeing, it must therefore be concluded that direct reduction of chromium(VI) by cystine is the main route by which chromium(III) is generated in the chrome dyeing process.

Hartley also proposed [34] that the process by which chromium(VI) is converted to chromium(III) is via chromium(IV) and chromium(II), and in a later publication [35] put forward Scheme 7.2. This reaction causes a loss of hydrogen ions from solution, which leads to the observed increase in pH during the wool chroming process. Although this sequence was determined on undyed wool, it must be assumed that a similar process occurs on dyed wool, although the final product will be either a chromium–dye complex or a mixed complex with wool.

Carr *et al.* [36] used electron spin resonance (ESR) and x-ray photoelectron spectroscopy (XPS) to show that chroming of wool leads to little modification of the fibre surface, in terms of oxidation of cystine or chromium content. The conclusion was that chromium is present in the trivalent form in the bulk (core) of the fibre.

The final stage in chrome dyeing, be it prechrome, metachrome or afterchrome, is the formation of a chromium(III)–dye complex. It has long been known for azo dyes [37] that hydroxyl and/or carbonyl groups in the $0'$-position enable complex formation with chromium to take place. It has latterly been recognised that amino groups might also

Chromium(VI)

↓ Oxidation of cystine

Chromium(VI)

↓ Oxidation of cystine,

methionine and tyrosine

Chromium(II)

↓ Wool-CO_2H

Wool-COO Cr(II)

↓ Rapid oxidation by air

Wool-COO Cr(III)

Scheme 7.2   *Process by which dichromate is reduced to chromium(III).*

be involved in complex formation. Typical structures for such complexes are shown in Structures **6** and **7**.

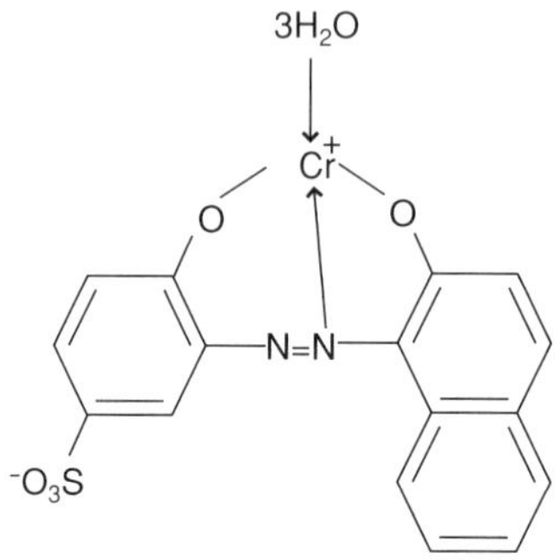

$$3H_2O$$

6

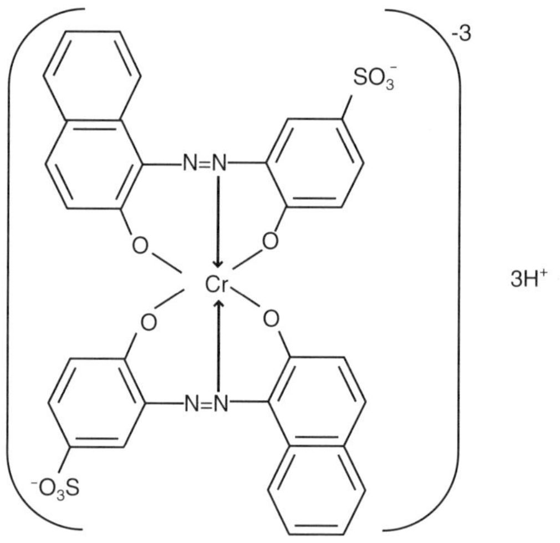

7

A great deal of investigative work has been conducted to determine whether dye–chromium(III) complexes on wool are 1:1 or 1:2 and it has been concluded [38] that the nature of the complex depends on:

- *Solubility of the 1:1 complex:* Low solubility favours formation of the 1:1 structure.
- *Solution pH:* Low pH favours formation of the 1:1 complex, being an internal 'zwitterion' with no net charge, and high pH favours the formation of a 1:2 complex, with an overall negative charge.

Having established the probable structure of the dye–chromium(III) complex, it is necessary to consider the nature of the forces that 'fix' it in the wool fibre. In this case, there are several proposals that are accepted, but none that is universal [21]. A schematic representation for the 1 : 1 complex, forwarded by Dobozy [23], indicates his view that chromium acts as a bridge between polypeptide chains of wool and the dye.

Additional views are that, once the complex is formed on the fibre, water molecules are replaced by coordinate links with oxygen and nitrogen atoms in the fibre's protein chains. Salt links formed by the sulphonic acid groups of the dye and chromium atom with side chains of the protein are also implicated. Hydrogen bonding between nitrogen or hydrogen atoms on the dye and hydrogen atoms on the wool have also been suggested. The high fastness of chrome dyes has been attributed in part to these linkages and to the formation from smaller components of a large molecule inside the fibre. It has been suggested that such large molecules will be difficult to remove from the fibre [24,25,39].

7.4.3   Low-Chrome Dyeing

The one main issue that restricts and is leading to a decline in the use of chrome dyes is the discharge of trivalent and hexavalent chromium from dyeing and rinsing baths. It has long been recognised that high levels of chromium in the environment are unacceptable, and for several decades research has been conducted to minimise contamination from chrome dyeing, even though the potential impact is very much lower than from the metal industry or leather processing. Handling of dichromate by operatives is also an issue and has been addressed by the use of alternative products or improved processes. Additionally, the increasing proliferation of 'eco' claims for textile products has meant that residual labile chromium on wool products is also an issue. Traditional chrome dyeing processes that resulted in significant levels of residual chromium(VI) on the fibre could lead to skin irritation. These are avoided with the techniques described in this section, but even the best methods cannot reduce the levels of extractable (via saliva) levels of chromium(III) sufficiently to meet the requirements of some eco standards, such as Öko-Tex 100.

The information that follows relates primarily to afterchrome dyeing, because it is this process that dominates commercial practice globally. Traditional methods of chrome dyeing used a large excess of dichromate to ensure full and uniform chroming of dyes. However, this approach also led to very high residues of chromium in the exhausted dyebath (typically 200–250 mg l^{-1} total chromium) and in rinse baths. Statutory limits on the discharge of chromium to sewer obviously vary globally, but typical values are:

- hexavalant chromium: 0.0–0.5 mg l^{-1};
- trivalent chromium: 2.0–5.0 mg l^{-1};
- total chromium: 2.0–5.0 mg l^{-1}.

In mixtures of chromium(III) and chromium(VI), the only simple method of determining chromium(III) content is to analyse for chromium(VI) and deduct this value from a measurement of total chromium. For this reason, many authorities set limits for chromium(VI) and total chromium only. The levels listed here are clearly very much lower than can be achieved by conventional chrome dyeing procedures, even after typical dilution with other dyehouse effluent. It has therefore been necessary to develop techniques for applying chrome dyes that can meet these regulations.

The first steps in minimising the discharge of chromium are to reduce the amounts added to the bath. Scheffner and Mosimann [40] recommended a simple formula, based on dye concentration, to calculate the minimum addition of dichromate. It should be noted that the pH of chroming has a significant effect on chromium residues and the range (pH 3.5–3.8) used by Scheffner and Mosimann provides the lowest values for both chromium(III) and chromium(VI). They also worked at 92 °C, and other workers have shown that cooling the exhausted dyebath to 90–80 °C can further increase exhaustion. With this technique it was possible to obtain values in the exhausted chroming bath of typically 0.0–1.0 mg l^{-1} chromium(VI) and 5–10 mg l^{-1} chromium(III). Thomas *et al.* [41] confirmed the effect of pH and demonstrated how different chrome dyes and substrate types (untreated and Hercosett-treated wool) could affect residual chromium levels. Their work also showed that rinsing and alkaline aftertreatment baths contributed significantly to discharged chromium levels. In fact, the residues from alkaline aftertreatment were similar to those from the original chroming bath.

It was found that reducing dichromate additions to levels that were close to those stoichiometrically required for complex formation led to unlevel or incomplete chroming of the applied dyes. This issue may be overcome by adding a salt, such as sodium sulphate, to the bath during the chroming cycle, which has the effect of liberating ionically bound dichromate on the fibre to provide greater uniformity of uptake [42]. This so-called 'Glauber's salt method' was developed by workers at Bayer and relied on GCr factors that the company provided for each of their chrome dyes. The GCr factors were the minimum proportions of dichromate required for each dye that enabled the total quantity of dichromate to be calculated for each recipe. The system worked well but was applicable only to Bayer (now DyStar) dyes for which the factors were available.

Addition of reducing and or complexing agents to the chroming bath was found to reduce residual chromium levels. Benisek [43] studied a wide range of products and found lactic acid to be the most effective. Spinacci and Gaccio [44] discovered that the addition of sodium thiosulphate increased the rate of chroming sufficiently to allow the process to be conducted at 80 °C with specific dyes. Further work [42] suggested that dyeing and chroming at 90 °C was possible for all dyes and provided significant benefits in terms of reduced fibre damage when compared with conventional dyeing and chroming at the boil. Sandoz (now Clariant) introduced a process that employed a specific reducing agent, Lyocol CR [45], which effectively reduced chromium concentrations in effluent, but the product is no longer available, as the company ceased supplying chrome dyes.

Xin and Pailthorpe [46] showed that using sulphamic acid to replace, either wholly or partly, formic acid in the chroming bath would significantly reduce chromium levels by automatically increasing the bath pH as the sulphamic acid decomposed to ammonium sulphate.

Maximising dyebath exhaustion was also considered to assist in the reduction of chromium residues, and this assumption was supported by marked reductions in chromium(VI) and total chromium residues if chroming was conducted in a fresh bath. The effect was thought to be related to a reduction of chromed, unexhausted dye in the bath [47]. However, this view was shown to be inaccurate, and soluble proteins that reduce chromium(VI) in the dyebath and possibly bind the resultant chromium(III) are a more likely source of residual chromium when using optimised dyeing processes [48]. This observation corroborates the work by Thomas *et al.* [41], because alkaline aftertreatment is known to remove a significant proportion of soluble proteins.

Alternative mordants were evaluated in conjunction with dichromate [49]. Rare earth salts (neodymium and cerium chlorides) were applied prior to dichromate and it was suggested that the dye–rare earth complex was slowly converted to the chromium complex, which allowed very low additions of dichromate to be used. The system led to very low residual concentrations of chromium.

To avoid the exposure of operatives to chromium(VI) and its presence in effluent, workers have looked at the possibilities of chroming with chromium(III) salts. Lewis and Yan [50] employed α-hydroxycarboxylic acid ligands to form anionic complexes with chromium(III), which were substantive to wool fibres and satisfactorily chromed a range of mordant dyes. Xing and Pailthorpe [51] similarly used a ligand that was not identified but was claimed to overcome some of the issues related to earlier work with chromium(III). The process was successfully adopted by commercial cashmere manufacturers in China and, because the fibre was not oxidised, as when dichrome was used as the mordant, measurable reductions in fibre damage were noted.

However, although this method virtually eliminated chromium(VI) residues, the level of total chromium was unaffected. Additionally, the use of chromium(III) as a mordant has a disadvantage when applying C.I. Mordant Black 9, because this dye requires oxidation to achieve the correct shade. Dichromate can act as the oxidising agent, whereas chromium(III) cannot. This fact may explain why addition of a green dye was necessary to achieve the correct shade when applying C.I. Mordant Black 9 by the new method.

The practical dyer must make a choice from the many available improved chrome dyeing methods, and this will be based on the requirements of the local regulatory authorities and the dyer's customers. If, as in some areas, discharge of chromium in effluent is prohibited completely, the dyer must either treat wastewater on site or use alternative dyes. In those areas where discharge is restricted, dyers must select an application process that enables them to meet the requirements. One of the optimised processes based on dichromate should provide the lowest chromium residues in the dyebath, but consideration must also be given to chromium that arises from rinsing and any aftertreatment processes. Fibre damage may also be a consideration in selecting a dyeing process, and details of low-temperature methods are provided later in this chapter. If handling of dichromate by operatives is an issue, the use of a liquid version that is dipensed automatically would be acceptable, or else one of the processes based on chromium(III) salts.

As noted previously, environmental pressure has led to a great deal of research and development effort being expended to find chrome dyeing processes that have less impact on our environment. The information presented in this section represents just a selection of the developments that are most relevant to practical chrome dyeing, but even the best of these will not allow chrome-dyed wool to meet some of the retail standards that have been introduced for textiles. For example, the use of chrome dyes on organic wool products is specifically prohibited; the human ecology standard Öko-Tex 100 further effectively prohibits their use by setting maximum levels of extractable chromium which cannot normally be achieved with chrome dyes. Retailers and brand owners too are increasingly imposing restrictions on the use of chrome dyes on their products, so that the future market for these dyes will probably continue to decline. However, the ease of application, economy of use and very high wet fastness are still attractive to some dyers. For some specific applications, such as woven piece dyeing to machine washable fastness, there are no acceptable alternatives yet available for heavy shades except for black and navy. Reactive dyes provide acceptable fastness, but the overall uniformity of shade is not acceptable to some customers.

7.5   Specific Dyeing Methods

Standard application methods for acid and mordant dyes were described earlier in this chapter, but there have been developments in dyeing techniques that are relevant to these and other wool dye classes and which can provide benefits for the wool dyer and/or in subsequent processing. The many investigations into pretreatments with a range of physical and chemical processes that are claimed to improve dyeing performance are not fundamentally the subject of this chapter, but it should be noted that their benefits, in terms of increased dyeing rate, can provide significant advantages in the wool-dyeing process. Some examples of recent publications are given in the list of references [52–57].

However, it should be noted for practical dyeing that these developments must be considered in terms of the added water, energy and time costs involved in carrying them out, which in some instances may outweigh any benefit to the dyeing process. Additionally, for milling dye application, an increase in the rate of dyeing may be detrimental with respect to levelness. A critical factor in the application of such dyes that have high fibre affinity is the control (retardation) of dye uptake to ensure uniformity of shade, and pretreatments that increase the rate of dyeing will potentially lead to unlevelness of shade or the need for higher levels of retarding agents. Such comments are obviously not relevant to the application of those dyes that can be levelled by extended dyeing at the boil, such as level-dyeing acid and many chrome dyes.

These pretreatments and other modified dyeing processes may however be employed to reduce the temperature of dyeing with acid, mordant and other dyes. If similar rates of dyebath exhaustion and fastness performance can be achieved at temperatures below a conventional boil (95–100 °C), significant reductions in fibre damage may be achieved. There may be similar benefits to be gained from adopting the pretreatments or modified processes in order to reduce dyeing times at the boil. In this case, both reduced fibre damage and improved production efficiency may be achieved.

The role of low-temperature dyeing auxiliaries in the application of acid dyes was covered in Chapter 3, but for chrome dyes an additional technique for low-temperature chroming is also needed. Figure 7.7 [58] illustrates one that is applicable to all chrome dyes, gives equivalent fastness to dyeings at the boil and reduces both wool damage and the dyeing time. Reduced fibre damage is beneficial when dyeing wool in all substrate forms, but for

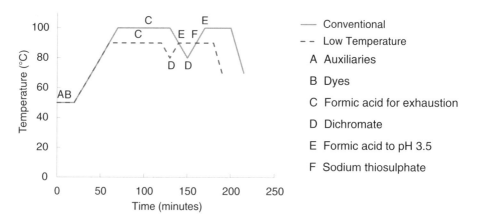

Figure 7.7   *Example of a low-temperature dyeing process.*

fibre dyeing it can provide significant efficiency advantages, and hence cost savings, in subsequent yarn-forming processes.

Researchers in Spain have developed what they term 'integration' dyeing [59], which provides a method of applying milling dyes with high fibre affinity in a uniform manner. The process relies on continuous dosing of dyes and/or auxiliaries, which obviously necessitates the use of relevant equipment to provide and control the dosing function. This method could be used for difficult substrate/dye class combinations (the development work was conducted on yarn packages) and is not limited to acid dyes.

References

[1] J R Easton, *Colour in Dyehouse Effluent* (Bradford, UK: Society of Dyers and Colourists, 1995).

[2] European Parliament Directive 2002/61/EC (July 2002).

[3] J A Bone, J Shore and J Park, *Journal of the Society of Dyers and Colourists*, **104** (1988) 12.

[4] K Parton, in *Wool: Science and Technology*, ed. W S Simpson and G H Crawshaw (The Textile Institute, 2002), 237.

[5] T Vickerstaff, *The Physical Chemistry of Dyeing*, 2nd Edn (London and Edinburgh, UK: Oliver & Boyd, 1954), 348.

[6] J Frauenknecht, P C Hextall and A Welham, *Textilveredlung*, **21** (1986) 331.

[7] H Flensberg, W Mosimann, H Salathe, *American Dyestuff Reporter.*, **73** (1984) 30.

[8] B Glover and H Pierce, *Journal of the Society of Dyers and Colourists*, **109** (1993) 5.

[9] D J Hill, *Review of Progress in Coloration and Related Topics*, **27** (1997) 18.

[10] G Dalby, *Natural Dyes, Fast or Fugitive* (Ashill Publications, 1989).

[11] C L Bird, *The Theory and Practice of Wool Dyeing*, 3rd Edn (Bradford, UK: Society of Dyers and Colourists, 1963).

[12] P Duffield, M Rushforth, K Lordan, A Ng, F Gruener and Y C Kim, *International Dyer*, (2010).

[13] R Nietski, *Journal of the Society of Dyers and Colourists*, **5** (1889) 175.

[14] C B Stevens, F M Rowe and J B Speakman, *Journal of the Society of Dyers and Colourists*, **59** (1943) 165.

[15] A P B Maasdorp, PhD Thesis, University of Port Elizabeth (1983).

[16] F R Hartley, *Australian Journal of Chemistry*, **21** (1968) 2277.

[17] F R Hartley, *Australian Journal of Chemistry*, **21** (1968) 2723.

[18] F R Hatley, *Australian Journal of Chemistry*, **22** (1969) 129.

[19] F R Hartley, *Australian Journal of Chemistry*, **22** (1969) 209.

[20] F R Hartley, *Australian Journal of Chemistry*, **23** (1970) 275.

[21] A P B Maasdorp, SAWTRI Special Publication, WOL 61 (November 1983).

[22] P Kay & E Bastow, *Journal of the Society of Dyers and Colourists*, **3** (1887) 118.

[23] O K Dobozy, *American Dyestuff Reporter*, **62** (1973) 36.

[24] J F Gaunt, *Journal of the Society of Dyers and Colourists*, **65** (1949) 429.

[25] P Fink, *Textil Rundschau*, **8** (1953) 279.

[26] N V Perrymann, *Proceedings of the Wool Textile Research Conference*, **E** (1946) 329.

[27] H F Bichsel, *Textil Rundschau*, **10** (1955) 471.

[28] J J Hammel and W M Gardner, *Journal of the Society of Chemical Industry*, **14** (1895) 452.

[29] L Benisek, *Dyer*, **156** (1976) 600.

[30] A P B Maasdorp, SAWTRI Technical Report, **538** (1983).

[31] J A Maclaren and B Milligan, *Wool Science, The Chemical Reactivity of the Wool Fiber* (Marrickville, NSW, Australia: Science Press, 1981).

[32] P W Carlene, F M Rowe and J B Speakman, *Journal of the Society of Dyers and Colourists*, **62** (1946) 329.

[33] E Race, F M Rowe, J B Speakman and T Vickerstaff, *Journal of the Society of Dyers and Colourists*, **54** (1938) 141.

[34] F R Hartley, *Wool Science Review*, **37** (1969) 54.

[35] F R Hartley, *Journal of the Society of Dyers and Colourists*, **86** (1970) 209.

[36] C M Carr, J C Evans and M W Roberts, *Textile Research Journal*, **57** (1987) 109.

[37] R Nietzki, *Farben-Zeitung*, **1**, **8**, (1989/1990) 205.

[38] W Beal, *Cirtel Conference, Paris*, **III** (1965) 223.

[39] E Race, F M Rowe and J B Speakman, *Journal of the Society of Dyers and Colourists*, **62** (1946) 372.

[40] K Scaffner and W Mosimann, *Textilveredlung*, **24** (1979) 12.

[41] H Thomas, R Kaufmann, R Peters, H Höker, M Lipp, J Goshnick and H-J Ache, *Journal of the Society of Dyers and Colourists*, **108** (1992) 186.

[42] P A Duffieldand K H Hoppen, *Melliand Textilber*, **68** (1987) 195.

[43] L Benisek, *Dyer*, **156** (1976) 600.

[44] P Spinacci and N C Gaccio, *12th Congress of IFATTC, Budapest*, **6** (1981) 10.

[45] A C Welham, *Journal of the Society of Dyers and Colourists*, **102** (1986) 126.

[46] J Xing and M T Pailthorpe, *Journal of the Society of Dyers and Colourists*, **108** (1992) 17.

[47] P A Duffield, J M Wimbush and P F A Demot, *Proceedings of the 8th International Textile Research Conference*, **IV** (1990) 97.

[48] D G King, *Journal of the Society of Dyers and Colourists*, **125** (2009) 161.

[49] J Xing and M T Pailthorpe, *Journal of the Society of Dyers and Colourists*, **108** (1992) 265.

[50] D M Lewis and G Yan, *Journal of the Society of Dyers and Colourists*, **111** (1995) 317.

[51] J Xing and M T Pailthorpe, *Journal of the Society of Dyers and Colourists*, **116** (2000) 91.

[52] A Negri, H J Cornell and D Rivett, *Journal of the Society of Dyers and Colourists*, **109** (1993) 297.

[53] T Wakida, M Lee, Y Sato and Y Yanai, *Journal of the Society of Dyers and Colourists*, **109** (1993) 393.

[54] T Wakida, M Lee, Y Sato, S Ogasawara, Y Ge and S Niu, *Journal of the Society of Dyers and Colourists*, **112** (1996) 233.

[55] J Shao, C J Hawkyard and C Carr, *Journal of the Society of Dyers and Colourists*, **113** (1997) 127.

[56] C W Kan, K Chan, C W M Yuen and M H Miao, *Journal of the Society of Dyers and Colourists*, **114** (1998) 61.

[57] A Riva, I Algaba and R Prieto, *Coloration Technology*, **118** (2002) 59.

[58] P A Duffield and R R D Holt, *Textilveredlung*, **24** (1989) 40.

[59] J Cegarra, F Enrich, M Pepio and P Puente, *Journal of the Society of Dyers and Colourists*, **115** (1999) 92.

8

Dyeing Wool with Metal-complex Dyes

Stephen M. Burkinshaw
School of Design, University of Leeds, UK

8.1   Introduction

As a means of overcoming the inherent disadvantages of using *mordant* dyes on wool (colour matching can be difficult to achieve and fibre tendering can occur during the prolonged, two-stage, pre- and after-chrome dyeing processes), dye makers found that by combining the dye and mordant (metal salt) together prior to dyeing, the resulting *metal complex* dyes could be applied to wool using a single-stage application process similar to that employed for non-metallised *acid dyes*. In metal-complex dyes, one metal atom, commonly chromium, is complexed with either:

(1) one, typically monoazo, dye molecule (1 : 1 metal-complex dye) such as C.I. Acid Red 183 (**1**); or

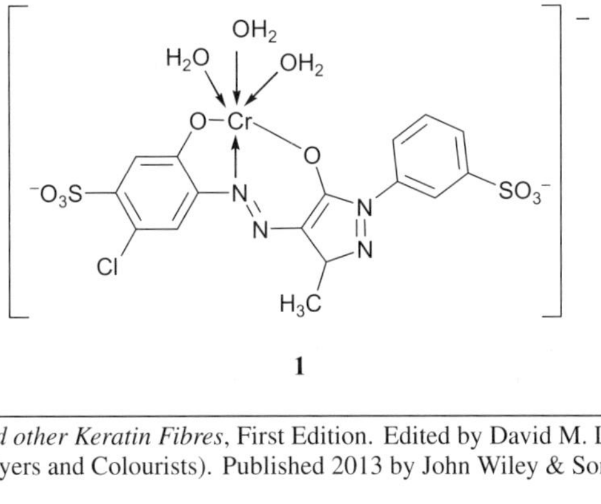

1

The Coloration of Wool and other Keratin Fibres, First Edition. Edited by David M. Lewis and John A. Rippon.
© 2013 SDC (Society of Dyers and Colourists). Published 2013 by John Wiley & Sons, Ltd.

(2) two, typically monoazo, dye molecules (1 : 2 metal-complex dye) such as C.I. Acid Violet 90 (**2**).

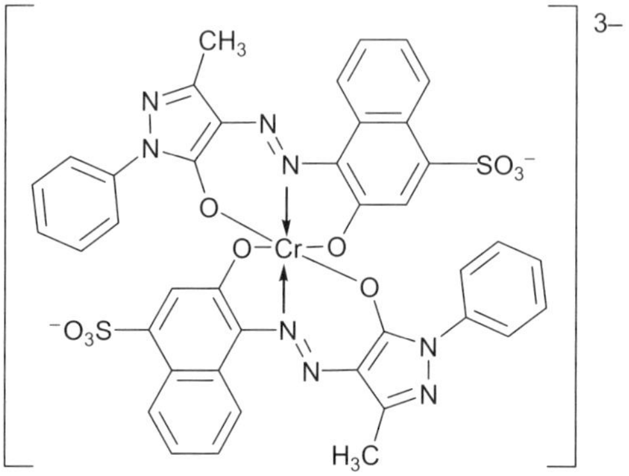

2

In each case, the dye contains nucleophilic groups, such as -OH, -COOH or –NH₂, which can coordinate with the metal atom.

In the *Colour Index* [1], metal complex dyes are included in the *Acid Dye* application classification (hence the dyes are also known as *pre-metallised acid dyes*) and provide shades on wool that are duller than those given by non-metallised acid dyes but which are slightly brighter than those obtained using mordant dyes. 1 : 1 and 1 : 2 metal-complex dyes resemble levelling and milling non-metallised acid dyes respectively, as regards the general application conditions used, and intermediate and intermediate/milling acid dyes respectively, in terms of their wet fastness properties on wool. Metal-complex dyes nowadays enjoy applications other than wool dyeing, including, for example, high-density memory storage (CD-R and DVD-R) [2,3], nonlinear optics [4], ink jet dyes [5,6] and thermal dye transfer printing [7].

While, in this chapter, commercial names of dyes and auxiliaries are mentioned and details of application procedures are given for commercial dyes, this is not intended to imply superiority of product but rather to serve as a guide. It is considered that much of the practical value of the chapter would be lost without the inclusion of such specific application details. There has been no intention to stress one particular manufacturer's products and methods to the detriment of another's.

8.2   Dye Structure

Metal-complex dyes for wool typically incorporate tridentate, bicyclically metallisable monoazo dye ligands, notably *o,o′*-dihydroxyazo (**3**), *o,o′*-carboxyhydroxyazo (**4**), *o,o′*-hydroxyaminoazo (**5**), *o,o′*-dihydroxyazomethine (**6**) and *o,o′*-carboxyhydroxyazomethine (**7**) dyes.

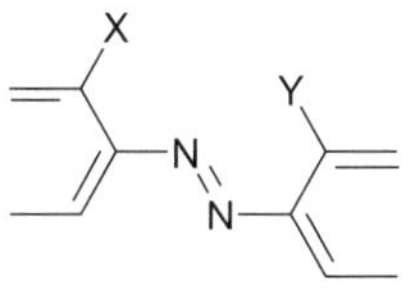

3 X=OH; Y=OH
4 X=COOH; Y=OH
5 X=OH; Y=NH$_2$

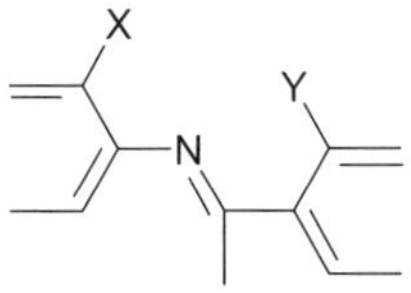

6 X=OH; Y=OH
7 X=COOH; Y=OH

Coordination of these trifunctional ligands with a metal ion, in which a nitrogen atom of the azo or azomethine group participates, involves the loss of a proton from each of the two *o*-substituted hydroxyl, amino or carboxyl groups of the azo or azomethine dye. This results in a structure comprising one five- and one six-membered ring (**8**), in the cases of dyes derived from Structures **3**, **5** and **6**, or two six-membered rings (**9**) in the case of a dye based on Structures **4** and **7**. More detailed accounts are available [6,8–11].

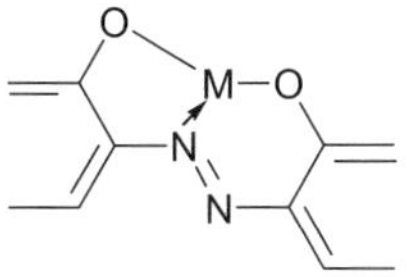

8 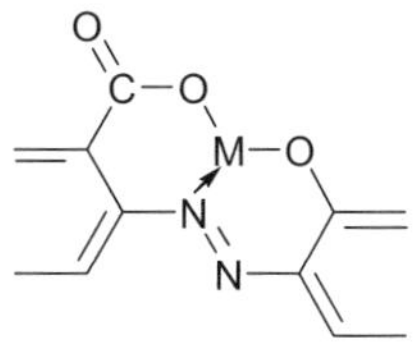   9

The metal ion used is most commonly trivalent and hexa-coordinate and is usually either Cr^{3+} (predominantly) or Co^{3+}, although Fe^{2+} has been recently explored [12–15] on environmental grounds. The coordination sphere of the metal ion is completed either by an additional three monodentate ligands, such as water (**10**, derived from **8**), or by a single tridentate dye ligand, such as an *o,o*-dihydroxyazo dye (**11**, also derived from **8**). **10** represents the fundamental structure of a 1 : 1 metal-complex dye and **11** that of a symmetrical 1 : 2 metal-complex dye; if in **11** the additional tridentate ligand were, for example, an *o,o'*-hydroxyaminoazo dye then an unsymmetrical 1 : 2 metal-complex acid dye would arise (**12**).

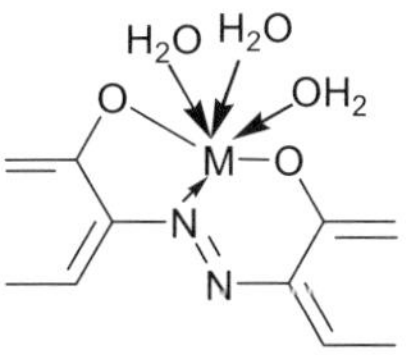

10 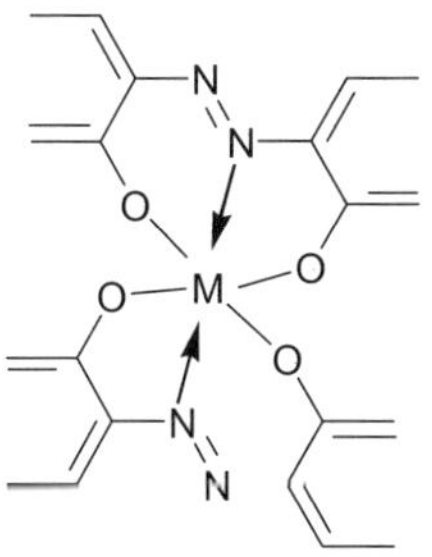   11 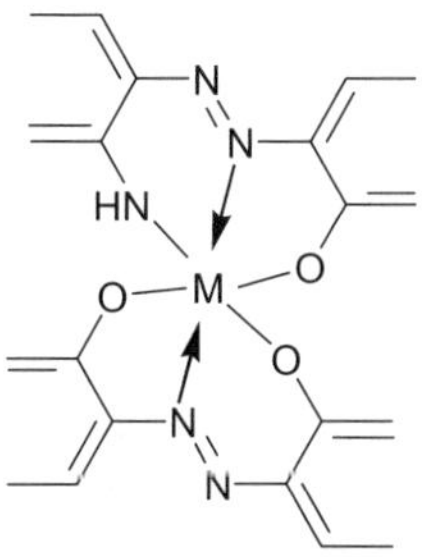   12

Since coordination of the dye ligand involves the loss of two protons from the substituents located *o,o'* to the azo or azomethine group, 1 : 1 complexes, such as **11**, carry a single positive charge, and 1 : 2 complexes, such as **12**, a single negative charge. However, the overall charge carried by a 1 : 1 metal-complex acid dye is determined by the nature of the monodentate ligands used (e.g. H_2O, NH_3) and both the nature and the number of ionic groups (e.g. $-SO_3H$, $-COOH$) present in the dye ligand. For 1 : 2 metal-complex dyes, the nature and number of ionic groups in the two contributing dye ligands determine the overall charge.

8.2.1 Electronic Structure

Although many transition-metal ions form complexes with metallisable azo dyes, those of trivalent cobalt and in particular trivalent chromium are the most used commercially, owing to their marked propensity to resist demetallisation during application or subsequent processing and/or use. The high stability of chromium(III) and cobalt(III) dye complexes is attributable to the particular electronic configurations of these trivalent ions; that of chromium is shown in Figure 8.1.

Both Cr^{3+} and Co^{3+} (and other transition-metal cations) readily form coordination complexes between a central metal cation and negative or neutral ligands containing lone pairs of electrons, because their vacant *d* orbitals can accommodate electron pairs donated by the ligands. The following discussion concerns the *3d* orbitals, which are involved in the complexes of trivalent chromium and cobalt, although the higher (the *4* and *5*) *d* orbitals have similar characteristics.

The five *d* orbitals differ, and there are two sets (Figure 8.2): the t_{2g} (or *dε*) orbitals lie between the *x*-, *y*- and *z*-axes and are denoted d_{xy}, d_{yz} and d_{zx} orbitals respectively, while the e_g (or *dγ*) (denoted $d_{x^2-y^2}$ and d_{z^2}) orbitals lie along the axes. In an isolated atom, the five *d* orbitals are degenerate (i.e. have the same energy). As discussed later, both trivalent chromium and cobalt form complexes (Figure 8.3) in which the metal cation lies at the centre of an octahedron and the ligands are at the six corners. Two theories have been proposed to describe metal–ligand bonding:

(1) *Electrostatic theory (or crystal field theory):* Bonding is described purely in terms of electrostatic attraction operating between the metal cation and the negative charge – or

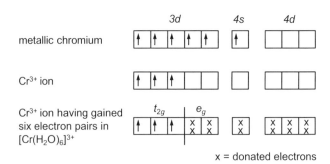

x = donated electrons

Figure 8.1 *Electronic configuration of chromium (the inner electron shells are completely filled).*

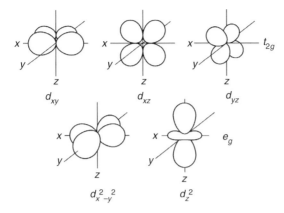

Figure 8.2 *The d orbitals of a transition metal cation.*

negative pole of the dipole (or induced dipole) – of the ligand; the crystal field splitting energy arising from *d* orbital splitting augments this electrostatic interaction (this is discussed later).

(2) *Molecular orbital theory:* Bonding is considered to be purely covalent, arising from the provision, by the ligand, of lone pairs of electrons that occupy bonding molecular orbitals in the complex.

The two theories are essentially similar, in that both consider the *3d* electrons to be accommodated in an upper set of two and a lower set of three orbitals that are separated by an energy difference, ΔE. In both cases, the lower set comprises the d_{xy}, d_{yz} and d_{zx} atomic orbitals; the upper set in the electrostatic approach comprises the $d_{x^2-y^2}$ and d_z^2 atomic orbitals, while in the molecular orbital theory, antibonding molecular orbitals arising from the combination of the $d_{x^2-y^2}$ and d_z^2 atomic orbitals with ligand orbitals make up the upper set. The combination of both these theories is known as the *ligand field theory*. The approach used in the following brief discussion of the principles of metal-complex formation combines the concepts of crystal field splitting of energy levels and the formation of sigma (σ) bonds. The reader is directed elsewhere for more detailed accounts of this large topic [16–20].

A ligand coordinates by means of a lone pair of electrons and has a negative charge (or the negative pole of a dipole) directed towards the central metal cation. As the negatively

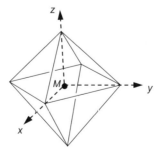

Figure 8.3 *The structure of an octahedral metal complex.*

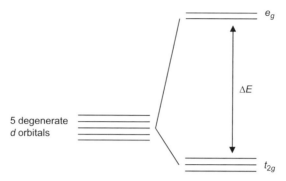

Figure 8.4   *Orbital splitting in a metal complex.*

charged ligand approaches the metal ion along the x-, y- and z-axes, the x-, y- and z-orientated d orbitals (i.e. the $d_{x^2-y^2}$ and d_{z^2} orbitals) of the ion are in a higher negative field than those that are non-axially oriented (i.e. the d_{xy}, d_{yz} and d_{zx} orbitals). Hence, the e_g orbitals are of higher energy than are the t_{2g} orbitals, and the degeneracy of the five $3d$ orbitals is split into two groups of different energy (Figure 8.4).

In the case of the Cr^{3+} cation, which has the electronic configuration shown in Figure 8.1, the three d electrons in the free ion have equal probability of occupying any three of the five degenerate $3d$ orbitals. However, in the presence of, for example, water ligands, the five d orbitals are not all equivalent, since the e_g orbitals are in regions that are closer to the ligands than are the t_{2g} orbitals. Thus, the three electrons of the cation will avoid entering the e_g orbitals and, therefore, occupy the lowest-energy, t_{2g} orbitals; the ligands thus utilise the e_g orbitals as shown in Figure 8.1. In the case of the $[Cr(H_2O)_6]^{3+}$ complex, the orbitals are split as shown in Figure 8.4, where ΔE (expressed as wave number or electron volts) is the energy difference between the e_g and t_{2g} orbitals. The magnitude of the splitting (ΔE) of the degenerate d orbitals depends on the size of the (electrical) ligand field, which also determines which d orbitals are occupied by the electrons of the metal cation. In occupying the t_{2g} orbitals, the three d electrons of the metal ion gain energy, termed *crystal field stabilisation energy (CFSE)*, relative to that which they would have possessed if splitting had not occurred. The electrons have parallel spins and the t_{2g} orbitals contain nonbonding electrons, while the e_g orbitals are used for bonding.

The Co^{3+} ion (Figure 8.5), in contrast to that of Cr^{3+}, has six d electrons; two alternative electronic configurations are possible, depending on the strength of the ligand field:

(1) *High-spin or weak-field configuration:* In a weak ligand field, four of the six d electrons of the Co^{3+} ion enter the t_{2g} level with one paired spin, while the two remaining ones

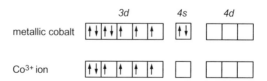

Figure 8.5   *Electronic configuration of cobalt (the inner electron shells are completely filled).*

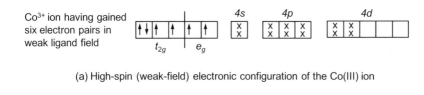

(a) High-spin (weak-field) electronic configuration of the Co(III) ion

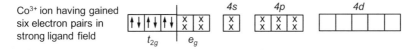

(b) Low-spin (strong-field) electronic configuration of the Co(III) ion

Figure 8.6   *Electronic configuration of the Co(III) ion.*

enter the e_g level. The lone pairs in the ligand occupy the *4s*, *4p* and two *4d* orbitals (Figure 8.6a).

(2) *Low-spin or strong-field configuration:* In a strong ligand field, all six *d* electrons of the Co^{3+} enter the t_{2g} level with paired spins; the t_{2g} level is completely filled and the ligand lone pairs occupy the e_g, *4s* and *4p* orbitals. The e_g orbitals are therefore used for bonding and the t_{2g} orbitals contain nonbonding electrons (Figure 8.6b).

The CFSE for the high-spin configuration is lower than that for the configuration of low-spin, and so adoption of the latter configuration represents a more stable arrangement. The inherent stability of half-filled Cr(III) and filled Co(III) *d* orbitals in octahedral complexes is attributable to the general symmetry of the electron clouds and the high exchange energy of such configurations, which relates to *Hund's Rule of Maximum Multiplicity*, that electrons which enter degenerate orbitals have, as far as possible, parallel spins; the low-spin configuration for Co(III) therefore gives rise to the most stable complexes.

Although metal-complex dyes containing metals other than chromium and cobalt have been the subject of research [9,12–15,21], relatively few are used on wool. The ligand contributes to dye–metal complex stability insofar as dye ligands that form five- and six-membered ring systems give more stable complexes than those that form seven-membered rings [22]; furthermore, in general, stability increases with increasing basicity of the ligand [21,22]. While tridentate dye ligands are used extensively in the preparation of dyes for wool, both bi- and tetradentate dye ligands have been the subject of research [8,9,21].

8.2.2   Colour and Light Fastness

Metallisation usually results in a bathochromic shift of the λ_{max} of the metallisable azo dye, this often being manifest in the dyeing of wool with mordant dyes; for example, C.I. Mordant Blue 1 when applied to wool has a red hue, but when the dyeing is afterchromed, a blue hue is secured. Bathochromicity is attributable [22,23] to perturbation of the π-electron density distribution of the dye chromogen. This perturbation has also been suggested to be responsible for the generally high light fastness properties of metallised azo dyes [22],

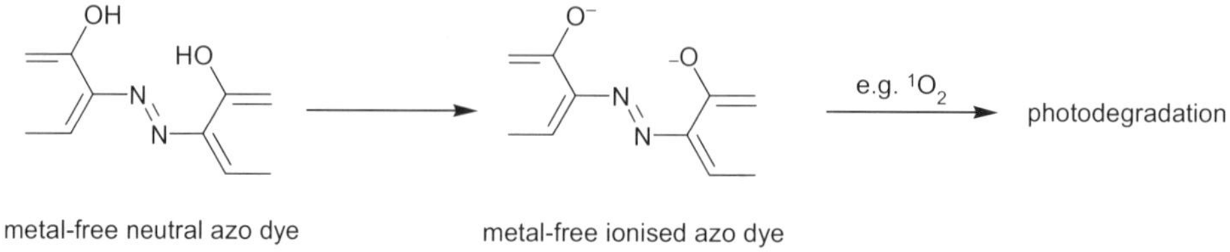

Scheme 8.1   *Attack at an azo group by electrophiles such as $^1O_2$.*

while the high light fastness of metal-complex dyes on wool has also been ascribed to dye aggregation within the fibre [24] and to the Cr or Co atoms shortening the lifetime of the triplet states of the dyes [25]. The mechanisms which underlay the photo fading of dyed textiles are complex, and various electrophilic species can be produced upon irradiation (e.g. singlet oxygen $^1O_2$, superoxide radicals $O_2^{\bullet-}$, etc.) which cause photodegradation [26,27]. In the case of azo dyes, attack occurs preferentially at or around the azo (or hydrazone) group of the *anionised* form of the dye, rather than the *neutral* dye [6,27] (Scheme 8.1), as the ionised dye is more electron rich than the neutral dye and is therefore more susceptible to attack by electrophiles such as $^1O_2$. The high light fastness of metal-complex dyes can be attributed [6] not only to their inability to form such anions but also to the presence of the metal cation, which decreases the electron density around the azo group and physically hinders the approach of electrophiles, as can be seen from structure **8**.

The characteristic dull shades produced by Cr(III) and Co(III) medially metallised azo dyes accrue from broadening of the visible absorption spectrum of the deprotonated azo dye, owing to transitions of the degenerate *d* orbital electrons in the metal ion [21]. For a series of 1 : 2 metal-complex dyes, bathochromicity was found to increase in the order $Ni^{2+} < Co^{3+} < Cr^{3+}$, whereas a series of Ni^{2+}, Co^{3+} and Cr^{3+} 1 : 1 metal-complex dyes exhibited similar spectral curves to each other, but absorbed at shorter wavelengths than the corresponding 1 : 2 metal-complex dyes [6]. The observation that the 1 : 1 metal-complex dyes displayed narrower absorption curves than both the 1 : 2 dyes and the metal-free precursor dyes was attributed [6] to metallisation, having increased the rigidity of the dye, which decreased the number of vibrational degrees of freedom and thereby reduced the number of vibronic transitions. That metal-complex dyes can comprise several isomeric forms in equilibrium is considered to be a major cause of their relatively dull shades, since many isomers display slightly different absorption spectra, leading to broader spectra [5,11,28].

8.2.3   Stereochemistry and Isomerism

The 1 : 2 complexes of dyes with Cr(III) or Co(III) are octahedral, with structural isomerism arising from different sources [29]:

(1) *The ligands can occupy different spatial arrangements around the central metal ion:*
    A total of 11 possible isomers are realised. The tridentate ligands can be arranged perpendicular to each other (*Drew–Pfitzner* or *meridial* ('mer') type; Figure 8.7a) [30],

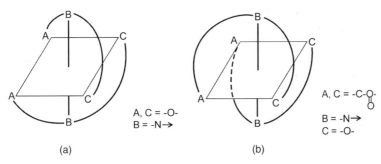

Figure 8.7 *Geometrical isomers of 1 : 2 dyes: (a) meridial ('mer') type; (b) facial ('fac') type.*

resulting in two enantiomers, which are adopted [31] when the two dye ligands (e.g. *o,o'*-dihydroxyazo) form a five- and a six-membered chelate ring (**6**). Alternatively, four enantiomeric pairs together with one centrally symmetric isomer result when the dye ligands are arranged parallel to each other in the *Pfeiffer–Schetty* or *facial* ('fac') orientation (Figure 8.7b) [32], which is adopted [31] in the cases of those dyes (e.g. *o*-carboxy,*o'*-hydroxyazo) that form two six-membered chelate rings. The mer and fac configurations possess different coloration properties insofar as Cr(III)–dye complexes with the mer configuration are deeper in hue and more water soluble than their fac isomeric counterparts; in addition, they display higher wet fastness but marginally lower light fastness properties on wool.

(2) *Either of the two N atoms of the azo group (N_α or N_β) of a ligand can coordinate with the metal ion:* In the case of unsymmetrical 1 : 1 complexes, two isomers are realised [33]; in a symmetrical 1 : 2 metal complex, three isomers (N_α, N'_α; $N_\alpha N_\beta$, $N_\beta N'_\beta$) are formed [11,34,35]; with mixed 1 : 2 complexes containing asymmetrical compounds, four isomers can be found [11].

(3) *The coordinating N atom in the hydrazone form of the azo dye is formally sp^3 hybridised, giving rise to different conformational possibilities:* In the case of five/six-membered ring systems, a total of six isomers and their mirror images can arise if azo-hydrazone tautomerism occurs (three enantiomeric conformers if both dye ligands exist in the hydrazone form, two conformers if one ligand coordinates in the hydrazone and the other in the azo form, and one pair of enatiomers if both ligands are in the azo form) [5,11,28].

As already mentioned, the fact that metal-complex dyes can comprise several isomeric forms in equilibrium results in the typically dull shades of the dyes, as many isomers display slightly different absorption spectra, leading to broad spectra [5,11,28]. Comparatively little research has attended the stereochemistry of 1 : 1 Cr(III)–dye complexes; it has been suggested that such octahedral dyes assume either a fac [36] or a mer [37] configuration.

8.2.4 1 : 1 Metal-Complex Dyes

Although BASF described the preparation of water-soluble 1 : 1 chromium complexes of sulphonated hydroxyanthraquinone dyes [38] and chromium complexes of azo mordant dyes [39] in 1912, the *Palatin Fast* (BASF) and *Neolan* (Geigy) ranges of 1 : 1 metal-complex acid dyes were not commercially introduced until 1919 [40], following the development of a suitable application method by Ciba in 1915 [41,42] and improvements in dye synthesis [9,21,43]. The metal ion in the almost exclusively Cr(III) complexes of (primarily) *o,o'*-dihydroxyazo and *o,o'*-hydroxyaminoazo dyes, as represented by C.I. Acid Red 183 (**1**) and C.I. Acid Green 12 (**13**), is coordinated with the single monoazo dye ligand and three other ligands, usually water, although the *Neolan P* (Huntsman) range of dyes, introduced in 1988, utilise colourless hexafluorosilicate ligands [44].

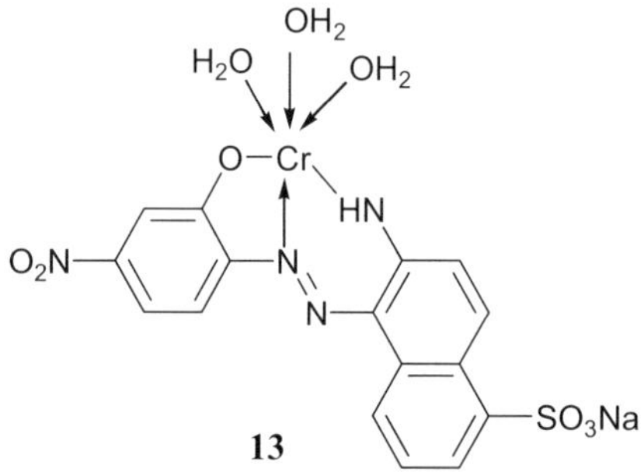

13

Solubility in water is typically conferred by the presence of one or more sulphonic acid group, although C.I. Acid Orange 76 (**14**) contains aminosulphone ($-SO_2NH_2$) groups as solubilising aids [1]. As discussed earlier, depending on the nature and number of the solubilising groups and the nature of the monodentate ligands present, the dyes are either effectively uncharged (**13**) or carry an overall negative charge (**1**) (because of the presence of the Cr(III) cation); dyes that contain no ionic solubilising group (e.g. **14**) have an overall positive charge. Blus [45] prepared a series of orange and red 1 : 1 copper-azo dyes derived from I-(3'-*N*-benzenesulphonamido)phenyl-3-methyl-*S*-pyrazolone, which, owing to the presence of sulphonamide groups, were suitable for dyeing wool from a weakly acidic dyebath.

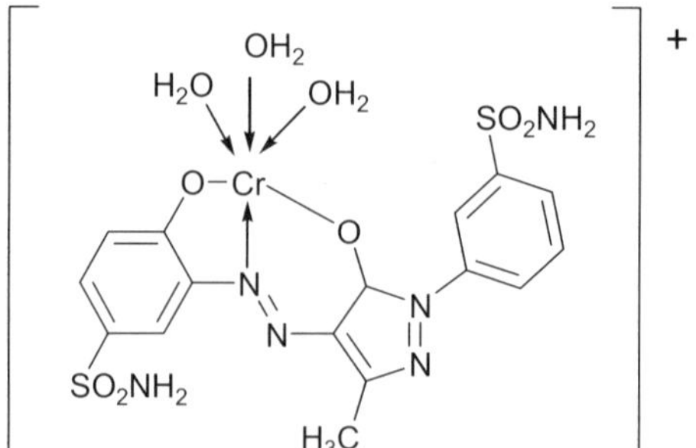

14

Preparation of 1 : 1 metal-complex dyes generally entails heating the chelatable monoazo dye under acidic conditions with an excess of a trivalent salt, such as chromium(III) fluoride or formate, under reflux or pressure [9,11,21].

8.2.5   1 : 2 Metal-Complex Dyes

Although the wet fastness properties of 1 : 1 metal-complex dyes on wool are lower than those of mordant dyes, their excellent migration and penetration character, ease of application, good light fastness and comparatively bright shades, secured their considerable use on wool for over 25 years, until the introduction of 1 : 2 metal-complex acid dyes in 1951. Commonly, two types of 1 : 2 metal-complex dye are differentiated, depending on the nature of the solubilising groups present in the dye ligands: *weakly polar* dyes, which contain no water-solubilising groups, and *strongly polar* dyes, which contain water-solubilising groups.

8.2.5.1   *Weakly Polar 1 : 2 Dyes*

Although studies of 1 : 2 metal-complex dyes devoid of ionic solubilising groups (e.g. $-SO_3H$) began as early as the 1920's [9], the first commercially available 1 : 2 metal-complex acid dye for wool, *Polar Grey BL* (later renamed *Irgalan Grey BL*; C.I. Acid Black 58; **15**), was introduced by Geigy in 1949. This was the forerunner of the first commercial range (*Irgalan*) of 1 : 2 metal-complex dyes for wool, marketed by Geigy in 1951, which contained no water-solubilising groups.

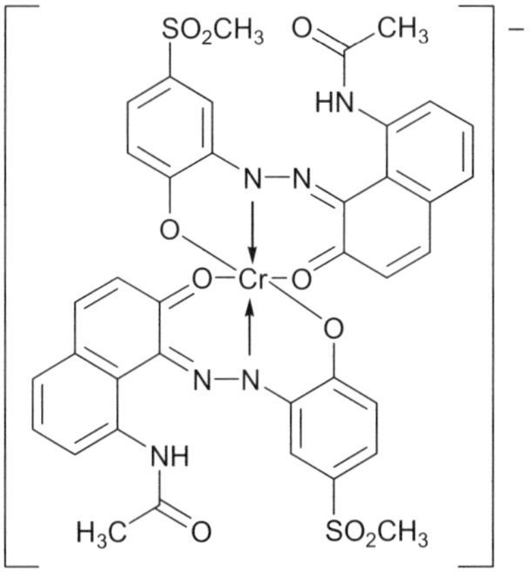

15

Typically, such weakly polar 1 : 2 dyes, typified by C.I. Acid Black 60 (**16**) and C.I. Acid Violet 78 (**17**), are symmetrical Cr(III) or Co(III) complexes that are free of strongly polar, ionic solubilising groups (e.g. $-SO_3H$ or $-COOH$), water solubility being conferred by the inherent anionicity of the 1 : 2 structure (arising from the loss of four protons from the two dye ligands) and the presence of nonionic, hydrophilic substituents (e.g. $-SO_2CH_3$).

Other solubilising aids that have been employed include mono- or di-alkyl-substituted sulphonamide, ethylsulphone and cyclic sulphone groups [9].

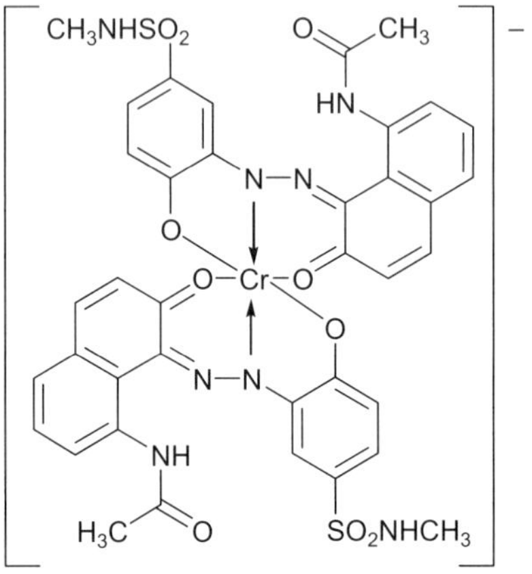

16

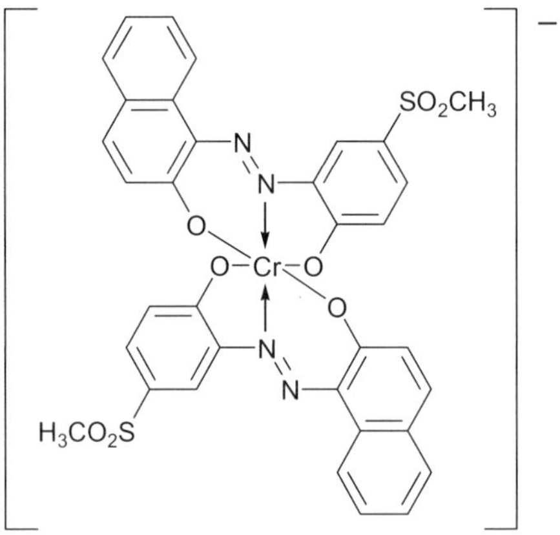

17

Different methods are used for the synthesis of Cr and Co 1 : 2 metal-complex dyes, details of which are outside the scope of this chapter (see [5,9,11,21,46]).

8.2.5.2   Strongly Polar 1 : 2 Dyes

Two types of strongly polar 1 : 2 metal-complex dye are available, namely *monosulphonated* types (e.g. structure **18** [11]) and *disulphonated* types, of which there are both asymmetrical (C.I. Acid Orange 148; **19**) and symmetrical (C.I. Acid Orange 142 (**20**) [47]) variants.

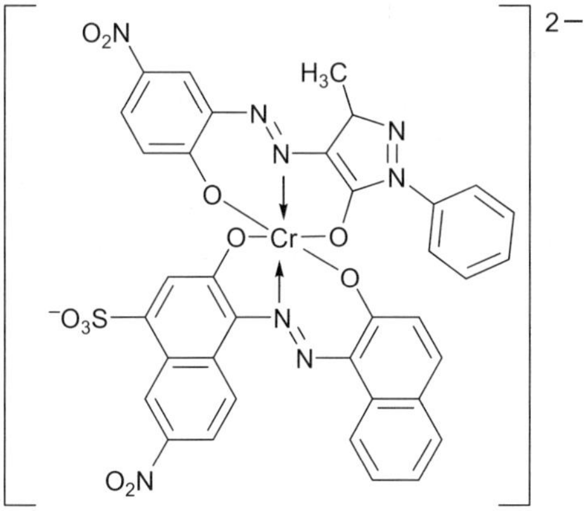
O2N
H3C
N
N
N
N
O
O
O Cr O
O
−O3S
N
N
O2N
2−

18

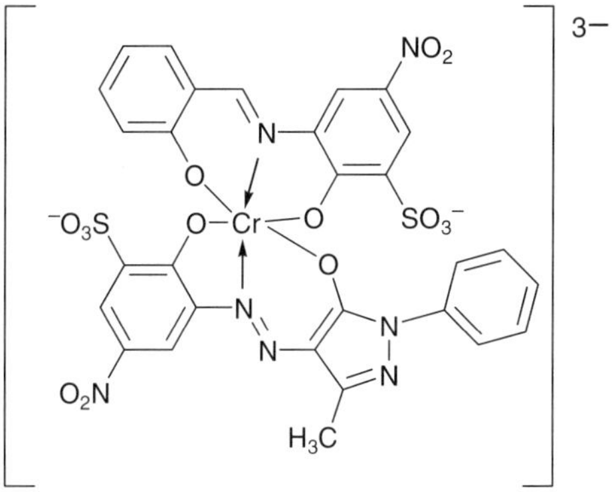
NO2
N
O
−O3S
O Cr O
SO3−
O
N
N
N
N
O2N
H3C
3−

19

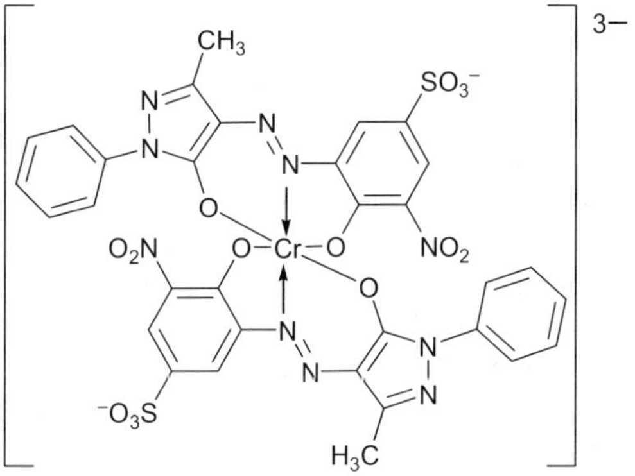
CH3
SO3−
N
N
N
N
O
O2N
O Cr O
NO2
O
N
N
N
N
−O3S
H3C
3−

20

Although symmetrical, disulphonated 1 : 2 dye–chromium complexes were patented as early as 1926 [48] and asymmetrical monosulphonated 1 : 2 complexes were discussed six years later [49], the synthesis and application of strongly polar 1 : 2 metal-complex dyes were widely studied only from the 1950's onwards. This resulted in the commercial intro-duction of asymmetrical, monosulphonated 1 : 2 chromium complex dyes in the 1960's. While symmetrical 1 :2 metal-complex dyes typically contain two sulphonate solubilising groups, as typified by **20**, one commercial range contained carboxylic acid solubilising groups (e.g. C.I. Acid Red 308; **21**). Current 1 : 2 metal-complex dyes include the mono-sulphonated *Lanasyn M* (Clariant) and *Isolan 1S* (DyStar) and the disulphonated *Isolan 2S* (DyStar) ranges.

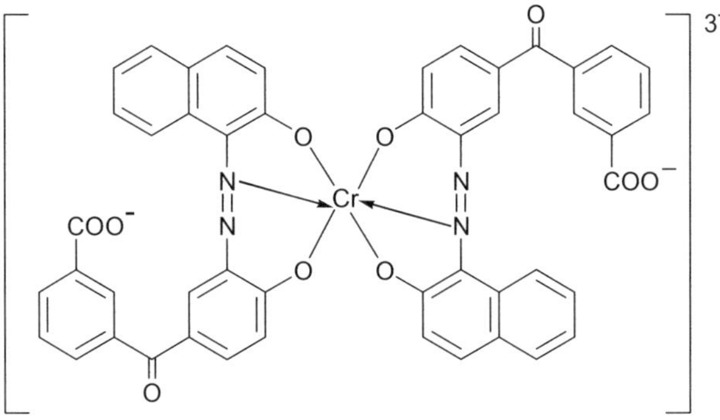

21

8.3   Dye Application

8.3.1   1 : 1 Metal-Complex Dyes

1 : 1 metal-complex dyes, as exemplified by the *Neolan* and *Neolan P* (Huntsman), *Chromene* (Townend) and *Palatin* (DyStar) ranges, enjoy use in the dyeing of loose stock and yarn for floor coverings, hand-knitting yarns and piece goods. They display excel-lent level-dyeing and penetration characteristics and are especially suitable for dyeing un-neutralised carbonised and acid-milled wool. The dyes have the ability to cover irreg-ularities in the substrate and yield dyeings on untreated wool of good to very good light fastness and moderate to good fastness to wet treatments, even in deep shades.

The dyes are commonly applied to wool from a strongly acidic (pH ~2) dyebath (hence they are sometimes referred to as 'acid dyeing' metal-complex dyes; see Figure 8.8), under which conditions they possess excellent migrating and levelling character. Since wool absorbs ~4% on mass fabric (o.m.f.) 96% H_2SO_4, an excess of acid is required in order to maintain a suitably acidic dyebath. The amount of excess sulphuric acid required depends on the dyeing method and liquor ratio used (Table 8.1).

When un-neutralised carbonised or acid-milled wool is dyed, the H_2SO_4 content of the fibre should be determined and the total quantity of acid used for dyeing adjusted accordingly. In view of the high concentration of sulphuric acid used, it is necessary to either neutralise or buffer the residual acid in the fibre at the end of dyeing. Excessively

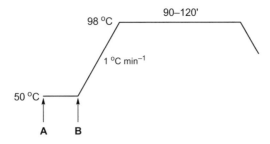

A 4% + 1 g l⁻¹ 96% H_2SO_4 (>1% o.m.f dye)
   4% + 0.7 g l⁻¹ 96% H_2SO_4 (<1% o.m.f dye)
   5% Na_2SO_4
   pH 1.9–2.1

B *Neolan* dye

Figure 8.8 *Application of* Neolan *dyes [50]. Reproduced by permission of Huntsman Advanced Materials (Switzerland) GmbH.*

hard water may affect the quantity of H_2SO_4 required; chelating agents are unsuitable due to demetallisation of some dyes. As prolonged boiling at such low pH can impart fibre damage, either reduced amounts of H_2SO_4 or a proprietary levelling agent (Figure 8.9) can be employed.

The *Neolan P* (Huntsman) range of dyes can be applied at pH 3.5, achieved using HCOOH rather than H_2SO_4, thus minimising fibre damage (Figure 8.10) [51]. The dyes exhibit excellent levelling character, very good fastness properties and high exhaustion [31] and are also suitable for dyeing wool/PAN and wool/PA blends [51]; because of the pH used for application, the need for neutralisation or buffering of the dyed material is alleviated. These particular dyes are 1 : 1 chromium complexes of sulphonated azo dyes that contain colourless hexafluorosilicate ligands [44] and are applied in conjunction with *Albegal Plus* (Huntsman), an amphoteric levelling agent that contains an ethoxylated fatty amine and ammonium hexafluorosilicate ((NH_4)$_2SiF_6$) [52,53].

Table 8.1 *Amount of 96% H_2SO_4 required for* Neolan *dyes [50]. Reproduced by permission of Huntsman Advanced Materials (Switzerland) GmbH.*

Liquor Ratio	% by Mass of H_2SO_4 (96%)	
	>1% o.m.f. Dye	<1% o.m.f. Dye
10:1	5	4.7
20:1	6	5.4
30:1	7	6.1
40:1	8	6.8
50:1	9	7.5
60:1	10	8.2

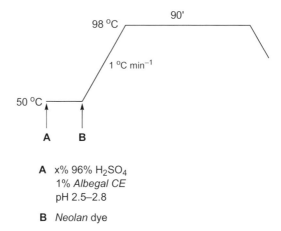

A x% 96% H_2SO_4
1% *Albegal CE*
pH 2.5–2.8

B *Neolan* dye

Figure 8.9 *Application of Neolan dyes at higher pH values in presence of auxiliary [50]. Reproduced by permission of Huntsman Advanced Materials (Switzerland) GmbH.*

8.3.1.1   Dye–Fibre Interaction

Although studied by several workers [34], the precise nature of dye–fibre interaction remains a matter of debate. Dyes show maximum exhaustion in the pH range 3–5, depending on the dye [54], but these conditions give rise to tippy dyeings of poor wet fastness. Under such pH conditions, the dye will interact with the fibre by virtue of:

- Electrostatic forces operating between the anionic dye and protonated amino ($^+NH_3-$) groups in the fibre;

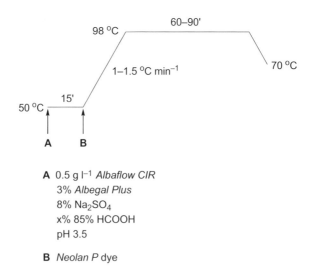

A 0.5 g l⁻¹ *Albaflow CIR*
3% *Albegal Plus*
8% Na_2SO_4
x% 85% HCOOH
pH 3.5

B *Neolan P* dye

Figure 8.10 *Application of Neolan P dyes [51]. Reproduced by permission of Huntsman Advanced Materials (Switzerland) GmbH.*

- Coordination of the chromium ion in the dye with appropriate ligands (such as carboxyl or imino groups) in the substrate;
- Ion–dipole, dipole–dipole and related forces.

Dye exhaustion decreases with decreasing pH of the dyebath below pH 3 [54]; as previously mentioned, under these strongly acidic conditions the dyes exhibit excellent migrating power at the boil. At such low pH (<3), the secondary amino (–NH–) groups in wool are protonated ($^+NH_2$–) and therefore cannot coordinate with the dye; the carboxyl groups in the substrate are un-ionised, however, and coordination of these groups with the dye is possible. Ender and Müller [55,56] proposed that the dyeing behaviour of the Palatin *Fast* (BASF) dyes depended on their anionicity and that such 1 : 1 metal-complex dyes coordinated with appropriate ligands in the fibre. Hartley [57,58] demonstrated that at low pH, 1 : 1 Cr(III)–dye complexes coordinated with carboxyl groups present in wool. However, if this dye–fibre coordination were to occur to any great extent during dyeing, the dyes would be strongly attached to the substrate and consequently would be expected to display poor migrating power, *which of course, they do not*. Giles *et al.* [59] suggested that coordination of 1 : 1 metal-complex dyes with carboxyl groups in the fibre plays no part in their adsorption by wool, the dyes behaving as simple, non-metallised acid dyes. Thus, it can be proposed that under the strongly acidic (approximately pH 2) conditions normally employed for the application of 1 : 1 metal-complex dyes to wool, dye–fibre substantivity arises predominantly by virtue of electrostatic interaction forces operating between the anionic dye and protonated primary and secondary amino groups ($^+NH_3$– and $^+NH_2$–) in the fibre; other forces of interaction, such as ion–dipole or dipole–dipole forces, may also be involved. The characteristic excellent migrating power displayed by 1 : 1 metal-complex dyes at low pH can therefore be attributed to electrostatic interaction, the metal-complex dyes behaving essentially as acid levelling, non-metallised acid dyes. The observed reduction in dye exhaustion that accompanies a decrease in application pH below 3 can be attributed to the corresponding absence of dye coordination with the amino groups in the fibre. In contrast, the presence of such coordination may account for the skitteriness that occurs during application at pH 3 and above, in the absence of levelling agent.

The question as to whether dye–fibre coordination occurs, and if so, to what extent, can also be discussed in terms of the wet fastness properties of 1 : 1 metal-complex dyes on wool. If dye–fibre coordination were a major factor, the dyes would, by virtue of this very strong attachment, be expected to exhibit very high wet fastness on wool fibre. However, the fastness of these dyes to wet treatments is only moderate – indeed, only slightly greater than that of acid levelling, non-metallised acid dyes, and similar to that displayed by intermediate-dyeing, non-metallised acid dyes on wool. Consequently, the contribution of dye–fibre coordination to dye–fibre interaction is, at most, very low. The fact that 1 : 1 metal-complex dyes possess slightly higher wet fastness than their acid levelling, non-metallised acid counterparts of similar molecular size may, as Giles *et al.* [59] have proposed, be due to a greater tendency of the 1 : 1 complexes to aggregate within the fibre. Peters [60] suggests that a 1 : 1 metal dye complex may, once adsorbed, revert to a 1 : 2 metal dye complex; however, if this were to occur, the wet fastness properties of 1 : 1 metal-complex acid dyes on wool might be expected to approach those of their 1 : 2 metal-complex dye counterparts, which they do not.

8.3.2   1 : 2 Metal-Complex Dyes

8.3.2.1   Weakly Polar 1 : 2 Dyes

Weakly polar 1 : 2 metal-complex dyes, as represented by the *Isolan* (DyStar) range, display very good/excellent light fastness and very good fastness to wet treatments on wool, in pale to medium depths. The dyes, which exhibit good levelling and penetration properties and typically yield non-skittery dyeing, are used on loose stock, slubbing, yarn and piece goods. Nowadays, the dyes are applied to untreated wool using a method that is also suitable for the application of strongly polar dyes (Figure 8.11).

8.3.2.2   Strongly Polar 1 : 2 Dyes

Dyes, such as C.I. Acid Orange 148 (**19**), have very good build-up properties, very good/excellent light fastness and very good fastness to wet treatments on wool, and generally display higher fastness to wet treatments than their weakly polar counterparts [62]. They are therefore generally suitable both for applications in which mordant dyes are used and for applications in which mordant dyes cannot be employed due to the prolonged dyeing times required. Their principal use is on loose stock, tops, yarn and piece goods; however, in general, strongly polar 1 : 2 dyes have poorer lower levelling properties than do their weakly polar counterparts. The dyes also have a greater tendency for skittery dyeing, with the disulphonated variants possessing the lowest levelling characteristics; these dyes are therefore rarely used on woven fabrics. Figure 8.11 shows typical application conditions for mono- and disulphonated dyes [61].

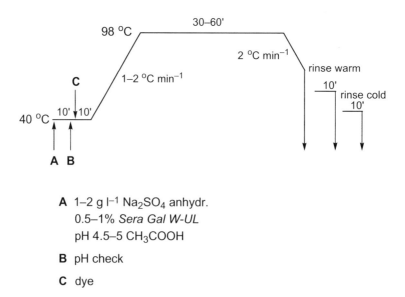

A   1–2 g l^{-1} Na$_2$SO$_4$ anhydr.
   0.5–1% *Sera Gal W-UL*
   pH 4.5–5 CH$_3$COOH

B   pH check

C   dye

Figure 8.11   *Application of mono- and disulphonated dyes [61]. Reproduced by permission of DyStar Colours Distribution GmbH, Frankfurt, Germany.*

8.3.2.3 Dye–Fibre Interaction

The nature of the interaction of both weakly and strongly polar 1:2 complexes with wool fibres has received relatively little attention. In contrast to their 1:1 metal-complex counterparts, in both weakly and strongly polar 1:2 metal-complex dyes, the metal atom is fully coordinated with the two dye ligands and, as a consequence, coordination of the chromium or cobalt ion of the dye with ligands such as amino or carboxyl groups in the fibre is not possible [63]. Since a weakly polar 1:2 metal-complex carries acid dye carries a single negative charge (due to the loss of four protons from the two component dye ligands), a monosulphonated 1:2 complex an overall negative charge of two and a disulphonated 1:2 complex an overall negative charge of three, then ion–ion interaction, operating between the dye anion and the protonated amino groups in the substrate, can be expected to contribute to dye–fibre substantivity. This electrostatic interaction can also be expected to be more pronounced in the case of the strongly polar dyes, which carry a comparatively greater – and localised – negative charge compared with the weakly polar dyes, which carry a nonlocalised, single negative charge. Indeed, the lower level-dyeing character of strongly polar 1:2 metal-complex acid dyes may be attributable to the correspondingly greater contribution that electrostatic interaction makes towards dye–fibre substantivity. Whilst the dyes are applied to wool under weakly acidic to near-neutral pH conditions, at such pH values the number of protonated amino groups in the substrate is small and electrostatic interaction will be small. Hence, the dyes in effect behave as large molecular size, non-metallised (i.e. milling) acid dyes in their adsorption characteristics on wool and, as with milling non-metallised acid dyes, forces other than electrostatic interactions will contribute to dye–fibre substantivity. Furthermore, although both weakly and strongly polar dye types are hydrophilic in as much as they possess anionicity, the dyes are predominantly hydrophobic in character, since this relatively small anionicity is present within the large, essentially hydrophobic 1:2 metal-complex structure. Therefore, hydrophobic interactions, operating between the dye and hydrophobic regions within the fibre, can be expected to make an important contribution to substantivity.

The characteristic high wet fastness properties of 1:2 metal-complex dyes on wool can be attributed to the nature of the various forces of interaction (such as ion–ion and hydrophobic interactions) operating between the dye and the fibre, as well as to the low diffusion coefficient of the large dye molecule within the substrate. Owing to the dye's structural characteristics, dye aggregation within the fibre will probably arise (by virtue of, for instance, hydrophobic interactions operating between the dye molecules); this will serve to further reduce the tendency of the dye to diffuse out of the fibre during wet treatments. In this context, it is therefore not surprising that the wet fastness properties of 1:2 metal-complex acid dyes closely resemble those of mordant dyes, which, when complexed with chromium, form 1:2 metal complexes [64] *in situ* in the fibre.

However, there are differences in the fastnesses of the different types of 1:2 metal-complex dyes. In a study of the effect of solubilising groups in metal-complex dyes and fastness properties on wool, Rexroth [62] observed that fastness to both potting and milling decreased in the order 1:2 (diSO$_3$H) — 1:1 > 1:2 (monoSO$_3$H) > 1.2 > 1.2 (SO$_2$NH$_2$). Similar findings were obtained using nine disulphonated, monsulphonated and unsulphonated 1:2 metal-complex dyes on nylon 6,6. It was observed [65] that the fastness of 2% o.m.f. dyeing to washing at 60 °C was markedly dependent upon the degree of sulphonation of the dyes, insofar as whilst the change in shade that occurred during washing at 60°C increased in the order (diSO₃H) ≫ (monoSO₃H) > (SO₂NH₂), the degree of staining of adjacent materials during washing increased in the order (SO₂NH₂) ≫ (monoSO₃H) > (diSO₃H). Elliott [66] found that the wash-fastness properties of 1 : 2 metal complexes solubilised by carboxylic acid groups were superior to those of conventional sulphonated 1 : 2 metal-complex dyes.

8.4   Environmental Aspects

Dye effluents contain, in addition to residual dyes, electrolytes, surfactants, dyeing auxiliaries etc., and, characteristically, dye wastewaters are remarkably recalcitrant towards biodegradation [67]. As such, it is understandable that the treatment and disposal of dye wastewater has attracted a large amount of interest for many years. In recent times, the decolourisation of dye effluents has been the subject of a seemingly endless number of publications. In the context of metal-complex dyes, such studies have included the use of low-cost alternative absorbents [68–70], biosorption [71–74], anaerobic advanced oxidation processes (AOPs) [75,76] and bacterial cells [77–79].

References

[1] SDC, Colour Index, Vol 4 online (Bradford, UK: Society of Dyers and Colourists).

[2] H Y Park, N H Lee, J T Je, K S Min, H J Hun and E R Kim, *Molecular Crystals and Liquid Crystals*, **371**(1) (2001) 305.

[3] S Wang, S Shen and H Xu, *Dyes and Pigments*, **44**(3) (2000) 195.

[4] S J Wu, W Qian, Z J Xia, Y G Zou, S Q Wang, S Y Shen and H J Xu, *Chemical Physics Letters*, **330**(5–6) (2000) 35.

[5] K Grychtol and W Mennicke, in *Ullmann's Encyclopedia of Industrial Chemistry* (Wiley-VCH Verlag, 2000).

[6] P Gregory, J A McCleverty and T J Meyer, in *Comprehensive Coordination Chemistry II* (Oxford, UK: Pergamon, 2003), 549.

[7] T Abe, S Mano, Y Yamaya and A Tomotake, *Journal of Imaging Science and Technology*, **43**(4) (1999) 339.

[8] F Beffa and E Steiner, *Review of Progress in Coloration and Related Topics*, **4**(1) (1973) 60.

[9] F Beffa and G Back, *Review of Progress in Coloration and Related Topics*, **14**(1) (1984) 33.

[10] K Grychtol and W Mennicke, in *Ullmann's Encyclopedia of Industrial Chemistry* (Wiley-VCH Verlag GmbH & Co KGaA, 2000).

[11] K Hunger, *Industrial Dyes* (Weinheim, Germany: Wiley-VCH, 2003).

[12] W Czajkowski, R Stolarski, M Szymczyk and G Wrzeszcz, *Dyes and Pigments*, **47**(1–2) (2000) 143.

[13] L Y Zhou, B T Tang, S F Zhang and J Z Yang, *Chinese Chemical Letters*, **20**(11) (2009) 1296.

[14] J Sokolowska-Gajda, H S Freeman and A Reife, *Textile Research Journal*, **64**(7) (1994) 388.

[15] J Sokolowska-Gajda, H S Freeman and A Reife, *Dyes and Pigments*, **30**(1) (1996) 1.

[16] J S Griffiths, *The Theory of Transition Metal Ions* (Cambridge, UK: Cambridge University Press, 1960).

[17] C F Bell, *Priciples and Applications of Metal Chelation* (Oxford, UK: Oxford University Press, 1977).

[18] L E Orgel, *An Introduction to Transition Metal Chemistry* (London, UK: Methuen, 1966).

[19] C J Jones, *d- and f-block Chemistry* (Cambridge, UK: The Royal Society of Chemistry, 2002).

[20] E A Moore, E W Abel and R Janes, *Metal-Ligand Bonding* (Cambridge, UK: The Royal Society of Chemistry, 2004).

[21] R Price, *The Chemistry of Synthetic Dyes*, **Vol. 3**, ed. K Venkataraman (New York, USA: Academic Press, 1970).

[22] P F Gordon and P Gregory, *Organic Chemistry in Colour* (New York, USA: Springer-Verlag, 1983).

[23] J S Griffiths, *Colour and Constitution of Organic Molecules* (New York, USA: Academic Press, 1976).

[24] N A Evans and I W Stapleton, *The Chemistry of Synthetic Dyes*, **Vol. 18**, ed. K Venkataraman (New York, USA: Academic Press, 1978).

[25] C H Giles, D G Duff and R S Sinclair, *Review of Progress in Coloration and Related Topics*, **12**(1) (1982) 58.

[26] J Oakes, *Review of Progress in Coloration and Related Topics*, **31**(1) (2001) 21.

[27] R J Chudgar and J Oakes, in *Kirk-Othmer Encyclopedia of Chemical Technology* (Chichester, UK: John Wiley & Sons, Ltd, 2000).

[28] H Zollinger, *Color Chemistry: Syntheses, Properties, and Applications of Organic Dyes and Pigments* (Weinheim, Germany: Wiley-VCH, 2003).

[29] R Price, in *Comprehensive Coordination Chemistry*, **Vol. 6**, ed. G Wilkinson, R D Gillard and J A McCleverty (Oxford, UK: Pergamon, 1987).

[30] H D K Drew and R E Fairbairn, *Journal of the Chemical Society (Resumed)*, (1939) 823.

[31] G Schetty, *Helvetica Chimica Acta*, **47**(4) (1964) 921.

[32] G Schetty and W Kuster, *Helvetica Chimica Acta*, **44**(7) (1961) 2193.

[33] H Pfitzner, *Angewandte Chemie, International Edition in English*, **11**(4) (1972) 312.

[34] G Schetty and E Steiner, *Helvetica Chimica Acta*, **57**(7) (1974) 2149.

[35] U Lehmann and G Rihs, in *Chemistry of Functional Dyes*, ed. Z Yoshida and T Kitao (Tokyo, Japan: Mita Press, 1989).

[36] M Idelson, I R Karady, B H Mark, D O Rickter and V H Hooper, *Inorganic Chemistry*, **6**(3) (1967) 450.

[37] G Schetty and F Beffa, *Helvetica Chimica Acta*, **50**(1) (1967) 15.

[38] BASF, GP 280505 (1912).

[39] BASF, GP 284856 (1912).

[40] G Schetty, *Journal of the Society of Dyers and Colourists*, **71**(12) (1955) 705.

[41] CIBA, GP Application G43123 (1915).

[42] CIBA, GP 416379 (1920).

[43] S M Burkinshaw, in *The Chemistry and Application of Dyes*, ed. G Hallas and D Waring (New York, USA: Plenum, 1990), 237.

[44] C Meulemeester, I Hammers and W Mosimann, *Proceedings of the 8th International Wool Textile Research Conference, Christchurch* (1990).

[45] K Blus, *Dyes and Pigments*, **25**(1) (1994) 15.

[46] G Hallas and D Waring, *The Chemistry and Application of Dyes* (New York, USA: Plenum, 1990).

[47] P Cesla, J Fischer, E Tesarová, P Jandera and V Stanek, *Journal of Chromatography A*, **1149**(2) (2007) 358.

[48] IG, GP 455277 (1926).

[49] CIBA, GP 600545 (1932).

[50] Huntsman, *Neolan Dyes for Wool Piece Dyeing* (2011).

[51] Huntsman, *Neolan P Optiflow Metal Complex Dyes* (2011).

[52] G Back and W Mosimann, EP 163608 A1 (1985).

[53] G Back and W Mosimann, EP 264346 A1 (1988).

[54] D R Lemin and I D Rattee, *Journal of the Society of Dyers and Colourists*, **65**(5) (1949) 221.

[55] W Ender and A Muller, *Melliand Textilberichte*, **19**(65) (1938) 181.

[56] W Ender and A Muller, *Melliand Textilberichte*, **19**(65) (1938) 272.

[57] F R Hartley, *Journal of the Society of Dyers and Colourists*, **85**(2) (1969) 66.

[58] F Hartley, *Australian Journal Chemistry*, **23**(2) (1970) 275.

[59] C H Giles, T H Macewan and N Mciver, *Textile Research Journal*, **44**(8) (1974) 580.

[60] R H Peters, *Textile Chemistry*, **Vol. 3** (London, UK: Elsevier, 1975).

[61] DyStar Textilfarben GmbH & Co Deutschland, *Supralan Pattern Card* (2007).

[62] E Rexroth, *Arbeitstagung Proceeedings (Aachen)*, **89** (1972).

[63] C H Giles and T H MacEwan, *Journal of the Chemical Society*, (1959) 1791.

[64] E Race, F M Bowe and J B Speakman, *Journal of the Society of Dyers and Colourists*, **62**(12) (1946) 372.

[65] R S Blackburn and S M Burkinshaw, *Journal of the Society of Dyers and Colourists*, **114**(3) (1998) 96.

[66] A Elliott, *Dyer*, **153** (1975) 481.

[67] S M Burkinshaw and O Kabambe, *Dyes and Pigments*, **83**(3) (2009) 363.

[68] M Özacar and I A Sengýl, *Biochemical Engineering Journal*, **21**(1) (2004) 39.

[69] M Özacar and I A Sengil, *Bioresource Technology*, **96**(7) (2005) 791.

[70] F Deniz and S Karaman, *Microchemical Journal*, **99**(2), 296.

[71] M C Ncibi, B Mahjoub, A M ben Hamissa, R ben Mansour and M Seffen, *Desalination*, **243**(1–3) (2009) 109.

[72] M C Ncibi, B Mahjoub and M Seffen, *Bioresource Technology*, **99**(13) (2008) 5582.

[73] Z Aksu and G Karabayir, *Bioresource Technology*, **99**(16) (2008) 7730.

[74] Y Yang, G A Wang, B Wang, Z L Li, X M Jia, Q F Zhou and Y H Zhou, *Bioresource Technology*, **102**(2) (2011) 828.

[75] M Bauman, A Lobnik and A Hribernik, *Ozone: Science and Engineering*, **33**(1) (2011) 23.

[76] W K Józwiak, M Mitros, J Kaluzna-Czaplinska and R Tosik, *Dyes and Pigments*, **74**(1) (2007) 9.

[77] L N Du, S Wang, G Li, Y-Y Yang, X-M Jia and Y H Zhao, *Water Science and Technology*, **63**(7) (2011) 1531.

[78] T Li and J T Guthrie, *Bioresource Technology*, **101**(12) (2010) 4291.

[79] Z Aksu and E Balibek, *Journal of Environmental Management*, **91**(7) (2010) 1546.

9

Dyeing Wool with Reactive Dyes

David M. Lewis
Department of Colour Science, University of Leeds, UK

9.1 Introduction

All classes of dye are absorbed by textile fibres through the operation of one or more of the following: hydrogen bonding, Coulombic interactions, van der Waals forces, London forces, dispersion forces, aromatic interactions and hydrophobic interactions. Dyes may be classified into two groups according to how the dyed fibre performs in response to exterior agencies such as subsequent washing. With dyes of the first group, which includes acid dyes, basic dyes, direct dyes and disperse dyes, the coloration process is reversible under conditions to which the coloured material may be exposed during its lifetime. In the second group, which includes sulphur dyes, vat dyes, mordant dyes, ingrain dyes, oxidation dyes, reactive dyes and resin-bonded pigments, the coloration process is irreversible under conditions normally encountered in use.

The reactive dyes are unique among all dyes in that they are covalently bonded to the substrate; that is, the dye and the fibre substrate form a bond of shared electrons. The energy required to split this bond is of the same order as that required to split carbon–carbon bonds in the substrate itself; hence the high degree of wet fastness observed with these dyes.

It comes as a surprise to find that the first reactive dye ever produced was Supramine Orange R (**1**; C.I. Acid Orange 30), marketed by IG Farben as a wool dye as long ago as 1932; the high wet fastness of this dye compared with that of its acetamido analogue was not at that time attributed to the lability of the ω-chlorine atom. A German patent [1] published in 1937 indicated that dyes could be firmly attached to wool by covalent bonds. From 1948 onwards, Hoechst concentrated on dyes containing either a vinylsulphone (VS) group or a VS precursor. The first reactive dye patent was, in fact, claimed by Heyna and Schumacher [2,3]; this work led in 1952 to the marketing of the Hoechst range of Remalan Fast dyes,

The Coloration of Wool and other Keratin Fibres, First Edition. Edited by David M. Lewis and John A. Rippon.
© 2013 SDC (Society of Dyers and Colourists). Published 2013 by John Wiley & Sons, Ltd.

which gave dyeings of high wet fastness on wool. It is unlikely that the patentees did not realise that these were in fact reactive dyes, and one may conclude that the sales division of the company did not wish this fact to be common currency. This groundbreaking work was recognised in 2000 by the posthumous award of the Society of Dyers and Colourists' Perkin Medal to both Heyna and Schumacher.

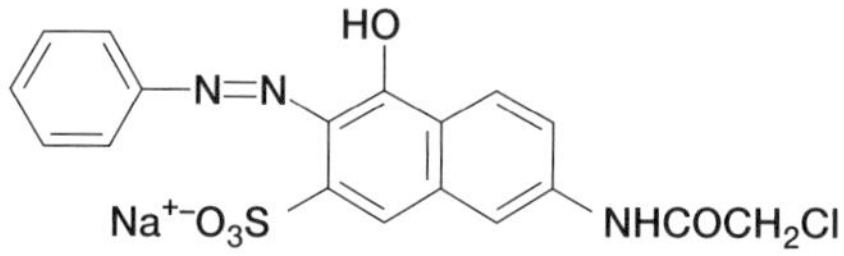

1

In 1954, Ciba introduced a range of bright wool dyes (Cibalan Brilliant dyes) with improved wet-fastness properties, which contained the chloroacetamido group originally employed in Supramine Orange R (**1**). Again, however, there was no published comment on their ability to form covalent bonds with the substrate during dyeing, and strangely enough neither Hoechst nor Ciba claimed the first reactive dye–fibre system.

Although much of the early work with reactive dyes was carried out on wool, initial market success was achieved on cellulose fibres. ICI Dyestuffs Division introduced the first commercial range of reactive dyes in 1956, with their Procion dichloro-*s*-triazine dyes for cellulose, developed by Rattee and Stephen [4]. These dyes were followed almost immediately by such ranges as Cibacron (Ciba), Remazol (HOE), Drimarene (S), Levafix (BAY) and Reactone (Gy); the Remazol range incorporated many of the previously mentioned Remalan dyes introduced for wool. In 1985, the reactive dye share of the wool-dyeing market was estimated to be about 5% [5], but a figure of 20% is currently estimated by the author.

9.2   Commercial Reactive Dyes for Wool

Three ranges of reactive dyes became available for use on wool during the period 1966–1980, but in the 1980s the Drimalan F and Hostalan ranges were withdrawn; subsequently, DyStar put together a range of selected sulphatoethylsulphone (SES) dyes (Realans). Currently, Lanasols (Huntsman) and Realans (Dystar) appear to be the only ranges marketed specifically for wool. Table 9.1 summarises the position.

These systems generally satisfy the following criteria:

(1) A high degree of dye–fibre covalent bonding is achieved at the end of the dyeing process, minimising the clearing treatment required to give maximum wet fastness.
(2) The rates of adsorption and of reaction are such that the former is always greater than the latter; otherwise, dyeing will be uneven. A dye that is too highly reactive will react rapidly with the fibre even at low temperatures, reducing the possibility for dye levelling or migration. Conversely, a dye that is of too low a reactivity will require extended dyeing times at the boil, to ensure adequate covalent bonding and optimum wet fastness.

Table 9.1   *Reactive dyes for wool.*

Commercial Name	Reactive Group	Year of Introduction
Lanasol (CGY)	$\overset{\displaystyle Br}{\underset{\displaystyle \alpha\text{-bromoacrylamido}}{-NHCO-\overset{\mid}{C}=CH_2}}$	1966
Drimalan F (S)	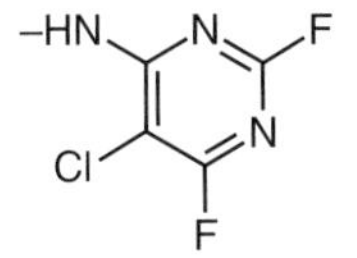 2,4-difluoro-5-chloro-pyrimidyl (FCP)	1969
Hostalan (HOE), Hostalan E (HOE)	$-SO_2CH_2CH_2-\overset{\displaystyle CH_3}{\overset{\mid}{N}}-CH_2CH_2SO_3H$ *N*-methyltaurine-ethyl sulphone (also $-SO_2CH_2CH_2OSO_3H$ β-sulphatoethyl sulphone)	1971
Realan (Dystar)	β-sulphatoethyl sulphone $-SO_2CH_2CH_2OSO_3H$ (also selected FCP dyes)	1997

9.3   The Chemistry of Reactive Dyes

In theory, any group that is capable of reacting with sites in the fibre – such as hydroxyl groups in cellulosic fibres, and amino, thiol and hydroxyl groups in wool – is a potential reactive system capable of incorporation in a reactive dye molecule. In practice, there are many restrictions on the type of reactive group employed, such as level of reactivity, stability to hydrolysis, stability of the dye–fibre bond and, not least, cost and ease of manufacture.

Reactive groups in commercially available reactive dyes used for wool dyeing are of two types: systems that react by nucleophilic substitution reactions and those that react by the Michael addition reaction; the latter are typically present in the Lanasol and Realan ranges.

9.3.1   Nucleophilic Substitution Reactions

These reactions can best be described in terms of the attraction of an electron-deficient carbon atom for the free lone pair of electrons on the nucleophile. In general, this reactive centre on the carbon atom is activated by electron-withdrawing groups adjacent to it (usually SO_2 or $C=O$). The reactive carbon atom is also attached to a leaving group, usually halogen, sulpho or quaternary nitrogen. For example, such a system may be described by the reaction of a chloroacetyl reactive dye with an organic amine, as depicted in Scheme 9.1 (where D represents the chromophoric residue). Since both the reactants are involved simultaneously in covalence change, this mechanism is termed 'bimolecular' and denoted by S_N2.

9.3.2   Michael Addition Reaction

The general reaction of dyes containing polarised, unsaturated carbon–carbon double bonds with nucleophiles can be considered to be a 1,2-*trans*-addition. The double bond is activated

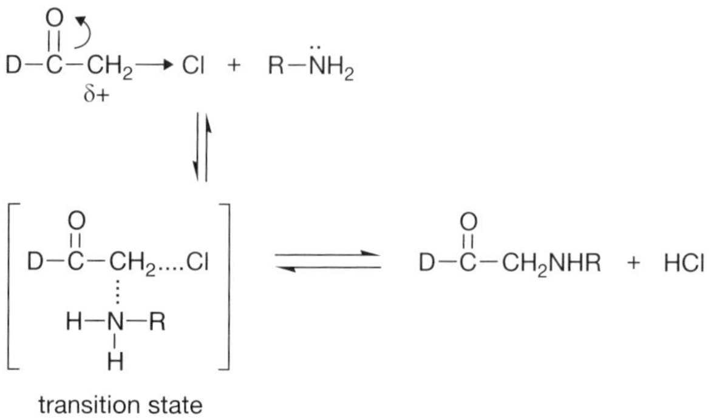

$$D-\overset{\overset{\displaystyle O}{\|}}{C}-CH_2 \rightarrow Cl \quad + \quad R-\ddot{N}H_2$$
$$\delta+$$

$$\left[\begin{array}{c} \overset{\overset{\displaystyle O}{\|}}{D-C-CH_2....Cl} \\[2mm] \vdots \\ H-N-R \\ | \\ H \end{array} \right] \quad \rightleftharpoons \quad D-\overset{\overset{\displaystyle O}{\|}}{C}-CH_2NHR \quad + \quad HCl$$

transition state

Scheme 9.1 *Bimolecular nucleophilic substitution reaction of a chloroacetyl dye with an amino residue.*

$$\delta+ \overset{\overset{\displaystyle H}{|}}{\underset{\underset{\displaystyle H}{|}}{C}}=CH-\overset{\overset{\displaystyle O}{\|}}{\underset{\underset{\displaystyle O}{\|}}{S}}-D \quad + \quad R\ddot{N}H_2 \quad \rightleftharpoons \quad RNH-\overset{\overset{\displaystyle H}{|}}{\underset{\underset{\displaystyle H}{|}}{C}}-\overset{\overset{\displaystyle H}{|}}{\underset{\underset{\displaystyle H}{|}}{C}}-\overset{\overset{\displaystyle O}{\|}}{\underset{\underset{\displaystyle O}{\|}}{S}}-D$$

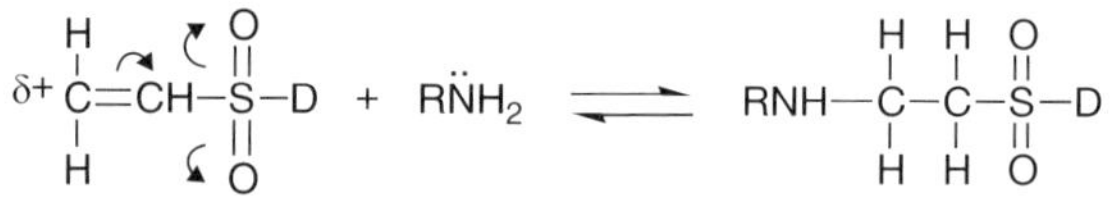

Scheme 9.2 *Reaction of a vinyl sulphone dye with an amino residue.*

by the presence of an electron-withdrawing substituent, such as a carbonyl or sulphonyl group. The reaction of a VS dye with an amino group in the wool may be represented as in Scheme 9.2.

9.3.3 Specific Reactive Dyes for Wool

9.3.3.1 *Lanasol (CGY, now Huntsman) Dyes*

Lanasol dyes, launched in 1966, are the most successful class of reactive dyes for wool. Of especial value has been the compatible trichromatic system based on Lanasol Yellow 4G, Blue 3G and Red 6G (respectively C.I. Reactive Yellow 39, Blue 69 and Red 84).

Bühler and Casty [6], Mosimann [7] and Mäusezahl [8] have reviewed important aspects of the chemistry of these dyes. Another useful review of Lanasol dye chemistry was published by Church *et al.* [9]. Mäusezahl [8] measured bromide ion liberation from various Lanasol dyes at 60, 77 and 100 °C in the presence and absence of wool. Typical results from Lanasol Blue 3R (C.I. Reactive Blue 50) are shown in Table 9.2. In the absence of wool, no bromide ion was detectable even after 15 hours at the boil, indicating the excellent stability of these dyes to hydrolysis under weakly acid conditions.

Working with model amines, Mäusezahl observed that both nucleophilic substitution and addition reactions occur, with the formation of aziridine derivatives (Scheme 9.3). Using nuclear magnetic resonance studies of model compounds, Mäusezahl [8] also demonstrated that the aziridine form (**4**) was produced in preference to Structure **2**.

The aziridine derivative is capable of further reaction with nucleophilic amino groups (Scheme 9.4). It is not necessary to postulate the formation of the aziridine ring (**4**) in order to propose further peptide reaction of this type; however, Structures **2** and **3** are

Table 9.2   *Bromide ion production and fixation of Lanasol Blue 3R.*

Temperature (°C)	Time (Minutes)	Exhaustion (%)	Bromide (%)	Fixation (%)
60	15	30	4	12
	30	52	8	19
	60	86	19	39
	180	96	39	–
	995	98	52	79
77	17	97	38	35
	32	98	47	43
	68	100	51	56
	187	100	64	67
	1277	100	74	88
100	5	95	37	43
	10	95	45	52
	30	96	67	71
	60	97	79	85
	180	98	92	92
	1350	97	94	95

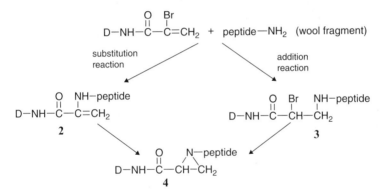

Scheme 9.3   *Bifunctionality of the α-bromoacrylamido dye in its reaction with wool amino residues.*

Scheme 9.4   *Final reaction of the aziridine–peptide to form a disubstituted product.*

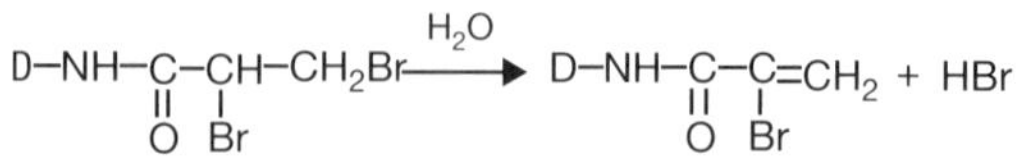

$$\text{D–NH–C–CH–CH}_2\text{Br} \xrightarrow{\text{H}_2\text{O}} \text{D–NH–C–C=CH}_2 + \text{HBr}$$

Scheme 9.5 *Formation of the α-bromoacrylamido dye from the α,β-dibromopropionamido dye.*

quite capable of reacting further with nucleophiles to give Structure **5**. These studies indicate that the α-bromoacrylamide dyes are essentially bifunctional reactive dyes, provided sufficient nucleophilic groups are available for reaction and that these reactions are not sterically hindered.

Patents [10,11] covering the Lanasol reactive dyes usually refer to α,β-dibromopropionylamide dyes, and it is believed that these are precursors of the α-bromoacrylamide dyes, being converted to the latter by simple elimination of hydrogen bromide on dissolving in water (Scheme 9.5).

Some members of the Lanasol range of dyes have two α-bromoacrylamido reactive groups. These include Lanasol Red G, Lanasol Red 2G, Lanasol Scarlet 3G and Lanasol Orange R (respectively C.I. Reactive Red 83, Red 116, Red 178 and Orange 68). The level of fixation of these dyes with the wool fibre is particularly high, leading to very high wet-fastness properties; in fact these latter dyes may be looked upon as tetrafunctional.

Aziridine dyes of type **4** have been reported in the patent literature [12], but their fixation ratio on wool is only 50–70%, indicating that reaction to the crosslinked form (**5**) is likely to be incomplete.

Zollinger *et al.* [13] and Mosimann *et al.* [14] studied changes in the solubility values obtained from various wool fibre solubility tests carried out on dyeings produced with Lanasol dyes. The bis-α-bromoacrylamide Lanasol dyes, which are theoretically tetrafunctional, showed marked evidence of fibre crosslinking, as evidenced by the reduced solubility of the dyed wool in the urea thioglycollate solubility test [15].

Lanasol dyes have been strongly promoted as environmentally friendly alternatives to chrome dyes; thus the metal-free Lanasol Black PV was launched in 1996 and shown to give similar hues on wool as the traditional chrome dye, Eriochrome Black PV; the price was also similar [16]. The black reactive system was improved even further by the launch of Lanasol Black CE in 1997; an important feature of this mixture dye was the incorporation of a bireactive orange component. Other Lanasol CE dyes included the trichromatic yellow, red and blue, plus a navy. These reactive dyes also offer an important fibre-protective effect, even when dyeing at 120 °C [16].

9.3.3.2 *Drimalan F (S, now Clariant) Dyes*

These dyes contain the 2,4-difluoro-5-chloro-pyrimidine (FCP) reactive group. They are regarded as bifunctional in their substitution reactions with the nucleophilic sites in wool, since both fluorine atoms are capable of reaction. Hildebrand and Meier [17] observed that the fluorine atom in position 4 reacts first, but that under dyeing conditions at the boil the fluorine atom in the 2 position, though less strongly activated, is also eliminated by nucleophilic substitution. This bifunctional character, coupled with exceptional resistance to hydrolysis in the pH region 5–7, leads to a very high degree of dye–fibre covalent bonding

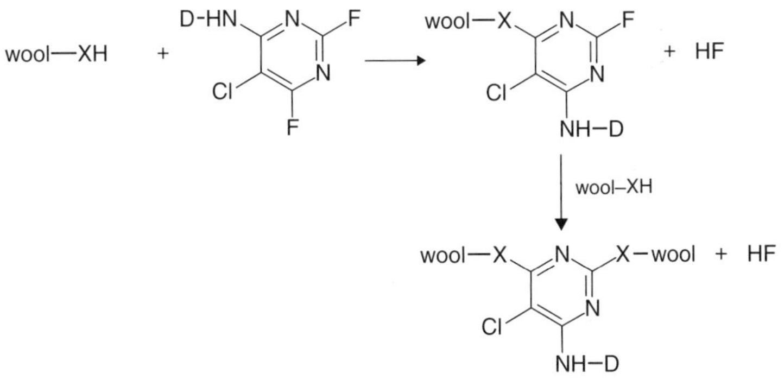

Scheme 9.6   *Reaction of FCP dyes with wool nucleophiles (wool–XH).*

and hence to very good wet-fastness properties of the dyeings. Again, using fibre solubility tests, Zollinger *et al.* [13] found evidence of fibre crosslinking with this type of reactive group. The high resistance to hydrolysis of the FCP reactive group may be explained by the inability of the pyrimidine ring system to absorb a proton under acid dyeing conditions, unlike the triazine or quinoxaline ring systems.

The Bayer range of FCP dyes for wool, the Verofix dyes, was withdrawn in 1987. Most of the Levafix EA (BAY) and Drimarene R/K (S) reactive dyes for cellulosic fibres are FCP dyes, and some dyes suitable for the dyeing of wool can be selected from these ranges. As indicated in Table 9.1, selected Realan (DyStar) dyes may be FCP types [18].

The reactions of FCP dyes with the wool fibre are summarised in Scheme 9.6, where wool–XH is the nucleophilic site in the wool fibre involved with the dye–fibre reaction (cystine thiol, histidine amino, lysine ϵ-amino or terminal α-amino groups).

9.3.3.3   Realan (DyStar) Dyes

These dyes, similar to the old, now unavailable Hostalan dyes (Hoechst), are blocked VS derivatives [19,20], which gradually activate to the reactive VS at elevated temperatures, even under slightly acidic conditions. The main advantage of such a system is an improvement in dye levelness, due to suppression of dye–fibre covalent bonding at temperatures below the boil.

The Hostalan dyes (*N*-methyltaurine-ethylsulphones and sulphato-ethylsulphones) were found to give good level dyeings, since they only form the fibre-reactive VS dye when the weakly acidic dyebath reaches the boil. Fuchs and Konrad [21] claimed that the Hostalan E dyes were based only on aromatic disulphonated chromophores.

The chemistry of the preparation and subsequent activation of the *N*-methyltaurine dyes (**6**) in the dyebath is summarised in Scheme 9.7 [19].

Osterloh [20] has published a detailed study of the dyeing mechanism of the *N*-methyltaurine-ethylsulphone dyes. It appears that the *N*-methyltaurine group is readily eliminated in dilute acidic aqueous solutions at pH 5.5 and at temperatures above 80 °C, resulting in the formation of the reactive VS species (Figure 9.1). The rate of conversion to the reactive VS is important in determining the levelness and final fixation of the system,

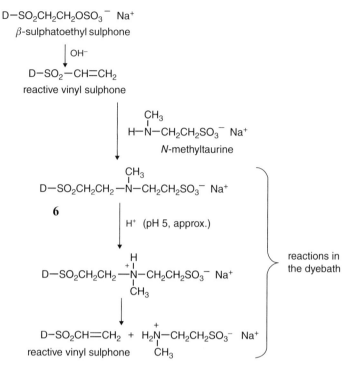

Scheme 9.7   *The formation of N-methyltaurine-ethylsulphone dyes and their elimination under wool dyeing conditions to reform the fibre-reactive vinyl sulphone dye.*

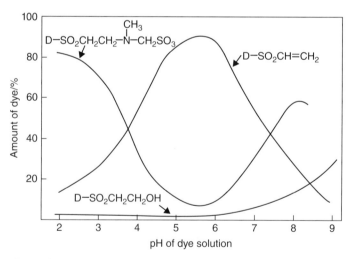

Figure 9.1   *Effect of pH on the conversion of Hostalan dyes to the reactive vinyl sulphone form and the hydrolysed form in 1 hour at 100 °C.*

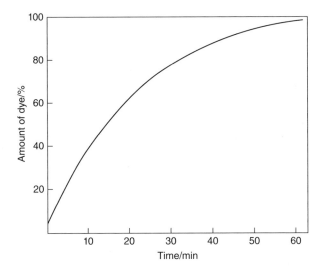

Figure 9.2   *Rate of conversation of Hostalan dyes to the reactive vinyl sulphone form at pH 5 and 100 °C.*

and Osterloh was able to demonstrate that conversion was complete after 60 minutes at the boil (Figure 9.2).

Osterloh [19] had previously carried out similar investigations with β-sulphatoethyl sulphone dyes, and found, in the case of Remazolan Red R, a maximum activation pH of 6.5 (Figure 9.3), confirming that it is possible that selected β-sulphatoethyl sulphone dyes might have been included in the Hostalan and Procilan E wool ranges.

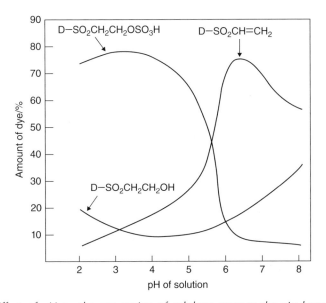

Figure 9.3   *Effect of pH on the conversion of sulphate ester to the vinyl and hydroxyethyl forms in 1 hour at 100 °C.*

9.4   Application Procedures

9.4.1   Auxiliary Agents

Reactive dyes for wool would probably not have been commercially successful without parallel developments in auxiliary products. Without the appropriate auxiliary, dyeings would have been skittery, if not grossly unlevel, and the concept of dyeing a shade with a trichromatic mixture unrealised. Ciba-Geigy introduced the amphoteric auxiliary product, Albegal B, at the same time as the Lanasol dyes were launched. This unusual levelling agent overcame the tippy dyeing properties of wool by forming dye–surfactant complexes which at low temperatures probably exhaust more evenly and extensively on to the surface of the wool fibre than does the dye alone. As the dyebath temperature is raised, the dye–surfactant complex breaks down, allowing the dye to penetrate and react with the fibre.

The amount of this type of auxiliary normally recommended for use is 1% add on mass fabric (o.m.f.), but for deep dyeings it is necessary to employ up to 1.5% o.m.f. If too much of this product is present in the dyebath, the phenomenon of 'reverse tippiness' (weathered fibre tips dyeing less deeply than the undamaged roots) is observed.

All manufacturers of reactive dyes for wool offer amphoteric, or weakly cationic, auxiliary products that promote dye uptake at low temperatures, indicating complex formation. Christoe and Datyner [22] have assigned Structure **7** to a typical amphoteric surfactant used for dyeing wool with reactive dyes; Table 9.3 summarises other available agents.

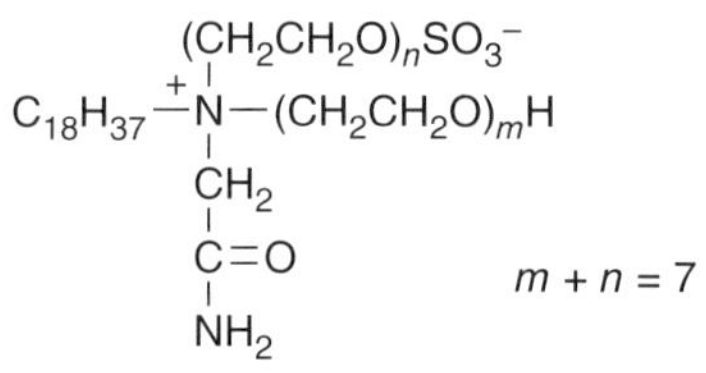

$$C_{18}H_{37} \overset{+}{\underset{\underset{\underset{\underset{NH_2}{|}}{C=O}}{\underset{|}{CH_2}}}{N}} {\overset{(CH_2CH_2O)_nSO_3^-}{\overset{|}{}}}{-}(CH_2CH_2O)_mH$$

$$m + n = 7$$

7

These agents actually promote dye uptake and hence dye fixation on the wool fibre. Typical examples of this effect, taken from a paper by Graham *et al.* [23], are reproduced in Table 9.4. On Hercosett-treated wool, these agents function rather differently and do not enhance dye uptake in this way (Table 9.5).

Table 9.3   *Amphoteric auxiliary products for use with reactive dyes.*

Range of Dyes	Auxiliary Product
Drimalan F (S)	Lyogen FN
Lanasol (CGY)	Albegal B
Hostalan (HOE)	Eganol GES
Verofix (BAY)	Avolan REN

Table 9.4   *Effect of Albegal B on reactive dye exhaustion/fixation (untreated wool) [18].*

Dye	Albegal B (% o.m.f.)	Exhaustion (%)			Fixation (%)		
		60 °C	80 °C	100 °C	60 °C	80 °C	100 °C
Lanasol	0	46	64	91	40	56	70
Scarlet 3G	2	96	99	100	71	83	99
Drimalan Red	0	34	44	73	28	38	70
F-2GL	2	70	94	100	53	77	98
Hostalan Red	0	18	48	57	14	32	48
E-G	2	52	78	97	13	44	87

9.4.2   Dyeing Processes Used with Reactive Dyes

9.4.2.1   *Untreated Wool*

Untreated wool is dyed with reactive dyes in loose stock, top, yarn and, to a lesser extent, piece form by exhaustion methods. Dyeing is usually carried out in a slightly acid dyebath; generally the pH is determined by the depth of the dyeing required. Thus full depths are dyed at pH 5.0–5.5, and paler depths at pH 5.5–6.0. Too high a starting pH will result in poor exhaustion of the dyebath; too low will produce unlevel dyeing, because dye uptake will be too rapid. With medium-depth dyeing, optimum fixation is generally achieved after boiling for 1 hour; paler dyeings may require only a short period at the boil, and deep dyeings could require prolonged boiling (1.5–2 hours) to bring about maximum covalent bonding. Most dyeing methods include a hold period of 20–30 minutes at 70 °C in order to achieve as much dye migration as possible and hence improve the levelness.

Lewis *et al.* [24,25] showed that the plot of pH versus uptake of the ω-chloroacetylamino dye (**1**) on wool was quite different from that of a simple mono-sulphonated acid dye of similar structure, the acid dye showing a marked drop in substantivity at pH values above 4 and the reactive dye not showing this reduction until pH 6; these results are reproduced in Figure 9.4. It is thus clear that covalent interactions are especially important in determining dye–fibre substantivity in the pH range 4–6.

Depending on the depth of the dyeing, an alkaline aftertreatment may be necessary in order to achieve maximum wet fastness; the alkaline conditions should remove most of

Table 9.5   *Effect of Albegal B on reactive dye exhaustion/fixation (Hercosett-treated wool).*

Dye	Albegal B (% o.m.f.)	Exhaustion (%)			Fixation (%)		
		60 °C	80 °C	100 °C	60 °C	80 °C	100 °C
Lanasol	0	89	97	100	57	85	98
Scarlet 3G	2	92	96	100	57	83	99
Drimalan Red	0	82	92	99	57	86	98
F-2GL	2	69	87	100	45	77	98
Hostalan Red	0	47	70	94	21	59	87
E-G	2	50	75	97	14	44	87

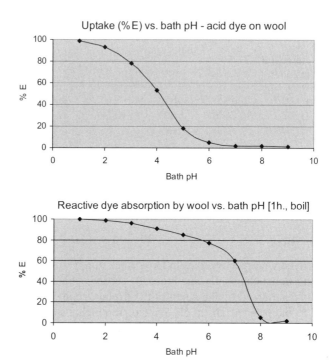

Figure 9.4   *Effect of pH on uptake of nonreactive and reactive dyes on wool.*

the noncovalently bonded dye forms, be they nonreacted reactive dye, hydrolysed dye, thiolysed dye or dye reacted with soluble protein fragments. The aftertreatment should be carried out in a fresh bath, with 1.5 g dm^{-3} ammonia (specific gravity 0.880) (pH 8.5–9.0) for 15 minutes at 80 °C; such a clearing treatment is very important and is considered in more detail in Section 9.4.2.3. Following rinsing, the goods should be neutralised by treatment in a bath containing acetic acid (pH 5–6).

9.4.2.2   Machine-Washable Wool

Dyers are currently dyeing wool that has been given a chlorine–polymer treatment so as to achieve machine washability. One of the commonest wool top treatments is the chlorine–Hercosett process developed by the Commonwealth Scientific and Research Organisation (CSIRO) and the International Wool Secretariat (IWS). The self-crosslinking polymer used in this process is cationic in character, and remains so even after curing or crosslinking; since reactive dyes are anionic, the treated wool absorbs dye more readily.

The advent of truly machine-washable wools has led to the development of specifications, originally controlled by the IWS, for garment stability (relaxation and felting shrinkage) and for colour fastness (light, washing (TM 193), perspiration (TM 174)). In the case of the wet-fastness tests, an important requirement is that the adjacent fabric used must have been pretreated by the chlorine–Hercosett process, which results in any loose dye being more readily absorbed. Consequently, an inferior rating for staining is recorded compared to when the usual SDC standard adjacent fabric is used. These factors explain the widespread use and success of reactive dyes for dyeing of this particular substrate. In addition, reactive dyes are often employed on wools which are subsequently chlorinated in garment form; when predyeing with nonreactive dyes, cross-staining can occur during the chlorination procedure.

Cockett and Lewis [26] studied the performance of reactive dyes when applied in full depths (5%o.m.f.) on Hercosett-treated wool. They examined the improvement in dye–fibre fixation achieved by prolonged dyeing at the boil and by dyeing for up to 2 hours at 110 °C. Significant differences were found from dye to dye: some dyes showed no improvement in fixation when boiling was prolonged for up to 3 hours, while others showed progressive improvement. In general, the best fixation values and wet-fastness ratings were obtained after dyeing for 30 minutes at 110 °C. Fastness to alkaline perspiration (IWS TM 174) generally declined as dyeing was continued beyond either 1 hour at the boil or 30 minutes at 110 °C; this phenomenon was attributed to the presence within the dyed material of coloured peptide material produced in significant amounts by fibre hydrolysis. The proposed method for the production of deep dyeings at 110 °C was regarded as highly promising, especially since there was little difference between the physical characteristics of the yarns and those of control samples dyed at the boil. It was therefore postulated that Hercosett-treated wool withstands dyeing at temperatures above the boil better than untreated wool. Other authors [27–29] have also demonstrated this positive protective nature of the Hercosett resin layer.

9.4.2.3   Aftertreatment of Dyeings

An alkaline aftertreatment is equally important in achieving excellent wet-fastness prop-erties with reactive dyes on both shrink-resisted and untreated wool, especially in deeper shades. Adequate fastness to a long-liquor washing test is normally readily achieved, but short-liquor staining, often exemplified in the alkaline perspiration test, is more critical, and is an important indicator of the performance of the dyeing in close-contact damp-wearing and laundering situations. The usual procedure is to aftertreat dyeings deeper than 1% o.m.f. with a dilute alkaline solution (usually of ammonia but occasionally of sodium bicarbonate, at pH 8.5) for 15 minutes at 80 °C. To be effective, this aftertreatment has to remove most of the dye that is not covalently bonded to the substrate.

The ammonium hydroxide aftertreatment does have damaging effects when dyeing loose wool and tops; it may also be ineffective in certain parts of the dye pack because ammonia promotes wool swelling, and should this be uneven, as is likely in tightly packed loose wool or top dyeing machines, channelling, will occur giving rise to uneven removal of unfixed dye. Evans [30] examined these shortcomings and proposed the use of 2.5% o.m.f. hexamethylenetetramine (hexamine), which decomposes at the boil to give ammonia and formaldehyde [31]. Two advantages were claimed for this compound:

(1) It does not cause the fibre swelling that occurs when ammonia is used; thus, good even penetration of the aftertreating agent is achieved, even through a tightly packed mass of wool.
(2) Unfixed dye is removed more efficiently at a lower pH (6.5 as against 8.5) and there is thus less risk of fibre damage and yellowing of the wool.

Some azo dyes are severely affected by the reducing action of the formaldehyde produced; dyeings of Lanasol Red 6G (C.I. Reactive Red 84), for example, are discoloured by this aftertreatment. Finnimore *et al.* studied pH changes in hexamine solutions (2.5 g dm^{-3}) and noted that when starting at pH 4.5, the pH rose to 6.5 after 30 minutes boiling; starting at pH 5.5, the corresponding rise was to pH 7.5 [32]. The use of hexamine is unusual and the simple explanation of its mode of action being due to a rise in pH deserves further investigation. Under the pH conditions employed, it may not hydrolyse completely to ammonia and formaldehyde, as this reaction is catalysed by strongly acidic conditions [31]. Since the reactive electrophile, formaldehyde, is produced at the boil, it is intriguing to note that crosslinking of hydrolysed dye or wool proteins can also be a factor in producing the improved wet-fastness properties following this aftertreatment; Lewis, Lei and Wang 33 demonstrated that hexamine, co-applied to wool with nonreactive, nucleophilic aminoalkylamino-*s*-triazine dyes (Scheme 9.8) at pH 5 and at the boil, covalently fixed the dye to the fibre; the dyes have an excellent opportunity to migrate and produce level dyeings as covalent fixation does not commence until the boil is reached, since it is only then that sufficient formaldehyde is produced. Figure 9.5A shows typical exhaustion/absorbed dye fixation/overall fixation efficiency values (E,F,T – see Equation 9.1

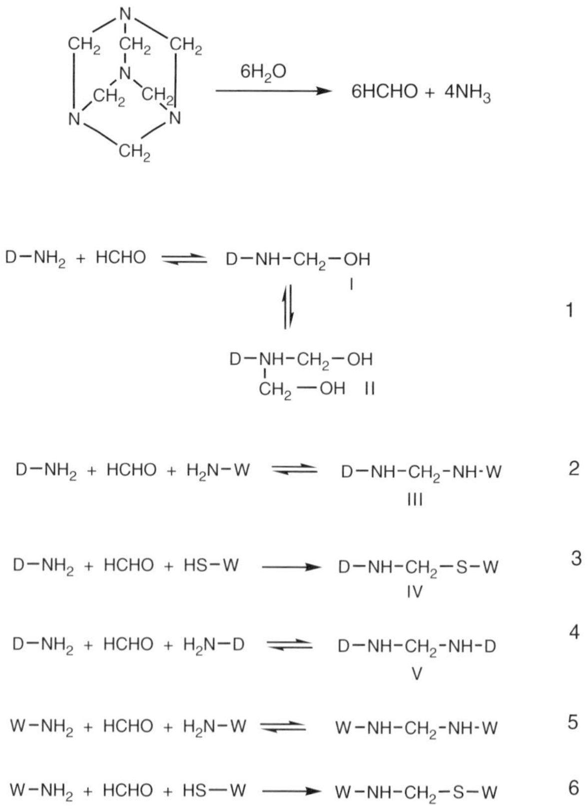

Scheme 9.8   *Hexamine as a reagent to covalently bond amino-dyes to wool.*

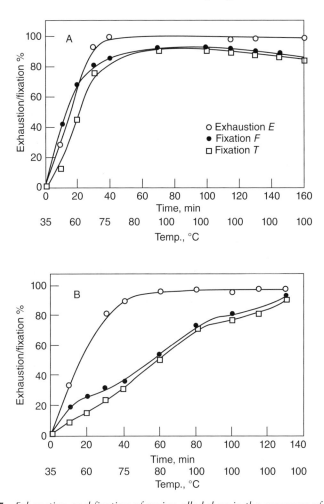

Figure 9.5   *Exhaustion and fixation of amino-alkyl dyes in the presence of hexamine.*

in Section 9.4.2.4) for dyeings produced with the parent monochloro-*s*-triazine reactive dye, and Figure 9.5B shows corresponding E,F,T values for dyeings produced with the aminoalkylamino-*s*-triazine derivative (**8**) in the presence of hexamine.

8

$$^-O_3S\text{-}D\text{-}SO_2\text{-}CH\text{=}CH_2 + NH_3 \longrightarrow {}^-O_3S\text{-}D\text{-}SO_2\text{-}CH_2\text{-}CH_2\text{-}NH_3{}^+$$

Scheme 9.9   *Reaction of vinyl sulphone dyes with ammonia, where D is a chromophore residue.*

Since hexamine has three tertiary amino groups, it may be an effective tertiary amine catalyst for enhancement of the reaction of unfixed reactive dye with nucleophilic groups in the fibre. (The effect of tertiary amines on improving reactive dye fixation on cotton [34] and on wool [35,36] had previously been noted.) Chloro-difluoropyrimidine dyes in particular are highly resistant to hydrolysis during the dyeing procedure, and the aftertreatment procedure is thus mainly designed to remove species still capable of reaction. Using chloro-difluoropyrimidine dyes, Finnimore *et al.* [35] studied a variety of aftertreatment techniques and concluded that aftertreatment at pH 7.0–7.5 with borate buffer for 15 minutes at the boil was very effective in clearing unfixed dye. Carlini *et al.* [37] extended this work to include red azo and blue anthraquinone FCP, α-bromoacrylamide and VS-reactive dyes; they also clearly showed that the improvement in wet fastness brought about by neutral to alkaline aftertreatments is solely due to an increase in fixation ratio (that is, the percentage of residual dye covalently bonded to the fibre following aftertreatment).

Little attention has been paid to the undoubted reaction between ammonia and reactive dyes; FCP dyes react readily with ammonia [35], and Lewis [38] has demonstrated rapid reaction of VS-reactive dyes with ammonia to give dyes that are markedly less water soluble. This decrease in solubility will inhibit desorption and may be caused by the formation of a zwitterion (Scheme 9.9).

More recently, Lewis and Smith [39] have shown that when dyeing wool with VS dyes, a nonreactive thioether dye is formed, through the reaction of the dye with hydrogen sulphide produced by the partial hydrolytic decomposition of cystine and cystine. This dye may be represented as $D\text{–}SO_2CH_2CH_2\text{–}S\text{–}CH_2CH_2SO_2\text{–}D$. Due to its increased molecular size, it will not readily desorb from wool.

Cockett and Lewis have proposed the use of sodium sulphite (1 g dm^{-3}) as an aftertreatment for wool dyed with reactive dyes; these workers suggest that the nucleophilic addition of sulphite to the reactive group decreases dye substantivity and increases aqueous solubility [26]. The reactivity of sulphite towards chlorotriazine [40] and VS [41] reactive dyes is well established: in the former case the sulpho-*s*-triazine dye formed is still reactive, but in the latter an inert sulpho-ethylsulphone dye is produced – a factor employed in the development of resist processes for cellulosic materials. Lanasol dyes also react readily with sodium sulphite [42], even at room temperature, to form inactive species. Inactivation of activated double bonds by bisulphite may be described by Scheme 9.10.

Clearly, the aftertreatment of wool dyed to full shades with reactive dyes in order to achieve maximum wet-fastness properties is an area that requires further research and development, paying particular attention to the possibility of increasing fibre damage under alkaline conditions.

$$D\text{—}SO_2CH\text{=}CH_2 + NaHSO_3 \longrightarrow D\text{—}SO_2CH_2CH_2SO_3{}^-Na^+$$

Scheme 9.10   *Reaction of vinyl sulphone dyes with bisulphite to form nonreactive dyes.*

$$CCl_3-COO^- \; Na^+ \xrightarrow{\;H_2O\;} NaOH + CHCl_3 + CO_2$$

Scheme 9.11 *Decomposition of sodium trichloroacetate acid in hot aqueous solution.*

Fuchs and Konrad [21] described the aftertreatment of wool dyed with reactive dyes using Hostalan Salt K. The procedure is to add 5% o.m.f. Hostalan Salt K to the dyebath 20 minutes before the end of the dyeing time; the dyebath pH then automatically shifts from 5.0–6.0 to 6.7–6.9. Hostalan Salt K is believed to be sodium trichloroacetate, which hydrolyses in boiling water (Scheme 9.11). The carbon dioxide and trichloromethane (chloroform) produced are volatile and the sodium hydroxide formed effectively raises the pH.

9.4.2.4 *Measurement of Dye Fixation*

Reported measurements of reactive dye fixation, on both wool and cotton substrates, are a source of some confusion because of the tendency to quote fixation as the percentage of the original dye applied that becomes covalently bound; this figure is in effect the overall fixed colour yield ($T\%$). As far as the wet-fastness properties are concerned, the important factor is the percentage of dye exhausted that is covalently bonded to the fibre ($F\%$). In pad-dyeing processes, the problem does not arise, as fixation can only be related to the dye initially applied to the fibre at the nip.

For long-liquor dyeing, F and T are related as follows:

$$T = \frac{FE}{100} \tag{9.1}$$

where E is the percentage exhaustion [24]. Zollinger [43] differentiated between these parameters by terming F the 'fixation quotient' and T the 'fixation ratio'.

Noncovalently bonded reactive dye remaining on the wool fibre after the dyeing process may take form of unreacted dye, hydrolysed dye, dye that has reacted with ammonia or dye that has been inactivated by reaction with soluble peptides, or even of amino acids liberated in small amounts from the wool fibre due to amide and/or disulphide bond hydrolysis. Recently, it has been shown [39] that reactive dyes of the VS type may also be inactivated by hydrogen sulphide liberated from the hydrolytic breakdown of cystine or cystine residues (Scheme 9.12). The β-elimination mechanism for hydrogen sulphide production shown in this scheme has not been entirely clarified, but every dyer is aware of the strong smell of hydrogen sulphide produced when boiling wool in water. Hydrogen sulphide is a powerful

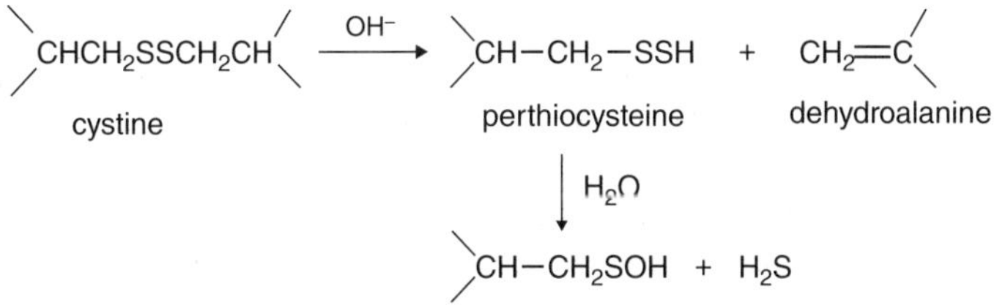

Scheme 9.12 *Hydrolysis of cystine to form hydrogen sulphide.*

$$D-SO_2CH=CH_2 + H_2S \longrightarrow D-SO_2CH_2CH_2SH$$

$$\Big\downarrow D-SO_2CH=CH_2$$

$$D-SO_2CH_2CH_2-S-CH_2CH_2-D$$

Scheme 9.13   *Reaction of hydrogen sulphide with vinyl sulphone dye.*

nucleophile and reacts with the VS dye with the formation of a thioether dye (Scheme 9.13).

The thioether cannot undergo β-elimination under mildly acidic conditions in order to reform the reactive VS, and thus must be regarded as a highly substantive acid dye. Dyeings from such a dye display significantly reduced wet-fastness properties compared with those produced from covalently bound reactive dye. Such an inactivated derivative has indeed been shown to be present on the fibre after dyeing with model VS-reactive dyes [39].

Stripping techniques have therefore been devised to determine the amount of reactive dye covalently bonded to the wool fibre following the reactive dyeing cycle. These are usually based on several short high-temperature extractions with aqueous mixtures of powerful solvents for acid dyes, such as urea or pyridine, which remove noncovalently bonded dye.

One of the earliest proposals [44] was to employ repeated extraction with a boiling solution of 50% urea and 1% Dispersol VL (an ethoxylated alkylamine auxiliary from ICI). Extraction was repeated until no more colour was removed, and the coloured extracts were then combined and the dye content, estimated spectrophotometrically. This method has been modified [45,46], mainly by reducing the extraction temperature with the urea–surfactant solvent to 60 °C.

Other workers have favoured systems based on the use of boiling 25% aqueous pyridine as an extraction solvent [24]; in some cases, improved reproducibility can be achieved by adjusting a 50% aqueous pyridine solution to pH 7 [47] and extracting for 2 hour at 80 °C. Acid aqueous pyridine extraction (10 parts pyridine, 20 parts 90% formic acid and 70 parts water) at the boil has also been recommended [48].

Asquith *et al.* [48] attempted to reconcile these widely differing views on extraction procedures by making a detailed study of their relative effectiveness and reproducibility when employed to determine the extent of fixation of the dichloro-*s*-triazine dye C.I. Reactive Red 1, following its application to wool fabric by a pad–batch procedure. These workers concluded that in this case, acid pyridine extraction at the boil gave the most accurate assessment.

Problems arising with the urea–surfactant method include the length of time required to carry out the repeated extractions until no more colour is removed, the tendency for urea to crystallise on the glass spectrophotometer cells from the stripped solutions and the lack of reproducibility with certain classes of reactive dye [25].

The present author has used repeated extraction with boiling 25% (by volume) aqueous pyridine solutions and finds the results to be reproducible; usually, four 1-minute extractions with intermediate water rinsing are sufficient to remove all uncombined dye, after which colourless extracts are obtained. The main criticism of this method is that the slightly alkaline pH of this extraction medium (7.5–8.0) may promote further dye–fibre covalent

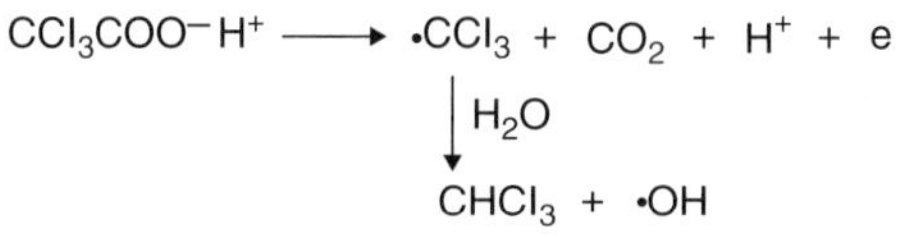

$$CCl_3COO^-H^+ \longrightarrow \bullet CCl_3 + CO_2 + H^+ + e$$

$$\downarrow H_2O$$

$$CHCl_3 + \bullet OH$$

Scheme 9.14   *Radical decomposition of trichloroacetic acid on heating.*

bonding; in practice, most dyeings with reactive dyes are aftertreated at pH 8.5 to achieve maximum wet fastness, and thus this argument is of minimal importance.

9.4.2.5   *Novel Processes for use with Reactive Dyes*

In order to achieve improved levelness, it would be desirable if reactive dye systems could be developed which do not fix to the wool fibre either until the bath is raised to the boil or until the bath is made alkaline during an aftertreatment. Lewis [49] has therefore proposed a two-stage method for dyeing wool with all types of reactive dyes. This provides for an initial level-dyeing period under acid conditions, during which reduced dye–fibre reaction occurs; then, as soon as a bath temperature of 100 °C is reached, the acid disappears from the system through a free radical decomposition reaction. Such a decomposition of the acid is advantageous in that it occurs in a progressive, level manner throughout the dyebath and fibre mass; practical experience has shown that neutralisation of an acid bath using alkalis leads to uneven results, due to unequal absorption of the alkali by the wool fibre.

Trichloroacetic acid decomposes in water at about 100 °C to give free radicals (Scheme 9.14). Since the chloroform and carbon dioxide produced are volatile, they are rapidly removed from the dyebath. Trichloroacetic acid is a very strong acid capable of giving dyebath pH values below 2; this strong acidity is attributed to the combined inductive effect of the three chlorine atoms (which gives almost complete dissociation of the proton from the carboxylic acid group). A summary of this effect in the related acetic acid and substituted acetic acid series is given in Table 9.6.

Various solutions of trichloroacetic acid (0.5–3.0 g dm^{-3}) were prepared and the rate of change of pH of these solutions with temperature in the presence of wool was studied. The results are shown in Figure 9.6.

Exhaustion (E)/fixation (F) profiles for dyeings of Lanasol Red G (C.I. Reactive Red 83) (2% o.m.f.), produced by the standard pattern card method (pH 5.5) and by dyeing in the presence of 3 g dm^{-3} trichloroacetic acid, are shown in Figure 9.7. These results show that

Table 9.6   *Ionisation of organic acids (20 °C).*

Acid	Structure	pK_a ($-\log K_a$)	Ionisation (%) (0.03 mol l^{-1} in Water)
Acetic	CH_3COOH	4.76	2.4
Chloroacetic	$ClCH_2COOH$	2.81	22.5
Dichloroacetic	$Cl_2CHCOOH$	1.29	70.0
Trichloroacetic	Cl_3CCOOH	0.08	89.5

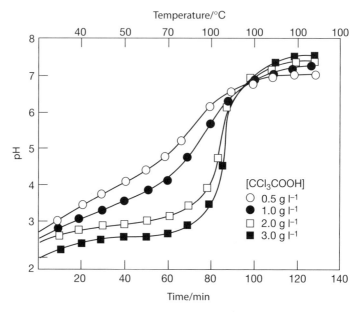

Figure 9.6   *Variation in pH of trichloroacetic acid wool dyebaths versus time and temperature.*

the system does indeed delay the rate of reaction of the dye with the fibre; only at the boil, when the pH starts to rise, does covalent bonding with the fibre become significant. Even though the rate of dyebath exhaustion is much higher for the trichloroacetic acid dyeing, the levelness of the dyeings produced is significantly better than that of the control dyeing, due to the increased opportunity for dye migration, which is, of course, hindered by a too-rapid fixation. The improved levelness of the system allows trichromatic shades to be produced with mixtures of wool-reactive dyes normally considered to be quite incompatible.

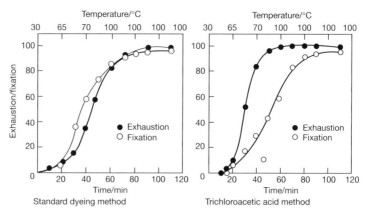

Figure 9.7   *Comparison of exhaustion/fixation values from different dyeing methods (for details of the methods, see text).*

Unfortunately, trichloroacetic acid is comparatively expensive and one of its decomposition products, chloroform, is regarded as potentially toxic. The further development of this concept thus requires the selection of a strong acid that readily hydrolyses to safe nonacidic compounds at the boil.

An alternative approach concerns the pad–batch dyeing of wool with reactive dyes. This has proved to be a very successful method of dyeing cotton with reactive dyes, offering clear advantages in terms of increased colour yield, minimum water and energy usage and increased productivity. Lewis and Seltzer [50] investigated the application of reactive dyes to wool fabric by such a system and observed that it was necessary to include fairly large amounts of urea in the pad liquor. An optimum aqueous pad liquor contains the following agents:

- reactive dye: x g kg^{-1};
- urea: 300 g kg^{-1};
- sodium metabisulphite: 0–20 g kg^{-1};
- acetic acid: to give pH 5;
- auxiliary agent: 10–20 g kg^{-1};
- thickener: 5–15 g kg^{-1}.

Urea additions to the pad liquor improve both dye fixation and the subsequent levelness of the resulting dyeings. At a concentration of 300 g kg^{-1}, urea effectively disaggregates dyes [51,52], thereby promoting dye penetration into the fibre at low temperatures. It also has profound effects on proteins, acting as a denaturing agent; with wool, this is manifested as increased swelling in water.

Using highly reactive dichloro-*s*-triazine dyes, not normally employed for the dyeing of wool by long-liquor processes due to their propensity for unlevel dyeing and hydrolysis, Lewis and Seltzer [50] observed high fixation levels following batching in the presence of 300 g dm^{-3} urea; their results for Procion Red M-5B (ICI; C.I. Reactive Red 2) are reproduced in Figure 9.8. The fixation level achieved following an ammonia rinse (1 g

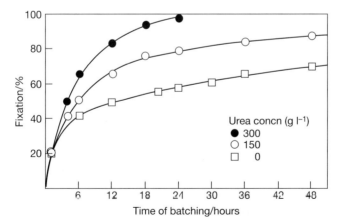

Figure 9.8 *Effect of concentration of urea on rate of fixation in cold pad–batch dyeing (all dyeings aftertreated with ammonia).*

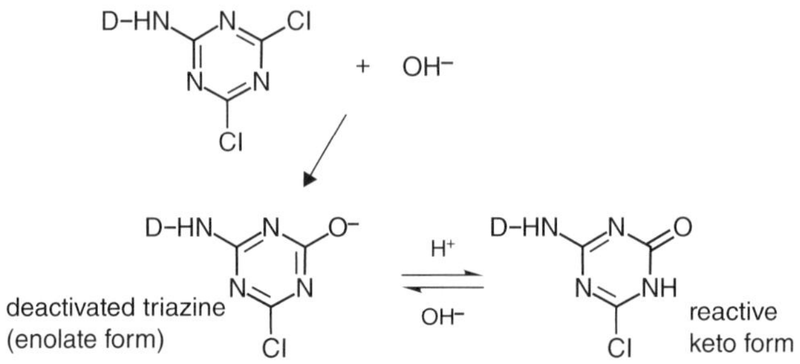

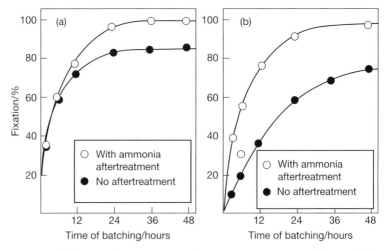

Scheme 9.15   *Alkaline hydrolysis of a dichloro-s-triazine dye.*

dm^{-3} ammonia (specific gravity 0.880), 40 °C for 15 minutes) is surprisingly high ($\geq 95\%$), indicating that the procedure is remarkably effective.

Dichloro-*s*-triazine reactive dyes hydrolyse relatively easily to the monochloro-monohydroxy-*s*-triazine but are relatively difficult to hydrolyse to the di-hydroxy-*s*-triazine [53] under alkaline conditions; under acidic conditions, however, the dihydroxy-*s*-triazine is readily formed. The reasons for this observation are made clear in Scheme 9.15.

Lewis and Seltzer [50] prepared the monochloromonohydroxy-*s*-triazine analogue of C.I. Reactive Red 2 by treating the parent reactive dye with sodium hydroxide solution (0.05 mol dm^{-3}) at 40 °C for 30 minutes, then neutralising. This dye was applied to wool fabric by the urea pad–batch (cold) technique and its rate of fixation was studied with and without the ammonia aftertreatment (any coloured solution removed by the ammonia aftertreatment was saved and combined with subsequent aqueous pyridine extracts in order to calculate the overall fixation value). The results obtained, along with corresponding results for the parent dichloro-*s*-triazine dye, are illustrated in Figure 9.9, which shows

Figure 9.9   *Rates of fixation of (a) Procion Red M-5B (C.I. Reactive Red 2) and (b) its mono-hydroxy analogue (cold pad–batch dyeing).*

that the role of the alkaline ammonium hydroxide aftertreatment is rather different than its role in clearing long-liquor reactive dyeing – in this case, it acts to further enhance fixation rather than to desorb unfixed dye; this is especially pronounced in the case of the 2-chloro-4-hydroxy-1,3,5-s-triazine dye. When applying this mono-hydroxy dye to cellulosic fibres under alkaline conditions, this form is inactive, and it is thus regarded as 'hydrolysed' dye and as being incapable of forming covalent bonds with the substrate; under the mildly acidic pad–batch conditions, in marked contrast, the 'hydrolysed' dye fixes efficiently to the wool fibre. In acidic conditions, the monochloromonohydroxy-s-triazine residue is in the reactive ketone form and reacts covalently with highly nucleophilic thiol (cystine) and amino (histidine) residues in wool keratin; when dyeing cellulosic fibres above pH 8, however, the mono-chloro-mono-hydroxy-s-triazine exists as a different tautomer (the cyanurate form), which deactivates the system to nucleophilic attack [54].

Apart from allowing the use of the highly reactive dichloro-s-triazine dyes – not normally seen as suitable for wool dyeing under normal long-liquor hot dyeing conditions, presumably because of acid-catalysed hydrolysis to the inactive dihydroxy-s-triazine and unlevel dyeing problems – the pad–batch system is also suitable for other reactive dyes. These include the Lanasol (Hunsman), Drimalan F and Drimarene R and K (Clariant) and Levafix E-A (DyStar) ranges. It is also suitable for the Remazol (DyStar) range, provided the β-sulphatoethyl sulphone dye is preactivated to the reactive VS form; usually a pretreatment at pH 11.5 and 20 °C for 30 minutes will suffice.

At an early stage in the development of the pad–batch process, it was noticed that the addition of sodium metabisulphite to the pad liquor had a remarkable effect on dye levelness and rate of fixation. On certain fabrics, such as wool double-jersey knits, pad–batch dyeing with reactive dyes gave an unacceptable striped effect unless sodium metabisulphite was included in the liquor. It is believed that irregular autoclave setting of yarns is the main variable that produces stripes in cold dyeing processes; bisulphite additions eliminate the variability due to their effect on the cystine disulphide linkages. Bisulphite reacts with the disulphide residues in wool through a reversible reaction (Scheme 9.16), the original disulphide bonds being reformed on rinsing away of the excess bisulphite. The cystine thiol residue generated is extremely reactive towards reactive dyes [55,56]. Wool has a relatively

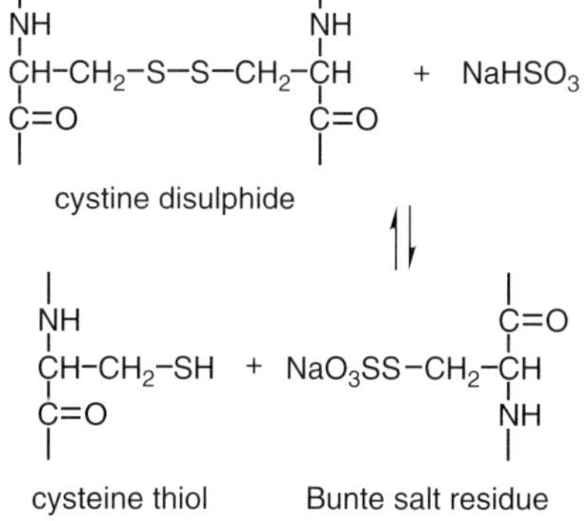

Scheme 9.16 *Reaction of cystine disulphide residues with bisulphite.*

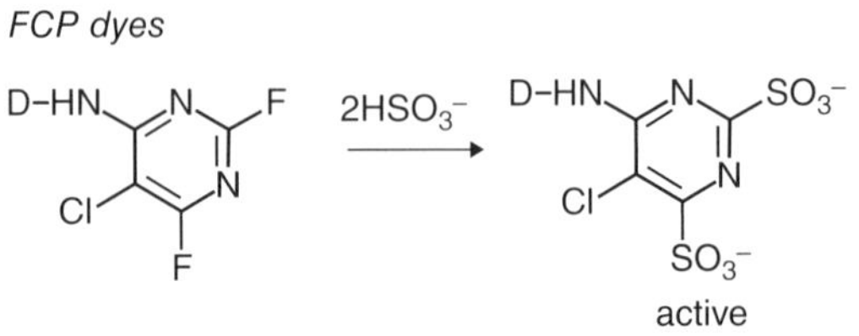

FCP dyes

Scheme 9.17 *Reaction of FCP dyes with bisulphite.*

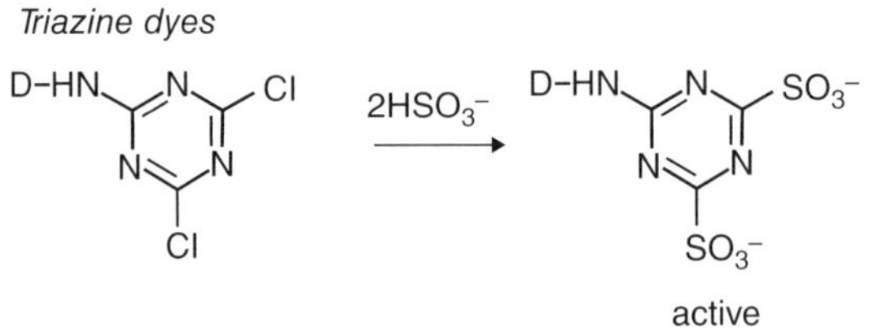

Triazine dyes

Scheme 9.18 *Reaction of halo-triazine dyes with bisulphite.*

high disulphide content but a low free thiol content; most analyses indicate a cystine value of 400–500 μmol g^{-1} fibre and a cystine (thiol) value of 20–40 μmol g^{-1} fibre [57].

The effect of bisulphite on increasing the rate of reaction is due not only to the formation of highly nucleophilic thiol residues but also to the increased fibre swelling produced by aqueous urea–bisulphite solutions. Not all reactive dyes can be used in the presence of bisulphite, however, as this anion is a fairly strong nucleophile and reacts with some reactive groups (Schemes 9.17, 9.18 and 9.19). These side reactions are of no significance in the case of dyes that react by nucleophilic substitution reactions, since the products of reaction are still fibre-reactive. However, with dyes that react by addition reactions at activated double bonds, such as Lanasol and Remazol dyes, the addition of bisulphite anions leads to the formation of a dye that is nonreactive. Since such dyes would be formed gradually during pad liquor storage, there would be the danger of a drop-off in colour yield during the padding process. A practical solution to the problem would be to use minimum-volume pad troughs and to add the bisulphite only at the very last minute before the cloth was padded; this step could be accomplished using the alkali mixing device developed for cellulosic fabric pad dyeing with reactive dyes. (In fact, the analogy between alkali instability problems in cellulose fibre dyeing and bisulphite instability with certain reactive dyes in wool dyeing is very close.)

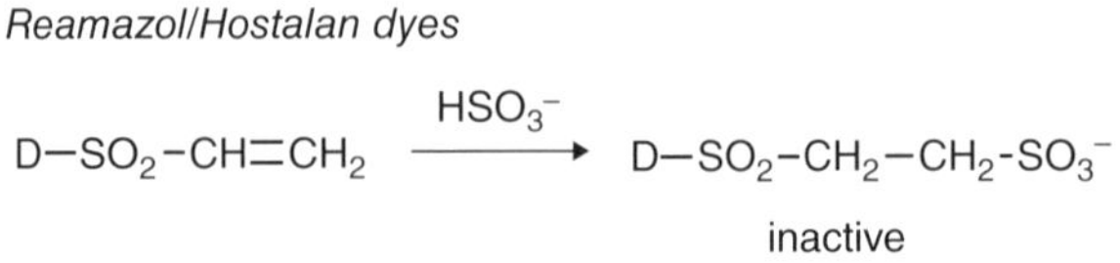

Reamazol/Hostalan dyes

$$\text{D}-\text{SO}_2-\text{CH}{=}\text{CH}_2 \xrightarrow{\text{HSO}_3^-} \text{D}-\text{SO}_2-\text{CH}_2-\text{CH}_2\text{-SO}_3^-$$

inactive

Scheme 9.19 *Reaction of vinyl sulphone dyes with bisulphite.*

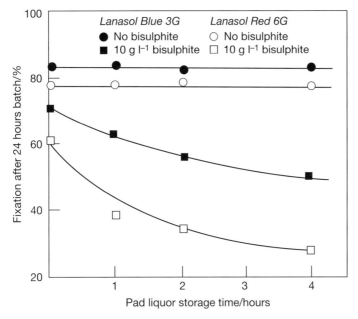

Figure 9.10   *Effect of bisulphite on Lanasol fixation in the cold pad–batch procedure.*

The effect of bisulphite on Lanasol dye pad liquor storage stability is demonstrated in Figure 9.10.

Apart from its effect on dye levelness and the fact it increases the rate of reaction during cold pad–batch dyeing, bisulphite also allows for the application of large-molecule dyes used for speciality shades. Of particular note is its effect in pad liquors containing copper phthalocyanine reactive dyes; the molecules of these dyes are relatively large and at room temperature do not adequately penetrate the fibre, even in the presence of 300 g kg^{-1} urea. In order to obtain desirable turquoise and emerald-green shades it is therefore essential to include sodium bisulphite in the pad liquor.

Lewis [58] also proposed a combined dye–polymer shrink-resist process for wool fabric, in which the pad liquor contained the reactive Bunte salt polyether prepolymer, Securlana (Cognis), along with reactive dye. The use of bisulphite in the pad liquor was essential to generate sufficient reactive cystine thiol residues to bring about disulphide crosslinking of the polymer during the batching process; Bunte Salt residues in the polymer react readily with wool thiols to form mixed disulphides, thus rendering the polymer water insoluble and imparting an excellent degree of shrink resistance.

The choice of auxiliary product in the pad–batch (cold) procedure plays a vital role in achieving adequate dye levelness. In the early days, pad dyeing wool fibres gave rise to 'frosty' dyeings, in which the centre of the fibrous mass contained more dye than the surface. The problem was resolved by careful selection of auxiliary products; suitable products perform several functions, acting as rapid wetting agents and forming strong films or coacervate complexes with the dye. They include lauroyl diethanolamide and anionic surfactants based on sodium di-iso-octyl sulphosuccinate.

9.4.3   Effect of Reactive Dyes on Fibre Properties

As described in Chapter 4, Section 4.7, permanent setting in wool fibres during the boiling process contributes to loss in wool fibre strength during dyeing [59–61]. The setting reactions are related to the ready elimination reactions at cystine and cystine [62–64] and to the subsequent formation of new, more hydrolytically stable crosslinks such as histidino-alanine, lysino-alanine and lanthionine [65,66]. Also of some importance in setting is the so-called thiol-disulphide interchange reaction [67,68].

As summarised in Chapter 4, control of setting in dyeing can be achieved by the addition of chemicals which scavenge hydrosulphide anions as they are liberated from cystine breakdown or which rapidly modify free cystine thiol residues to prevent the elimination reaction. Køpke's crease-angle method is usually used to quantify set in dyeing [69]; typically, blank dyeings of wool fabric in pH 5 buffer for 1 hour at the boil, without antisetting agent present, give set values of ~70%, whereas dyeing under the same conditions but with an effective antisetting agent gives a set value of ~30%.

Several workers have indicated that reactive dyes protect wool against damage in dyeing. Steenken *et al.* [70] observed that when dyeing wool/cotton blends with reactive dyes, considerable protection of the wool fibres against the alkaline conditions required to fix the cotton-reactive dye (pH 10, 40 °C) was obtained if the wool component was dyed first with a wool-reactive dye. Dittrich and Blankenburg [71] also noted that when wool tops were dyed under industrial conditions with reactive and chrome dyes and their strength measured by the bundle test (IWTO (E)-5-73), the dyeings with reactive dyes showed consistently higher strength values. Rouette [72] employed the bifunctional reactive dye Hostalan Black SB (**9**; C.I. Reactive Black 5) for the safe dyeing of deep shades on severely damaged carbonised wools. Rouette ascribed the fibre protection afforded by using this dye to its ability to crosslink adjacent peptide chains.

9

It is now well understood that selected classes of reactive dyes, applied in moderate to heavy depths of shade, actively prevent damage in wool dyeing [73–75], the effect being strongest with those dyes which contain activated carbon–carbon double bonds (these include acrylamido dyes and VS dyes). The study by Lewis [73] quantified the damaging effect of dyeing wool fabric with reactive dyes, using the wet burst strength test. These dyeings were produced at pH 4, in the presence of Albegal B (1% o.m.f.) and ammonium sulphate (4% o.m.f.), for 2 hours at 98 °C (liquor ratio 30 : 1). Figure 9.11 shows that the magnitude of this effect increased with increasing amounts of reactive dye applied, being optimum at about 3% dye o.m.f. The importance of this effect when dyeing wool fabric at pH 4 at 100, 105, 110 and 120 °C for various times in the presence and absence of 4% o.m.f. Lanasol Red 6G is demonstrated in Figure 9.12.

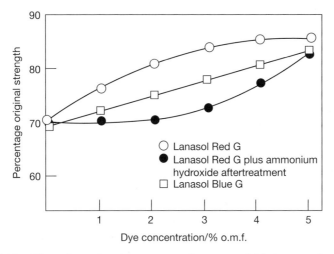

Figure 9.11   *Effect of reactive dye concentration on wool fabric strength retention.*

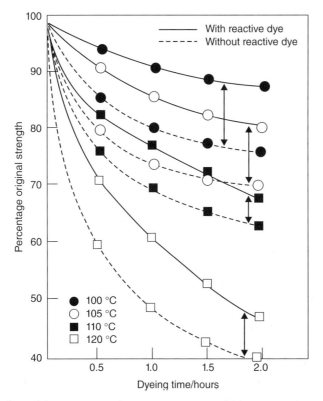

Figure 9.12   *Effect of dyeing time and temperature on wool fabric strength retention with and without reactive dye.*

This damage-limiting effect in full depths could be used in the field of high-temperature dyeing. Dyeing at 120 °C, as is often proposed for wool/polyester fabrics, is seen to be a very damaging procedure for the wool fibre, despite the protection afforded by the reactive dye. Levels of protection with full-shade dyeings of reactive dyes are equivalent to or even better than those obtained by Liechti for the fibre-protective agent Irgasol HTW [76].

It is interesting to reflect upon why reactive dyes based on reactive halogenated heterocy-cles, which react with wool fibre nucleophiles via a nucleophilic substitution reaction, are less effective in controlling wool damage in dyeing than are the activated carbon–carbon double bond-type reactive dyes. From the discussions in this section, it is clear that suc-cessful control of damage and level of set go hand in hand, and it is thus necessary to look carefully at the reactivity/stability of the reactive dye–cysteinyl residue covalent bond. Thioether derivatives of triazine or pyrimidine heterocycles will react further with amines to form bonds of greater stability; the leaving group in this reaction is the substituted thiol [77]. The thioether formed from reaction with an activated carbon–carbon double bond is, however, resistant to nucleophilic attack or β-elimination under the mildly acidic conditions pertaining in wool dyeing [78].

When set was measured on wool fabrics dyed at pH 5 with 3% o.m.f. of the activated halogenated heterocycle-type reactive dye Drimalan Red F-2G (Clariant), a value of 74% was obtained. In contrast, when this dye was replaced with a reactive dye containing an activated carbon–carbon double bond, Lanasol Red 6G (Huntsman) (3% o.m.f.), a set value of 41% was obtained [79]. The reactions responsible for these differences are summarised in Scheme 9.20.

Since the fibre-protective effect of reactive dyes is of real significance only at depths of shade greater than 3% o.m.f., Lewis and Wool Development International [80] developed colourless fibre-reactive compounds that can be included in wool dyebaths (pH 4–7) and reduce fibre damage in dyeing, acting in the same manner as reactive dyes. It is believed that Mirolan HTP and Mirolan Q from Huntsman function in this manner [81]; they might possibly be maleic acid derivatives.

Lewis and Smith [82] studied the nature of the fibre-protective effect when using modified VS-based reactive dyes and concluded that two mechanisms were involved.

First, reactive dyes interfere with the thiol–disulphide interchange reaction, restricting the extent of wool fibre setting in boiling dyebaths – provided there is sufficient reactive dye present to block cystine thiol groups and scavenge hydrogen sulphide. Confirmation of this hypothesis for dyeings produced with VS dyes was obtained by demonstrating the presence of the thioether dye in residual dyebaths and on the fibre. Further support for this mechanism comes from the work of Lee-Son and Hester [83], who used resonance Raman spectroscopy to examine wool dyed with α-bromoacrylamide dyes; evidence of a thi-irane ring was obtained at 695 cm^{-1} (excitation wavelength 488 nm for C.I. Reactive Red 84). It is possible that the thi-irane ring is formed according to Scheme 9.21.

Second, since reactive dyes react with wool proteins, they may end up immobilised in different morphological regions of the wool fibres when compared with the nonreactive acid dyes. Lewis and Smith [82] therefore suggested that reactive dyes react preferentially with the nonkeratinous regions of the cell membrane complex and the endocuticle; these proteins are easily accessible and are generally accepted to be the dyeing pathway into the fibre [84–86]. Selective modification of these proteins by covalent reaction with reactive

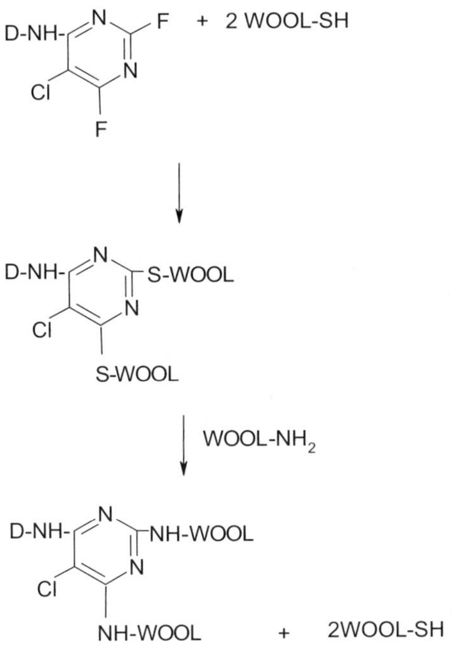

D-NH-CO-C(Br)=CH$_2$ + 2 WOOL-SH

↓

D-NH–CO-CH(S-WOOL)-CH$_2$-S-WOOL

↓ WOOL-NH$_2$

NO REACTION

(b)

Scheme 9.20 *Interference in wool setting reactions by reactive dyes: (a) activated heterocycle dye; (b) activated carbon–carbon double-bond dye.*

dyes may well reduce their solubility, thereby further maintaining the physical integrity of the fibre.

Fluorescent reactive vinylsulphonyl chlorotriazinyl (**10**), β-sulphatoethyl sulphonyl triazinyl derivatives (**11**) and a nonreactive analogue (**12**) were prepared [82], and the distribution of these fluorescent whitening agents (FWAs) within the fibre was examined by fluorescence microscopy. The photomicrographs obtained are reproduced in Figure 9.13 and show that reactive dyes selectively stain the outermost regions of the wool, the endocuticular and intercellular cuticular regions being particularly heavily dyed.

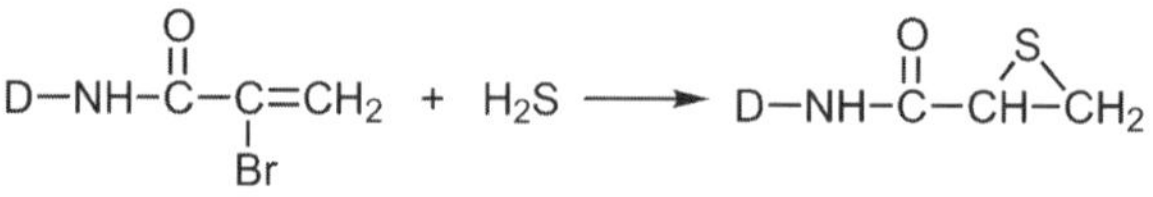

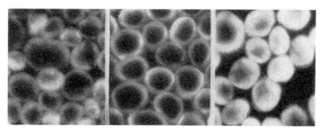

Scheme 9.21   *Reaction of α-bromoacrylamido dyes with hydrogen sulphide.*

Figure 9.13   *Photomicrographs of cross-sections from merino 21.5 μm wool staples dyed with (left to right) model compounds **10** and **11** (reactive FWAs) and **12** (nonreactive FWA) (0.5% o.m.f.) after 45 minutes at the boil [82].*

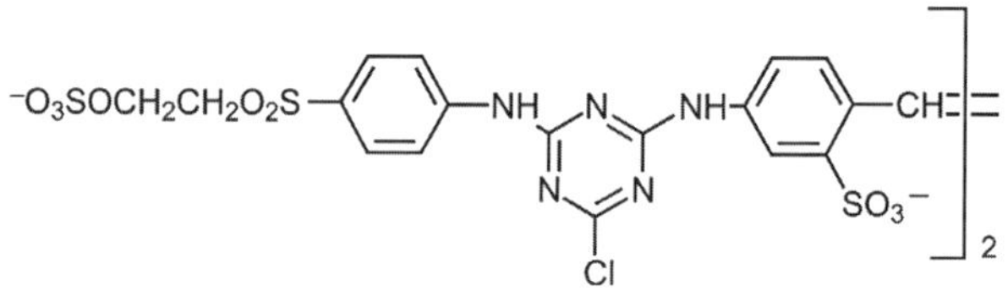

10

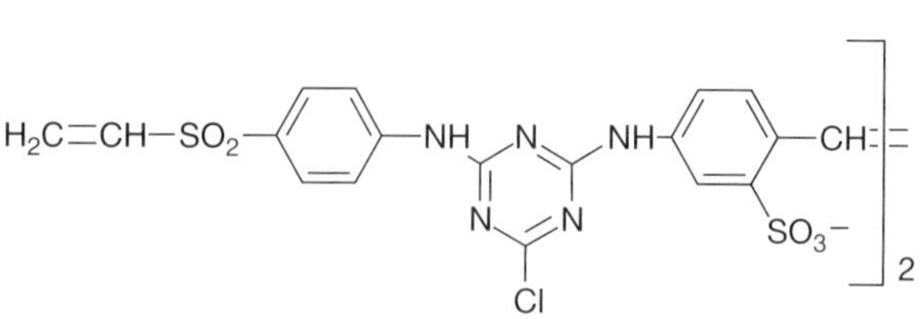

11

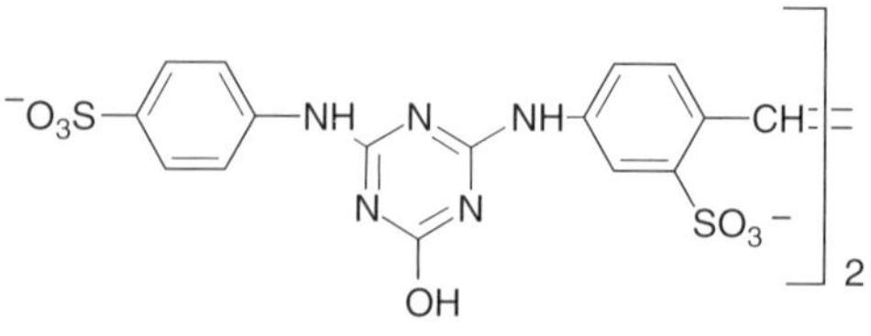

12

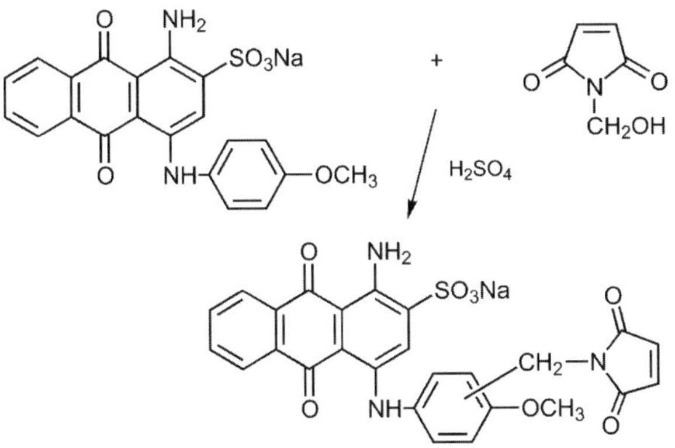

Scheme 9.22 *Tscherniac–Einhorn reaction to prepare fibre-reactive N-hydroxymethyl-maleinimide dyes.*

These observations have been supported by studies of similar dyeings using the Zeiss scanning photometer microscope [87]. This technique clearly demonstrates that cross-sections of the dyed fibres are predominantly 'ring-dyed' in the case of reactive dyes, in marked contrast to the good fibre penetration shown by the nonreactive dyes.

9.5   Novel Reactive Dye Systems for Wool

Several novel reactive systems have been proposed as reactive dyes for wool. The most promising groups are considered in this section.

9.5.1   Maleinimides

This system was proposed by Altenhofen and Zahn [88]. The proposed reactive dyes could be easily prepared from *N*-hydroxymethylmaleinimide and acid dyes using the Tscherniac–Einhorn reaction. Scheme 9.22 gives an example of this preparation to produce a blue reactive dye. The dye reacted very effectively with the fibre at pH 4 and 80 °C, giving a fixation value of 96.0%. At pH 5, these reactive dyes gave much reduced bath exhaustion.

9.5.2   Isocyanate and Isothiocyanate Bisulphite Adducts

Kirkpatrick and Maclaren [89,90] studied the preparation (Schemes 9.23 and 9.24) and reaction with wool of dyes with carbamoyl sulphonate and thiocarbamoyl sulphonate groups. The main promise of these dyes lay in the observation that application at pH 2 gave very little dye–fibre covalent bonding, while a simple aftertreatment of these dyeings with

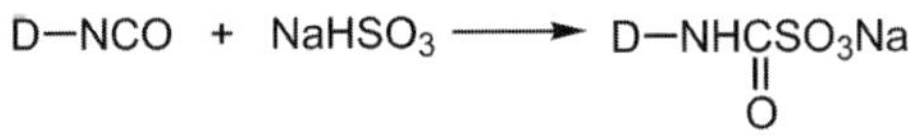

$$D{-}NCO + NaHSO_3 \longrightarrow D{-}NHCSO_3Na$$

Scheme 9.23   *New reactive dyes based on carbamoyl sulphonates.*

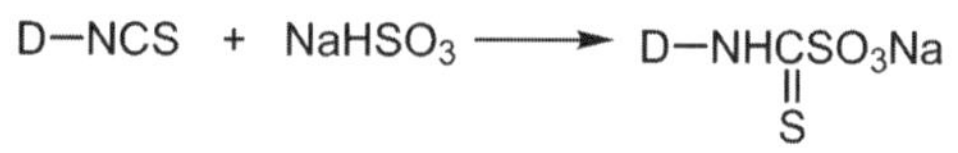

$$D-NCS \;+\; NaHSO_3 \longrightarrow D-NHCSO_3Na$$
$$\overset{\parallel}{S}$$

Scheme 9.24   *New reactive dyes based on thiocarbamoyl sulphonates.*

$$\overset{S}{\overset{\parallel}{D-CSCH_2COOH}} \;+\; wool{-}NH_2 \longrightarrow \overset{S}{\overset{\parallel}{D-CNH}}{-}wool \;+\; HSCH_2COOH$$

Scheme 9.25   *Reaction of carboxymethyl-carbodithioate dyes with amino groups in wool.*

sodium carbonate gave a high degree of fixation. Unfortunately, the carbamoyl sulphonate or thiocarbamoyl sulphonate group was the only solubilising group in the dye molecule; it was thus impossible to attribute high wet fastness solely to a high degree of covalent bonding with the fibre, since products of hydrolysis would be completely water insoluble and would therefore also show good wet fastness.

9.5.3   Carboxymethyl Carbodithioate Dyes

These dyes were described in a Unilever patent [91]. They show good resistance to hydrolysis and fix with amino groups in wool, according to Scheme 9.25. Van den Broek [92] studied their application to wool and observed a high degree of reaction with thiol and amino groups. Some yellowness was associated with thioamide bond formation; it faded rapidly on exposure to light, due to carbonamide formation by photooxidation of the thioketone.

9.5.4   Trifunctional Reactive Dyes Prepared from Bis-(chloroethyl-sulphonylethyl)amine [P-3] Reaction with a DCT Dye

Cho *et al.* [93] reported the following synthesis of a model trifunctional reactive dye, starting with a specially synthesised DCT dye analogous to Procion Red MX8B; this simple DCT dye was reacted (at 20 °C and pH 5) with 1 mol of the specially synthesised intermediate, *N,N′*-bis-{2[(2-chloroethyl)sulphonyl]ethyl}-amine reactive intermediate [P-3]. The reaction is shown in Scheme 9.26.

This dye was applied to wool serge at 3% o.m.f. along with other wool-reactive dyes for comparison, and their exhaustion and fixation efficiency (T) profiles versus dyebath pH

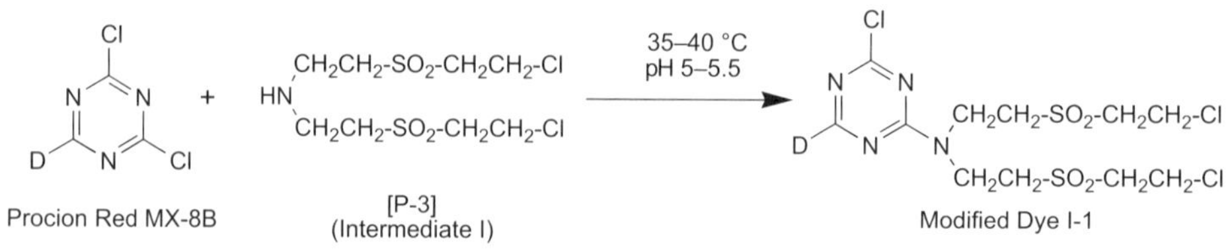

Scheme 9.26   *Preparation of a trifunctional reactive dye for wool.*

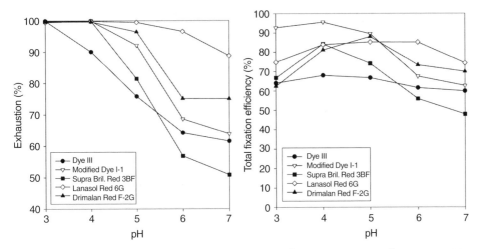

Figure 9.14 *Exhaustion/fixation efficiency (T) values for various reactive dyes versus pH.*

were determined (Figure 9.14). The trifunctional reactive dye showed remarkably improved dyeing properties when compared with the conventional wool-reactive dyes Lanasol Red 6G and Drimalan Red F-2G, especially in the application pH range of 3–4; additionally, its fibre protection properties were excellent [93].

9.5.5   Crosslinking Agents to Covalently Fix Acid Dyes to Wool

In Section 9.4.2.3, on hexamine aftertreatments of wool-reactive dyeings, the use of the liberated formaldehyde to covalently fix dyes containing nucleophilic side chains was fully described [31].

Van Beek and Heertjes [94] showed that acid dyes with the *o*-hydroxyazo structure could be fixed covalently to the fibre, following dyeing, by an aftertreatment with an alcoholic solution of a carbodi-imide such as *NN'*-dicyclohexylcarbodi-imide (0.20 mol l^{-1} solution).

Padhye and Rattee [95] have also studied this interesting route to fixing acid dyes, but they applied the carbodi-imide from aqueous alcoholic solutions. They claimed that the main effect of the carbodi-imide was to crosslink amino and carboxylic acid residues in the protein, thus entrapping the dye.

The earliest commercial success in using crosslinking agents to covalently fix dyes was achieved by BASF with the Basazol system [96,97]. In this case, dyes containing pendant nucleophilic groups were fixed to cellulose hydroxyl residues using the trifunctional crosslinking agent 1,3,5–triacroylaminohexahydro-*s*-triazine (**13**; Fixing Agent P). The dyes employed contained sulphonamide residues that were sufficiently nucleophilic above pH 10.5 to undergo Michael addition with the activated double bonds of the crosslinker. In fact, many of the aminosulphonyl dyes selected as Basazol dyes were already available as 2 : 1 premetallised dyes for the dyeing of wool (Ortolan – BASF). Although this concept was originally developed for cellulosic fibres, Lewis *et al.* [98] found that wool could be

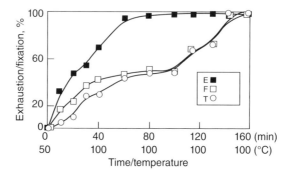

Figure 9.15 *Exhaustion/fixation curves from the bis-aminoalkyl dye/FAP dyeing method.*

dyed from long-liquor baths set at pH 6 and containing aminoalkylamino-*s*-triazine dyes and Fixing Agent P. Very high exhaustion values were obtained and total fixation efficiency T values of 95% were recorded. This system had the advantage of low reactivity until the boil was reached, giving the dye an excellent opportunity to level before covalent bonding became significant. Typical results for a 4% o.m.f. shade of the bis-aminoalkyl dye (**14**) fixed with 3% o.m.f. FAP are shown in Figure 9.15.

13   **14**

Disodio-2-chloro-4′,6-di(aminobenzene-4-sulphatoethylsulphone)-*s*-triazine (**15**; XLC) has been investigated as a crosslinking agent by which to covalently fix amino-nucleophile-containing dyes to wool [99], and very high values of covalent fixation have been recorded, either by co-applying dye and crosslinker at pH 6 (100 °C) or by dyeing first and then adding the crosslinker to the exhausted bath.

NaO₃SO-CH₂CH₂SO₂ — ⟨benzene⟩ — NH — triazine — NH — ⟨benzene⟩ — SO₂CH₂CH₂-OSO₃Na

15

Table 9.7   *Side chains involved in dye–fibre reaction.*

Amino Acid	Reactive Side Chain
Lysine	$-CH_2CH_2CH_2CH_2NH_2$ ϵ-amino
Cystine	$-CH_2SH$ β-thiol
Histidine	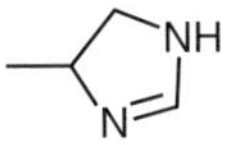 imizadole amine
Threonine	$\overset{\displaystyle OH}{\underset{\textstyle }{-CH_2CHCH_3}}$ secondary aliphatic hydroxyl
Serine	$-CH_2OH$ primary aliphatic hydroxyl
Tyrosine	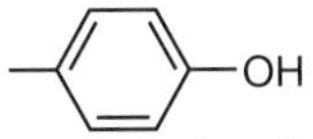 phenolic hydroxyl
Methionine	$-CH_2SCH_3$ thioether
N-terminal amino	$-\overset{\displaystyle O}{\overset{\displaystyle \|}{C}}-CHRNH_2$ α-amino

9.6   Identification of the Reaction Sites in the Fibre

Although mainly of theoretical interest, the identification of amino acid side chains reacting with reactive dyes has been the subject of many investigations. Shore [100] has reviewed the techniques for identifying these sites; the reactive side chains involved are listed in Table 9.7. Using model compounds, Shore [101] showed that for monochlorotriazine dyes the following groups would be involved in nucleophilic substitution (in order of relative reactivities on an equimolar basis): cystine thiol, N-terminal amino, histidine imidazole, lysine ϵ-amino, serine hydroxyl, tyrosine phenolic hydroxyl, arginine guanidino and threonine hydroxyl.

Baumgarte [102] and Corbett [103] confirmed the high reactivity of acrylamide and of VS dyes respectively towards cystine thiol groups. Lewis [48] studied the reaction of chloroacetyl dyes with amino acids and various modified amino acids, and noted reaction with cystine thiol at pH >2, histidine at pH >6, lysine at pH >5, glycine and valine at pH >7, tyrosine at pH 7.5 and methionine at pH >6.5.

The main problem in identifying the actual site of reaction in wool is that during dissolution of the protein into its constituent amino acids, many of the dye–amino acid linkages are also hydrolysed. The extent to which this occurs varies from dye to dye. In particular, the bond between Remazol Brilliant Blue R (C.I. Reactive Blue 19) and amino acids is resistant to acid hydrolysis; moreover, the blue chromophore is also unaffected. Thus N-ϵ-Remazol Brilliant Blue R–lysine, S-Remazol Brilliant Blue R–cystine

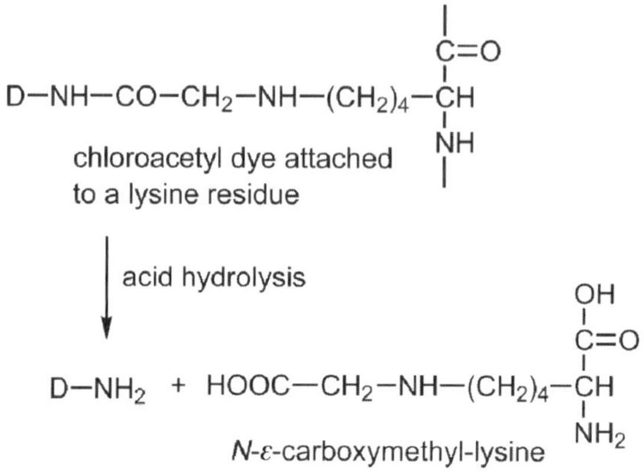

$$
\begin{array}{c}
\qquad\qquad\qquad\qquad\qquad\qquad\qquad | \\
\qquad\qquad\qquad\qquad\qquad\qquad\qquad C{=}O \\
\qquad\qquad\qquad\qquad\qquad\qquad\qquad | \\
\text{D}{-}\text{NH}{-}\text{CO}{-}\text{CH}_2{-}\text{NH}{-}(\text{CH}_2)_4{-}\text{CH} \\
\qquad\qquad\qquad\qquad\qquad\qquad\qquad | \\
\qquad\qquad\qquad\qquad\qquad\qquad\qquad \text{NH} \\
\qquad\qquad\qquad\qquad\qquad\qquad\qquad |
\end{array}
$$

chloroacetyl dye attached
to a lysine residue

⬇ acid hydrolysis

$$
\begin{array}{c}
\qquad\qquad\qquad\qquad\qquad\qquad\qquad\qquad OH \\
\qquad\qquad\qquad\qquad\qquad\qquad\qquad\qquad | \\
\qquad\qquad\qquad\qquad\qquad\qquad\qquad\qquad C{=}O \\
\qquad\qquad\qquad\qquad\qquad\qquad\qquad\qquad | \\
\text{D}{-}\text{NH}_2 \;+\; \text{HOOC}{-}\text{CH}_2{-}\text{NH}{-}(\text{CH}_2)_4{-}\text{CH} \\
\qquad\qquad\qquad\qquad\qquad\qquad\qquad\qquad | \\
\qquad\qquad\qquad\qquad\qquad\qquad\qquad\qquad \text{NH}_2
\end{array}
$$

N-ε-carboxymethyl-lysine

Scheme 9.27   *Acid digestion of wool dyed with chloracetylamido reactive dyes.*

and N-imidazole–Remazol Brilliant Blue R–histidine have all been isolated from dyed wool [104].

Some reactive dyes, despite being destroyed during the hydrolysis procedure, chemically alter the amino acid residue to which they were attached; thus Lewis [48] showed that wool dyed with chloroacetyl reactive dyes yielded carboxymethylated amino acids, and Derbyshire and Tristram [105] showed that with wool dyed with acrylamide reactive dyes, carboxyethyl derivatives were obtained. These modified amino acids are formed according to Schemes 9.27 and 9.28, and could readily be identified using the Moore and Stein automatic technique for amino acid analysis. When dyeing with chloroacetyl amino dyes under acidic dyeing conditions (pH 4–6), large amounts of S-carboxymethylcysteine and small amounts of N-carboxymethylhistidine were detected, indicating that cystine and histidine residues are important sites (cystine is progressively hydrolysed to cystine under hot aqueous conditions). On the other hand, at pH 7 significant amounts of N-ε-aminocarboxymethyl-lysine could be detected.

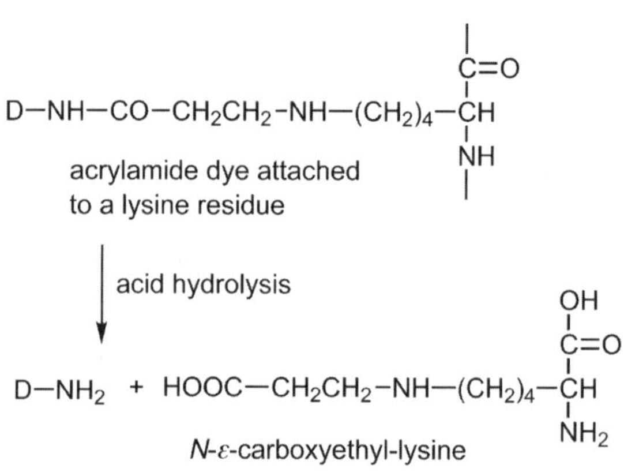

$$
\begin{array}{c}
\qquad\qquad\qquad\qquad\qquad\qquad\qquad | \\
\qquad\qquad\qquad\qquad\qquad\qquad\qquad C{=}O \\
\qquad\qquad\qquad\qquad\qquad\qquad\qquad | \\
\text{D}{-}\text{NH}{-}\text{CO}{-}\text{CH}_2\text{CH}_2{-}\text{NH}{-}(\text{CH}_2)_4{-}\text{CH} \\
\qquad\qquad\qquad\qquad\qquad\qquad\qquad | \\
\qquad\qquad\qquad\qquad\qquad\qquad\qquad \text{NH} \\
\qquad\qquad\qquad\qquad\qquad\qquad\qquad |
\end{array}
$$

acrylamide dye attached
to a lysine residue

⬇ acid hydrolysis

$$
\begin{array}{c}
\qquad\qquad\qquad\qquad\qquad\qquad\qquad\qquad OH \\
\qquad\qquad\qquad\qquad\qquad\qquad\qquad\qquad | \\
\qquad\qquad\qquad\qquad\qquad\qquad\qquad\qquad C{=}O \\
\qquad\qquad\qquad\qquad\qquad\qquad\qquad\qquad | \\
\text{D}{-}\text{NH}_2 \;+\; \text{HOOC}{-}\text{CH}_2\text{CH}_2{-}\text{NH}{-}(\text{CH}_2)_4{-}\text{CH} \\
\qquad\qquad\qquad\qquad\qquad\qquad\qquad\qquad | \\
\qquad\qquad\qquad\qquad\qquad\qquad\qquad\qquad \text{NH}_2
\end{array}
$$

N-ε-carboxyethyl-lysine

Scheme 9.28   *Acid digestion of wool dyed with acrylamido reactive dyes.*

9.7 Conclusions

Reactive dyes for wool have achieved a significant market share. In particular, Ciba (now Huntsman) have developed excellent navy and black formulations to match the shade and fastness properties of chrome dyes on untreated wool. With current and foreseeable environmental restrictions on the discharge of heavy metals potentially affecting the usage of chrome and premetallised dyes, an increasing trend towards using reactive dyes is likely; their benefits in terms of colour reproducibility and fibre protection have also been recognised.

Major problems still requiring solutions include the need for improved level-dyeing properties, making these dyes suitable for hank and winch dyeing, and elimination of the alkali aftertreatment to achieve maximum wet fastness. Reactive dyes tend to act as the best possible 'indicators' for preparatory variations. Thus, until either fabric and garment preparations become virtually uniform or the level-dyeing properties of reactive dyes approach those of acid levelling dyes, their usage in this type of machinery will be restricted.

References

[1] German patent 721231 (1937).
[2] German patent 965902 (1949).
[3] H-U Von der Eltz, *Textilveredlung*, **7** (1972) 297.
[4] I D Rattee and W E Stephen, British patents 772030, 774925, 781930.
[5] I D Rattee, *Journal of the Society of Dyers and Colourists*, **101** (1985) 46.
[6] A Bühler and R Casty, *Melliand Textilber*, **48** (1967) 693.
[7] W Mosimann, *Textile Chemist and Colorist*, **1** (1969) 282.
[8] D Mäusezahl, *Textilveredlung*, **5** (1970) 839.
[9] J S Church, A S Davie, P J Scammells and D J Tucker, *Review of Progress in Coloration and Related Topics*, **29** (1999) 85.
[10] CIBA BeP 683734 (1965).
[11] CIBA DP 6609467 (1965).
[12] CGY, D Mäusezahl and A Wohlkönig, DOS 2155464 (1970) and British patent 1351976 (1971).
[13] P Ball, U Meyer and H Zollinger, *Textile Research Journal*, **56** (1986) 447.
[14] W Mosimann and H Flensberg, *Proceedings of the 7th International Wool Textile Research Conference, Tokyo*, **5** (1985) 39.
[15] J B Caldwell, S J Leach and B Milligan, *Textile Research Journal*, **36** (1966) 1091.
[16] 16 K Hannemann, *Proceedings of the 8th International Wool Textile Research Conference, Aachen*, CD of Papers DY-6 (2000) 1.
[17] D Hildebrand and G Meier, *Textil Praxis*, **26** (1971) 499.
[18] R S Blackburn, S M Burkinshaw and K Ghandhi, *Advances in Coloration Science and Technology*, **2** (1999) 109.
[19] F Osterloh, *Melliand Textilber*, **49** (1968) 1444.
[20] F Osterloh, *Textil Praxis*, **26** (1971) 164.
[21] H Fuchs and H Konrad, *Melliand Textilber*, **55** (1974) 458.

[22] J R Christoe and A Datyner, *Applied Polymer Symposium.*, **18** (1971) 447.
[23] J F Graham, R R D Holt and D M Lewis, *Proceedings of the 5th International Wool Textile Research Conference, Aachen*, **5** (1975) 200.
[24] D M Lewis, I D Rattee and C B Stevens, *Proceedings of the 3rd International Wool Textile Research Conference, Paris*, **3** (1965) 305.
[25] D M Lewis, PhD Thesis, Leeds University (1966).
[26] K R F Cockett and D M Lewis, *Journal of the Society of Dyers and Colourists*, **92** (1976) 141.
[27] J-H Dittrich and H-J Henning, *Textilveredlung*, **9** (1974) 227.
[28] N Werkes, *Melliand Textilber*, **70** (1989) 52.
[29] R Jerke, E D Finnimore, J Kopecky, J-H Dittrich and H Höcker, *Schriftenreihe DWI, Aachen*, **102** (1987) 49.
[30] D G Evans, *Textile Journal of Australia*, **46** (1971) 20.
[31] E M Smolin and L Rapport, *The Chemistry of Heterocyclic Compounds; s-triazines and Derivatives* (New York, USA: Interscience, 1959).
[32] E D Finnimore, U Meyer and H Zollinger, *Journal of the Society of Dyers and Colourists*, **94** (1978) 17.
[33] X-P Lei, D M Lewis and Y-N Wang, *Journal of the Society of Dyers and Colourists*, **108** (1992) 383.
[34] T L Dawson, *Journal of the Society of Dyers and Colourists*, **80** (1964) 134.
[35] A Datyner, E D Finnimore and U Meyer, *Journal of the Society of Dyers and Colourists*, **93** (1977) 278.
[36] T J Abbott, R S Asquith, D K Chan and M S Otterburn, *Journal of the Society of Dyers and Colourists*, **91** (1975) 133.
[37] F Carlini, U Meyer and H Zollinger, *Melliand Textilber*, **60** (1979) 587.
[38] D M Lewis, *Melliand Textilber*, **67** (1986) 717.
[39] D M Lewis and S M Smith, *Proceedings of the 8th International Wool Textile Research Conference, Christchurch*, (1990).
[40] ICI, British patent 849772; S, French patent 1246743.
[41] M Noda, H Kajita and K Kuwuyana, German patent 1619606 (1973); Dethloth and Klein, German patent 2326522 (1976).
[42] D M Lewis and I Seltzer, *Journal of the Society of Dyers and Colourists*, **88** (1972) 327.
[43] H Zollinger, *Textilveredlung*, **6** (1971) 57.
[44] H R Hadfield and D R Lemin, *Journal of the Textile Institute*, **51** (1960) T1351.
[45] J R Christoe, A Datyner and R L Orwell, *Journal of the Society of Dyers and Colourists*, **87** (1971) 231.
[46] A Datyner, E D Finnimore, B Furrer and U Meyer, *Proceedings of the 5th International Wool Textile Research Conference, Aachen*, **3** (1975) 542.
[47] H K Rouette, J F K Wilshire, I Yamase and H Zollinger, *Textile Research Journal*, **41** (1971) 518.
[48] R S Asquith, W-F Kwok and M S Otterburn, *Textile Research Journal*, **48** (1978) 1.
[49] D M Lewis, *Journal of the Society of Dyers and Colourists*, **97** (1981) 365.
[50] D M Lewis and I Seltzer, *Journal of the Society of Dyers and Colourists*, **84** (1968) 501.
[51] H Niederer and P Ulrich, *Textilveredlung*, **3** (1968) 337.

[52] R S Asquith and A K Booth, *Textile Research Journal*, **40** (1970) 410.

[53] S Horrabin, *Journal of the Chemical Society*, (1963) 4130.

[54] D M Lewis and E L Gillingham, *Proceedings of the ISF Conference, Yokahama*, (1994) 26.

[55] J F Corbett, *Proceedings of the 3rd International Wool Textile Research Conference, Paris*, **3** (1965) 321.

[56] U Baumgarte, *Melliand Textilber*, **43** (1962) 1297.

[57] J A Maclaren and B Milligan, *Wool Science: The Chemical Reactivity of Wool* (Marrickville, NSW, Australia: Science Press, 1981), 6.

[58] D M Lewis, *Journal of the Society of Dyers and Colourists*, **93** (1977) 105.

[59] A J Farnworth and J Delmenico, *Permanent Setting of Wool* (Watford, UK: Merrow, 1971).

[60] J R Cook and J Delmenico, *Journal of the Textile Institute.*, **62** (1971) 27.

[61] P G Cookson, K W Fincher and P R Brady, *Journal of the Society of Dyers and Colourists*, **107** (1991) 135.

[62] I Steenken and H Zahn, *Journal of the Society of Dyers and Colourists*, **102** (1986) 269.

[63] B Milligan and J A Maclaren, in *Wool Science: The Chemical Reactivity of Wool*, ed. J A Maclaren and B Milligan (Marrickville, NSW, Australia: Science Press, 1981).

[64] H Zahn, *Proceedings of the 9th International Wool Textile Research Conference, Biella*, **I** (1995) 1.

[65] L R Mizell and M Harris, *Journal of Research of the National Bureau of Standards*, **30** (1943) 47.

[66] J B Speakman, *Journal of the Society of Dyers and Colourists*, **52** (1936) 335.

[67] F Sanger, A P Ryle, L F Smith and R Kitai, *Proceedings of the International Wool Textile Research Conference, Melbourne*, **C** (1955) 49.

[68] A P Ryle, F Sanger, L F Smith and R Kitai, *Biochemical Journal*, **60** (1955) 541.

[69] V Kǿpke, *Textile Research Journal*, **61** (1970) 361.

[70] I Steenken, I Souren, U Altenhofen and H Zahn, *Textil Praxis*, **39** (1984) 1146.

[71] J H Dittrich and G Blankenburg, *Textil Praxis*, **38** (1983) 466.

[72] P F Rouette, *Textil Praxis*, **27** (1972) 722.

[73] D M Lewis, *Journal of the Society of Dyers and Colourists*, **106** (1990) 270.

[74] P Ball, U Meyer and H Zollinger, *Proceedings of the 7th International Wool Textile Research Conference, Tokyo*, **5** (1985) 33.

[75] H Flensberg and W Mosimann, *Proceedings of the 7th International Wool Textile Research Conference, Tokyo*, **5** (1985) 39.

[76] P Liechti, *Journal of the Society of Dyers and Colourists*, **98** (1982) 284.

[77] D Brock, T Youseff and D M Lewis (1999), WO 99/51684, WO 99/51686 and WO 99/51687, to Procter and Gamble.

[78] D M Lewis and S M Smith, *Dyes and Pigments*, **29** (1995) 275.

[79] D M Lewis, *Advances in Coloration Science and Technology*, **7** (2004) 1.

[80] D M Lewis and Wool Development International, EP application 85306256 (1985).

[81] K Hannemann, *Proceedings of the 11th International Wool Textile Research Conference, Leeds*, CD of Papers CCF (2005) 27.

[82] D M Lewis and S M Smith, *Journal of the Society of Dyers and Colourists*, **107** (1991) 357.

[83] G Lee-Son and R E Hester, *Journal of the Society of Dyers and Colourists*, **106** (1990) 59.

[84] J D Leeder, J A Rippon, F E Rothery and I W Stapleton, *Proceedings of the 7th International Wool Textile Research Conference, Tokyo*, **5** (1985) 99.

[85] J D Leeder, J A Rippon and I W Stapleton, *Proceedings of the 8th International Wool Textile Research Conference, Christchurch, Christchurch*, **4** (1990) 227.

[86] J A Rippon, in *Wool Dyeing*, ed. D M Lewis (Bradford, UK: Society of Dyers and Colourists, 1992), 1.

[87] G Blankenburg, K Laugs and A Thiessen, *Textilveredlung*, **24** (1989) 1.

[88] U Altenhofen and H Zahn, *Textilveredlung*, **12** (1977) 9.

[89] A Kirkpatrick and J A Maclaren, *Journal of the Society of Dyers and Colourists*, **93** (1977) 272.

[90] A Kirkpatrick and J A Maclaren, *Australian Journal of Chemistry*, **30** (1977) 897.

[91] N H Leon and J A Swift, Unilever, British patent 1309743 (1973).

[92] D Van den Broek, PhD Thesis, Leeds University (1974).

[93] H J Cho, D M Lewis and B H Jia, *Coloration Technology*, **123** (2007) 86.

[94] H C A Van Beek and O M Heertjes, *Melliand Textilber*, **44** (1963) 987.

[95] R M Padhye and I D Rattee, *Proceedings of the 7th International Wool Textile Research Conference, Tokyo*, **5** (1985) 59.

[96] U Baumgarter, *Melliand Textilberichte*, **49** (1968) 1432.

[97] G Lützel, *Journal of the Society of Dyers and Colourists*, **82** (1966) 293.

[98] D M Lewis, Y N Wang and X P Lei, *Journal of the Society of Dyers and Colourists*, **111** (1995) 12.

[99] X P Lei, D M Lewis, X M Shen and Y N Wang, *Dyes and Pigments*, **30** (1996) 271.

[100] J Shore, *Journal of the Society of Dyers and Colourists*, **84** (1968) 408.

[101] J Shore, *Journal of the Society of Dyers and Colourists*, **84** (1968) 413.

[102] U Baumgarte, *Melliand Textilber.*, **43** (1962) 1297.

[103] J F Corbett, *Proceedings of the 3rd International Wool Textile Research Conference, Paris*, **3** (1965) 321.

[104] H Baumann and H Zahn, *Textilveredlung*, **3** (1968) 241.

[105] A N Derbyshire and G R Tristram, *Journal of the Society of Dyers and Colourists*, **81** (1965) 584.

10

Dyeing Wool Blends[1]

David M. Lewis
Department of Colour Science, University of Leeds, UK

10.1 Introduction

The use of wool in admixture with other natural fibres dates from early times. The Old Testament prohibition on wearing garments woven of wool and linen suggests that the use of such mixture fabrics was customary among the neighbours of the ancient Hebrews [1]. By the 19th century, a wide range of fabrics made from mixtures of wool with silk or cotton was being produced. With the introduction of synthetic fibres over the past 70 years, blends of wool with polyester, nylon and acrylic fibres have assumed major importance. The combination of two or more different fibres into a blend makes it possible to produce textile articles with properties that cannot be obtained by the use of a single fibre. In addition to expanding aesthetic properties such as texture, lustre and colour, the blending of non-wool fibres with wool can lead to articles with physical properties superior to those of the pure wool goods. For example, improved resistance to wear is obtained by incorporating nylon into woollen socks and carpets, and inclusion of a suitable level of polyester in a worsted fabric enables it to be permanently pleated by a heat-setting operation. For some end uses, the incentive to blend is economic: cheaper products can be made by partly substituting wool fibres with others that are less expensive.

Although synthetic fibres were designed to overcome certain deficiencies of natural fibres, they are not without their own limitations [1]. The inability of most synthetics to take up moisture can lead to the build-up of static electricity, and harshness of handle is

[1] Some information in this chapter is based on that published in Chapter 9 of *Wool Dyeing*, ed. David M. Lewis, published SDC, 1992 and is used with permission from the Society of Dyers and Colourists.

The Coloration of Wool and other Keratin Fibres, First Edition. Edited by David M. Lewis and John A. Rippon.

Table 10.1   *Consumption of principal wool-containing goods[a] (1988) [2].*

Product Area	Consumption (million kg)		
	All-Wool	Wool-Rich	Wool-Poor
Men's outerwear	80.6	27.8	46.9
Women's outerwear	98.2	72.4	66.5
Adults' knitwear	56.1	45.3	42.2
Carpets	139.9	60.3	12.1

[a]In Belgium, France, West Germany, Italy, Japan, the Netherlands, the UK and the USA.

often a problem. The inclusion of wool as the minor component in a synthetic fabric can significantly improve properties such as comfort, handle and drape.

Statistics relating to the consumption of wool and wool blends in 1988 are given in Table 10.1. These indicate the significance of blends. For men's and women's outerwear and adults' knitwear, blends constituted 56% (on a mass basis) of the market for the countries considered, and 34% of wool-containing carpets were blends. In outerwear markets, blends with polyester are the most important [2]. In 1987, the estimated world consumption of wool/polyester blends was 450 million kg (excluding the People's Republic of China and Eastern Europe) [3]; estimated consumption in 1977 was 360 million kg (excluding Eastern Europe) [4]. In the carpet sector, wool/nylon blends have become increasingly popular. The bulk characteristics of the acrylic fibre make it suitable for incorporation in blends with wool, especially in the areas of knitwear and hand-knitting yarns. An increasing consumer awareness of the desirable aesthetic qualities of fabrics and garments made completely from natural fibres has been reflected in a resurgence of interest in blends of wool with cotton and with silk.

Although blends can be produced from different fibres that have been dyed separately, there is increasing pressure on the dyer to dye each fibre in the presence of the other(s), as in the cases of spun-blended yarns and fabrics, and yarn-blended fabrics. Fibre mixtures may be dyed in any of the following ways [5,6]:

- Each fibre is dyed to the same shade and depth (solid).
- Each fibre is dyed to a different depth of the same shade (shadow or tone-on-tone).
- One fibre is dyed, while the other is left undyed (reserve).
- Each fibre is dyed to a different shade (contrast).

Certain wool blends may be dyed using the same dyes for each fibre. In these cases, the partitioning of dye between the two fibres is of utmost importance, requiring judicious selection of dyes and dyeing procedures in order to achieve the desired effects. Although better control can be achieved when different classes of dye are used for each component, staining of the wool fibre by the non-wool dye is invariably a problem, and careful optimisation of dyeing methods is important. Care must also be taken to ensure that the conditions used for dyeing of the non-wool component do not lead to excessive damage to the wool fibre.

The dyeing of (binary) blends of wool with cotton, silk, polyamide (nylon), polyester and polyacrylonitrile (acrylic) fibres is dealt with in this chapter. For an understanding of blend dyeing, it is essential to appreciate the different properties of both the wool and the non-wool components, and how these properties affect dyeing behaviour. The dyeing of wool is covered in previous chapters, and both fundamental and practical aspects of the dyeing of the non-wool fibres are also presented in this chapter. Where appropriate, problems relating to wool damage due to extremes of pH and/or the temperature required to dye the non-wool component are discussed.

10.2   Wool/Cotton

Blends containing similar proportions of wool and cotton have been of considerable importance for many years, in spite of the inroads made by manmade fibres [5]. Union fabrics made from a cotton warp and a woollen weft have enjoyed some popularity for shirtings, pyjamas and similar end products, but these have now been largely replaced by synthetic-fibre blends with lighter constructions and superior laundering properties. Khaki shirtings for service wear have been produced from intimate blends of wool and cotton, and the union of cotton warp and wool weft has been used for gaberdine raincoating. Another large outlet has been union fabrics for blazers and coatings, in which the woollen wefts are derived from reprocessed wools.

Wool/cotton blends are used for articles such as shirts, blouses, knitting yarns, women's outerwear (where viscose is often used instead of cotton), general leisurewear and sportswear [7]. The use of wool/linen blends has also been reported [8,9]. A demand for items such as shirts, blouses and knitwear, which stand up to the rigours of domestic washing, has led to a need for better wet-fastness properties [10]. In addition, for blends containing more than 30% wool, a shrink-resist treatment is usually necessary if the final product is to have a satisfactory stability to washing. Shrinkproofing can be achieved by blending cotton with chlorine–Hercosett machine-washable wool, or by the pad application to the untreated wool/cotton fabric of a suitable resin, such as the blocked isocyanate Synthrappret BAP (Dohmen) or Securlana K (Cognis), a Bunte salt-terminated polyether applied by a pad–batch method [11,12]. The latter process was applied to Viyella wool/cotton (45:55) fabrics for many years.

10.2.1   Dyeing of Cotton

Of all the naturally occurring cellulosic fibres that have textile applications, cotton is the most important [13]. Other fibres include linen, which is obtained from flax, and hemp, jute and ramie, which are used mainly for industrial textiles. Regenerated fibres such as viscose are also of major importance.

For full details of the structure of cotton and its basic chemistry, the reader is referred to several excellent texts [14–17].

Natural and regenerated cellulose fibres adsorb dyes at different rates and to different saturation levels [13]. For example, exhaustion and fixation values of reactive dyes on viscose are usually higher than on cotton or linen. Fibres from different sources will often differ in their dyeing behaviour, and the scouring and bleaching treatments used for natural fibres may also modify their dyeability.

The cotton fibre contains small amounts of ionised carboxylate groups, which arise from oxidative bleaching, and, under alkaline conditions, alcoholate residues, which give rise to a negative surface charge, providing a barrier to the adsorption of anionic dyes. Electrolyte is thus added to dyebaths in order to reduce the magnitude of the surface potential, facilitating the approach of the dye anion to the range within which Van der Waals forces become effective [18–20]. In addition, electrolyte increases dye activity in solution and reduces the osmotic work that must be done in bringing sodium ions into the fibre during the adsorption of dye [19]. The use of dyes of large molecular area aids interactions between fibre and dye [18].

Direct, reactive, vat, azoic and sulphur dyes are the types most commonly used for the dyeing of cotton [13,20,21]. Direct dyes are normally applied to cotton under neutral conditions. All other classes of dye are applied under alkaline conditions, which poses a potential problem for wool/cotton blends, because of the sensitivity of wool towards alkali damage. Direct or reactive dyes are normally used to dye wool/cotton blends. Procedures are also available for the application of azoic and vat dyes [22]. The chemistry of direct, vat, azoic and reactive dyes for cellulosic fibres and their application processes to cellulosics has been comprehensively reviewed and will not be covered in detail here [13,23].

Direct dyes have relatively poor wet-fastness properties, and often require an aftertreatment with an agent to achieve satisfactory levels of fastness. Recommended procedures for the dyeing of the cotton portion of wool/cotton blends with direct dyes normally involve aftertreatment with cationic products. The effectiveness of cationic aftertreatments can be explained in terms of an interaction of these compounds with the free sulphonate groups of the dye molecules, leading to the formation of insoluble dye complexes that are more resistant to laundering [24].

When dyeing cotton with reactive dyes, alkaline conditions are usually required to bring about covalent reaction between dye and fibre; these reactions occur preferentially with the primary C6 hydroxy group in the cellulose fibre. In contrast, reactive dyes are applied to wool under neutral or slightly acidic conditions, and reaction occurs mainly with cystine thiolate anions $(-S^-)$ and amino groups $(-NH_2)$ [25]. Hydrolysis of reactive dyes is promoted at higher pH values [18], and the maximum fixation level expected for reactive dyes on cotton is about 80% [13], but in deep shades is often of the order of only 50% [26]. Significantly higher fixation levels (>90%) are obtained for reactive dyeings on wool, even in full depths [25]. The dyes listed in Table 10.2 are recommended

Table 10.2   *Cotton reactive dyes for wool/cotton blends.*

Dye Range	Structural Type
Remazol (HOE)	β-sulphatoethyl sulphone
Basilen M (BASF)	Dichlorotriazine
Procion MX (ICI)	Dichlorotriazine
Levafix E-A (BAY)	Monochlorodifluoropyrimidine
Drimarene K (S)	Monochlorodifluoropyrimidine
Cibacron F (CGY)	Monofluorotriazine

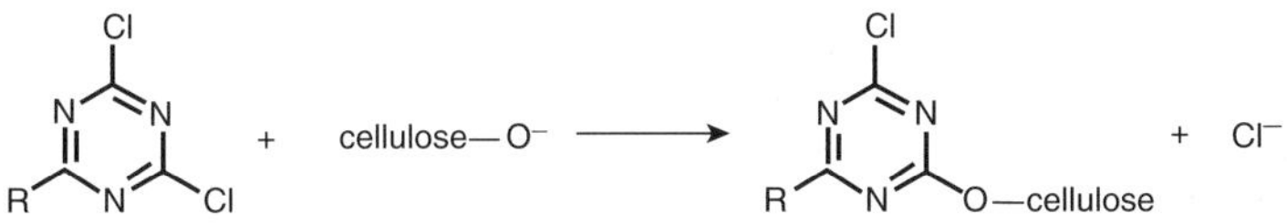

Scheme 10.1 *Nucleophilic substitution reaction of dichloro-s-triazine with cellulose.*

for the dyeing of the cotton portion of wool/cotton blends (see Section 10.2.2); of these, only the monochlorodifluoropyrimidine and β-sulphatoethyl sulphone (especially the *N*-methyltaurine adduct) types are recommended for wool [25]. It is interesting that the Lanasol α-bromoacrylamido dyes are not recommended for use on cellulosic fibres, presumably due to potential alkali instability of the linking amide group.

The exhaustion dyeing of cellulose with reactive dyes is normally carried out in the following three stages:

(1) Initial exhaustion in the presence of electrolyte under neutral conditions at temperatures usually ranging from 20 to 40 °C. Dye is adsorbed in a manner similar to that used for direct dyes, with uptake generally ranging from 25 to 60% [21]. During this stage, very little dye–fibre covalent bonding occurs, and the dye molecules can achieve uniform distribution in the cellulose fibres because they are free to migrate; the use of levelling agents is normally unnecessary. In the case of wool, however, reactive dyes can generally react with the fibre during adsorption, and suitable auxiliaries are necessary to reduce the problem of unlevel dyeing [25]. Scheme 10.1 describes the nucleophilic substitution reaction of a dichloro-*s*-triazine reactive dye with cellulose, and Scheme 10.2 describes the Michael addition mechanism of a vinylsulphone (VS) reactive dye with cellulose.
(2) Addition of alkali (pH 8–12) to promote chemical reaction of adsorbed dye with the fibre and further dye uptake. For those dyes suitable for dyeing of the cotton portion of wool/cotton blends, fixation temperatures normally range from 20 to 60 °C.
(3) Rinsing and soaping of the dyed material to remove electrolyte, alkali and unfixed dye. The removal of unfixed dye, which is present in relatively large amounts because of hydrolysis, is of the utmost importance if adequate fastness properties are to be ensured. Hydrolysed dye normally has low substantivity for cellulose and can be effectively removed during washing-off at the boil. Reactive dyes have been deliberately developed so that the hydrolysed species have low substantivity [21], but when dyeing wool/cotton, hydrolysed reactive dye has a higher substantivity for wool where it is adsorbed as an acid dye, and thus its removal is somewhat more difficult [18].

An alternative approach has been proposed, which avoids a separate alkali fixation step for reactive dyes on cotton [27,28]; this depends on using temperature to ionise the cellulose hydroxyl groups. A study of the neutral fixation (pH 7) of a variety of reactive dyes on

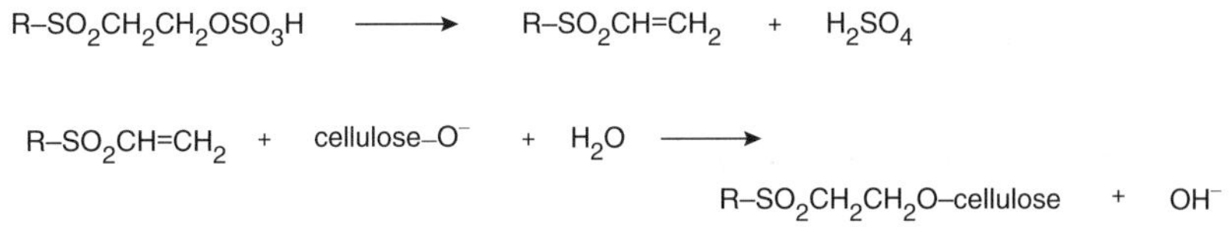

$$R–SO_2CH_2CH_2OSO_3H \longrightarrow R–SO_2CH=CH_2 + H_2SO_4$$

$$R–SO_2CH=CH_2 + cellulose–O^- + H_2O \longrightarrow$$
$$R–SO_2CH_2CH_2O–cellulose + OH^-$$

Scheme 10.2 *Michael addition of vinylsulphones with cellulose.*

cotton demonstrated that promising overall total fixation efficiencies can be obtained by increasing the electrolyte concentration, by working at the lowest possible liquor ratios and by dyeing at 100 °C for fluoro-triazine and VS derivatives and at 140 °C for the less reactive monochloro-triazine derivatives. Because of the neutral dyeing conditions, the risk of hydrolysis of dyes can be minimised. This new dyeing process requires that the temperature and the concentration of electrolyte used are higher than in the conventional dyeing method. However, because of the low liquor ratio employed, less energy is used to heat the dyebath solution and lower amounts of chemicals are used. The smaller volume of water used for dyeing and the reduction in waste effluents potentially reduce the cost of wastewater treatment after dyeing. The neutral exhaust application of reactive dyes to cellulosic fibres gives dyeings with excellent reproducibility, levelness and penetration. Moreover, since all the chemicals, such as dyes, sodium sulphate and sodium acetate, can be put into the dyeing solution at the same time, it is possible to simplify the dyeing process and save operation costs. The most suitable dyes, giving fixation values similar to those achieved in the conventional alkali fixation process, are the bis-monofluoro-*s*-triazines and the VS types. Clearly, this method will have use when dyeing wool/cotton blends, since the damaging effect of alkaline conditions can be avoided.

Reactive dyes can be applied to cellulosic fabrics using padding procedures. Of particular importance for wool/cotton blends is the cold pad–batch operation, which involves the following sequence of operations [13]:

(1) Impregnation of the fabric with a cold solution of dye and alkali.
(2) Uniform expression of the surplus liquor by padding.
(3) Storage of the batched roll of wet fabric at ambient temperature for a predetermined fixation period (between 2 and 48 hours).
(4) Washing-off.
(5) Drying.

In order to minimise dye hydrolysis in the padding trough, liquor-feed devices have been developed whereby separate dye (including auxiliaries) and alkali solutions are brought together immediately before the mixed padding solution comes in contact with the fabric. In comparison with exhaustion processes, the relatively high concentration of dye means that the concentration of electrolyte must be reduced, because of the hazards associated with precipitation of dye; urea may be added to the pad bath to improve dye solubility. After batching, fabric should be covered with plastic sheeting to prevent moisture evaporation and rotated to avoid seepage. When caustic soda is used as the alkali, there is some risk of poor fixation at the selvedges due to a drop in pH through reaction with carbon dioxide. This problem can be largely overcome by using sodium silicate, which produces a strong buffering action on caustic soda [29]. Silicate also increases the stability of the pad bath, but does not adversely affect the rate of dye fixation. The use of a pad–batch (rather than exhaustion) procedure to dye cotton leads to a more efficient reaction between dye and fibre, with consequent savings in dye costs [30].

10.2.2   Exhaustion Dyeing of Wool/Cotton Blends

The normal alkaline conditions used as a preparation for cotton goods cannot be employed for wool/cotton blends due to the damage which would be caused to the wool. Scouring and setting operations can be carried out in a similar manner to those used for pure wool. The procedures for exhaustion dyeing of blends fall into four major categories, depending on whether direct, reactive, azoic or vat dyes are used to colour the cotton component [12].

10.2.2.1   Direct Dyes

Both the wool and the cotton portions of a blend can be dyed in a single dyebath using a mixture of direct dyes and wool dyes to generate a wide range of shades. In certain instances, depending on the dye structures, direct dyes alone can be used for both fibres. When producing heavy shades with direct dyes, it is necessary to use large quantities of Glauber's salt (up to about $60\,g\,l^{-1}$) in order to achieve a satisfactory build-up of colour on the cotton. Cooling of the dyebath to $60\,^{\circ}C$ prior to rinsing can further improve uptake of the direct dye on the cotton component.

In order to minimise cross-staining of the wool by direct dyes, the use of a blocking agent is usually recommended. These auxiliaries are normally large-molecular-weight aromatic sulphonates which exhaust on to the wool and restrict the uptake of direct dye. Similar compounds are used to reduce the rate of uptake of acid dyes by nylon (see Section 10.4.1). Especially for heavy shades, aftertreatment with a polycationic product such as Indosol E-50 (Clariant) or Tinofix ECO (Huntsman) is essential in order to achieve satisfactory levels of fastness.

10.2.2.2   Reactive (Cotton) Dyes

When generating bright shades on wool/cotton blends, the cotton can be dyed with reactive dyes and the wool with reactive or selected nonreactive dyes. Because of the different pH and temperature requirements for the dyeing of each fibre, dyeings must be carried out in two stages.

The dyes and conditions recommended for the dyeing of deep (solid) shades on the cotton portion of an intimately blended $55:45$ wool/cotton yarn are shown in Table 10.3. Conventional wool-dyeing methods are used to dye the wool component in a separate dyebath. For shades composed of mixtures of sulphonated metal-complex wool dyes and fluoro-chloro-pyrimidine (FCP) reactive dyes, it has been recommended that the wool portion of the yarn is dyed first. For other dyes, the cotton is dyed first, and it has been suggested that this approach is essentially no more complicated than that used to dye pure cotton, since the wool may be regarded as being dyed during a normal 'soaping-off' process [31]. For pale and medium shades, the concentrations of electrolyte and the dyeing times can be reduced for all dyes. With the exception of Cibacron F (now Novacron F dyes (Huntsman)), where $2\,g\,l^{-1}$ soda ash (pH 9–10) is used for all depths of shade on wool/cotton blends [32,33], the amount of alkali may also be decreased. Reductions in alkali concentration at higher liquor-to-goods ratios are advised for dyeings with Procion MX (DyStar) dyes. In order to achieve good tone-on-tone dyeing of both fibres in a blend using reactive dyes, dye recipes should be restricted to binary combinations on each fibre [34].

Following dyeing of the cotton portion of a blend with certain Drimarene K (Clariant) dyes and replacement of the dyebath without intermediate rinsing, it is possible to dye the wool component with unfixed dye remaining on the fibres [7]. This seems to be an odd recommendation, since the nonfixed dye is a mixture of unreacted and hydrolysed dye;

Table 10.3   *Cotton reactive dyes for wool/cotton yarn (liquor ratio 15 : 1, deep shades) [12].*

Dye	Temperature (°C)	Alkali	Electrolyte	Time (minutes)
Remazol	20	$5 g l^{-1}$ soda ash + 4.5 ml l^{-1} caustic soda (38° Bé)	$80 g l^{-1}$ sodium sulphate	90
Basilen M	30	$10 g l^{-1}$ soda ash	$60 g l^{-1}$ sodium chloride	60
Procion MX	30	$12 g l^{-1}$ soda ash	$55 g l^{-1}$ sodium chloride	60
Drimarene K	40	$5 g l^{-1}$ soda ash	$60 g l^{-1}$ sodium sulphate	90
Levafix E-A	40	$9 g l^{-1}$ soda ash	$60 g l^{-1}$ sodium sulphate	90
Cibacron F	60	$2 g l^{-1}$ soda ash	$60 g l^{-1}$ sodium sulphate	45

absorption of the hydrolysed dye component by wool would compromise the wash-fastness properties of the dyeing. Drimarene K dyes contain the monochlorodifluoropyrimidine group, which reacts with wool or cotton via nucleophilic substitution of one or both fluorine atoms [25]. A similar procedure, called the Leva-metering process, was promoted for the Levafix E-A dyes [35,36]. The cotton is dyed first using caustic soda at 40 °C, then, after adjusting the pH, the wool is dyed in the same bath. The addition of alkali is regulated by a metering pump, and the pH is controlled so that the optimum colour yield is obtained on the cotton, with only minimal damage being caused to the wool. A modified procedure with Cibacron F dyes has been developed, in which the pH is adjusted to 6–7 at the end of the cotton-dyeing cycle [33]. The hydrolysed dye remaining in the bath behaves like an acid dye and exhausts on to the wool as the temperature is raised to 93 °C. In some cases, it is necessary to adjust the shade of the wool using Lanaset dyes (CGY), which can be added to the dyebath at the beginning of the cotton-dyeing operation.

10.2.2.3   Vat Dyes

When applying vat dyes [22,37] to 50 : 50 wool/cotton blends, the optimum conditions are $4 g l^{-1}$ caustic soda and $4 g l^{-1}$ sodium hydrosulphite at 40 °C for 45 minutes. Under more severe conditions, the wool portion of the blend is unacceptably damaged. Oxidation is carried out at 40 °C for 15 minutes with $3 g l^{-1}$ of sodium perborate, percarbonate or persulphate. Finally, the goods are given an acid sour and then soaped at the boil. If the dyeing is off-shade, the wool component can be adjusted in shade through the addition of selected acid dyes to the soap-off process.

   Use of this procedure imposes severe limitations on the choice of vat dyes, since many of the commonly used dyes require larger quantities of caustic soda to ensure sufficient stability in the leuco form, and higher dyeing temperatures to promote an adequate rate of sorption of dye by the fibres. The following dyes form a basis on which a wide variety of colours, over a range of depths, may be obtained, with both the wool and the cotton being dyed to a similar shade:

- C.I. Vat Oranges 3 and 15;
- C.I. Vat Red 10;
- C.I. Vat Violets 14 and 17;
- C.I. Vat Blues 14 and 16;
- C.I. Vat Greens 1 and 3;
- C.I. Vat Browns 3, 38 and 49;
- C.I. Vat Black 27.

10.2.2.4   *General Comments*

Azoic dyes have become scarcer in the market, possibly due to environmental and safety issues; their chemistry and use are thus not covered in this chapter.

A major problem in the exhaustion dyeing of wool/cotton blends is the inadequate build-up of colour on the cotton component when dyeing heavy shades, especially when using reactive dyes. This is brought about by the following factors:

- The alkaline scouring conditions required to break down the primary wall of the cotton fibre prior to dyeing are too severe to be used for a blended yarn or fabric, because of the problem of damage to the wool fibre. The need to use relatively mild pretreatments means that the dyeability of the cotton fibre will be less than optimum. In addition, when reactive dyes are used for the cotton, the optimum dyeing conditions with regard to both the amount of alkali required and the dyeing temperature cannot always be employed, because of wool damage. For a yarn-blended fabric, it would be possible to alkaline-scour or mercerise the cotton yarn prior to weaving and subsequent piece dyeing.
- Compared with wool dyeing, the uptake of cotton dyes (especially direct and reactive dyes) is often poor. This is further exacerbated in the dyeing of, for example, a 55 : 45 wool/cotton blend, where a liquor-to-goods ratio of 15 : 1 means that the liquor-to-cotton ratio is as high as 33 : 1.
- Cross-staining of the wool by the cotton dyes means that hydrolysed reactive dye is present on the wool, and the wash fastness and perspiration fastness of such a dyeing is thus suspect.

For a substrate such as an intimately blended yarn, it may be feasible to overcome a lack of depth on the cotton by adjusting the wool-dye recipe to give the correct overall depth and shade. For heavy, dull shades, direct dyes should be used in preference to reactive dyes for the cotton component. For certain deep shades, selected vat dyes are available as alternatives, the methods of application usually being restricted to jig dyeing of pieces and to yarn dyeing in pressure-circulating machinery.

As the level of wool damage increases in a blend, cross-staining of the wool by direct, reactive (cotton) and vat dyes increases. This is especially true for blends containing chlorinated wool, where the wool is often dyed to a heavier shade than the cotton when using cotton dyes alone. Although the use of blocking agents [10] corrects the problem to some extent, solidity of shade may be difficult to achieve. The post-chlorination of a dyed wool blend is not normally recommended, due to problems regarding the fastness to chlorine of the dyes on the non-wool fibre. Colour-woven styles for wool/cotton blends are often produced by the separate application of fast dyes to the cotton and chlorinated wool fibres before blending [10].

For most shades of medium or greater depth on wool/cotton blends dyed with direct or reactive (cotton) dyes, reasonable fastness to machine washing can be obtained if an aftertreatment with a suitable cationic product is carried out. In addition to improving the fastness of the dyes on the cotton fibre, these auxiliaries improve the fastness of the dyes – including hydrolysed reactive (cotton) dye – on the wool fibre. Because of this, selected nonreactive dyes can be used for the wool component in order to achieve the levels of fastness normally expected from reactive dyes. The presence of the shrinkproofing polymer Securlana (Cognis) on a wool/cotton fabric has little effect on dyeing quality when dyeing with mixtures of Lanasyn S (Clariant) and Indosol SF dyes; shrinkproofing before dyeing is therefore a realistic option.

Cai *et al.* [38] have studied the dyeing of a Hercosett wool/cotton blend (30 : 70) with reactive dyes. They concluded that the usual blocking agents based on Syntans should be avoided as they reduce light fastness and give rise to discoloration; a better blocking agent, which overcame these deficiencies, was the anionic surfactant sodium dodecylbenzene-sulphonate.

10.2.3 Pad Dyeing of Wool/Cotton Blends

Procedures similar to those used for pure cotton can be adopted for the pad–batch dyeing of the cotton portion of wool/cotton fabric with reactive (cotton) dyes [12]; recommended alkali concentrations and batching times (at ambient temperature) are given in Table 10.4. Following a thorough washing-off operation to remove alkali and unfixed dye, it is usually necessary to exhaust-dye the wool; in most instances the unmodified wool fibre is only lightly stained by the pad–batch operation. A pad–batch/exhaustion procedure can be used to generate heavy shades and full bright shades on wool/cotton fabric. It is possible to shrinkproof fabric with Securlana K (Cognis) prior to dyeing. Although the polymer-treated wool fibre in a blend is generally stained more heavily than the untreated fibre, it is still necessary to dye the wool in a separate operation.

For those blend fabrics in which the wool has been chlorinated, it is possible to dye both fibres by the pad–batch application of reactive (cotton) dyes, since the chlorinated wool fibre readily takes up dye. Pad–batch dyeing of piece-chlorinated wool/cotton fabric has been practised commercially, but the nonuniformity and irreproducibility of the chlorination step

Table 10.4 *Pad–batch dyeing of the cotton portion of wool/cotton fabric.*

Dye	Alkali	Batching Time (hours)
Remazol	Caustic soda (38°Bé): 19.5 ml l^{-1} (up to 30 g l^{-1} dye) or 27 ml l^{-1} (over 30 g l^{-1} dye); + sodium silicate (37–40° Bé): 145 ml l^{-1}	24
Procion MX	Soda ash: equal to mass of dye, up to a maximum of 30 g l^{-1}	4
Drimarene K	Soda ash: 10–40 g l^{-1}, depending on depth of shade	24
Cibacron F	Caustic soda (70° Tw): 2–12 ml l^{-1}, depending on depth of shade, + sodium silicate (No. 1): 60 ml l^{-1}	24

have caused dyeing problems. In addition, for certain shades, it is often difficult to avoid significant differences in hue on the two fibres. Solid shades on wool/cotton fabric have been obtained by the pad–batch application of selected reactive (cotton) dyes on to fabric that has been pretreated with hydrogen peroxide [39].

Sandoz (now Clariant) has recommended a pad–roll operation using a mixture of Solar 3L (direct) and Sandolan MF dyes, or of Indosol SF and Lanasyn S dyes, to dye both fibres [7]. Fabric is padded at 20–50 °C (according to the depth of shade) and pH 5.5–6.5. Fixation is achieved by batching in a chamber for 3 hours at 85 °C. The fabric is then washed-off and, if necessary, aftertreated with a cationic product. Heavy shades with excellent fastness properties are reported.

10.2.4   Wool Damage during Dyeing

As previously discussed, because of the susceptibility of wool to alkaline damage, the use of alkaline conditions to dye the cotton portion of a blend poses a potential problem for the wool fibre. To model the effect on wool when long-liquor dyeing wool/cotton blends in the presence of alkalis, wool was treated at 40 °C with different levels of soda ash, and the physical effects of these conditions were evaluated [40–43]; these conditions are relevant to the application of FCP dyes, such as Levafix E-A and Drimarene K, to the cotton portion of a blend. A treatment employing sodium carbonate at a concentration of $10\,g\,l^{-1}$ leads to a loss of cystine residues and the formation of dehydroalanine and lanthionine residues; a reaction pathway involving β-elimination is proposed (Scheme 10.3) [43]. The loss of cystine residues occurs only after a reaction time of 50 minutes, whereas significant amounts of lanthionine and dehydroalanine residues are formed within 30 minutes; the formation of the latter from serine, cystine and threonine residues is therefore likely for short reaction times. After 70 minutes the dehydroalanine content falls off, and after prolonged treatment (over 90 minutes) peptide hydrolysis (Scheme 10.4) is the major degradation route.

Following blank dyeings carried out with soda ash at 40 °C on a wool/cotton blend (wool sliver intertwined with a cotton yarn), it has been proposed that the loss in cystine (+ cystine) content and the decrease in urea–bisulphite solubility can be used to measure the extent of wool degradation [40]. The effects of the dyeing conditions on these properties are shown in Figure 10.1. Decreases in tensile strength have been found to be negligible

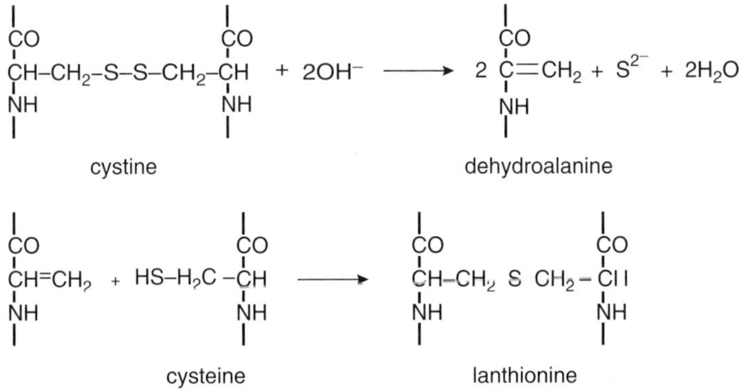

Scheme 10.3   *Elimination reactions of wool cystine.*

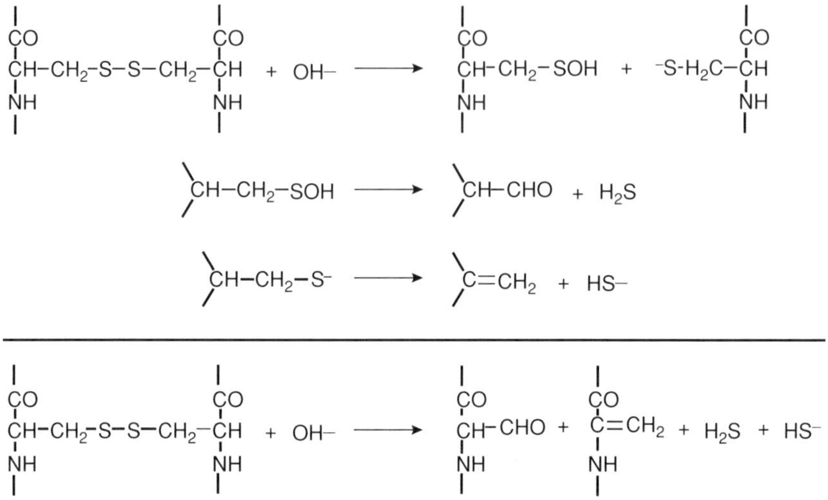

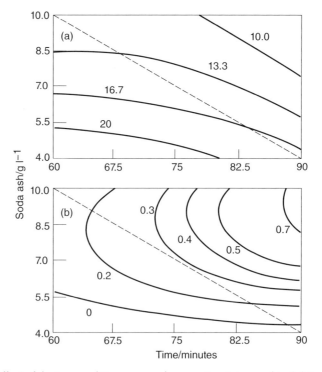

Scheme 10.4   *Elimination reactions of wool cystine: severe alkaline treatment.*

Figure 10.1   *Effect of dyeing conditions on wool properties: (a) urea–bisulphite solubility (%); (b) absolute cystine decrease (%); for the undyed fibre, urea–bisulphite solubility = 56% and cystine content = 11.55%.*

in most instances, even under the most severe operating conditions, and it has been suggested that this indicates that the peptide bonds are more important than disulphide bonds in determining the mechanical properties of the wool fibre. Dyeing conditions which lie beneath the diagonal lines in Figure 10.1 are said not to lead to any unacceptable changes to the wool fibre.

In another study, using a 50 : 50 wool/cotton union fabric, the wool was dyed first in the normal manner at 98 °C, then the cotton was dyed with $10\,g\,l^{-1}$ soda ash at 40 °C for 90 minutes and finally the goods were aftertreated with ammonia at pH 8.5 (80 °C) [44]. Although substantial reductions in alkali solubility (14.8 → 8.7%), urea–bisulphite solubility (48.1 → 7.9%) and cystine content (11.6 → 10.9%) occurred, it was suggested that the solubility data were of limited value and that the cystine content was within acceptable limits for dyed wool. A decrease in the dry tensile strength of the wool of 6.7% is considered to be insignificant.

Recommended procedures for the dyeing of wool/cotton yarn with vat dyes [22] and reactive/milling dyes [12] cause little deterioration in the breaking load and extension at break. Although the abrasion resistance of a 50 : 50 wool/cotton fabric dyed with azoic dye has been found to be similar to that of the undyed material, fabric dyed with a vat dye shows a significant reduction (up to 50%) [22]. It appears that abrasion resistance is a more sensitive test for the presence of alkali-damaged wool in a blend than is tensile strength. There is evidence to suggest that the wool in a blend is protected by the cotton, which can absorb relatively large quantities of alkali [40].

For the pad–batch dyeing of wool/cotton, it has been found that damage to the wool is not serious for batching times of up to 48 hours if the pH of the padding liquor is kept below 11.3 [40]. The combined operations of chlorination and pad–batch dyeing (with Remazol dyes), when applied to a pure wool fabric and a 50 : 50 wool/cotton fabric, cause little damage to the wool, as assessed by changes in alkali solubility, urea–bisulphite solubility, yarn strength and extension, abrasion resistance and cystine loss [45].

10.3 Amination of Cellulosic Fibres

Practical processes have been developed to chemically aminate both cotton and viscose fibres in order to render them dyeable with wool dyes under wool-dyeing conditions (pH 4–5, 1 hour at the boil). At one time, it was thought that a cheap, convenient way of modifying cotton to make it readily dyeable with reactive dyes under neutral to mildly acidic conditions was to pretreat it, from a long liquor, with a reactive cationic polymer under alkaline conditions [46]. Such treatments imparted a very high neutral substantivity for reactive dyes, in the absence of electrolyte, and gave dyeings of good wet fastness. To date, such approaches have not met with commercial success, as the treated fabric dyed to duller shades than those produced by the conventional salt/alkali process on untreated cotton and a significant drop in light fastness of 1–2 points was noted [47]. Lenzing, however, had significant commercial success with its Rainbow fibre, which was produced by adding a cationic polymer to viscose dope; this cationised viscose could be blended with wool and dyed in one bath with wool dyes such as milling and premetallised dyes [48].

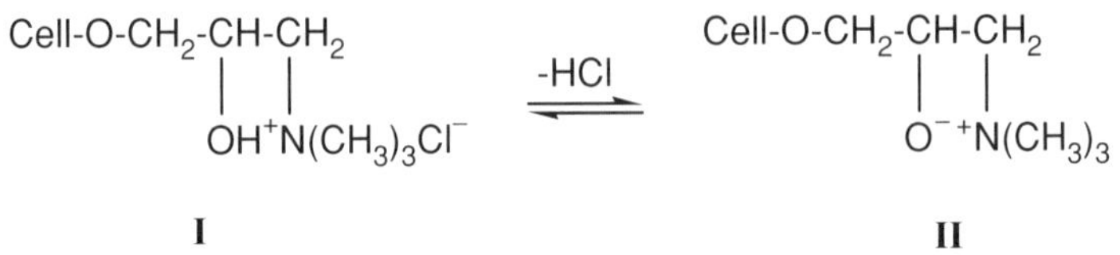

Cell-O-CH$_2$-CH-CH$_2$ Cell-O-CH$_2$-CH-CH$_2$

OH$^+$N(CH$_3$)$_3$Cl$^-$ -HCl   O$^-$ $^+$N(CH$_3$)$_3$

I   II

Scheme 10.5 *Different forms of cotton modified with Glytac A.*

Bright, light-fast dyeings can, however, be obtained by using a variety of methods that incorporate amino residues in cellulosic substrates, and some of the most promising techniques will be discussed here.

Rippon found that an exhaust pretreatment with a cationic polymer, chitosan, improved the dyeing properties of cotton, in particular the dye coverage of immature fibres [49]. Rupin *et al.* [50] and Dvorsky and Cerovsky [51] studied the dyeability of cellulose substrates modified with glycidyl-trimethyl-ammonium chloride (Glytac A from Protex) or its precursor 3-chloro-2-hydroxy-*N,N,N*-trimethyl propanaminium chloride. This modification can be carried out by an alkaline pad–bake (200 °C) [52] or by an alkaline pad–batch procedure. Unfortunately, the product has insufficient substantivity to allow its application by long-liquor methods. The modified substrate may be considered to have the structure shown in Scheme 10.5, although the possibility of such treatments leading to the introduction of oligomeric polyether chains bearing cationic functionality has not been established.

The ease of neutral dyeing of this substrate with reactive dyes led Lewis *et al.* [53] to speculate that the deprotonated form II predominates even when dyeing at pH 7. Thus, the anionic sulphonated reactive dye is initially absorbed on to the fibre by a powerful ionic attraction to the quaternary ammonium residue and fixes to the adjacent ionised (nucleophilic) hydroxyl group. Typical dyeing conditions that could be employed were pH 7, with *no salt*, the bath being raised to the boil over 30 minutes and dyeing continued at the boil for a further 60 minutes. Dyeing with a 2% on mass fabric (o.m.f.) shade of C.I. Reactive Red 5 (dichloro-*s*-triazine dye), for example, gave exhaustion values of 85%, and even more interestingly, 99% of the absorbed dye was apparently covalently fixed; thus, soaping-off of the dyeing gave rise to hardly any colour removal. Colour yields compared to the conventional salt/alkali method on untreated cotton were doubled.

Due to perceived health-and-safety problems in handling the epoxide form, recent efforts to modify cotton in this manner have concentrated on a pad–batch procedure, using a mixture of the chlorohydrin analogue 3-chloro-2-hydroxy-*N,N,N*-trimethyl propanaminium chloride and sodium hydroxide. The high alkalinity converts the agent, during the batching procedure, to the corresponding epoxide, which then fixes covalently to the fibre. In the USA, the commercial development of the Virkatone ECO system (Virkler) is worth noting, and there is undoubtedly mill production of cotton modified by this method. This treated cotton can be dyed with selected reactive dyes in the absence of salt; dyebath exhaustion is 99–100%, even in full depths of shade, and fixation values also equal or approach 100%.

An alternative method of introducing different aliphatic amino groups into cotton involved a pad–bake pretreatment of the fabric with *N*-methylol acrylamide (NMA), which reacted covalently with cellulose by a Lewis acid-catalysed baking reaction [54] (Scheme 10.6). Reaction of this substrate with appropriate amines gave a series of

Cell-OH + HOCH$_2$-NH-CO-CH=CH$_2$

$\downarrow$

Cell-O-CH$_2$-NH-CO-CH=CH$_2$........ **I**

Scheme 10.6 *Reaction of NMA with cotton.*

Cell-O-CH$_2$-NH-CO-CH$_2$CH$_2$-NH$_2$ ················· **II**
Cell-O-CH$_2$-NH-CO-CH$_2$CH$_2$-NHCH$_3$ ··············· **III**
Cell-O-CH$_2$NH-CO-CH$_2$CH$_2$-N(CH$_3$)$_2$ ············· **IV**
Cell-O-CH$_2$-NH-CO-CH$_2$-$^+$N(CH$_3$)$_3$ ············· **V**

Scheme 10.7 *Amino derivatives of NMA-treated cotton.*

amino-substituted cellulosic substrates (Scheme 10.7). Substrates II–V were dyed at pH 5 in the absence of electrolyte with 2% o.w.f. C.I. Reactive Red 5; the bath was raised to the boil and boiled for 1 hour. The results, in terms of colour yield (K/S) before and after soaping-off, are illustrated in Figure 10.2. A study of this figure reveals that optimum exhaustion/fixation was achieved on Substrates II and III; Substrate IV, containing the tertiary amine, gave surprisingly poor fixation results.

10.4 Wool/Silk

Silk is formed by solidification of the viscous fluid excreted from the glands of the silkworm. *Bombyx mori* is the species of greatest commercial importance [55]. Significant quantities of silk are also produced by silkworms living 'wild'; the most important of these is known as tussah. The raw silk fibre consists of two continuous filaments of a protein (fibroin),

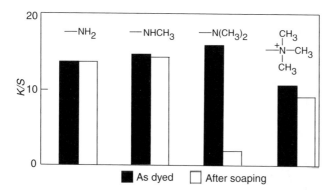

Figure 10.2 *Comparison of colour yields obtained for dyeings of C.I. Reactive Red 5 (2% o.w.f.) on various aminopropionamido-methylene-oxycelluloses (Cell-O-CH$_2$NH-CO-CH$_2$CH$_2$NR$_1$R$_2$R$_3$) [54].*

cemented together and surrounded by a gum (sericin), which normally accounts for up to 25% of the mass of the fibre. The lustre and softness normally associated with silk appear only after the gum has been removed, which may be done by treating yarn or fabric with a soap solution (with added alkali) at 95 °C for at least 2 hours [20,56]. The degummed filaments are smooth and translucent, with a mean fibre diameter of about 13 μm [55]. The mass loss that results from degumming can be compensated for by allowing the silk to absorb various inorganic salts, but this procedure has become less common [55,56].

10.4.1   Dyeing of Silk

Silk fibroin, like wool keratin, is formed by the condensation of α-amino acids into polypeptide chains [55–57]. The fibroin protein is composed of 17 amino acids; in contrast to wool, the chemically simpler species such as glycine, alanine and serine are present in abundance, and there is very little cystine (Table 10.5). Because of the virtual absence of sulphur in fibroin, the long-chain molecules are not linked together by disulphide bridges as they are in wool.

The polypeptide chains in silk run parallel to the fibre axis, with neighbouring chains running in opposite directions and being hydrogen-bonded to form a sheet (Figure 10.3) [57]. This results in a fibre that is very strong, because the resistance to tension is borne directly by the covalent bonds of the polypeptide chains. It is not appreciably extensible, because the chains are already extended as far as they can go without breaking the hydrogen

Table 10.5   *Amino acid compositions of wool (merino) and silk fibroin* (Bombyx mori) *[58].*

Amino Acid	Residues per 1000	
	Wool	Silk
Glycine	83	439
Alanine	53	289
Serine	111	119
Tyrosine	39	51
Valine	55	24
Aspartic acid	61	16
Threonine	66	13
Glutamic acid	114	13
Phenylalanine	27	8
Isoleucine	31	6
Leucine	74	6
Arginine	67	5
Proline	67	4
Lysine	26	4
Tryptophan	4	2
Histidine	7	2
Methionine	5	
[Cystine]$_{1/2}$	113	2

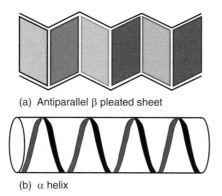

(a) Antiparallel β pleated sheet

(b) α helix

Figure 10.3   *Structures of (a) silk and (b) wool.*

bonds that hold the sheets together, but it is quite flexible, as the sheets are packed together (Figure 10.4), with only Van der Waals forces of attraction between them. In silk from *Bombyx mori*, approximately 60% of the protein is crystalline. There is no room in the regularly packed, crystalline structure for bulky side chains such as tyrosine; ordered regions alternate with disordered regions, which contain, in addition to the three primary residues, all the large side chains.

In silk fibroin, a relatively small number of amino acids contain strongly basic residues, such as lysine, histidine and arginine (Table 10.5) [56,58], and the maximum amount of acid that the fibre can adsorb is somewhat less than that for wool (Table 10.6). Since most acid dyes are the salts of strong acids [18], these dyes combine with wool and silk in a similar manner to hydrochloric acid. The adsorption of C.I. Acid Blue 45 by wool, silk and nylon is shown in Figure 10.5 [59]. Although the number of amino groups in silk is only about 20% of that in wool, there is still a sufficient number of dyeing sites available

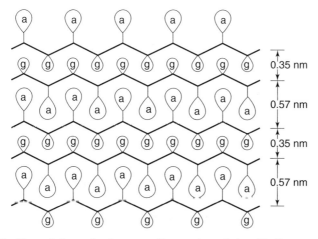

Figure 10.4   *Packing of sheets in the crystalline region of the silk fibre; g = glycine, a = alanine.*

Table 10.6 *Basicity of wool and silk [19].*

Fibre	Equivalent Number of Basic Groups (kg^{-1})	Amount of 1 mol l^{-1} Hydrochloric Acid Adsorbed (ml kg^{-1})	
		Expected	Experimentally Observed
Wool	0.82	820	850
Silk	0.15	150	120–220

for practical purposes. Over the pH range at which wool is normally dyed, the combining power of wool for acid dyes is greater than that for hydrochloric acid (at any given pH), because of the relatively high affinity of the dye anion (compared with Cl$^-$) for the fibre due to nonionic, aromatic and electrostatic interactions [60,61]; similar considerations also apply when analysing the nature of the interactions of acid dyes with silk.

10.4.2   Dyeing of Wool/Silk Blends

Although relatively small in volume, wool/silk blends are becoming more important because of the increased demand for high-quality apparel. Natural silk is often blended in equal proportions with wool, and contributes lustre and strength to the final product [5]. Wool/silk blends are important for goods such as kimonos, and are finding increasing usage in dress fabrics and knitwear [62]. Silk yarn is also used as an effect thread in worsted fabrics [5].

Dyeing of wool/silk blends is normally carried out on yarn, and to a lesser extent on woven fabric, with solid shades being the most important [62]. Degumming of the silk should be carried out prior to blending in order to avoid alkaline damage to the wool (see Section 10.2.4). Like wool, silk can be dyed using milling and 1 : 2 metal-complex dyes [63]; both portions of a blend can therefore be dyed simultaneously in a single dyebath. The main problem for the dyer lies in the distribution of dye between the two fibres, with dyeing temperature being a crucial parameter in controlling dye partition. The silk fibre lacks the impervious scale structure of the wool fibre, and dyes are more readily adsorbed

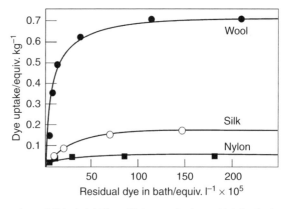

Figure 10.5 *Adsorption of C.I. Acid Blue 45 (pure dye) at pH 1.6; dyeings were carried out at 85 °C for sufficient time to attain equilibrium: 6 hours for wool and silk, 15 hours for nylon.*

and desorbed by silk [6]. At low temperatures, dye is taken up preferentially by the silk in a blend and later migrates to the wool at higher temperatures. Under normal wool-dyeing conditions, more acid dye is adsorbed by wool than by silk.

The use of strongly acidic dyebaths favours dyeing of the silk, although an acceptable partition of dye between the two fibres to yield a solid shade may be obtained [5,6]. The fastness to wet treatments is lower with silk [6,20]. Dyeing temperatures close to the boil may adversely affect the tensile strength and lustre of the silk fibre [20].

Improvements in wet-fastness properties are obtained by dyeing the silk portions of blends with reactive dyes. The fixation of reactive dyes on silk requires alkaline conditions, and occurs through reaction with hydroxyl-containing side chains such as serine and tyrosine [56,63]. As for wool/cotton blends, mild alkaline conditions must be used to minimise any damage to the wool (see Section 10.2.4).

10.4.2.1   Milling and 1 : 2 Metal-Complex Dyes

Combinations of milling and premetalllised dyes are dyed on 50 : 50 wool/silk blends: the dyebath is set at pH 4.5–5.0 and raised to 70 °C, at which temperature the silk is dyed in preference to the wool; the dyebath is then raised to 90 °C, where dye uptake by the wool increases. The goods are held at 90 °C for sufficient time to yield a solid shade. Lanaset (Huntsman) dyes can be applied to both of the fibres in a blend by dyeing at pH 5 (95 °C) [64]. The amount of Glauber's salt used influences the uptake of dye on each fibre. Only Lanaset Red 2B, Blue 2R and Green B are unsuitable for achieving similar shades on both fibres. Sandoz (now Clariant) has found that the following dyeing temperatures (and pH values) ensure optimum distribution of nonreactive dyes between wool and silk [62]:

- Sandolan MF: 70–80 °C, pH 4.5–5.5;
- Sandolan Milling N: 80–85 °C, pH 4.5–6.5;
- Lanasyn S: 85–90 °C, pH 4.5–6.5.

Sandolan MF dyes are used mainly for pale to medium shades on piece goods, because of their somewhat poorer wet-fastness properties on silk and their good migration characteristics; the addition of 10–20% Glauber's salt improves the initial levelness of the strike on each fibre. Sandolan Milling N and Lanasyn S dyes are used for medium and heavy shades.

10.4.2.2   Reactive Dyes

For higher levels of wet fastness, a two-bath system using reactive dyes has been recommended [65]. The silk is dyed first, with monofluorotriazine (MFT) dyes at pH 8.5 (1–2 g l^{-1} soda ash) in the presence of 20–80 g l^{-1} Glauber's salt (or common salt) at 60 °C. After rinsing, soaping and acid souring, the wool portion of the blend is then dyed with Lanasol dyes. Intermediate soaping at 80 °C is necessary to remove the 20–40% of the MFT dye that does not fix covalently to the silk. A slightly different approach has been adopted by Sandoz (now Clariant), for the application of FCP dyes on to both of the fibres under neutral or slightly acidic conditions, followed by alkaline fixation in the same bath [62]. Two procedures can be used:

- Dyeing at pH 7 (90 °C) with 20–60 g l^{-1} Glauber's salt.
- Dyeing at pH 4.5–5.5 (70 °C) with 5 g l^{-1} Glauber's salt.

In each case, fixation of dye on the silk is carried out by reducing the bath temperature to 40 °C and adding 2–5 g l^{-1} sodium bicarbonate. Better results, especially with combination dyeings, are obtained by dyeing at 90 °C under neutral conditions, but there is greater danger of chafe marks in piece dyeing (caused by mechanical damage to the silk fibres [63]) at higher temperatures. With deep shades, an additional soaping (at 70 °C) can bring about an improvement in fastness properties. The distribution of dye between the two fibres depends on the individual dye, and the following ternary combinations are recommended when dyeing 50 : 50 wool/silk blends:

- Drimalan Golden Yellow F-3RL;
- Drimarene Orange K-GL;
- Drimalan Red F-BR;
- Drimarene Brilliant Red K-8B;
- Drimalan Blue F-GRL/F-B;
- Drimalan Blue F-GRL.

The alkaline conditions required for the dyeing of the silk portion of a blend with reactive (cotton) dyes are milder than those required for the cotton portion of a wool/cotton blend (Table 10.4), and so there should be less alkaline damage to the wool for blends with silk. The silk fibre is somewhat less susceptible to alkaline damage than is the wool fibre [55], because of its very low cystine content.

10.5   Wool/Nylon

The term 'nylon' was originally a DuPont brand name, but has now become a generic term for any synthetic linear polyamide fibre. Nylon fibres are renowned for their outstanding tensile strength and resistance to abrasion. The inclusion of polyamide fibres (often in only small proportions) in blends with wool is a particularly useful way of improving the performance properties, such as tensile strength and abrasion resistance, of wool textiles. Wool/nylon blends containing 5–50% nylon are widely used, the most important blend being 80 : 20 wool/nylon, for use in carpets [66]. Other areas of use include hosiery, woollen outerwear fabrics (which often contain small quantities of nylon, mainly in the warp yarns) and automobile upholstery fabrics.

10.5.1   Dyeing of Nylon

10.5.1.1   Fibre Structure

Polyamides are so named because they contain amide groups as integral parts of their polymer chains, and they are thus chemically similar to the natural polymers wool and silk. Nylon 6.6 and nylon 6 are the two most important polyamide fibres, and have similar properties.

Nylon 6.6, so designated because the two starting materials each contain six carbon atoms, is formed from the reaction of hexamethylenediamine and adipic acid [67,68]. A condensation reaction takes place initially (Scheme 10.8), and two molecules of this condensate, then react together (Scheme 10.9). This self-condensation continues, eventually giving rise to a long polymer molecule. The molecules of nylon are long and straight, and

$$NH_2(CH_2)_6NH_2 + HOOC(CH_2)_4COOH \longrightarrow NH_2(CH_2)_6NHCO(CH_2)_4COOH + H_2O$$

Scheme 10.8　*First stage of condensation to form nylon 6.6.*

$$2NH_2(CH_2)_6NHCO(CH_2)_4COOH \longrightarrow$$
$$NH_2(CH_2)_6NHCO(CH_2)_4CONH(CH_2)_6NHCO(CH_2)_4COOH + H_2O$$

Scheme 10.9　*Self-condensation to form nylon 6.6.*

there are no side chains or cross-linkages. The process of polymerisation must not be allowed to go on indefinitely, but must be stopped at a given chain length; that is, the polymer must be 'stabilised'. Stabilisation of nylon 6.6 can be brought about by using equimolar quantities of diamine and diacid in the reaction mixture and adding (at the appropriate time) a monofunctional reagent such as acetic acid. Acetylation of the amino groups prevents any further reaction with carboxylic acid groups, and the polymerisation is stopped. Nylon 6.6 polymers have an average relative molecular mass (r.m.m.) in the range 12 000–20 000 Da.

Extrusion of the molten polymer through spinnerets leads to the formation of filaments [67,68]. At this stage, the polymer molecules are in a state of random orientation, and the fibre is weak and opaque. The filaments are now stretched to four or five times their original lengths, in a process known as drawing. This causes the molecules to straighten and pack together in a regular manner with hydrogen bonding between adjacent chains, resulting in a fibre with a high degree of crystallinity (Figure 10.6) [69]. As the extent to which the molecules are orientated along the fibre axis increases, the affinity of dyes for the fibre decreases; in addition, nonuniformity in the degree of orientation can lead to an uneven uptake of dye.

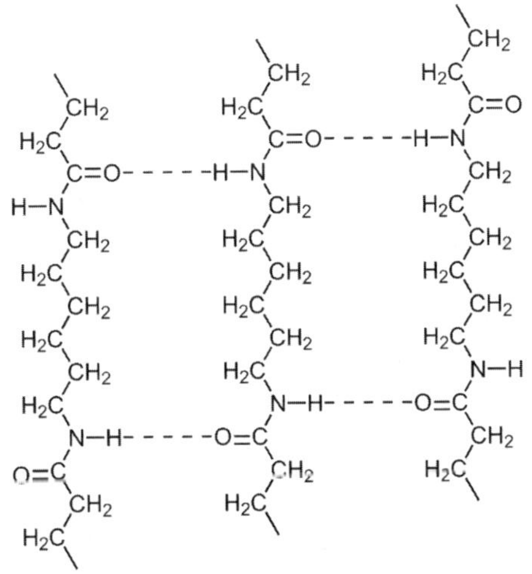

Figure 10.6　*Crystalline structure of nylon 6.6.*

Table 10.7   *Polar groups in nylon 6.6.*

Group	Concentration (equiv. kg^{-1})
Amino (–NH$_2$)	0.036
Acetamido (–NHCOCH$_3$)	0.090
Carboxylic acid (–COOH)	0.063
Amido (–CONH–)	8.85

The nylon 6.6 fibre has a glass transition temperature (T_g) of 57 °C [70]. Being thermoplastic, the fibre can be set by a heating operation followed by cooling, a procedure known as heat setting [68]. Heat setting may be carried out before or after dyeing and can be used, for example, to stabilise yarn twist and to ensure the dimensional stability of fabrics during processing and wear [68]. Dry heat (210–220 °C), saturated steam under pressure (120–130 °C) or hot water under pressure (130 °C) may be used to heat-set nylon 6.6. The specific conditions and temperatures used have significant effects on the rate of subsequent dye adsorption [71].

The polar groups in the nylon 6.6 polymer are amino, acetamido, carboxyl and amido groups; the quantities typically present are shown in Table 10.7 [21]. There are fewer amino groups than carboxylic acid groups because of the stabilisation process.

Nylon 6 is produced from the self-condensation of 6-aminocaproic acid (Scheme 10.10) [67,68]. Further self-condensation proceeds in a manner similar to that for nylon 6.6.

10.5.1.2   Acid Dyes

Acid dyes are used widely to dye nylon 6 and nylon 6.6, which have the same fundamental dyeing behaviour [21]. As a result of their dyeing properties on wool, acid dyes are classified into categories such as acid levelling and milling dyes. Similar classifications are less useful for nylon, where the dye affinities are higher, and as a consequence higher pH conditions are recommended for dyeing.

The saturation level for dye uptake by nylon is less than that for either wool or silk (Figure 10.5). Whereas uptake of dye by wool reaches a saturation value of 30–60% (concentration of commercial dye), saturation on nylon occurs with only 1.5–3.0% of dye [19]. The cause of this poor build-up on nylon lies in the limited number of sites available for dye adsorption [19,21]; nylon 6.6 contains only 0.036 equiv. kg^{-1} of primary amino groups (Table 10.7), compared with values for wool and silk of 0.82 and 0.15 equiv. kg^{-1}, respectively (Table 10.6). The inability to achieve heavy shades with many dyes under practical dyeing conditions is a major problem in the dyeing of nylon.

$$2NH_2(CH_2)_5COOH \longrightarrow NH_2(CH_2)_5CONH(CH_2)_5COOH + H_2O$$

Scheme 10.10   *Preparation of nylon 6.*

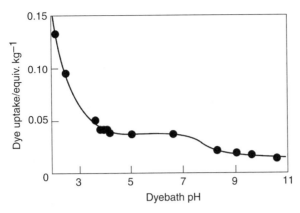

Figure 10.7   *Adsorption of C.I. Acid Orange 7 (pure dye) by nylon. Dyeings carried out at 60°C for 4–7 days (to attain equilibrium) with a large excess of dye; pH adjusted with sulphuric acid [72].*

The relationship between dye uptake and dyebath pH is shown in Figure 10.7 for dyeings of nylon yarn with relatively concentrated solutions of the acid levelling dye C.I. Acid Orange 7 [72]. The following important points emerge from this study:

(1) At pH 10.5, there is significant dye sorption.
(2) As the pH of dyeing is lowered to 6.5, dye uptake increases to a point corresponding to a dye concentration of 0.037 equiv. kg^{-1}.
(3) Over the pH range 6.5–4.0, dye sorption increases only very slightly.
(4) At pH values below 4, dye uptake increases dramatically as the pH of dyeing is lowered. Dye uptake in excess of the number of amino groups is a phenomenon referred to as 'overdyeing'.

The occupancy of amino sites in nylon by dye molecules cannot normally be estimated when using highly concentrated dyebaths (as in Figure 10.7), because of overdyeing. In order to determine the extent to which the amino groups are saturated with dye, dye adsorption (at pH 3.2), as a function of the concentration of dye in the bath at equilibrium, has been determined for a range of acid dyes; the results are presented in Figure 10.8 [73]. The values of the intercepts on extrapolation of the linear parts of the curves in Figure 10.8 to zero concentration are given in Table 10.8 for an extended range of dyes. Any value which is significantly less than 0.035 equiv. kg^{-1} indicates that, for the particular dye, not all the amino sites are occupied.

To examine the overdyeing properties, nylon was acetylated so that its amine end-group concentration was decreased to 0.0029 equiv. kg^{-1} and was dyed at pH 3.2 using dye concentrations of 0.004 mol l^{-1}. Dye adsorption figures are listed in Table 10.8, and overdyeing is indicated by values in excess of 0.0029 equiv. kg^{-1}.

Dye (anion) affinities have been determined by desorption experiments at pH 3.2, and are given in Table 10.8 [73]. Dye affinities may also be assessed from the pH at which the amino groups are half-saturated with dye [19]. For any given acid dye, this point occurs at a much higher pH with nylon than with wool; that is, acid dyes have a higher affinity

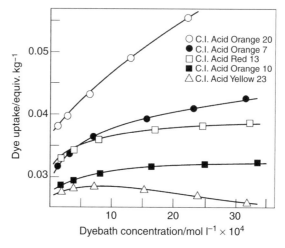

Figure 10.8 *Adsorption of (pure) acid dyes by nylon.*

for nylon than for wool,; this accounts for the superior fastness properties and inferior migration behaviour of dyes on nylon.

10.5.1.3 Factors Influencing the Dyeing Behaviour of Nylon

(1) *Dye adsorption:* In aqueous solution at neutral pH, nylon possesses a zwitterion structure ($^+NH_3$–nylon–CO_2^-) [74]. Under acidic conditions the carboxylate anions are protonated, and the dye anions associate with the amino groups (Scheme 10.11) in a

Table 10.8 *Acid dyes[a] for nylon 6.6 (see text for experimental details).*

C.I. Acid	Number of Sulphonate Groups	Intercept on Adsorption Curve[b] at Zero Concentration (equiv. kg^{-1})	Adsorption by Acetylated Nylon (equiv. kg^{-1})	Dye Anion Affinity (kJ mol^{-1})
Red 88	1	0.035	0.0466	-32.1
Orange 20	1	0.037	0.0411	-25.8
Orange 8	1	0.036	0.0230	-25.2
Orange 7	1	0.035	0.0172	-24.4
Red 13	2	0.035	0.0110	-27.3
Red 1	2	0.034	0.0083	-21.4
Yellow 17	2	0.034	0.0055	-22.2
Blue 45	2	0.030		-21.1
Blue 69	2	0.030	0.0054	-19.6
Orange 10	2	0.030	0.0052	-21.0
Red 18	3	0.037	0.0030	-24.6
Yellow 23	2[c]	0.028	0.0025	-21.0

[a]Listed in decreasing order of tendency to overdye, as assessed by adsorption by acetylated nylon.
[b]See Figure 10.7.
[c]This dye also contains a carboxylate group and effectively behaves as a trisulphonate.

$$^+NH_3–nylon–COOH + Na^+dye^- \rightleftharpoons dye^- \,^+NH_3–nylon–COOH + Na^+$$

Scheme 10.11 *Ionic interactions of acid dyes with nylon.*

similar manner to dye adsorption by wool. The maximum quantity of dye capable of being adsorbed on to these groups is approximately equivalent to the number of amino groups in the fibre. For a dye with a relatively high affinity, such as C.I. Acid Red 88, significant dye adsorption occurs at high pH values [72]. Electrostatic interactions between dye and fibre are less important when dyeing under neutral or alkaline conditions, and the adsorption of dye anions (together with sodium or hydrogen ions, to maintain electrical neutrality) is facilitated by hydrophobic interactions between fibre and dye.

(2) *Low site occupancy by polysulphonated dyes:* Not all the amino sites are occupied at saturation for polysulphonated dyes of low (anion) affinity (e.g. C.I. Acid Yellow 23, C.I. Acid Orange 10, C.I. Acid Blues 45 and 69), as indicated by the low intercept values in Table 10.8 [73]. For certain tri-, tetra- and pentasulphonated dyes, dye uptake at saturation corresponds to only 65, 40 and 10%, respectively, of the available amino sites (Table 10.9) [75]. The affinity of dye anions for both nylon and wool is governed by the same principles: any structural change which increases the hydrophobic nature or polarisability of a dye increases its affinity for the fibres. For example, increases in dye affinity for nylon of $6–9\,kJ\,mol^{-1}$ have been observed when a benzene nucleus is replaced by a naphthalene group [21,73]; corresponding differences of $5–6\,kJ\,mol^{-1}$ have been observed for a series of dyes on wool [72,76,77]. The introduction of an additional sulphonate group into a dye molecule causes a decrease in affinity for dyes of about $4\,kJ\,mol^{-1}$ on both nylon and wool [73]. The actual pattern of substitution of sulphonate groups in a dye molecule can have a profound effect on the ability of a dye to saturate all the available amino sites on nylon, as shown in Table 10.9 [75]. For a series of disulphonated dyes, affinity for nylon has been observed to increase with the distance

Table 10.9 *Occupancy of amino sites for dyes based on α-naphthylamine → β-naphthol[a].*

Sulphonate Substitution Pattern	Amino Site Occupancy (%)
4,3′,6′	100
4,6′,8′	100
8,6′,8′	65
3,6,8,6′	85
3,6,8,8′	40
3,6,8,3′,6′	30
3,6,8,6′,8′	10

[a]Calculated from results for the dyeing of fibre (containing $0.062\,equiv.\,kg^{-1}$ of amino end groups) at pH 3 (90 °C) using a 10% excess of dye.

between the sulphonate groups [21]. Despite problems with accommodating polybasic dyes of low affinity on to all the available sites in the nylon fibre, monobasic and polybasic dyes alike are taken up by the wool fibre to the extent of 0.8–0.9 equiv. kg^{-1} at saturation [21]; that is, each of the amino sites in wool is occupied by one dye molecule.

(3) *Overdyeing:* Once all the available sites are occupied by dye molecules, an increase in the concentration of dye in the bath, or a decrease in dyebath pH, can lead to further sorption of dye. Although overdyeing occurs to a lesser or greater extent at all pH values, dye uptake in excess of amino groups is most marked at low pH. The extent of overdyeing is greatest for monobasic dyes with high affinities, such as C.I. Acid Oranges 7, 8 and 20 and C.I. Acid Red 88 (Table 10.8) [73]. For dyes of high affinity, the availability of amino sites is not necessary for dye adsorption, as indicated by the dyeings of acetylated fibre. Sorption of excess dye, possibly as undissociated dye acid, will take place through nonpolar interactions, and this can occur at relatively high pH values; for example, at pH 4.5 for the milling dye C.I. Acid Blue 138 [78]. For dyes of low affinity, overdyeing occurs at relatively low pH values; for example, below pH 2.5 for C.I. Acid Blue 45 [72]. At pH 1, sorption of C.I. Acid Blue 45 and C.I. Acid Red 88 is 0.076 and 0.36 equiv. kg^{-1}, respectively. At low pH values, mineral acids are taken up by nylon in quantities significantly greater than the amino group content, as shown in Figure 10.9 for hydrochloric acid [19]. This adsorption of excess hydrogen ions has been explained in terms of protonation of the amide groups; the sorption of excess low-affinity dye can be rationalised in terms of ionic interactions with these positively charged sites. Under prolonged exposure to even mildly acidic conditions, amides will, however, hydrolyse to the corresponding protonated primary amine and carboxylic acid, which in the case of polyamides reduces the degree of polymerisation (DP). Thus the observed excessive 'overdyeing' at low pH values may be simply due to this increasing tendency to hydrolyse. Hydrolysis will make the fibre more accessible and lead to the production of further protonated amine dye sites [69]. In the presence of dyes (especially those of high affinity) at low pH values, however, there is strong evidence to suggest that hydrolysis occurs through random scission of the polymer chains, leading to an increase in the number of amino groups.

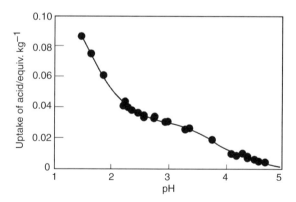

Figure 10.9   *Adsorption of hydrochloric acid by nylon.*

10.5.1.4   Practical Considerations

Acid dyes are normally applied to nylon at temperatures close to the boil. Because of the relatively high affinity of acid dyes for nylon, levelling properties below 100 °C are poor [21]. The initial dye application must be as uniform as possible, and this is best achieved by control of the temperature. The particular pH required to achieve optimum levelling and exhaustion depends upon the substrate as well as on the dyes that are used. Nylon 6 is generally dyed more readily than nylon 6.6, and texturised or staple yarns have a higher rate of dyeing than do flat, continuous-filament yarns.

Restraining agents are often used to overcome coverage problems on nylon by reducing the rate of dyeing [20,21,79–82]. These may be anionic compounds (blocking or reserving agents) or cationic compounds (levelling or retarding agents). Blocking agents are essentially colourless dyes that are preferentially adsorbed by the nylon, thus decreasing (reserving) the uptake of dye. Alkylbenzene-sulphonates, alkylnaphthalenesulphonates, alkanol sulphates and sulphated castor oil are all used. Levelling agents (used for both nylon and wool) are weakly cationic (and sometimes amphoteric) products that combine with the sulphonic acid groups of the dyes in the dyebath. The complexes so formed are maintained in dispersed states by the addition of a nonionic agent.

Because of the limited dye uptake by the nylon fibre, there is competition between acid dyes in mixtures for the available sites [21]. Dyes of higher affinity are taken up at the expense of dyes with lower affinity; the shade obtained on the fibre may bear little relationship to the proportions of dyes in the dyebath. This competition is most apparent when dyes of different basicities are used; very few dye combinations are free of these problems.

Afterchrome dyes can be applied to nylon in a similar way to how they are applied to wool [20]. They are applied, in the first stage, in a manner similar to that used for acid dyes. Because of the relatively hydrophobic nature of the nylon fibre, the uptake of chromium is slower than with wool, and chroming is carried out for 1 hour at the boil. The formation of a complex with the dye requires a reduction of the chromium from the hexavalent state to the trivalent state; keratin is capable of acting as the reducing agent, but nylon is not. It has been suggested that the dye on the nylon can catalyse the reduction of dichromate by the formic acid in the dyebath [83]. Many dyers accomplish the $Cr^{VI} \rightarrow Cr^{III}$ reduction by adding sodium thiosulphate to the chroming bath.

While reactive dyes can yield bright shades with good fastness to washing, they are not widely used on nylon because of their poor migration properties (when applied below the reaction limit), sensitivity to chemical variations in the substrate and difficulties in correcting or stripping faulty dyeings [81].

Attempts to dye nylon fibres with reactive dyes have been frustrated by the paucity of nucleophilic sites available for reaction; typically, nylon 6.6 from DuPont contains only 0.036 mol of free amine per kilogram of fibre, which contrasts greatly with wool (0.820 mol amine kg^{-1}) and silk (0.150 mol amine kg^{-1}). This factor alone has meant that achieving build-up of sulphonated reactive dyes even in moderate depths of shade is impossible, since every dye molecule covalently fixed means that, depending on the chromophoric component, one, two or even three strongly anionic sulphonate groups become fixed at the same time, resulting in a build-up of negative charge on the fibre which acts as a resist to further anionic dye uptake.

In the light of this, it is not surprising that the first range of reactive dyes for nylon was based on disperse dyes containing a pendant reactive group; these dyes were marketed as Procinyl dyes by ICI in 1959, and contained a variety of reactive groups, all but the yellow requiring alkali activation to form a more reactive residue. The dyes were: Procinyl Yellow GS (monochloro-*s*-triazine); Procinyl Scarlet GS (chloroethylaminosulphonyl); Procinyl Orange G (chloroethylaminosulphonyl); Procinyl Red GS (chloroethylaminosulphonyl); and Procinyl Blue RS (bis-chlorohydrin). All were applied to nylon at the boil for 30 minutes under neutral conditions; the pH was then raised to 10.5 by the addition of sodium carbonate and boiling continued for a further 30 minutes. The alkaline processing step achieved two objectives: first, the amino end groups in nylon were not significantly protonated, giving maximum nucleophilicity; second, the chlorohydrin groups in Procinyl Blue RS were converted to the highly reactive epoxide form or the chloroethylaminosulphonyl groups in Procinyl Scarlet GS, Procinyl Orange G and Procinyl Red GS were converted to the highly reactive aziridinylsulphone form. Scott and Vickerstaff [84] described the interesting chemistry behind these dyes and the chemistry of the application process, and demonstrated that in full shades of Blue RS or Scarlet GS, more than the theoretical amount of dye – based on an amino end group content of 0.0426 mol per kilogram of fibre – could be covalently bonded to the substrate. In particular, results with Scarlet GS were very surprising: applying 20% o.m.f. dye gave 0.1280 mol dye per kilogram of fibre as covalently bonded dye. To explain this excess fixation given the paucity of amino groups, double addition of the dye at each amino residue is clearly possible, as is reaction with ionised hydroxyl nucleophiles in the dye itself [85]. Procinyl Blue RS was used in some markets as a brilliant blue for dyeing 100% wool and gave very good level dyeings and excellent fixation; this was confirmed in a study by Stapleton and Waters [86].

Some work [87,88] has also been reported on the dyeing of nylon with VS disperse dyes. Good fixation was achieved without the need for an alkaline fixation stage. More recently, Lo [89] described the dyeing of nylon at pH 8 with sulphatoethylsulphone dyes in the presence of dispersing agent. As the bath temperature increased, the dye gradually changed to the active disperse VS and thus became covalently fixed to the fibre (Scheme 10.12).

The use of the VS-type dye, C.I. Reactive Blue 19, for the dyeing of nylon has been studied in detail [90]; it has been shown that this dye in the β-sulphatoethylsulphone form does not react efficiently with the available amino nucleophiles when applied at pH 5. When the dye was preconverted to the VS form before dyeing, however, much better fixation efficiencies were observed.

A cationic reactive dye of the general formula $H_2C=HC-SO_2-D-Q^+$, where D is an azo chromophore and Q is a quaternary ammonium residue, was prepared by Hinks *et al.* [91]

$$D-SO_2-CH_2CH_2-OSO_3^-Na^+ \quad \text{(water soluble form)}$$

$$\downarrow \text{NaOH}$$

$$D-SO_2-CH=CH_2 \quad + \quad Na_2SO_4 \quad + \quad H_2O$$

(disperse dye)

Scheme 10.12   *Temporary soluble disperse dyes.*

and shown to dye nylon efficiently at pH 9. The substantivity of this cationic dye for nylon is high under alkaline conditions, as the nylon is overall negatively charged; in addition, the amino groups are more nucleophilic, since they are likely to be deprotonated.

Sun and Lewis [92,93] synthesised blocked VS dyes by reaction of a VS disperse dye with 2-thiol-choline, so that the water-solubilising group was the trimethylammonium cation. When applied to nylon at pH 10, the dyes showed excellent substantivity, and yet at the boil the thiol-choline residue was eliminated to form the VS dye, which reacted readily with the deprotonated amines (this desirable ionic situation occurs under alkaline conditions) in nylon. A 2% dyeing (1 hour at the boil, at pH 10) with the pure cationic thiol-choline dye gave a total efficiency value (T) of 93%, which is impressive as anionic VS dyes give a maximum T value of 85% at 2% o.m.f. dye. Such cationic reactive dyes will also dye wool under alkaline conditions, giving similarly impressive fixation values.

10.5.2   Dyeing of Wool/Nylon Blends

Most wool/nylon blends are dyed to solid shades with acid levelling, milling or 1 : 2 metal-complex dyes. The procedures used are generally similar to those adopted for pure wool, with the exception that a restraining agent is often required to control the partitioning of dye between the two fibres. The problems encountered in the dyeing of nylon with dye mixtures are exacerbated in the dyeing of wool/nylon blends by the differing affinities of dyes for wool and nylon. The distribution of dye between nylon and wool depends on factors such as dye structure, applied depth, pH, blend ratio and the quality of the component fibres, and solid shades are more difficult to achieve than with most other wool blends. Good solidity can be achieved with staple fibre blends, as used in carpet yarns, where generally there is only a minor proportion (about 20%) of nylon fibre present. For blends containing higher proportions of nylon, however, such as fabric woven from a stretch-nylon warp and a wool weft for use in leisurewear, solidity is more difficult to achieve.

Although nylon fibres have a saturation value for the uptake of acid dye under practical dyeing conditions, wool shows no such limit (Figure 10.10) [80,81]. The rate of dyeing of nylon, particularly at 60–80 °C, is higher than that of wool, and pale dyeings on wool/nylon show a preferential dyeing of the nylon (in the absence of a restraining agent). As the applied depth increases, a critical depth is reached at which solidity is obtained. The critical depth is specific for the dye and is higher for a monosulphonated than for a disulphonated dye. Application of dye at heavier depths leads to preferential dyeing of the wool, and there is no auxiliary product that can control this effect and make possible the production of solid shades with acid dyes alone.

Wool/nylon blends can normally be prepared for dyeing in a manner similar to that used for pure wool. Any scouring or processing auxiliaries must be thoroughly removed, since their presence may cause interference with restraining agent in the dyebath.

10.5.2.1   Acid Levelling Dyes

For pale shades (less than about 1% dye), the nylon usually dyes to a darker shade than the wool. Any disparities in pale and medium shades can be overcome to some extent by using mixtures of monosulphonated and disulphonated dyes of similar hue [81].

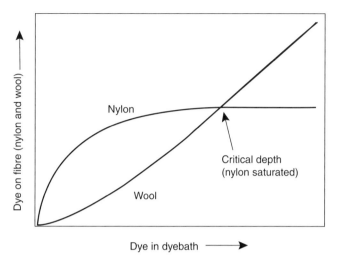

Figure 10.10   *Idealised representation of the partition (between wool and nylon) of a typical acid levelling dye.*

Monosulphonated dyes, because they are taken up by nylon in preference to poly-sulphonated dyes, exert a blocking effect. This can be used to advantage by varying the proportions of the two dyes (with little effect on the dyeing of the wool) so that the depth of shade on the nylon is matched to that on the wool. In order to avoid damage to the nylon fibre from the acid becoming 'dried in' after drying, formic acid should be used instead of sulphuric acid to achieve the required pH [80].

An appropriate amount of blocking agent must also be added to the dyebath to control partitioning between the nylon and wool [80]. When used with acid levelling dyes, levelling agents are effective in preventing dye uptake by the nylon fibre in the early stages of dyeing, but as boiling continues there is a constant migration of dye from the wool to the nylon; after prolonged boiling times, a distribution is obtained which is little different from that obtained in the absence of blocking agent.

Monosulphonated acid levelling dyes are widely used as the basis for selecting compatible binary and ternary combinations with high critical depths on wool/nylon blends [81]. Preferred dyes are as follows:

- C.I. Acid Yellows 25, 29, 49, 172, 196 and 219;
- C.I. Acid Orange 145;
- C.I. Acid Reds 42, 57, 266, 361 and 396;
- C.I. Acid Blues 40, 62, 72, 78, 258 and 277;
- C.I. Acid Brown 248.

On-tone partitioning between wool and nylon 6.6 is obtained for a wide range of hues, with solidity being achieved by the use of blocking agents.

Disulphonated acid levelling dyes can be used only for pale and medium shades on wool/nylon 6.6 [81]. The saturation limit is higher on nylon 6 fibres, and disulphonated dyes are often used to dye wool/nylon 6 blends. A compatible combination of disulphonates can be selected from the following dyes:

- C.I. Acid Yellows 17 and 44;
- C.I. Acid Reds 1 and 37;
- C.I. Acid Blues 23 and 45.

10.5.2.2 *Milling and 1 : 2 Metal-Complex Dyes*

These dyes offer improved wet-fastness properties compared with acid levelling dyes, at the expense of poorer levelling characteristics [80]. Because of this latter property, it is not always feasible to select a three-dye combination of bright primary hues for all shades, since the level-dyeing properties of such a combination would be inadequate for many applications and many valuable dyes of duller shade would be excluded. Blocking or levelling agents (or combinations of both) are used to optimise the partitioning of dye between nylon and wool [20,21,79–82]. In contrast to the behaviour shown by acid levelling dyes, the distribution of dye between the two fibres is largely maintained as the boiling continues, because of the poor migration properties of the dyes.

Based on dyeings of wool/nylon blends with 1 : 2 metal-complex dyes, it has been found that the activity of a blocking agent (Nylofixan P Liquid, Clariant) is inhibited when used in the presence of levelling agents that are substantive to the dyes, or with amphoteric or nonionic surfactants [94]. The problem can be minimised by modifying the dyeing procedure, either by delaying the addition of levelling agent to the dyebath or by carrying out a short pretreatment with blocking agent at 95 °C and then cooling the bath before the addition of levelling agent and dye. If the pH is lowered at the outset of the dyeing when the blocking agent is added, the reserve of the nylon is improved. Conversely, an increase in dyebath pH to the weakly acid region during dyeing at the boil with blocking agent present will lead to heavier dyeing of the nylon component.

So-called 'acid donors' can provide a means to bring about a desired pH change throughout the dyeing cycle. These products, which were originally developed for pure nylon, are generally hydrolysable esters which liberate free acid as the dyeing proceeds, gradually lowering the pH of the dyebath and facilitating level dyeing and good dye exhaustion.

Milling dyes are commonly used for bright shades. Colours are often based on a single dye, with a second dye as a shading component. They are also used to brighten shades based on 1 : 2 metal-complex dyes. Milling dyes are occasionally selected on the basis of their dyeing properties, such as in the piece dyeing of fabrics in which penetration and levelling problems are likely to be encountered with 1 : 2 metal-complex dyes, which have poorer migration properties; dyeing of leisurewear fabric containing texturised nylon (in the warp), which is prone to barré problems, is a common example. A compatible combination of milling dyes, with levelling properties superior to those of most 1 : 2 metal-complex dyes, can be selected from the following:

- C.I. Acid Oranges 116 and 127;
- C.I. Acid Red 299;
- C.I. Acid Blues 264 and 280.

These are monosulphonated dyes, with excellent buildup on nylon and good wet-fastness properties.

1 : 2 metal-complex dyes are used widely for wool/nylon blends, especially in applications requiring high fastness to light and washing. Their dyeing behaviour on wool/nylon blends is very dye-specific, and differences in affinity for wool and nylon are even more pronounced than with acid levelling or milling dyes [94,95]. Although the saturation limit on nylon for 1 : 2 metal-complex dyes decreases with an increasing level of sulphonation, the rate of dyeing can vary significantly for dyes of the same degree of sulphonation [94]. Unsulphonated dyes build-up to depths greater than 3/1 standard depth on 50 : 50 wool/nylon blends in the absence of blocking agent. Monosulphonated dyes reach depths ranging from 1/1 to 3/1 standard depth, whereas disulphonated dyes have saturation limits around 1/1 standard depth. Unsulphonated dyes preferentially dye the nylon component and require high levels of blocking agent to achieve solid shades. Monosulphonated dyes display a more balanced partition, which can be easily regulated by the use of blocking agent. Disulphonated dyes, with a relatively low saturation limit on nylon, tend to dye the wool preferentially. Unsulphonated dyes are useful for shading nylon and disulphonated dyes are suitable as shading elements for the wool.

Specific recommendations for the use of milling and 1 : 2 metal-complex dyes relate to the popular 80 : 20 wool/nylon blend; thus Lanaset (Huntsman) dyes are applied at pH 4.5–5.0 on carpet yarn [96]. Lanasyn S (Clariant) dyes (monosulphonated 1 : 2 metal-complex) have been recommended for the dyeing of wool/nylon blends in the form of loose fibres, hand- and machine-knitting yarns, furnishing and upholstery fabrics, outerwear and sportswear, and carpet yarns [97]. Preferred ternary combination elements are Lanasyn Yellow S-2GL, Lanasyn Red S-WP 479 and Lanasyn Black S-RL (or Lanasyn Grey S-BL), with brightening elements from the Lanasyn Brilliant dye range; the dyeing conditions are determined by the depth of shade and the dye combinations used. The Lanasyn S and Lanasyn Brilliant dyes have been classified into five groups on the basis of their restraining-agent requirements, and this is instructive with respect to their behaviour in combined application [94,97]. The use of Supranol and Alizarin dyes and of Isolan K dyes (unsulphonated 1 : 2 metal complex) at pH 4.5 has been recommended [95,98]. Isolan S dyes (monosulphonated 1 : 2 metal complex) can also be used at pH 4.5–5.5; better results are obtained using single dyes, such as a grey or a brown, than with a three-colour combination [99].

10.5.2.3   Afterchrome Dyes

It is difficult to achieve satisfactory solidity on wool/nylon blends with afterchrome dyes, because the wool is dyed more readily than the nylon [81]. Some solid shades can be based on selected chrome and 1 : 2 metal-complex dyes. Chroming of the dye on wool/nylon is even more difficult than on nylon alone, because the chromium tends to be adsorbed preferentially by the wool. The addition of a reducing agent (such as sodium thiosulphate) to the dyebath in the later stages of dyeing, the use of increased amounts of formic acid and the extension of the chroming time are techniques which can be used to ensure that optimum chroming takes place. Increasing the amount of dichromate is also effective, but is not favoured because of the undesirable environmental consequences. Afterchrome dyes that give acceptable solidity on wool/nylon are mainly monosulphonated monoazo types, which are used for orange, brown, red and black shades; at the depths applied, no restraining agent is required.

10.5.2.4 Reactive Dyes

The use of reactive dyes on wool/nylon blends is restricted. Not only are the shades often different on the two fibres but the saturation level on nylon is rather low because these dyes usually possess two or three sulphonic acid groups. Even in medium shades, nylon is dyed lighter than wool. This tendency is even more pronounced if Hercosett-treated wool is used, because the cationic Hercosett polymer at the fibre surface absorbs dye very rapidly at the beginning of the dyeing process and no subsequent migration occurs. Nevertheless, reactive dyes are used for the dyeing of blends of chlorine–Hercosett-treated wool with nylon; these blends are popular for machine-washable knitwear, especially hosiery yarns containing 20–30% nylon. A restraining agent is usually unnecessary, and dyeings are conducted as for the corresponding pure-wool articles.

In the absence of a restraining agent, 1 : 2 metal-complex dyes at low application levels partition strongly in favour of the nylon [95]. With careful selection, they are suitable for shading of the nylon component of machine-washable hosiery in pale and medium shades. In such cases, the wet-fastness properties of the metal-complex dyes on nylon adequately match those of the reactive dyes on the chlorine–Hercosett-treated wool. For heavier shades, where the partition of the metal-complex dye shifts towards the wool, solid shades are more difficult to achieve and the wet fastness is reduced.

10.6   Wool/Polyester

The polyester fibre, so called because it is a polymeric ester, is the most important of all synthetic fibres; its production overtook that of nylon in the early 1970s [3]. Polyester is renowned for its high tensile strength, excellent durability to wear and resistance to attack by chemicals. The ability to heat-set thermoplastic synthetic fibres (see Section 10.5.1) is exploited for the production of permanent pleats in polyester skirts.

Blends of polyester with wool represent an attempt to achieve a combination of desirable features of both fibres, allowing the production of fabrics with good wear properties and dimensional stability, yet which retain an attractive handle and drape reminiscent of pure wool. The most important end uses for wool/polyester blends are for outerwear, especially men's suitings, as well as women's suitings, dresses and skirts. The popular 45 : 55 wool/polyester blend is normally made up of warp and weft yarns of the same blend ratio. In the USA, the 20 : 80 wool/polyester blend is the most important; it is constructed from a texturised polyester warp and a 45 : 55 wool/polyester weft. The 80 : 20 wool/polyester blend, which is uncommon outside western Europe, can be woven from a 45 : 55 wool/polyester warp and a pure wool weft.

10.6.1   Dyeing of Polyester

10.6.1.1 Fibre Structure

Production of the polyester fibre is based on the condensation of ethylene glycol (ethanediol) and dimethyl terephthalate (or terephthalic acid), followed by polymerisation (Scheme 10.13) [67,68]. During polyester production, short chains known as oligomers

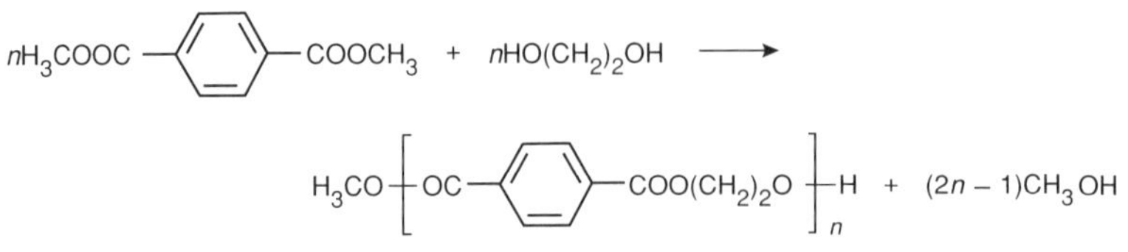

$$n\text{H}_3\text{COOC} - \!\!\left\langle \text{=} \right\rangle\!\! - \text{COOCH}_3 \ + \ n\text{HO(CH}_2)_2\text{OH} \longrightarrow$$

$$\text{H}_3\text{CO} \!\left[\!\text{OC} - \!\!\left\langle \text{=} \right\rangle\!\! - \text{COO(CH}_2)_2\text{O} \right]_n\!\!\!\! -\!\text{H} \ + \ (2n-1)\text{CH}_3\text{OH}$$

Scheme 10.13 *Condensation polymerisation to form polyesters.*

(consisting of only a few monomer units) are formed. All polyester fibres contain small quantities of oligomer, which can diffuse to the fibre surface during dyeing and form grey deposits.

As with nylon (see Section 10.5.1), extrusion of the molten polymer, followed by drawing, leads to a highly crystalline fibre (Figure 10.11) [100], which is noted for its high tensile strength [67,68]. The absence of ionic groups in the polyester molecule and its high degree of crystallinity inhibit the uptake of both water (Table 10.2) and anionic dyes. This problem has been overcome by using nonionic disperse dyes (originally developed for cellulose acetate), which are applied to polyester from an aqueous dispersion; examples of such dyes include C.I. Disperse Yellow 3 (**1**) and C.I. Disperse Blue 1 (**2**) [20]. The close packing of the polyester molecules and the rigidity of the aromatic rings [100] are responsible for a relatively high glass transition temperature ($T_g = 79\,^\circ\text{C}$) [54]. This leads to difficulties in dye adsorption, and necessitates the use of dyeing temperatures of up to $135\,^\circ\text{C}$. Polyester can also be produced by copolycondensation: small amounts of another acid (comonomer) are added to the terephthalic acid, and condensation takes place between this mixture and the glycol. Copolymers produced in this way differ from the homopolymer in properties such as thermoplasticity and tensile strength. In most cases, this modification

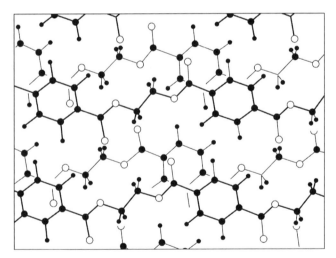

Figure 10.11 *Crystalline structure of polyester (larger dots, carbon; smaller dots, hydrogen; open circles, oxygen).*

reduces the state of crystallinity of the fibre, leading to an increased rate of dyeing with disperse dyes. Modified fibres have less tendency to pill, because of a reduction in tensile strength [101]. Comonomers containing anionic groups enable the polyester fibre to adsorb basic dyes.

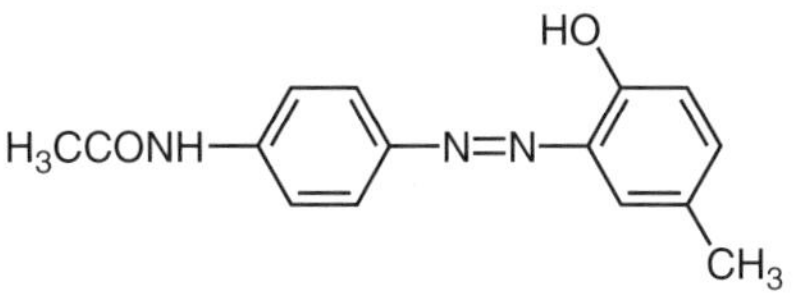

Because of the possible structural variations in the polyester fibre, the dyeing rates of different polyesters vary considerably. In order to quantify this behaviour so that appropriate dyeing parameters can be selected, a simple technique has been developed for determining the dyeing rate constant, V [102,103]. Goods are dyed at 95 °C for 30 minutes with 1% Resolin Blue FBL (BAY, now DyStar) (C.I. Disperse Blue 56) and then, in order to exhaust the bath, an equivalent quantity of the substrate is dyed at 125 °C for 1 hour. The value of V is determined by comparison of the two dyed fabrics with a prepared set of samples. A value of $V = 1$ is assigned to a standard fibre (Dacron 54 (Du Pont) [104]), and a higher rate of dye uptake is indicated by a higher value of V. Subject to variations of $\pm 10\%$, the value of V is independent of dye, dye concentration and dyeing temperature (in the range 95–130 °C) [104].

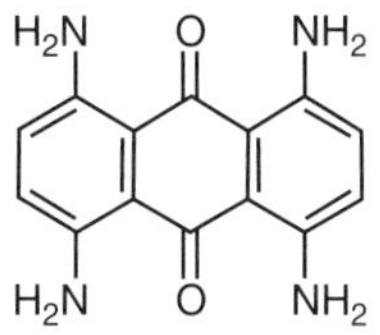

10.6.1.2   Adsorption of Disperse Dyes

Although disperse dyes are essentially insoluble in water, a small proportion of dissolves in the dyebath and is adsorbed from solution on to the surface of the fibre; the adsorbed dye then diffuses slowly into the fibre [81]. Dye adsorption occurs through π-bonding (involving aromatic nuclei) and hydrogen bonding with the fibre [18]. Planar dye molecules, which are able to come into close contact with the polymer chains and so facilitate the formation of bonds, generally have high substantivities. Attachment of undissolved dye particles to the fibre surface may also occur [102], but this is undesirable as it can create problems with levelness and fastness. The rate of solution of the dye in the liquor, and its diffusion into the fibre, depends on the dye, the fibre characteristics, the auxiliaries present and the dyeing temperature; the various processes involved interact in a complex manner during the dyeing operation [101].

In contrast to the behaviour of acid dyes on wool, the rate of dyeing of polyester with disperse dyes is slow. Whereas acid dyes start to build-up on wool from 40 °C, with the rate of dyeing doubling for every 10 °C rise in temperature, disperse dyes have virtually no substantivity for the polyester fibre below 85 °C, but then the rate of dyeing doubles for every 5 °C rise in temperature; the rates of dyeing for the two fibres are approximately the same at 120 °C [5,105]. At higher temperatures, the polyester molecules become more

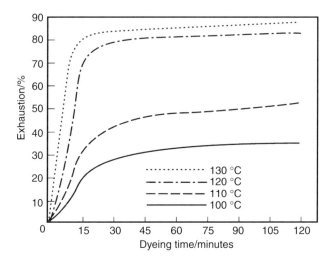

Figure 10.12   *Exhaustion of a typical disperse dye on polyester.*

mobile, leading to an increase in the rate of dye diffusion. An important point is that the 'reduced' dye uptake at lower temperatures is due to kinetic, rather than thermodynamic, effects; if dyeing at a lower temperature is continued for long enough, the equilibrium uptake of dye will be essentially the same as that at a higher temperature [106].

The dyeing behaviour of a typical disperse dye on polyester at temperatures ranging from 100 to 130 °C is shown in Figure 10.12 [105]. The need to dye at higher temperatures in order to achieve a high level of dye adsorption (at least in a reasonable period of time) is clearly indicated. Although the dyeing behaviour shown in Figure 10.12 is described as 'typical', the dyeing characteristics of disperse dyes vary considerably, and are largely influenced by molecular structure. Dyes of low molecular size (r.m.m. about 250–300 Da) diffuse readily into the fibre and may be applied at the boil; they display good coverage and level-dyeing behaviour, but have poor fastness to heat treatments and are inadequate for modern requirements [21]. Improvements in fastness are obtained by using dyes of higher r.m.m., but the migration properties and rate of adsorption are often adversely affected.

Disperse dyes have been classified into different groups according to their dyeing properties and fastness to sublimation [81,101,103]. Dyes with good fastness to sublimation are often referred to as 'high-energy dyes', and those with poor fastness to sublimation are called 'low-energy dyes'. It is usually advisable to dye at pH 4–5, because strongly acid, neutral and alkaline conditions are liable to impair the shade (or depth of shade) of some dyes.

It is considered that disperse dyes, in mixtures, do not generally influence one another with regard to adsorption rate [104]. For a mixture of dyes of similar properties, the one which is dyed at the highest concentration will exhaust most slowly, and there is no combination of disperse dyes which is completely compatible in varying concentrations. Independence in exhaustion also means that full shades are more easily obtained with a combination of several dyes than with a single one.

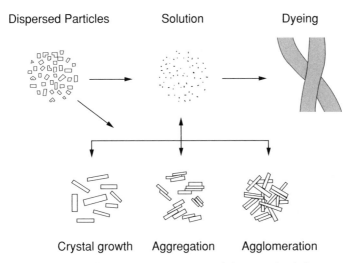

Dispersed Particles   Solution   Dyeing

Crystal growth   Aggregation   Agglomeration

Figure 10.13   *Schematic representation of disperse dye behaviour.*

10.6.1.3   *Dispersing Agents*

The particles of a disperse dye are normally in the size range 0.5–1.0 μm. Although disperse dyes are formulated with dispersants, additional dispersing agent must be added to the dyebath to prevent agglomeration and crystallisation of the dye (Figure 10.13). Poor dispersion stability leads to dye precipitation, manifested as spotting, slow exhaustion, poor colour yield, unlevel dyeings, poor reproducibility and soiled dyeing machines.

Anionic products such as sulphonic acid salts of naphthalene/formaldehyde condensates (**3**) and lignosulphonates are used widely as dispersing agents [107]. The hydrophobic group of the dispersant is adsorbed on to the surfaces of the dispersed dye particles, and this effectively encapsulates the dye. The sulphonate groups are solvated by the water (ensuring water solubility of the agent), with the formation of an electrical double layer. Mutual repulsion of the negatively charged dye particles inhibits breakdown of the dispersion. Because of differences in the formulation procedures and dispersing agents used, 'chemically equivalent' dyes can vary in dispersion characteristics, and dyes with the same C.I. number but from different manufacturers may differ in their dyeing behaviour [107,108].

3

Dispersion stability is adversely affected by high dye concentrations and dyeing temperatures, long dyeing cycles and a high degree of mechanical energy in the dyebath. The addition of electrolyte is not recommended, since it impairs dispersion stability, probably by adsorption of cations, which neutralise the charge of the dye complexes. At pH values between 4 and 6, dispersion stability is virtually independent of pH. Crystallisation is favoured by a high level of dye purity, a wide distribution of particle size, auxiliaries that increase the solubility of the dye and periodical heating and cooling of the liquor.

Yarn preparations and sizes often show inadequate emulsion stability under high-temperature conditions [101]. Breakdown of any emulsions can lead to a high concentration of dye in the form of droplets and cause specky dyeings. Adequate preparation of polyester goods prior to dyeing is therefore essential.

10.6.1.4   Levelling Agents

Dyeing levelness depends on the rate at which the dye is taken up by the fibre and the extent of migration at the maximum dyeing temperature. Since the rate of dyeing of polyester is relatively slow, there are often no problems with regard to the dyeing levelness [20]. There is very little migration, however, and it may be difficult to correct any faults by prolonging the dyeing time. Levelling agents in commercial use are, or contain, surface-active agents of the nonionic type [107,109,110].

Poly(ethylene glycol) derivatives such as $C_{16}H_{33}(CH_2CH_2O)_nCH_2CH_2OH$ (n = about 16) have been found to be useful [107]. The effectiveness of these products has been explained in terms of interactions with the dyes in the bath, leading to an increase in the solubility of the disperse dyes in water [110]. This results in better levelness due to increased migration, but generally gives rise to a decreased uptake of dye. Many nonionic agents appear to act by reducing the rate of dye exhaustion, with little effect on migration properties [109].

Nonionic levelling agents can have adverse effects on dispersion stability [107]. The solubility of these products decreases with increasing temperature; the temperature at which they become insoluble is known as the *cloud point*. If the cloud point of an added nonionic agent is below the dyebath temperature, a precipitate will form, resulting in a sticky, coloured deposit on the substrate being dyed. The cloud point can be influenced by the presence of other products in the liquor: electrolytes such as common salt depress it, while anionic surfactants raise it, as do anionic dispersing agents present in commercial dyes.

10.6.1.5   Carriers

The non-existence of high-temperature dyeing machinery was a serious impediment to the commercial adoption of the polyester fibre when it was first developed. Of crucial importance to polyester was the development of carriers. These dyeing auxiliaries are normally low-r.m.m. aromatic compounds such as biphenyl, dichlorobenzene and benzylphenol. Most carriers are insoluble in water and are formulated with surfactants to ensure emulsification in the dyebath [107,111]. Carriers increase the rate of dyeing of the polyester fibre (Figure 10.14), and this allows polyester to be dyed with disperse dyes of higher fastness at lower temperatures and, importantly, in commercially acceptable dyeing times [20].

The term 'carrier' is a misnomer. It originated from the idea that these compounds form complexes with the dyes which are 'carried' into the fibre [18]. It is now accepted that carriers modify the structure of the fibre, thus increasing the mobility of the polymer chains and facilitating the diffusion of dyes into the fibre [18,112]. It has also been suggested that water-insoluble carriers form a film on the fibre surface; the high solubility of disperse dyes

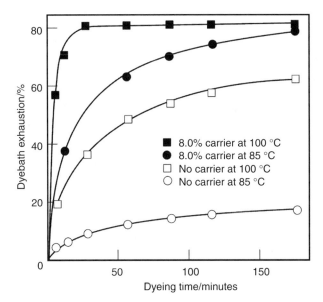

Figure 10.14   *Effect of a biphenyl-based carrier (and of temperature) on the rate of dyeing of C.I. Disperse Red 1.*

in this film aids dye uptake by the fibre [113]. Normally, carrier is added to the dyebath after the addition of dispersing agent and before the addition of dye.

Carriers are relatively toxic, and their presence in a dyehouse is normally signalled by their characteristic pungent odours. They can be broadly divided into two groups, depending on whether they are highly volatile or only mildly volatile in steam. Products belonging to the first group (such as di- and trichlorobenzene) are used mainly in enclosed machines, and have the advantage that they have little or no effect on the light fastness of dyeings as they can be readily removed from the fibre. In winches, jigs and partially filled jets, highly volatile carriers can condense on to the cooler roof of the machine and then drip on to the fabric. A locally high carrier concentration causes deeper dyeing, leading to the appearance of dark stains. In these types of machine, where the lid cannot be heated, or in open machines, the use of mildly volatile carriers such as *o*-phenylphenol and benzylphenol is recommended. To avoid light-fastness problems, carrier residues must be removed after dyeing by fabric stentering at 160–170 °C. Fibre shrinkage with carriers can be a problem too, especially in package dyeing, and stains may also be caused by breakdown of the carrier emulsion [3].

Carrier action can be described by the acceleration factor *a*; this is defined as the ratio of the amount of dye exhausted with auxiliary present to the amount of dye exhausted without auxiliary [110]. Most of the active substance of a carrier is taken up by the fibre to a degree that depends on the type of carrier. For a given carrier pick-up (by mass) on the fibre, different carriers have been observed to have the same acceleration factor. With increasing quantity of carrier, *a* increases until a maximum is reached (a_{max}), and further carrier brings about a reduction in *a* (Figure 10.15); this is due to a decrease in the equilibrium uptake of dye. No apparent relationship has been found between the dyeing rate constant *V* and

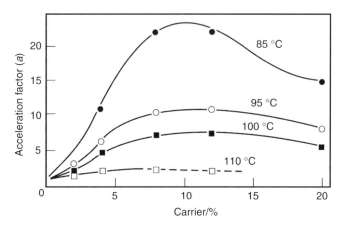

Figure 10.15   *Acceleration factor of an aromatic ester carrier on Dacron 54 at different temperatures as a function of carrier concentration (using 3% C.I. Disperse Blue 139).*

a_{max}. For values of $a < 8$, the relationship between a and carrier concentration is similar for different fibres [114].

With increasing temperature, a decreases (Figure 10.15) [110]. Values of a_{max} tend to be higher for dyes of lower diffusion coefficient, and differences in a for different dyes follow the same pattern for all carriers and fibres. If there are any variations in the structure of the polyester substrate to be dyed, carriers penetrate more rapidly into those parts of the fibre that take up the dye more readily. This results in a higher acceleration effect for such parts of the fibre, and accentuates the uneven uptake of dye during the exhaustion phase. In spite of the promotion of dye migration by the carrier, the levelness at the end of the dyeing may be unacceptable. Better results can be obtained by adding carrier after the initial exhaustion of dye.

10.6.1.6   Machinery

Temperatures above the boil are generally required for the dyeing of polyester fibres, necessitating the use of high-pressure machines in which temperatures as high as 140 °C can be attained [101]. In all dyeing machines operating at atmospheric pressure, the temperature must be maintained as close as possible to the boil. This applies particularly to machines in which the substrate is not constantly immersed in the liquor, such as jigs and winches.

10.6.1.7   Continuous Dyeing

Polyester fabric can be dyed continuously by the Thermosol process [20]. This procedure involves the following steps:

(1) The fabric is padded with a liquor containing disperse dye.
(2) The material is then dried, causing a film containing the dye to adhere to the surfaces of the fibres.
(3) The dyes are fixed by heating at 180–220 °C for 30–60 seconds; during this step, dye is transferred to the fibre interior.

Dyes with high vapour pressures (low-energy dyes) volatilise the most easily and give the best colour yields. If the vapour pressure is too high, however, dye can be deposited on to the machinery.

10.6.1.8   Heat Setting

Internal tensions are present in the polyester fibre as a result of production and processing [101]. Release of these stresses in subsequent wet and dry treatments can result in shrinkage of the substrate; polyesters shrink by about 7% in boiling water and even more at higher temperatures [20]. As with nylon (see Section 10.5.1), heat setting is required to improve dimensional stability. In addition, piece goods that have not been heat-set have a poor handle, and failure to carry out heat setting prior to piece dyeing may result in creases that are almost impossible to remove.

Yarns that have a tendency to shrink can be treated in the form of cops in saturated steam prior to dyeing [101]. Woven and knitted polyester fabrics are stabilised by setting at temperatures 30–40 °C above the highest temperature to which they will subsequently be subjected [20]; this is normally carried out in a stenter at 200–230 °C for 30–60 seconds.

Heat setting changes the dyeing properties of the fibre. In general, equilibrium dye uptake decreases as the heat-setting temperature increases, reaching a minimum at about 180 °C; above this temperature, dye uptake increases [20,21,115,116]. The initial fall in dye affinity has been attributed to an increase in crystallisation [115]. At temperatures above 180 °C, a two-phase structure tends to develop, in which the highly crystalline and amorphous zones increase at the expense of a zone of medium crystallinity. The resultant enlarged amorphous zone is a determining factor in increasing dye uptake.

10.6.1.9   Practical Considerations

Attention needs to be given to the following factors if polyester is to be dyed under optimal conditions [103]:

- starting temperature;
- temperature range for the initial adsorption of dye;
- rate of heating;
- amount of carrier required;
- dyeing time required at maximum temperature.

These factors are governed by the adsorption characteristics of the dyes, the dyeing rate constant of the substrate and the liquor circulation.

The following systematic approach to the dyeing of polyester has been developed by Bayer (all temperatures are in degrees Celsius) [103]. The starting temperature T_s can be calculated by:

$$T_s = T_r(\text{min}) + \Delta T_f - \Delta T_c \qquad (10.1)$$

T_r is the reference temperature at which the rate of dye exhaustion per minute is equivalent to about 1% of dye available when the dyeing rate constant V is equal to unity. This is determined by the adsorption characteristics of the dye and by dye concentration, and is calculated from published data; for a mixture of dyes, the minimum value of T_r is used. ΔT_f is a correction factor for fibres for which V is not unity. ΔT_c is a correction factor

Table 10.10   *Correction factors (ΔT_f) for different fibre types.*

V (dyeing rate constant of substrate)	ΔT_f (°C)
0.25	10
0.50	5
2	−5
4	−10
8	−15

Table 10.11   *Correction factors (ΔT_c) when carrier is used.*

a (acceleration factor)	ΔT_c (°C)
2	5
3	8
4	10
6	13
8	15

which applies when a carrier is added (see Tables 10.10 and 10.11). The amounts of carrier required to obtain a given acceleration factor are presented in Table 10.12. T_r varies from 75 °C for low concentrations of dyes with high rates of adsorption to about 120 °C for high concentrations of dyes with low rates of adsorption. In practice, starting temperatures as high as 80 °C are commonly used.

The upper temperature for the initial exhaustion phase (T_{end}) is calculated by [103]:

$$T_{end} = T_r(\text{max}) + \Delta T_f - \Delta T_c + 25 \tag{10.2}$$

$T_r(\text{max})$ is the maximum value of T_r for the dyes used. A rapid increase in temperature is permissible below and above the temperature range for the dyeing phase. Within that range, however, the temperature must be raised in uniform steps, with the rate of heating governed by the dyeing apparatus and the goods to be dyed.

Table 10.12   *Carrier addition required to obtain a given acceleration factor[a].*

a (acceleration factor)	Carrier (ex-Bayer) (% o.m.f.)			
	Levegal PT	Levegal D	Levegal TBE	Levegal OPS
2	1.0	1.0	1.5	1.5
3	1.5	1.5	2.2	2.6
4	2.2	2.0	3.0	3.7
6	3.2	3.1	4.0	5.8
8	4.0	4.0	5.0	7.5

[a]Values are for a liquor ratio of 20 : 1; carrier addition can be reduced by 10% for a liquor ratio of 10 : 1 and increased by 10% for a liquor ratio of 40 : 1.

The maximum temperature of the dyeing (T_{max}) should be higher than or equal to T_{end}. The dyeing time at T_{max} (in the range 120–135 °C and in the absence of carrier) is calculated from published information for the different dye groups, and should be sufficiently long to ensure good fibre penetration. The dyeing time at T_{max} should be multiplied by a factor of $1/V$ for a substrate with a dyeing rate constant other than 1, and should be multiplied by a factor of $1/a$ when carrier is used; a time of less than 10 minutes at maximum temperature is not advised.

10.6.2 Dyeing of Wool/Polyester Blends

Most of the dyeing information available for wool/polyester blends is related to wool-poor blends (in particular 45 : 55 wool/polyester), since these are of the greatest commercial importance. The polyester fibres normally used in blends are staple fibres (normal type, minimum-pilling type and also textured yarns) with relatively high levels of crimp [101]. They normally have a low level of stretch (and therefore a low level of crystallinity); this results in decreased tensile strength but a relatively high affinity for disperse dyes. The two main areas of concern in the dyeing of wool/polyester blends are:

(1) *Staining of the wool by disperse dyes:* Most disperse dyes have some affinity for wool, but on this substrate exhibit poor fastness to light, sublimation, wet treatments, rubbing and solvent treatments [101].
(2) *Appropriate conditions for dyeing of the polyester:* The normal high-temperature dyeing methods (120–135 °C) cannot be used because of the susceptibility of the wool fibre to damage (see Section 10.6.2.2).

Wool/polyester blends are generally prepared for dyeing in a similar manner to pure wool. It is important that any processing or scouring auxiliaries are removed before dyeing as these can interfere with the stability of the dye liquor. Wool-rich fabrics (especially lightweight fabrics) may require crabbing in order to stabilise the wool component and prevent the formation of running marks or creases during dyeing in a winch or jet/overflow machine. The crabbing operation can usually be omitted for wool-poor blends, as heat setting of the polyester component provides sufficient stabilisation of the fabric. Heat setting of wool/polyester fabrics is normally carried out prior to dyeing under less severe conditions (e.g. 180 °C for 30 seconds) than those used for pure polyester, in order to minimise yellowing of the wool. Fabrics which are to be beam-dyed do not require crabbing because the dyeing operation itself has a setting effect similar to that of crabbing.

10.6.2.1 Dyeing Procedures

Blends may be dyed by either a two-bath or a one-bath method [101]. In a two-bath operation, the polyester fibre is dyed first. This is followed by an intermediate wash or treatment with a reducing agent to remove disperse dye that has been taken up by the wool. Since no wool dyes have been added at this stage, a relatively intensive treatment can be given. This permits the selection of a broad range of disperse dyes, and leads to high levels of fastness. The wool is dyed subsequently in a fresh bath by a conventional method. In a one-bath operation, both fibres are dyed simultaneously with a mixture of disperse and wool dyes, and the goods are washed after dyeing. Disperse dye selection is somewhat more critical in order to minimise staining of the wool. It is possible to produce a wide range of shades with fastness properties that are scarcely inferior to those of two-bath dyeings. The one-bath dyeing method has established itself in practice because it saves time and ensures minimum tendering of the wool fibre.

Uptake of disperse dye by the wool fibre at lower temperatures (60–70 °C) is followed by migration on to the polyester at higher temperatures, an effect which is reversible [3,101]. During the wool-dyeing step in a two-bath process, therefore, dye can migrate from the dyed polyester fibre to the wool, thus negating to some extent the effects of the intermediate clearing treatment. Nevertheless, the two-bath process yields superior fastness properties for heavy shades. In a two-bath operation, problems may be encountered with certain anthraquinone dyes, which stain the wool and which, on reduction, give rise to coloured degradation products with hues that are different to those of the original dyes. Wool/polyester blends that have been dyed in a one-bath operation often cannot be given a reduction clear because of the susceptibility of wool dyes to decomposition. For deep shades (particularly blacks), however, an intensive clearing treatment can be used for material that is dyed with selected 1 : 2 metal-complex dyes, such as C.I. Acid Black 194.

10.6.2.2   Dyeing Conditions

Wool/polyester blends are dyed at temperatures ranging from the boil (95–98 °C) up to 120 °C. There are some advantages to dyeing wool above the boil, in that dyeing equilibrium and the attainment of maximum fastness properties are achieved in a shorter time, and increased migration promotes better levelness [117–119]. At temperatures above 110 °C, earlier practice was to include formaldehyde or formaldehyde derivatives such as DHDMEU in order to restrict the level of wool damage [5,120–125]. A rough guide to the appropriate dyeing conditions, based on recommendations from dye manufacturers, is given in Table 10.13.

The more rapidly the disperse dye is taken up by the polyester, the less the degree of staining of the wool [101]. Rapid uptake is favoured by a high dyeing temperature, addition of carrier and the use of readily dyeable types of polyester. Wool staining also depends on the type of carrier and the amount used. A high starting temperature decreases wool staining, but certain limits must be observed in order to ensure a level dyeing. Dyeing at higher pH values significantly reduces the extent of wool staining and improves the levelling characteristics of the wool dyes.

Dyeing at the boil is the only option if pressurised equipment is unavailable. The rate of disperse dye uptake by the polyester is low and staining of the wool by disperse dyes

Table 10.13   *Dyeing conditions for wool/polyester blends.*

Maximum Temperature (°C)	Time at Maximum Temperature (minutes)	Carrier[a]	Fibre Protection Agent	pH
95–98	60–120	yes	no	4.5–5.5
106	45–60	yes	no	4.5–5.5
110	30–60	yes	yes	4.5–5.5
120	15–30	yes	yes	4.5–5.5

[a]Less carrier is required at higher temperatures.

is relatively heavy, impairing the fastness of the dyed goods [121]. Dyeing under these conditions is inefficient as large quantities of carrier and long dyeing times are generally required. Variations in the affinity of the polyester for the disperse dyes are barely compensated for, or not at all. Wool/polyester goods are often dyed at about 106 °C [120,121]; this is more cost effective than dyeing at the boil, since dyeing times are considerably shorter and less carrier is required.

Dyeing at 110–120 °C can offer the following advantages: shorter dyeing times, better dye penetration of the polyester fibres, a more level uptake of dye by the polyester, higher colour yields, reduced problems associated with carriers and less staining of the wool [121]. However, such high temperatures produce significant wool damage, which has necessitated the development of so-called wool protective agents, which reduce the level of damage. Early studies described the use of 3–5% o.m.f. of a 30% solution of formaldehyde for a blend such as 45 : 55 wool/polyester. Later, to avoid the use of free formaldehyde, auxiliaries such as Irgasol HTW (Huntsman) and Lanasan PW Liquid (Clariant), based on formaldehyde/urea condensates, were promoted. Dyeing in the range 110–115 °C, rather than at 120 °C, is generally preferable, because the dyer has a far longer safe period before the onset of serious wool damage [126].

There is now a consensus that formaldehyde and formaldehyde-producing chemicals are not a modern solution to this damage-prevention problem. Over the past few years there have been significant advances in wool protective agents which do not give rise to the formaldehyde problem. Formaldehyde (all physical forms) was highlighted by the National Institute of Environmental Health Sciences (NIEHS) for reclassification as a potent carcinogen. This opinion was based on the 2004 review by the International Agency for Research on Cancer (IARC 2006), which concluded that there was sufficient evidence for the carcinogenicity of formaldehyde in humans; nasal tumours and brain tumours are implicated with formaldehyde exposure [127].

There is a direct relationship between setting and damage in the wool-dyeing process: the higher the degree of set imparted, the greater the fibre damage [128,129]; thus, antisetting agents are added to wool dyebaths in order to control damage. This procedure is clearly very necessary when dyeing wool/polyester blends above the boil.

10.6.2.3 *The Role of Oxidants in Preventing Setting in Dyeing*

One of the most useful antisetting systems for use in wool dyeing was developed by workers at the Commonwealth Scientific and Research Organisation (CSIRO), International Wool Secretariat (IWS) and BASF [130]. This system, offering improved fibre physical properties, was based on a mixture of hydrogen peroxide and a special auxiliary Basolan AS; the latter auxiliary performed two functions:

- Inhibition of the degradative effect of hydrogen peroxide on some wool dyes.
- Enhancement of the stability of the oxidant in the boiling dyebath.

In the BASF/IWS/CSIRO process [130], it is recommended that wool is dyed in the presence of hydrogen peroxide (35%) at a level of $1 \, ml \, l^{-1}$ (minimum 2% o.m.f.) and Basolan AS at $0.5 \, g \, l^{-1}$ (minimum 1% o.m.f.).

10.6.2.4   Dye Selection

A major problem in the dyeing of wool/polyester blends is the choice of appropriate *disperse dyes* that do not undergo reduction in the presence of wool [131]. Disperse dye reduction cannot be improved or corrected by simple pH adjustments or dyebath additives. Certain disperse dyes, especially blues and navy blues based on azo chemistry, are highly susceptible to reducing agents, and in the presence of wool may be degraded to give colourless compounds. When more than 20% wool is used in a blend, disperse dyes must be selected that are not sensitive to reduction.

Other important factors governing disperse dye selection are the temperature at which the substrate is to be dyed and the extent to which the disperse dye stains the wool [3]. The higher the dyebath temperature, the wider the choice of disperse dye, particularly for medium and heavy shades. For goods which are to be heat-set after dyeing, such as colour-woven fabric, the sublimation fastness of the disperse dyes is also an important consideration.

Dyeing at the boil with carrier in anything above pale to medium depths requires the use of low-energy dyes that show satisfactory build-up [3]. Only pale shades can be dyed at the boil with medium-energy dyes, and medium depths of shade require dyeing at 106 °C or above. Exhaustion and levelling of medium-energy dyes are inadequate at the boil, and wool staining is unacceptable. Dyeing at the boil is largely confined to woven piece goods in enclosed atmospheric winches, and the range of suitable disperse dyes is restricted. As the goods are invariably heat-set prior to dyeing, sublimation fastness is rarely a consideration. Combination shades are commonly selected from the following dyes:

- C.I. Disperse Yellows 23, 54, 93 and 218;
- C.I. Disperse Orange 25;
- C.I. Disperse Reds 50, 60 and 65;
- C.I. Disperse Blues 35, 56 and 81.

The selection of disperse dyes for piece goods in pressurised machinery (beams or jet/overflow machines), which can operate at between 106 and 120 °C, is made easier by the availability of a further group of dyes which exhaust well at these temperatures:

- C.I. Disperse Oranges 29 and 45;
- C.I. Disperse Reds 54 and 349;
- C.I. Disperse Violets 33 and 95;
- C.I. Disperse Blues 73 and 148.

Wool/polyester yarn is usually dyed on crosswound packages in radial-flow machines at 106 °C. Package-dyeing machines may have drainage systems that are configured to allow dumping of the dyebath at the end of the hold time at maximum temperature. This procedure minimises the risk of oligomers recrystallising in the cooling dyebath and depositing on the yarn surface through filtration effects. Suitable combination dyes for yarn dyeing include:

- C.I. Disperse Yellows 64, 201 and 210;
- C.I. Disperse Oranges 29, 66 and 97;
- C.I. Disperse Reds 82 and 106;
- C.I. Disperse Violets 33 and 40;
- C.I. Disperse Blues 73 and 333.

Several dyes suitable for the blue component of binary or ternary combinations are mixtures and are therefore not identifiable by C.I. numbers.

The selection of a wool dye for dyeing of the wool component is somewhat less restricted [3]. Levelling and coverage properties of the wool dyes are generally adequate under the conditions required to obtain a satisfactory dyeing of the polyester component, and wool dyes can usually be selected on the basis of fastness requirements. 1 : 2 metal-complex dyes, supplemented by selected milling dyes for brighter shades, are widely used.

Since 1965, mixtures containing disperse and wool dyes have been available [3]. The proportions of dyes in these mixtures are based around the dominant 45 : 55 wool/polyester blend, and it is claimed that blends with higher proportions of polyester can be dyed without the additional need for disperse dye. Some doubts have been expressed regarding the compatibility aspects with ternary combinations of mixture dyes. Although the possibility of dyeing wool and polyester with a single class of dye has been investigated, the results appear to be of little practical significance.

10.6.2.5   Carrier Selection

In addition to the requirements for an ideal carrier for the dyeing of polyester, a carrier for use in the dyeing of wool/polyester blends should not increase uptake of disperse dye by the wool. The ideal carrier which fulfils *all* the desired requirements for the dyeing of blends does not exist [3,132]. The dyer is therefore forced to compromise and select the most suitable carrier (or carriers) for the prevailing conditions.

o-phenylphenol (OPP) has been used widely for the dyeing of wool/polyester blends at temperatures close to the boil [3,107,111,132–134]. It is relatively nonvolatile, has a weak odour and can be used in open machines. Use of OPP can lead to staining problems due to crystallisation at temperatures below 60 °C. It may have an adverse effect on light fastness if not adequately removed; a drying temperature of 140–160 °C is therefore recommended. Relatively high concentrations of carrier are required due to its slight solubility in water and its absorption by wool. OPP is not readily biodegradable, has no influence on levelling and increases staining of the wool by disperse dyes [107,111,133]. In spite of these problems, and its high cost, OPP has remained popular and has been described as one of the most efficient carriers available for dyeing at the boil. The effects of different concentrations of an OPP-based carrier on the colour yield of the polyester portion of a 45 : 55 wool/polyester blend, dyed with a mixture of disperse dyes, are shown in Figure 10.16 [133].

Di- and trichlorobenzene are particularly effective carriers at low concentrations, giving good colour yields on polyester and minimal disperse dye staining on the wool, along with good economy in usage [3,107,111,132,133]. Their use is restricted to closed machinery, however, because of their volatility and toxicity; furthermore, they are not readily biodegradable. Data corresponding to those in Figure 10.16 for dyeings with a carrier based on trichlorobenzene at 100 and 105 °C are shown in Figure 10.17 [133]. At carrier concentrations greater than about 3 g l^{-1} at 100 °C, a decrease in colour yield is observed, caused partly by the slight dissolution of dyes in the carrier. The major factor in this retarding effect is the presence of the emulsifying agent in the carrier, which can restrict the uptake of both disperse dye and wool dye. The retarding effect of the emulsifier is not apparent for the same shade dyed at 105 °C.

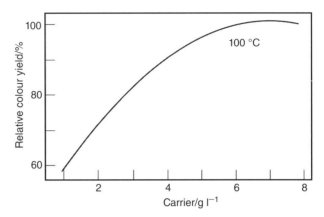

Figure 10.16   *Relationship between colour yield (on the polyester component of a 45:55 wool/polyester blend) and carrier concentration for an OPP-based derivative at 100°C.*

Like OPP, methylnaphthalene is useful for the dyeing of blends at the boil [3,107,132]. It has low volatility in water vapour and therefore there is little risk of carrier stains. Problems regarding the use of methylnaphthalene include its strong odour and its adverse effect on light fastness if the recommended drying conditions (e.g. 180 °C for 30 seconds) are not observed. It has low toxicity to animal, fish and plant life, however, and is fairly readily biodegradable [108].

Biphenyl has been widely used for wool/polyester blends, especially piece goods. It is relatively insensitive to variations in liquor-to-goods ratios, is cheap, has little or no effect on light fastness and minimises staining of wool by disperse dyes [111]. It is prone to crystallisation at temperatures below 80 °C, however, and is highly steam-volatile and therefore liable to cause condensation problems. These properties, together with its strong, unpleasant odour and low biodegradability, have led to a marked decline in its use.

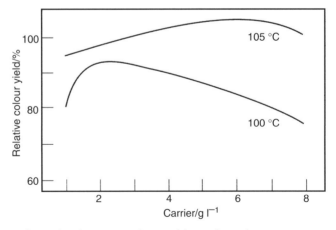

Figure 10.17   *Relationship between colour yield (on the polyester component of a 45:55 wool/polyester blend) and carrier concentration for a trichlorobenzene-based derivative.*

Methyl cresotinate is an efficient carrier for the dyeing of blends, but is more expensive than many other products [111]. It is best used in enclosed machinery, because of its high volatility and strong odour. This product has low toxicity and good biodegradability, can be readily removed by drying at 140–150 °C and minimises uptake of disperse dye by the wool.

N-alkylphthalamide derivatives have also been used [111]. They have little effect on light fastness, give a good wool reserve, have relatively little odour, are of low toxicity and are biodegradable. Other compounds that have been used include methyl salicylate and methyldichlorophenoxy acetate. Mixtures of carriers have been recommended in certain circumstances, allowing a combination of desirable properties to be achieved [3].

From a knowledge of the dyeing rate constant of the polyester component of a 45 : 55 wool/polyester blend, determined after dissolving out the wool component with caustic soda, the required concentrations of several carriers (for dyeing temperatures in the range 96–120 °C) can be calculated from published information.

The wrinkle recovery of the polyester portion of a blend can be affected by different carriers [135], and the extent of the reduction in wrinkle recovery is dependent on the type and amount of carrier retained by the polyester fibre [136].

10.6.2.6   *Reactive Disperse Dyes for Wool/Polyester Blend Dyeing*

The reactive disperse dyes described for the dyeing of wool/nylon blends (see Section 10.5.1) should be useful for the dyeing of wool/polyester blends with a single dye class [86–89]. This system has not yet been adopted by the industry. The most recent work was reported by Moussa *et al.* [137], who prepared two different VS precursors, Dye 1 and Dye 2 (Scheme 10.14). The importance of functionality on dye uptake by the wool and polyester components is illustrated in Figure 10.18a and b; clearly the bifunctional dye, Dye 1, performs the best.

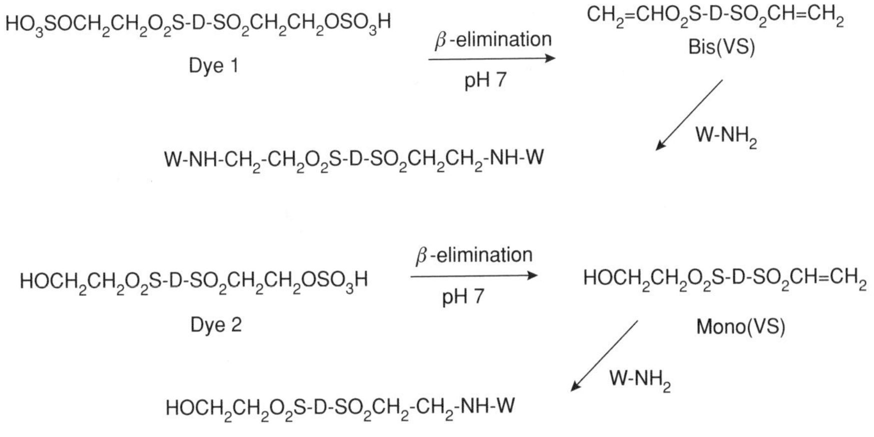

$HO_3SOCH_2CH_2O_2S$-D-$SO_2CH_2CH_2OSO_3H$

Dye 1

$\xrightarrow[\text{pH 7}]{\beta\text{-elimination}}$

$CH_2{=}CHO_2S$-D-$SO_2CH{=}CH_2$

Bis(VS)

↙ W-NH$_2$

W-NH-CH$_2$-CH$_2$O$_2$S-D-SO$_2$CH$_2$CH$_2$-NH-W

$HOCH_2CH_2O_2S$-D-$SO_2CH_2CH_2OSO_3H$

Dye 2

$\xrightarrow[\text{pH 7}]{\beta\text{-elimination}}$

$HOCH_2CH_2O_2S$-D-$SO_2CH{=}CH_2$

Mono(VS)

↙ W-NH$_2$

HOCH$_2$CH$_2$O$_2$S-D-SO$_2$CH$_2$-CH$_2$-NH-W

D = dye chromophore

Scheme 10.14   *Reaction of temporary water-soluble reactive disperse dyes with wool. Reproduced from A A Moussa et al., Coloration Technology, **127**, 304 (2011) with permission of John Wiley & Sons Ltd.*

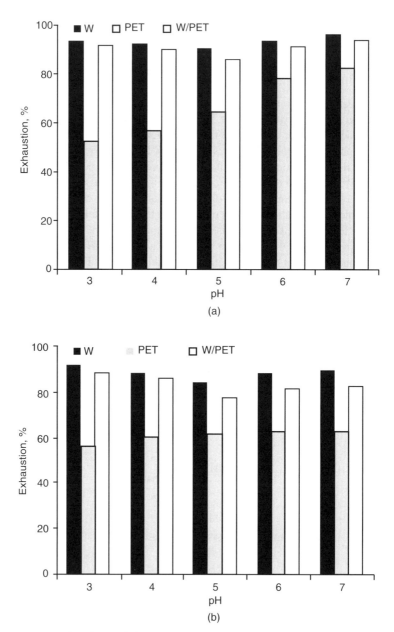

Figure 10.18 *Uptake of (a) a bifunctional sulphatoethylsulphone dye and (b) a monofunctional sulphatoethylsulphone dye on wool, wool/polyester and polyester. Reproduced from A A Moussa et al.,* Coloration Technology, **127***, 304 (2011) with permission of John Wiley & Sons Ltd.*

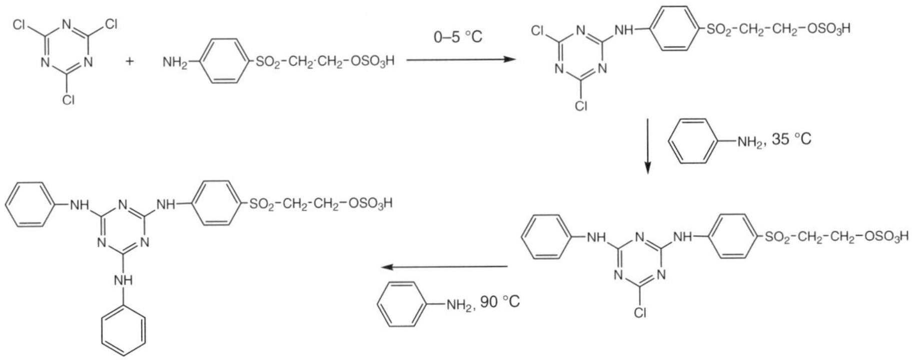

0–5 °C

—NH₂, 35 °C

4

—NH₂, 90 °C

Scheme 10.15   *Preparation of FAA 200 (**4**).*

10.6.2.7   *Novel Arylating Agents for Facilitation of the Dyeing of Wool and Polyester Fibres with Disperse Dyes*

Work carried out by Lewis *et al.* [138] achieved a one-bath, one-dye system for the dyeing of wool/polyester blends. These workers developed a water-soluble fibre-reactive agent for wool which was applicable from long liquor, was compatible with commercial disperse dye dispersions and formed covalent bonds with the fibre in the region of pH 5–6. In addition, the agent protected the wool fibre from damage even at 120 °C, a temperature used to ensure that the disperse dye penetrated the polyester component. The agent was coded FAA 200 (**4**) and was prepared as shown in Scheme 10.15. This agent is fibre-substantive and when absorbed reacts covalently with wool fibre nucleophiles (Wool-XH, where X is NH or S) to give Wool-X-(Ar)n (where (Ar)n represents a residue of FAA 200). It was applied to 100% wool along with selected disperse dyes at pH 6.0 at 120 °C; for comparison, the same disperse dye selection was dyed on 100% polyester fabric. The exhaustion of disperse dye from the wool dyebaths approached 100% when 10% o.m.f. FAA 200 was present, but in its absence very little disperse dye was taken up by wool. When dyeing wool/polyester blends (60 : 40), effective disperse dye uptakes were produced using 5% o.m.f. FAA 200 in the dyebaths. Tables 10.14 and 10.15 summarise the results achieved.

See Chapter 4 for more on this topic.

10.7   Wool/Acrylic

Of all the synthetic fibres, acrylic fibres have the most wool-like character, and knitted goods made of wool/acrylic are similar in appearance, handle and wear properties to articles made of pure wool [139]. The strength and bulk properties of acrylic fibres are enhanced by the water-sorption and aesthetic properties of wool fibres when the two fibres are combined in an intimate blend. The wool content of the most common blends ranges from 30 to 60%; blends containing 60% or more acrylic fibres have a high level of washability.

Table 10.14 *Colour yields for selected disperse dyes applied to all-wool fabric and polyester fabric (PES).*

Disperse Dye	Colour Yield (f_k) Values			
	Wool			PES
	No FAA	5% FAA	10% FAA	
C.I. Disperse Red 50	50	120	170	160
C.I. Disperse Orange 25	23	42	83	93
C.I. Disperse Violet 26	14	21	28	39
C.I. Disperse Red 167 : 1	48	106	128	127
C.I. Disperse Blue 79 : 1	37	81	117	136
C.I. Disperse Red 60	15	21	32	72
C.I. Disperse Yellow 119	12	51	60	73

10.7.1 Dyeing of Acrylic Fibres

10.7.1.1 Fibre Structure

Polyacrylonitrile fibres are produced by polymerisation of acrylonitrile [67,68] (Scheme 10.14). Comonomers are generally incorporated in the synthesis to modify the properties of the fibre. Polyacrylonitrile fibres containing at least 85% acrylonitrile units are described as *acrylic* fibres, whereas those containing 35–84% acrylonitrile units are described as *modacrylic* fibres [140]. Modacrylic fibres have had limited success, occupying a more specialised market position than acrylic fibres. Acrylics are produced almost exclusively in the form of staple fibres [141].

Table 10.15 *Colour yield values from dyeings produced with selected disperse dyes on 60:40 wool:PES blend fabric.*

Disperse Dye	Colour Yield (f_k) Values		
	Wool/PES		PES
	No FAA	5% FAA	
C.I. Disperse Red 202	70	109	139
C.I. Disperse Orange 30	64	80	120
C.I. Disperse Yellow 114	28	54	69
C.I. Disperse Violet 26	35	47	78
C.I. Disperse Red 167 : 1	73	121	127
C.I. Disperse Yellow 218	50	79	78
Dispersol Navy XF	72	129	150
Dispersol Navy G PC	62	130	110
Dispersol Rubine 3B PC	48	125	106
C.I. Disperse Red 60	19	23	40

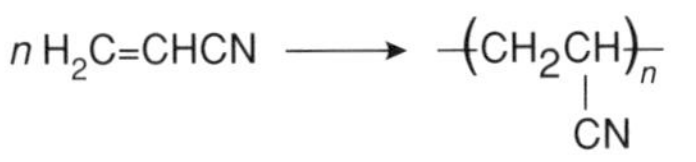

$$n \, H_2C{=}CHCN \longrightarrow (CH_2CH)_n$$
$$\overset{|}{CN}$$

Scheme 10.16 *Polymerisation of acrylonitrile.*

Acrylonitrile polymerises (Scheme 10.16) readily in aqueous solution in the presence of a suitable catalyst [20,67,68]. Melt spinning of polyacrylonitrile polymers is not practical because they decompose before melting. Filaments must therefore be wet- or dry-spun from a solution. In wet spinning, which is preferred for staple fibres, the polymer is dissolved in a solvent such as dimethylformamide and pumped through spinnerets into a bath containing a liquid with which the solvent is miscible, but in which the polymer is insoluble. The jets of solution coagulate into fine filaments, forming a tow, which is washed, dried, drawn, oiled and crimped. It may then be heated to relax the fibre before being cut into staple form. In dry spinning, the polymer solution is extruded through spinnerets into a chamber heated to about 400 °C; evaporation of the solvent produces solid filaments of polymer. An important property of acrylic fibres is that when they are hot-stretched by 15–30% of their original lengths, the elongation is not stable and they will relax in steam. If stretched and unstretched staple are spun together and subsequently steamed, only the stretched fibre will contract; this causes the low-shrinkage, unstretched component to buckle or bend outwards, giving a yarn with greatly increased bulk.

Polymerisation of acrylonitrile is usually achieved in the presence of peroxydisulphate, plus sulphite or thiosulphate as activator [142]. Sulphite and sulphate radicals are formed and act as initiators of the polymerisation of the acrylonitrile. Additional acrylonitrile molecules are added and the radical character of the growing molecular chain is always maintained. The polymerisation reaction eventually produces a giant polyacrylonitrile molecule, its radical state being finally saturated by the addition of a sulphite or sulphate radical. The polyacrylonitrile polymer produced by this process contains sulphate and sulphonate end groups [143,144].

It has been postulated that Van der Waals forces exist between the polymer chains, and an isotactic structure (**5**) has been assigned to the crystalline sections of the macromolecular micelle, and a syndiotactic structure (**6**) to the amorphous sections [143,144]. It has also been suggested that the fibre structure may be mesomorphic, however, and that it contains regions with pronounced lateral arrangements of the chain molecules, together with small crystalline regions with a three-dimensional periodic arrangement of chain segments that have a syndiotactic structure [144].

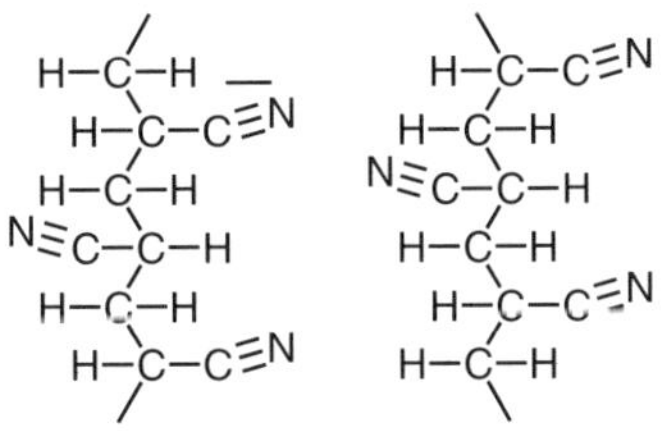

5 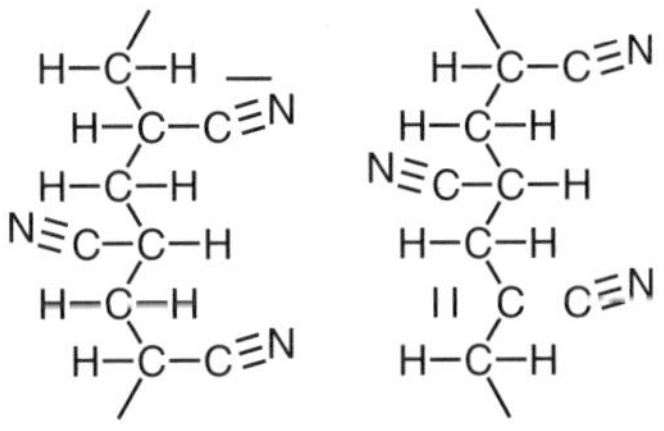   **6**

Table 10.16   *Acidic group content of acrylic fibres.*

Fibre	Acidic Group Content (equiv. kg^{-1})	
	Strongly Acidic	Weakly Acidic
Acribel (Fabelta)	0.028	0.030
Acrilan 16 (Monsanto)	0.031	0.021
Beslan (Toyo Rayon)	0.070	0.044
Courtelle E (Courtaulds)	0	0.154
Dralon (Bayer)	0.048	0.053
Orlon 42 (Du Pont)	0.046	0.017

An acrylic fibre (r.m.m. 60 000–120 000 Da) based almost entirely on acrylonitrile has several undesirable properties, including poor solvent solubility and a relatively high glass transition temperature ($T_g = 105\,°C$), making it extremely difficult for dyes to be adsorbed below 100 °C [143,144]. These problems have been overcome by copolymerisation of polyacrylonitrile with one or more comonomers (such as methyl methacrylate or vinyl acetate), leading to a less crystalline fibre with a T_g of 55–60 °C. Dyes' diffusion rates into the fibre are increased so that they can be applied at below 100 °C in conventional equipment. The temperature at which the fibre begins to adsorb dye depends on the comonomers used. Because acrylic fibres produced from different comonomers can have significantly different dyeing properties, polyacrylonitrile fibres are more difficult to consider as a general class than are nylon or polyester fibres.

Early acrylic fibres contained only small concentrations of anionic groups, and this precluded the dyeing of heavy depths and caused problems in shade matching [143,144]. These problems were solved by the use of acidic comonomers, most of which have weakly acidic constituents, and by the use of different catalysts to introduce strongly acidic groups. Concentrations of acidic groups in commercial acrylic fibres generally lie in the range 0.050–0.150 equiv. kg^{-1}; strongly and weakly acidic groups may be present in approximately equal proportions, or one of the two types may predominate (Table 10.16).

Another factor affecting the dyeing properties of acrylic fibres is the method of spinning, which controls the number and size of voids in the fibre [143]. These voids influence the rate of diffusion of dyes and, in many cases, the equilibrium exhaustion. In general, dry spinning produces fewer voids than does wet spinning, and spinning in an aqueous bath produces more than solvent spinning. The temperature of treatment of the fibre during spinning, in either the wet or the dry state, has a considerable influence over the microstructure of the acrylic fibre, and hence its dyeing properties.

10.7.1.2   Cationic Dyes

Most acrylic fibres are dyed with cationic dyes (traditionally referred to as 'basic dyes'), which dissociate in aqueous solution to give positively charged, coloured ions [20,143,144]. Those dyes especially developed for acrylics can be divided into three groups:

(1) *Dyes with a delocalised positive charge, such as C.I. Basic Violet 7 (7):* The dyes in this group are the brightest and have the highest tinctorial yield, but many have poor fastness to light and/or insufficient stability to hydrolysis.

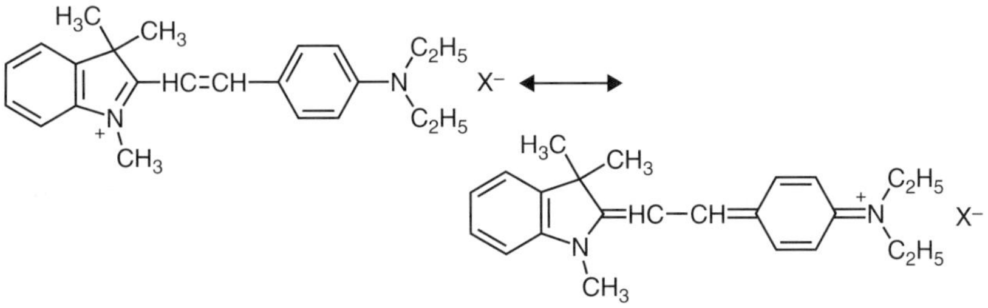

7

(2) *Dyes with a localised positive charge, such as C.I. Basic Red 18 (8):* The positive charge on substituent groups in these dyes is not conjugated with the chromophore. They have good fastness to light and good stability at high pH.

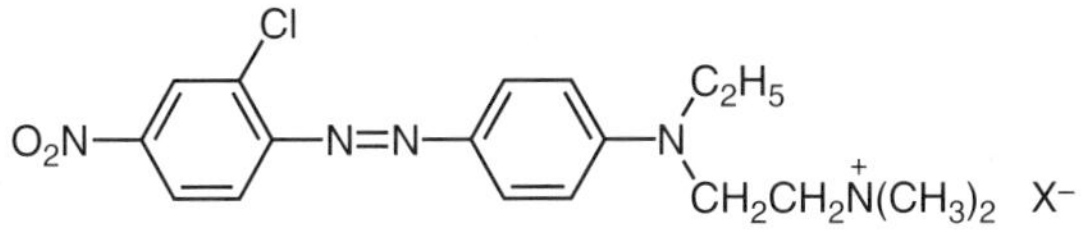

8

(3) *Dyes with a heterocyclic ring containing a quaternary nitrogen atom which does not form an integral part of the chromophore (9):* The tinctorial strength of these dyes is better than that of the second group, enabling them to be used on fibres with a low acidic group content. These dyes have relatively poor stability at high pH.

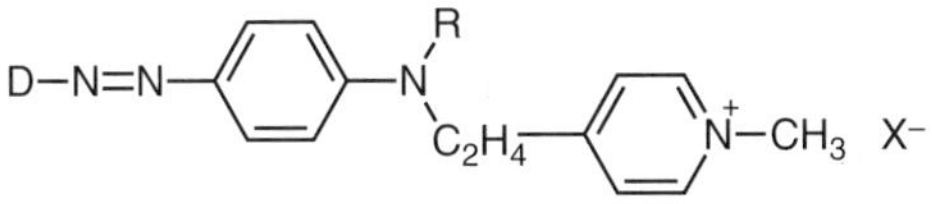

9

The interaction between cationic dyes and acrylic fibres is predominantly ionic (Scheme 10.17, in which $M = Na$ or H and X^- is the anion associated with the dye) [143,144]. The acrylic fibre has been considered to act like an ion-exchange resin during dyeing, taking up dye cations, which displace colourless cations (usually sodium) already

$$\text{fibre–M} \rightleftharpoons \text{fibre}^- + \text{M}^+$$

$$\text{dye–X} \rightleftharpoons \text{dye}^+ + \text{X}^-$$

$$\text{fibre}^- + \text{dye}^+ \rightleftharpoons \text{fibre–dye}$$

Scheme 10.17   *Interaction of cationic dyes with acrylic fibres.*

in the fibre. The number of dye sites available depends on the number of acidic groups and on whether these groups are accessible to enable interaction between dye and fibre to occur. The dye must be dissociated before it can react, and factors that influence the degree of ionisation of the fibre or the dye will affect dye uptake. The amount of dye adsorbed by the fibre is often slightly higher than that expected from the number of acidic groups. This anomaly has been attributed to dissolution of a small amount of dye in the fibre.

The number of dye sites available per unit mass of fibre has been termed the *fibre saturation value* [144]. This is defined as the quantity of pure (hypothetical) dye of r.m.m. 400 Da (calculated as % o.m.f.) that yields 90% dyebath exhaustion when applied for 4 hours at $100\,^{\circ}$C (pH 4.5) and with a liquor-to-goods ratio of 100:1. Similarly, *dye saturation factors* have been determined for commercial dyes. These fibre and dye characteristics can be used to determine the maximum amount of dye that can be bound to the acrylic fibre [139].

As the dyebath pH increases, dye uptake also increases [143]. This is more noticeable for fibres containing only weakly acidic groups than for those with strongly acidic groups. The difference in behaviour arises because dissociation of the strongly acidic groups is scarcely affected by increasing the pH, whereas the dissociation of weakly anionic groups increases significantly. Careful control over pH is necessary to ensure shade reproducibility; slightly acidic conditions are required to avoid the possibility of dye decomposition that occurs in hot alkaline solutions.

When an acrylic fibre is immersed in water, a zeta potential of $-44\,$mV is established between its surface and the water [143,144]. This negative charge is responsible for the adsorption of dye by the fibre surface. At low dye concentrations, the fibre gradually loses its negative potential and becomes slightly positive because of the accumulation of dye cations at the surface. The dye diffuses into the fibre, the potential becomes negative again and the cycle is repeated. A further increase in dye concentration above the amount required to neutralise the zeta potential changes the potential very slightly, indicating that the amount of dye adsorbed at the surface remains nearly constant. It has been suggested that other intermolecular forces (such as dispersion or dipole forces) are also involved in dye adsorption. The extent of dye uptake is independent of the liquor-to-goods ratio and the rate of dyeing, and also the dyeing temperature when this is below the glass transition temperature of the fibre. It does depend, however, on the nature of the acrylic fibre, the basicity of the dye and the pH of the dye liquor; the effect of pH is due to the competition between dye cations and hydrogen ions for the negatively charged sites on the fibre surface.

Since the formation of salt links between cationic dyes and the acrylic fibre is virtually irreversible, the wet-fastness properties of the dyes are high but the migration and levelling characteristics are poor [143,144]. In practice, cationic dyes are applied to acrylic fibres at a pH of 4.0–5.5 in the presence of an electrolyte and a retarding agent. The levelling effects of both pH control and additions of electrolyte are limited by the fibre type and the dye selection, so that temperature control and the addition of retarders are of the greatest practical benefit. Because of the high activation energies of dyeing for cationic dyes (about $293 \, kJ \, mol^{-1}$), temperature differences within the dyebath can have a marked influence on the rate of dyeing; an increase in temperature of just $1 \,°C$ can increase the rate of dyeing by about 30%.

Cationic retarding agents, such as quaternary ammonium salts with long aliphatic chains and aromatic groups, have been widely used to promote the level dyeing of acrylic fibres [144]. These water-soluble organic cations compete with the dye cations for the anionic dye sites, both at the fibre surface and within the fibre, thus reducing the rate of dye exhaustion. Because there is no direct interaction between a cationic retarding agent and a cationic dye, the compatibility values of the dyes are unaffected. The amount of retarder used, however, must be carefully calculated for each dyeing in order to avoid problems caused by 'oversaturation' or 'blocking'. Polymeric cationic retarding agents, containing up to several hundred cationic groups per molecule, have also been developed. These do not diffuse into the fibre and have little effect on migration, but the rate of dyeing is dramatically reduced.

Anionic retarding agents containing sulphonic acid groups that interact with cationic dyes can form dye complexes that have little affinity for the fibre [144]. A nonionic auxiliary is used with the anionic compound to keep the cationic–anionic complex in a suitable state of dispersion. At temperatures below the boil, the complex is loosely bound to the fibre, permitting some levelling on the fibre surface. As the temperature rises, the complex breaks down, allowing cationic dye to accumulate at the fibre surface and diffuse into the fibre. The concentration of free dye cation falls well below the value required for saturation of the fibre surface. In addition, the rate of dyeing and equilibrium exhaustion are reduced, the effects being largely dependent on the dyes used. With dyes of high affinity, the equilibrium favours complex formation, so that the compatibility values of the dyes are changed, enabling dyes whose rates of dyeing are normally different to be absorbed at a similar rate.

The rate of dyeing of cationic dyes is reduced by the addition of electrolytes, such as sodium sulphate, to the dyebath [143,144]. The more mobile sodium cations are preferentially adsorbed by the fibre and subsequently displaced by the dye cations. Electrolytes also reduce the magnitude of the surface potential of the fibre, thereby decreasing the rate of dyeing. At a level of 10%, sodium sulphate is a moderate retarder and enables a reduction of 20–30% in the amount of organic-based retarder required.

Compatibility in dye mixtures is extremely important for the level dyeing of cationic dyes [144]. A standard test has been instituted for the characterisation of cationic dyes according to a *compatibility value K*, which ranges from 1 to 5, the dyes with lower K values exhausting more rapidly. Dyes with the same K value are compatible in combination on acrylic fibres under practical exhaust-dyeing conditions, except in the presence of anionic dyes or auxiliaries. The value of K is primarily related to affinity, so that the effect of cationic retarders and electrolytes is more pronounced with dyes of higher K value.

Migrating cationic dyes, such as the Maxilon M range (CGY), have been developed. These are hydrophilic dyes of small molecular size that are taken up rapidly by the fibre, but which also migrate to promote level dyeing [144]. These dyes can be applied using sodium sulphate and a cationic retarding agent.

10.7.2   Dyeing of Wool/Acrylic Blends

10.7.2.1   *Preparation*

Blended yarns whose polyacrylic component contains a shrinkage fibre must be bulked before dyeing [139]. The choice of shrinking method depends on the desired yarn quality, as well as cost considerations. Discontinuous bulking at the boil can be carried out on yarn in the dyeing machine. If the shrinkage bath does not contain detergent, it can be cooled to the required initial dyeing temperature and used as the dye liquor. Alternatively, vacuum steamers can be used for the batch treatment of yarn muffs (cylindrical, crosswound soft packages). Yarn bulked in this manner has a greater volume than yarn bulked in water, and a further advantage is that the dyeing machine can be used to its maximum capacity for dyeing only. If steamed batches which are subsequently combined to form one dye lot have not been steamed under identical conditions, problems during dyeing may arise. Continuous package-to-package steaming units have been developed in which yarns can be bulked and wound on to the appropriate packages for subsequent use.

Dyeing of yarn hanks is economically justifiable with lower-count yarns such as hand-knitting yarns, which have superior bulk when dyed in this form. Normal yarns are dyed in the form of conventional crosswound packages. Despite preshrinking, high-bulk yarns are unable to develop their full bulk with this type of winding, and it is not possible to make adequate allowance for any residual shrinkage. Both high-bulk and normal yarns can be dyed in the form of soft packages. A small proportion of wool/acrylic blends are dyed as single or double jersey, mainly on beam or jet/overflow machines. Fully fashioned articles are dyed in paddle machines, or drum machines in the case of socks.

10.7.2.2   *Dyeing Procedures*

Mixtures of anionic dyes (milling, 1 : 2 metal-complex or reactive) and cationic dyes are generally recommended for use in a one-bath operation [139,145]. The degree of cross-staining of acrylic fibres (of the types which are generally used in blends with wool) by wool dyes is very slight, and is not a problem in blend dyeing [145]. Since wool contains carboxylic acid residues which are ionised under the mildly acidic conditions used when dyeing wool/acrylic blends, sites are available for the adsorption of cationic dyes [20,145].

Resisting of the wool by cationic dyes depends on the dye, dyeing time and temperature, concentration of electrolyte, pH and auxiliaries used [139]. A suitable cationic dye must have the following properties: good solubility, minimal staining of the wool, medium affinity for the acrylic fibre, good diffusibility, stability over a wide pH range, compatibility with other cationic dyes in mixture shades, good fastness to light and water, and adequate thermal stability (for goods which are to be decatised).

Lemin *et al.* carried out an extensive investigation into the dyeing behaviour of a 50 : 50 wool/acrylic (Orlon 42) blend in a one-bath procedure using cationic dyes [145]. One of the points to emerge from this study concerned the extent of cross-staining of the wool by cationic dyes. In most cases, there is considerable uptake of dye by the wool in the early stages of dyeing (up to 85 °C), and this adsorption is enhanced by the presence of a cationic retarder. By the time the temperature reaches the boil, however, the bulk of the dye transfers from the wool to the acrylic fibre, and this process continues during the boiling operation. This demonstrates the ability of wool to adsorb cationic dyes, and indicates that at temperatures above the glass transition temperature for the acrylic fibre, the substantivity of the dyes for the acrylic fibre is much higher than that for wool. The degree of cross-staining of the wool varies with the actual acrylic fibre in the blend, and this becomes particularly important when the depth approaches the saturation limit of the acrylic fibre.

The same study [145] showed that the use of a cationic retarding agent markedly reduces the rate of dyeing of milling dyes on wool. This is consistent with the known behaviour of cationic levelling agents that are specifically recommended for wool.

The rates of dyeing of mixtures of milling and cationic dyes were also studied [145]. Dyeings were carried out, in the presence (or absence) of a cationic retarder, with C.I. Acid Green 27 in admixture with either C.I. Basic Red 18 : 1 or C.I. Basic Red 22. Retarding agent reduces the rates of dyeing of both fibres. This effect is greater for the milling dye than for either of the cationic dyes, and is especially apparent for dyeing times of less than 30 minutes. This suggests that in the initial stages of dyeing, the cationic agent is either being adsorbed by the wool or forming a loose complex with the acid dye, which has a lower affinity for the wool than has the uncomplexed dye.

Dyeings of further combinations on three different 50 : 50 wool/acrylic blends (wool/Orlon 42, wool/Acrilan 16 and wool/Courtelle; see Table 10.16) have also been examined [145]. Dyeings were carried out in the presence of retarding agent. C.I. Acid Blue 140 and C.I. Basic Yellow 28 were adsorbed at similar rates on the Acrilan blend, but on the other two blends the exhaustion of the wool dye was significantly depressed, and was less than 50% after 1 hour at the boil. Using a combination of C.I. Acid Yellow 70 and C.I. Basic Blue 41, the wool dye was adsorbed more rapidly than the cationic dye on the Acrilan blend, and more slowly on the Orlon blend. With a combination of C.I. Acid Green 27 and C.I. Basic Red 22, the cationic dye was taken up more slowly than the acid dye on all three blends. Obviously, the dyeing of wool/acrylic blends is rather complex, and is influenced by the dyes used, the nature of the fibres present and the amount of retarder.

This work also showed that the same amount of retarder can be used for blends containing Orlon 42, Acrilan 16 or Courtelle [145]. Dyes with K values of 5 require less retarder than dyes with K values of 3, and this parallels the effect on pure acrylic goods. Dyes with K values of 5 are particularly useful when dyeing pale colours, and dyes with K values of 3 are recommended for medium and heavy dyeings. The compatibility of cationic dyes is often unaffected by the presence of wool, but when wool dyes are added to the system there is an adverse effect on compatibility for certain dye combinations. The following combinations of dyes with K values of 5 were recommended:

- C.I. Basic Yellow 59 (no longer made);
- C.I. Basic Red 22;
- C.I. Basic Blues 22 or 101.

For dyes with K values of 3, there are wider choices for three-colour combinations, and it has been reported that the following dyes are widely used:

- C.I. Basic Yellow 28 or C.I. Basic Orange 30 : 1;
- C.I. Basic Reds 18 : 1 or 46;
- C.I. Basic Blue 41.

The following procedure will produce satisfactory results when dyeing blends of wool with acrylic fibres such as Orlon 42, Acrilan 16 and Courtelle [145]. The dyebath is set at 50 °C (pH 5.5) with 0–1% cationic retarding agent and 0.5–1.0% antiprecipitating agent; since acid and cationic dyes have different ionic characters, an antiprecipitating agent is used to prevent co-precipitation, which can occur under certain circumstances and may result in unlevel dyeings with impaired fastness properties. After addition of the wool dyes, the dyebath is heated to 80 °C over 30 minutes and the cationic dyes are added. The temperature is then raised to 95 °C over 15 minutes, held there for a further 15 minutes, raised to the boil over 10 minutes and held there for 1 hour. The initial rate of heating of 1 °C per minute, followed by holding at 95 °C for 15 minutes before raising to the boil and continuing to dye at the boil, provides an approximately linear rate of dye adsorption which is satisfactory for yarn or piece dyeing in conventional machines. Under these conditions, at least 25% of the cationic dye is removed from the dyebath by the time the temperature reaches 80 °C, and most of this dye is adsorbed by the wool. After 30 minutes at the boil, there is virtually complete transfer of cationic dye from the wool to the acrylic fibre. Essentially, the wool exerts a levelling effect on the adsorption of cationic dye by the acrylic fibre.

The following details for the dyeing of wool/acrylic blends have been reported by Bayer (now DyStar) [139]. Especially when dyeing in heavy shades, a final dyeing temperature of at least 98 °C (60–90 minutes), and preferably 102–104 °C (45–60 minutes), is recommended. At lower temperatures, dye uptake is lower, and this cannot always be compensated for by longer boiling times; in addition, at temperatures below 98 °C the cationic dyes stain the wool more heavily, leading to poorer fastness properties. Electrolyte has a levelling effect on both fibre components, but the amount should not exceed $2 \, g \, l^{-1}$ Glauber's salt; larger quantities increase the staining of the wool by the cationic dyes. Wool/acrylic blends are best dyed at pH 4.5–5.0, as in this range the cationic dyes are the most stable, and damage to the wool is minimised. A nonionic auxiliary has been recommended to prevent dye precipitation. Since the acrylic fibre forms the major component in many blends, the retarding effect of the wool component is insufficient to obviate the need for a suitable cationic auxiliary. From a knowledge of the blend ratio, the dyeing rate constant V and the saturation value of the acrylic fibre and the cationic dye recipe, the required amount of retarder can be calculated. The following dyeing procedures have been recommended:

- One-bath, one-step process in which the wool and cationic dyes are added to the dyebath at 50 °C.
- One-bath, two-step process in which the wool dyes are added at 50 °C and the cationic dyes at 85 °C.
- One-bath, two-step process in which the acrylic fibre is dyed first with cationic dyes and the wool is then dyed in the normal way, once the bath is cooled to 85 °C. This procedure is used for heavy shades where the total dye concentration is greater than 2%.

When cationic dye is added to the dyebath at 85 °C, the rate of heating should be about 0.5 °C per minute. For blends in which the wool has been chlorinated to achieve shrink resistance, the starting temperature must be lowered, the rate of dyebath heating reduced and the pH increased in order to reduce the rate of dye adsorption by the wool. Blends containing wool treated with Hercosett should be dyed by the one-bath, two-step process in order to ensure good levelness.

Huntsman has recommended the use of mixtures of Lanaset and selected Maxilon M dyes (Yellow M-4GL, Yellow M-3RL 200%, Red M-4GL 200%, Red M-RL 200% and

Blue M-2G 200%) for the dyeing of 45 : 55 wool/Dralon blends in a one-bath, one-step operation. The dyebath is initially set at 50 °C with 0.7–1.0% Albegal SET, 5% Glauber's salt, 1 g l^{-1} sodium acetate and pH 4.5 with acetic acid. After 5 minutes, Lanaset dye is added; after a further 10 minutes, Maxilon M dye and Tinegal MR (retarding agent, 0.3% per 1% of cationic dye) are added. After another 10 minutes, the dyebath is heated to 98–103 °C at 1 °C per minute and then held at maximum temperature for 30–60 minutes.

Recommendations from the major dye manufacturers for wool/acrylic blends indicate that compatible combinations of cationic dyes are as follows:

- *K values of 3:*
  - C.I. Basic Yellow 28;
  - C.I. Basic Red 46;
  - C.I. Basic Blue 41.
- *K values of 5:*
  - C.I. Basic Oranges 29 and 43;
  - C.I. Basic Red 22;
  - C.I. Basic Blue 22.

10.7.2.3   Novel Dyeing Techniques for Wool/Acrylic Blends

An interesting recent paper proposed a simple method to chemically modify acrylic fibres through their reaction with hydroxylamine hydrochloride in order to incorporate primary amino groups in the fibre. Such modified fibres could be blended with wool and dyed with acid or reactive dyes using application conditions suitable for the dyeing of the wool component of a blend [146,147].

10.8   Conclusions

Wool blends have established important positions in both apparel and non-apparel markets over many years, and their importance is unlikely to diminish. The demands of quick-response manufacturing will lead to an increase in dyeing at the yarn and fabric stages. For blends, this will mean that there will be an even greater trend towards dyeing component fibres together in blend form, rather than separately.

Environmental concerns will place an increasing pressure on dyers to minimise the discharge of dyes to the effluent. This will apply to all fibres and blends, and be of critical importance for cotton and its blends, where dyebath exhaustion is often poor. The use of carriers is of particular importance for the dyeing of wool/polyester blends (especially wool-rich blends), because they enable dyeings to be carried out at temperatures sufficiently low to avoid serious wool damage. Increasing restrictions on the use of certain carriers, such as biphenyl and halocarbons, will intensify the need for the continued development of environmentally acceptable products [3,148]. For wool/nylon blends (and also pure wool) used in carpets where the application of an insect-resist agent is required, treatment methods will have to be developed which give virtually no discharge of the agent to the effluent, or which produce an effluent of such low volume that it can be economically contained and disposed of safely [149].

References

[1] Anon, *Ciba Review*, **12**(141) (1960).

[2] S M Doughty, *Review of Progress in Coloration and Related Topics*, **16** (1986) 25.

[3] R Walter, *Dyeing of Polyester/Wool Mixtures* (Leverkusen, Germany: Bayer, 1987).

[4] W Beckmann, H Flosbach, F Hoffmann, M Papmahl and R Walter, *Chemiefasern/ Textilindustrie*, **29/81** (1979) 339.

[5] R C Cheetham, *Dyeing Fibre Blends* (London, UK: Van Nostrand, 1966).

[6] C L Bird, *Theory and Practice of Wool Dyeing* (Bradford, UK: Society of Dyers and Colourists, 1972).

[7] E Engeler, *Dyeing Blends of Wool and Cellulosic Fibres for Washable Articles* (Basel, Switzerland: Sandoz, 1987).

[8] I Steenken, I Funken and G Blankenburg, *Textilveredlung*, **21** (1986) 128.

[9] G Kratz, A Funder, H Thomas and H Höcker, *Melliand Textilber*, **70** (1989) 128.

[10] ICI Sales Aid 267, Wool/Cotton Blends.

[11] D M Lewis, *Textile Research Journal*, **52** (1982) 580.

[12] P G Cookson, *Wool Science Review*, **62** (1986).

[13] C Preston, *The Dyeing of Cellulosic Fibres* (Bradford, UK: Dyers' Company Publications Trust, 1986).

[14] P G Cookson and F J Harrigan, in *Wool Dyeing*, ed. D M Lewis (Bradford, UK: Dyers' Company Publications Trust, 1992).

[15] D Klemm, B Philipp, T Heinze, U Heinze and W Wagenknecht, *Comprehensive Cellulose Chemistry* (Chichester, UK: Wiley-VCH, 1998).

[16] H R Krässig, *Cellulose: Structure, Accessibility and Reactivity* (New York, USA: Gordon and Breach, 1993).

[17] R A Young and R M Rowell, *Cellulose: Structure, Modification and Hydrolysis* (Chichester, UK: John Wiley & Sons, 1986).

[18] A Johnson, *The Theory of Coloration of Textiles*, 2nd Edn (Bradford, UK: Dyers' Company Publications, 1989).

[19] T Vickerstaff, *The Physical Chemistry of Dyeing*, 2nd Edn (London, UK: Oliver and Boyd, 1954).

[20] E R Trotman, *Dyeing and Chemical Technology of Textile Fibres* (High Wycombe, UK: Charles Griffin, 1984).

[21] R H Peters, *Textile Chemistry*, **Vol. 3** (Amsterdam, The Netherlands: Elsevier Scientific, 1975).

[22] D R Lemin and J K Collins, *Journal of the Society of Dyers and Colourists*, **75** (1959) 421.

[23] J Shore, *Blends Dyeing* (Bradford, UK: Dyers' Company Publications Trust, 1998).

[24] C C Cook, *Review of Progress in Coloration and Related Topics*, **12** (1982) 73.

[25] D M Lewis, *Journal of the Society of Dyers and Colourists*, **98** (1982) 165.

[26] F J Douthwaite, N Harrada, and T Washimi, *Proceedings of the IFATCC Conference, Vienna*, (1996) 447.

[27] D M Lewis and T T L Vo, *Coloration Technology*, **123** (2007) 306.

[28] D M Lewis and P J Broadbent, *AATCC Review*, **35** (2008) 35.

[29] H-U von der Eltz and R Klein, *International Textile Bulletin, World Edition Dyeing/Printing/Finishing*, **4** (1973) 2.

[30] S Hamby, *The American Cotton Handbook*, **Vol. 2** (New York, USA: John Wiley & Sons, 1966).

[31] T Robinson and W B Egger, *Textile Chemist and Colorist*, **15** (1983) 189.

[32] Ciba-Geigy, Dyeing of Wool/Cotton Blends (Basel, Switzerland).

[33] N E Hauser, *Textile Chemist and Colorist*, **18**(3) (1986) 11.

[34] H R Hadfield and D R Lemin, *Journal of the Textile Institute*, **51** (1960) T1351.

[35] D Hildebrand, *Proceedings of the 7th International Wool Textile Research Conference, Tokyo*, **5** (1985) 239.

[36] Bayer, D Hildebrand, The Dyeing of Wool/Cotton Blends by a One-bath Two-step Procedure (Leverkusen, Germany).

[37] ICI Technical Information (Dyehouse) No. 497, Application of Vat and Azoic Dyes to Wool/Cellulosic Fibre Blends.

[38] J Y Cai, J S Church and S M Smith, *Proceedings of the 11th International Wool Textile Research Conference, Leeds*, CD of Papers 26 CCF (2005).

[39] F G Dean, J W A Matthews and D S Orchard, Project TD4, Textile Technology Department, Australian Wool Corporation (1983).

[40] I Steenken, I Souren, U Altenhofen and H Zahn, *Textil Praxis International*, **39** (1984) 1146.

[41] I Steenken and H Zahn, *Proceedings of the 7th International Wool Textile Research Conference, Tokyo*, **5** (1985) 49.

[42] I Steenken and H Zahn, *Textile Research Journal*, **54** (1984) 429.

[43] I Steenken and H Zahn, *Journal of the Society of Dyers and Colourists*, **102** (1986) 269.

[44] H E Charwat, *Bayer Farben Review*, **32** (1981) 14.

[45] H Putze and G Dillmann, *Textilveredlung*, **15** (1980) 457.

[46] Sandene Process, Courtauld's Research Brochure (1989).

[47] J M Taylor, *International Dyer*, **173** (1988) 30.

[48] P Sulek, M Crnoja-Cosic and J Schlangen, *Proceedings of the 1st International Textile, Clothing and Design Conference* (2002).

[49] J A Rippon, *Journal of the Society of Dyers and Colourists*, **100** (1984) 298.

[50] M Rupin, G Veaute and J Balland, *Textilveredlung*, **5** (1970) 829.

[51] D. I. Dvorsky and I Cerovsky, *IFATCC Conference, Budapest* (1982).

[52] D. M. Lewis and X. P. Lei, *Textile Chemist and Colorist*, **21** (1989) 23.

[53] P J Broadbent, X P Lei and D M Lewis, in *Cellulosics: Materials for Selective Separations and Other Technologies*, ed. J F Kenendy, G O Phillips and P A Williams (New York, USA: Ellis Horwood, 1993), 297.

[54] D M Lewis and X P Lei, *Journal of the Society of Dyers and Colourists*, **107** (1991) 102.

[55] J G Cook, *Handbook of Textile Fibres, Vol. 1: Natural Fibres* (Shildon, UK: Merrow, 1984).

[56] R S Asquith, *Chemistry of Natural Protein Fibers* (New York, USA: Plenum Press, 1977).

[57] R E Dickerson and I Geis, *The Structure and Action of Proteins* (Menlo Park, CA, USA: W.A. Benjamin, 1969).

[58] E V Truter, *Introduction to Natural Protein Fibres: Basic Chemistry* (London, UK: Paul Elek, 1973).

[59] B G Skinner and T Vickerstaff, *Journal of the Society of Dyers and Colourists*, **61** (1945) 193.

[60] D M Lewis, in *Advances in Wool Technology*, ed. N A G Johnson and I M Russell (Cambridge, UK: Woodhead, 2009).

[61] M T Pailthorpe, in *Wool Dyeing*, ed. D M Lewis (Bradford, UK: Dyers' Company Publications Trust, 1992).

[62] Sandoz, Wool/Silk Blends (Basel, Switzerland).

[63] K Y Chu and J R Provost, *Review of Progress in Coloration and Related Topics*, **17** (1987) 23.

[64] Ciba-Geigy, Dyeing of Wool/Silk Blends with Lanaset Dyes (Basel, Switzerland).

[65] Ciba-Geigy, Dyeing of Wool/Silk Blends with Lanasol/Cibacron F Dyes (Basel, Switzerland).

[66] T L Dawson, *Review of Progress in Coloration and Related Topics*, **15** (1985) 29.

[67] R W Moncrieff, *Man-made Fibres* (London, UK: Newnes-Butterworths, 1975).

[68] J G Cook, *Handbook of Textile Fibres, Vol. 2: Man-made Fibres* (Shildon, UK: Merrow, 1984).

[69] R H Peters, *Textile Chemistry*, **Vol. 1** (Amsterdam, The Netherlands: Elsevier Scientific, 1963).

[70] E Snider and R J Richardson, in *Encyclopedia of Polymer Science and Technology*, **Vol. 10** (New York, USA: John Wiley & Sons, 1968), 347.

[71] H W Peters and T R White, *Journal of the Society of Dyers and Colourists*, **77** (1961) 601.

[72] R H Peters, *Journal of the Society of Dyers and Colourists*, **61** (1945) 95.

[73] E Atherton, D A Downey and R H Peters, *Textile Research Journal*, **25** (1955) 977.

[74] A R Mathieson, C S Whewell and P E Williams, *Journal of Applied Polymer Science*, **8** (1964) 2009.

[75] H J Palmer, *Journal of the Textile Institute*, **49** (1958) T33.

[76] G A Gilbert and E K Rideal, *Proceedings of the Royal Society, London*, **182A** (1944) 335.

[77] G A Gilbert, *Proceedings of the Royal Society, London*, **183A** (1944) 167.

[78] M Greenhalgh, A Johnson and R H Peters, *Journal of the Society of Dyers and Colourists*, **78** (1962) 315.

[79] Sandoz pattern card 1540.00.86, Nylosan: Dyes and Chemicals for Polyamide Fibres and Blends.

[80] ICI, Dyeing of Wool/Polyamide Fibre Unions, Technical Information D1235.

[81] D M Nunn, *Dyeing of Synthetic-Polymer and Acetate Fibres* (Bradford, UK: Dyers' Company Publications Trust, 1979).

[82] J Shore, *Colorants and Auxiliaries, Vol. 2: Auxiliaries* (Bradford, UK: Society of Dyers and Colourists, 1990).

[83] H R Hadfield and D N Sharing, *Journal of the Society of Dyers and Colourists*, **64** (1948) 381.

[84] D F Scott and T Vickerstaff, *Journal of the Society of Dyers and Colourists*, **76** (1960) 104.

[85] D M Lewis and D G Marfell, in *The Dyeing of Synthetic Fibres*, ed. C J Hawkyard (Bradford, UK: Dyers' Company Publication Trust, 2003).

[86] I W Stapleton and P J Waters, *Journal of the Society of Dyers and Colourists*, **97** (1981) 56.

[87] M Dohmyan, Y Shimizu and M Kimura, *Journal of the Society of Dyers and Colourists*, **106** (1990) 395.

[88] S M Burkinshaw and G W Collins, *Dyes and Pigments*, **25** (1994) 31.

[89] SS Lo, PhD Thesis, Leeds University (1991).

[90] D M Lewis and W C MacDougall, *Amerian Association of Textile Chemistry and Coloration, International Conference and Exhibition, Nashville, Book of Papers* (1996) 284.

[91] D Hinks, S M Burkinshaw, A H M Renfrew and DM Lewis, *Amerian Association of Textile Chemistry and Coloration, International Conference and Exhibition, Nashville, Book of Papers* (1999) 454.

[92] D M Lewis and L J Sun, *Coloration Technology*, **119** (2003) 286.

[93] D M Lewis and L J Sun, *Coloration Technology*, **119** (2003) 327.

[94] D Schwer, H Ritter and K Zesiger, *Textilveredlung*, **16** (1981) 479.

[95] Bayer, H Martel, Problems in Dyeing Fibre Blends of Wool and Polyamide (Leverkusen, Germany).

[96] Ciba-Geigy pattern card TS 5/84, Lanaset Dyes on Carpet Yarn WO/PA (80:20).

[97] Sandoz pattern card 2023/83, Dyeing Polyamide/Wool Fibre Blends with Lanasyn S Dyes as Well as Selected Lanasyn and Lanasyn Brilliant Dyes.

[98] Bayer pattern card Sp 454, Dyeings on Carpet Yarns from Wool and Wool Polyamide Blends.

[99] Bayer pattern card Le 1670 d-e-f-s-I, Isolan S-farbstoffe.

[100] R de P Daubeny, C W Bunn and C J Brown, *Proceedings of the Royal Society, London*, **226A** (1954) 531.

[101] BASF, Dyeing and Finishing of Polyester Fibres, Publication B363e/6.75.

[102] Bayer pattern card Sp 458 e, Resolin Dyestuffs for Textured Polyester.

[103] Bayer pattern card Sp 456 e, with supplement 650 e, Resolin S Process.

[104] W Beckmann, *Textile Chemist and Colorist*, **2** (1970) 350.

[105] H Baumann, *Textilveredlung*, **14** (1979) 515.

[106] E Waters, *Journal of the Society of Dyers and Colourists*, **66** (1950) 609.

[107] A Murray and K Mortimer, *Journal of the Society of Dyers and Colourists*, **87** (1971) 173.

[108] P Richter, *Melliand Textilber*, **55** (1974) 882.

[109] A N Derbyshire, W P Mills and J Shore, *Journal of the Society of Dyers and Colourists*, **88** (1972) 389.

[110] W Beckmann and H Hamacher-Brieden, *Textile Chemist and Colorist*, **5** (1973) 118.

[111] A Murray and K Mortimer, *Review of Progress in Coloration and Related Topics*, **2** (1971) 67.

[112] M J Schuler, *Textile Research Journal*, **27** (1957) 352.

[113] T Vickerstaff, *ICI Hexagon Digest*, **20** (1972).

[114] W Beckmann, *Textile Chemist and Colorist*, **2** (1970) 350.

[115] H-U von der Eltz, *Symposium of the Society of Dyers and Colourists of Australia and New Zealand, Melbourne* (1972).

[116] R Schroth and H Henkel, *Faserforschung und Textiltechnik*, **22** (1971) 2173.

[117] L Drijvers, *American Dyestuff Reporter*, **41** (1952) 533.

[118] J J Iannarone, H F Clapham and R J Thomas, *American Dyestuff Reporter*, **42** (1953) P666.

[119] G A Coutie, D R Lemin and H Sagar, *Journal of the Society of Dyers and Colourists*, **71** (1955) 433.

[120] H Baumann, H Müller, L Möchel and P Spiegelmacher, *Melliand Textilber*, **58** (1977) 420, 495.

[121] H H. Konrad and K Türschmann, *Textil Praxis International*, **33** (1978) 932.

[122] V Prchal and G Schröder, *Textiltechnik*, **30** (1980) 48.

[123] P Liechti, *Journal of the Society of Dyers and Colourists*, **98** (1982) 284.

[124] G Römer, *Textilveredlung*, **14** (1979) 332.

[125] G Römer, H-U Berendt, J B Feron, H Fierz and A Lauton, *Textilveredlung*, **15** (1980) 465.

[126] D M Lewis, *Review of Progress in Coloration and Related Topics*, **19** (1989) 49.

[127] 2010 Report from US Department of Health and Human Services, Public Health Service, National Toxicology Program, Research Triangle Park, NC, USA.

[128] P G Cookson, K W Fincher and P R Brady, *Journal of the Society of Dyers and Colourists*, **107** (1991) 135.

[129] H J Cho, D M Lewis and B H Jia, *Coloration Technology*, **123** (2007) 86.

[130] P G Cookson, P R Brady, K W Fincher, P A Duffield, S M Smith, K Reincke and J Schreiber, *Journal of the Society of Dyers and Colourists*, **111** (1995) 228.

[131] W T Sherill, *Textile Chemist and Colorist*, **10** (1978) 210.

[132] K Türschmann, *Textilbetrieb*, **93**(4) (1975) 49.

[133] G Römer, *Teinture et Apprets*, (1974) 203.

[134] A S Fern and H R Hadfield, *Journal of the Society of Dyers and Colourists*, **71** (1955) 277.

[135] I B Angliss and J D Leeder, *Journal of the Society of Dyers and Colourists*, **93** (1977) 387.

[136] R L Hayes and D G Phillips, *Journal of the Textile Institute*, **69** (1978) 364.

[137] A A Moussa, Y A Youssef, R Farouk, T M Ayeish and M H Arief, *Coloration Technology*, **127** (2011) 304.

[138] P J Broadbent, M Clark, S N Croft and D M Lewis, *Coloration Technology*, **123** (2007) 29.

[139] H Flosbach, R Walter and W Zimmermann, *Bayer Farb. Rev.*, **34** (1982) 18.

[140] J S Ward, *Review of Progress in Coloration and Related Topics*, **14** (1984) 98.

[141] I Holme, *Review of Progress in Coloration and Related Topics*, **13** (1983) 10.

[142] BASF publication S397e/5.72 (JHB), Dyeing and Finishing of Acrylic Fibres, Alone and in Blends with Other Fibres.

[143] J Cegarra, *Journal of the Society of Dyers and Colourists*, **87** (1971) 149.

[144] I Holme, *Chimia*, **34** (1980) 110.

[145] D R Lemin, *Journal of the Society of Dyers and Colourists*, **91** (1975) 168.

[146] RM El-Shistawy and N S E Ahmed, *Coloration Technology*, **121** (2005) 139.

[147] RM El-Shistawy, M M el-Zawahry and N S E Ahmed, *Coloration Technology*, **127** (2011) 28.

[148] D M Lewis, *Journal of the Society of Dyers and Colourists*, **105** (1989) 119.

[149] J. Barton, *International Dyer*, **185** (2000) 14.

11

The Coloration of Human Hair

Robert M. Christie[1] and Olivier J.X. Morel[2]
[1]School of Textiles & Design, Heriot-Watt University, UK
[2]Xennia Technology Ltd., UK

11.1 Introduction

Hair has always possessed powerful, symbolic and evocative properties. It is one of the most distinctive elements of our appearance. Hair coloration is among the oldest acts of human adornment [1], reflecting our common dissatisfaction with our natural colour. It is used to enhance appearance by concealing the greying process as we grow older, and also as a means of expressing individuality, for example as a fashion statement or by projecting a particular image. Hair dye manufacture has developed into a multinational, multibillion dollar industry, with an extensive range of dyes available [2]. As a result of an increasingly ageing, and thus greying, population, demand for the products has been increasing rapidly, a trend that seems likely to continue into the future [3]. Indeed, the demand for hair dyes may well accelerate due to individual expectations in the developing global economies.

Human hair is a protein fibre with certain physical and chemical similarities to the animal fibres that are used in textiles, as discussed in other chapters. However, hair coloration processes have unique features that distinguish them from textile dyeing, and consequently completely different dye types and dyeing processes are required. As the process is carried out in contact with the head, hair cannot be dyed above a temperature of ~40 °C, with a dyeing time generally not exceeding ~40 minutes. The process uses a low liquor to fibre ratio (1–2 : 1), the ratio of the volume of the wet hair-dyeing formulation to the weight of hair to which it is applied. Dyeing should aim to avoid or at least minimise damage to the hair and impairment of its natural texture or gloss, and it should not stain the scalp [4]. The colour produced should be stable to air, light, friction, perspiration and chlorinated water, and should remain unaffected by other hair treatments. Most critically, the dyeing

The Coloration of Wool and other Keratin Fibres, First Edition. Edited by David M. Lewis and John A. Rippon.
© 2013 SDC (Society of Dyers and Colourists). Published 2013 by John Wiley & Sons, Ltd.

process must be toxicologically safe, both to the individual undergoing treatment and to the hairdresser who is applying the formulation.

Hair coloration has been practised for thousands of years. Until synthetic dyes appeared around the middle of the 19th century, natural colouring materials of vegetable or mineral origin were used [5]. The best-known natural hair colouring product is *henna*, whose colouring qualities are due to lawsone (**1**), a naphthoquinone derivative released from the leaves of *Lawsonia inermis* (cultivated mostly in India, Tunisia, Arabia and Iran) when they are pulverised with an acidic liquid such as lemon juice or with aqueous sodium bicarbonate. Analysis of the hair cuticle of Rameses II revealed the presence of lawsone. Today, around 5000 years later, hair coloration with henna is still practised to a certain extent, especially in regions of Asia. Products based on henna suffer from a number of disadvantages, including difficulties in application, variability and a limited variety of shades. Used alone on white, grey or blond hair, lawsone from henna gives auburn shades. A wider colour range is produced when henna is mixed with other natural dyes, including pyrogallol (**2**) (from walnut shells), the yellow apigenin (**3**) (from chamomile), indigo (**4**) (from *Indigofera tinctoria*) and the reddish-brown haematein (**5**) (from logwood). Natural dyes such as these have also been used alone, but they are generally less effective than henna in terms of colour strength and stability. With the exception of henna, natural dyes have been virtually completely replaced by synthetic hair dye products, as these are easier to apply, more versatile and predictable and have a much wider shade range available.

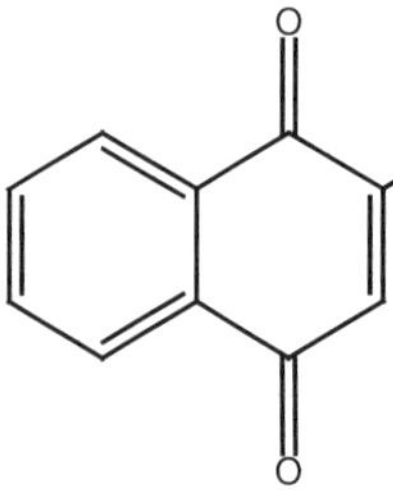

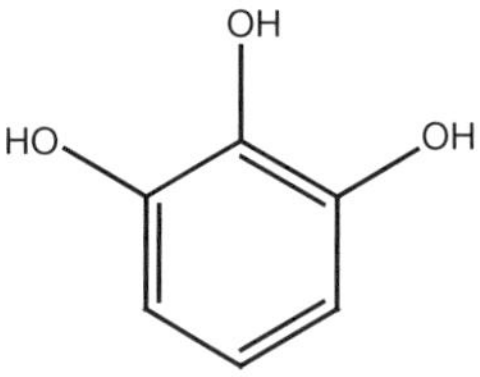

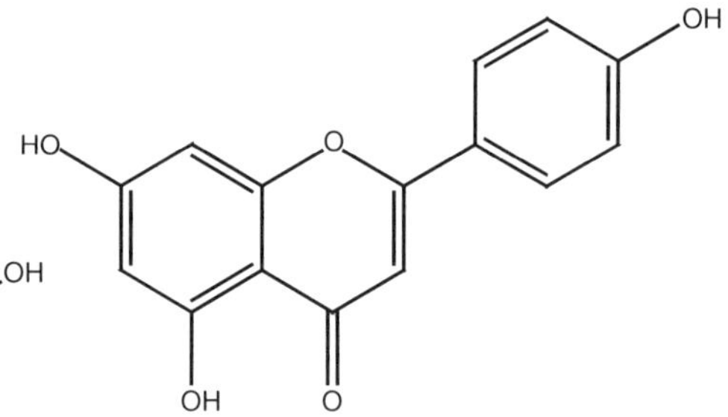

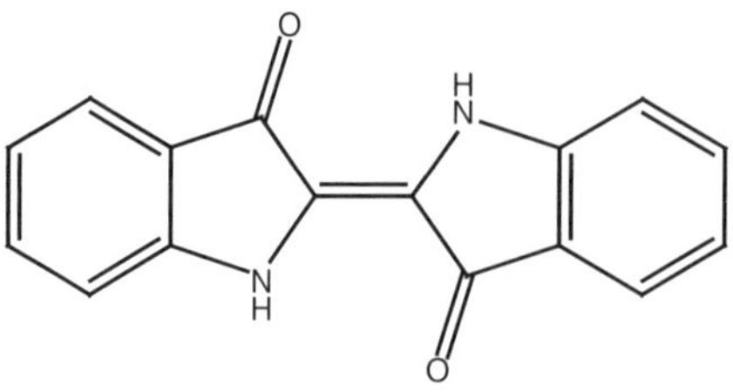

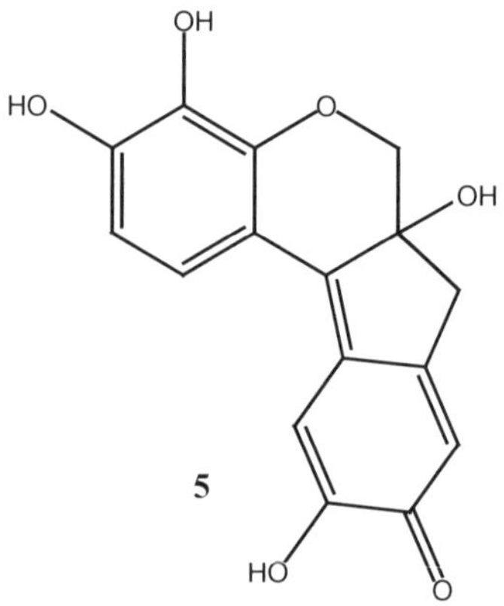

Processes which use lead salts obtained from mineral sources to generate lead oxides and sulphide on hair, leading to black colours, were also used in the past, but again these are rarely sold today, except in specific countries, for obvious toxicological reasons. Lightening the hair, another perennial preoccupation, was formerly achieved by exposure to sunlight in the presence of alkali.

Modern synthetic hair dye products may be categorised according to either the chemistry involved or the level of permanence of the colour. In chemical terms, they may be classed as involving either *oxidative* or *nonoxidative* processes. Oxidative dyeing products constitute around 70% of the hair colorant market. This group is dominated by the so-called 'permanent hair dyes', which owe their popularity to their long-lasting effect, ease of application and versatility in the range of colours achievable. Human hair grows around 0.3 mm each day and the growth cycle lasts about 3 years, at which point the hair falls out. Thus, the designation of the longest-lasting hair dyes as 'permanent' must be qualified, since additional treatment is required every few weeks to cover new growth. Besides the permanent hair dyes, there are two further subcategories of oxidative product: demipermanent and autooxidative. The former gives little or no lightening of the hair and less thorough grey coverage, and the dyeing time is shorter than with permanent hair dyes and the colours fade more rapidly with time. The latter offers users, particularly males, colour development over a period of time [6]. Nonoxidative hair dye products may be categorised as either semipermanent [7] (surviving around 4–6 washes) or temporary ('wash-in/wash-out') products. In addition to the colouring agent and a range of ingredients to assist in application, hair-dyeing products, especially of the permanent type, often contain a bleaching agent (hydrogen peroxide), which removes some of the natural hair colour. Dyes may be applied to the hair as one overall colour, although a modern trend is to use several colours to produce streaks or gradations on top of either the natural colour or a base colour. The literature accounts of aspects of hair dyeing [8–13] include a recent comprehensive review of the chemistry of permanent hair coloration by the authors of this chapter [14].

11.2   Structure and Morphology of Human Hair

Human hair grows from elongated sacs, the follicles or roots, which are generated from cells that multiply under the skin and move upwards until the hair shaft pierces it. These embryonic processes continue throughout the hair growth cycle [15]. Three types of hair may be identified, differing in macroscopic structure. Caucasian hair is generally fine and straight to curly, with a nearly circular cross-section (ellipticity 1.25). Ethiopian hair is coarse and wavy/woolly, with a more oval cross-section (ellipticity 1.75). Mongolian hair is coarse and straight to wavy, with cross-section similar to Caucasian (ellipticity 1.35). A fully-formed hair fibre contains four structural units: the cuticle, cortex, medulla and cell membrane complex. The cuticle has a protective outer layer of flat, overlapping scales which completely surround the hair [16–20] and an internal structure as illustrated in Figure 11.1. Each cell contains a thin outer membrane, the proteinaceous epicuticle, on the surface of which is a fatty acid layer (F-layer); this is connected to cystine residues in the underlying protein layer [21–25]. This feature explains the apparent hydrophobic character of the hair fibre. Beneath this membrane are three other major layers: the A-layer and B-layer of the exocuticle, and the endocuticle. Cuticle cells surround the

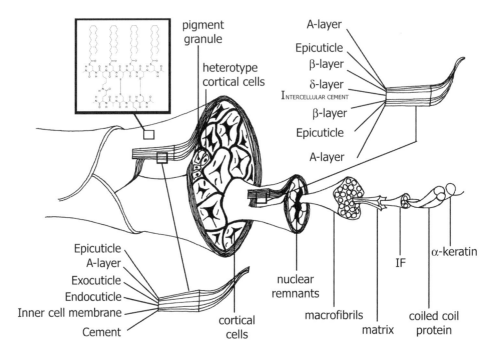

Figure 11.1   *Structure of human hair [29].*

cortex, which contains most of the fibre mass. Cortical cells contain pigment granules and nuclear remnants. They have a substructure of macrofibrils composed of intermediate filaments (IFs), highly organised units embedded in a less structured matrix. The IFs contain short sections of coiled helical proteins [26,27], approximating to an α-helical form, which is why hair proteins are often referred to as 'α-keratin'. The matrix, although often referred to as the 'amorphous region', has some limited structural organisation. Thin human scalp hair consists of cuticle and cortical cells only. Hairs of larger cross-section have a third type, the medullary cells, which can serve as a pigment reservoir, located in the centre of the fibre. The cell membrane complex consists of a pair of proteinaceous membranes bound together by adhesive material. On the hair surface, the proteinaceous membrane forms the epicuticle. The inner layer between the cuticle cells, referred to as intercellular cement, consists of a δ-layer of lightly crosslinked proteins sandwiched between two lipid-containing β-layers [28]. The cell membrane complex and the endocuticle are believed to provide important pathways for molecular diffusion into the hair fibre.

11.3   Natural Colour of Hair

Human hair is naturally coloured, the colour commonly reflecting an individual's geographic or ethnic origin [6]. Nature's abundance of hair colours provides shades ranging from light blond to black, the latter being characteristic of Asians, Arabics, Africans and southern Europeans. The hair of mid and northern Europeans shows a wide variety of

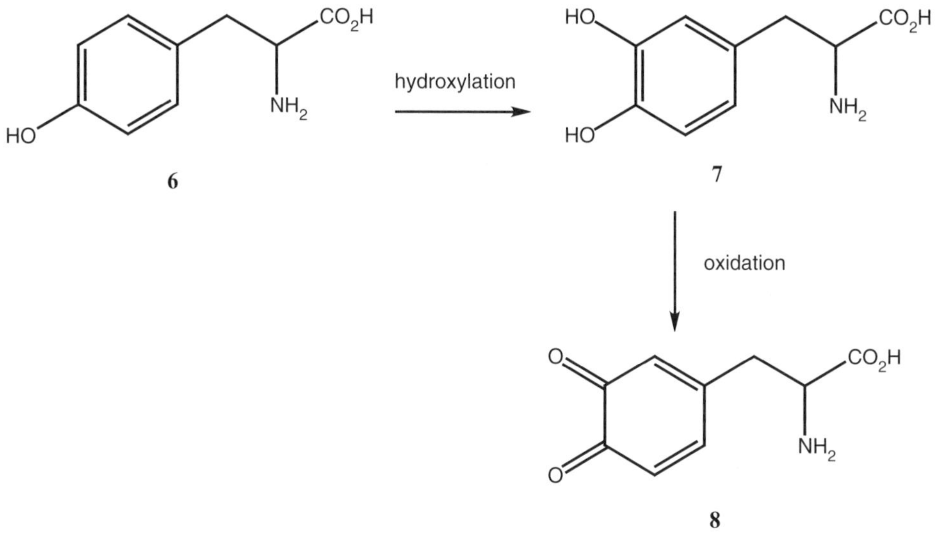

Scheme 11.1   *Oxidation of tyrosine (6) to dopaquinone (8) via DOPA (7).*

browns, with red hair originating in Celtic regions. Natural hair colours occupy a small segment of CIELAB colour space [30], corresponding to dominant absorption wavelengths in the range 586–606 nm, while lightness (L^*) varies over a wide range from 1.8 to 90%. Natural hair colour is due to the pigment melanin in the cortex and medulla, which occurs as granules ~1 μm long and 0.3 μm in diameter [31]. Melanin is formed in pigment-producing cells (melanocytes) in the follicle. Initially, these granules (melanosomes) consist primarily of a protein (melanoprotein). As they mature, the protein binds increasing amounts of

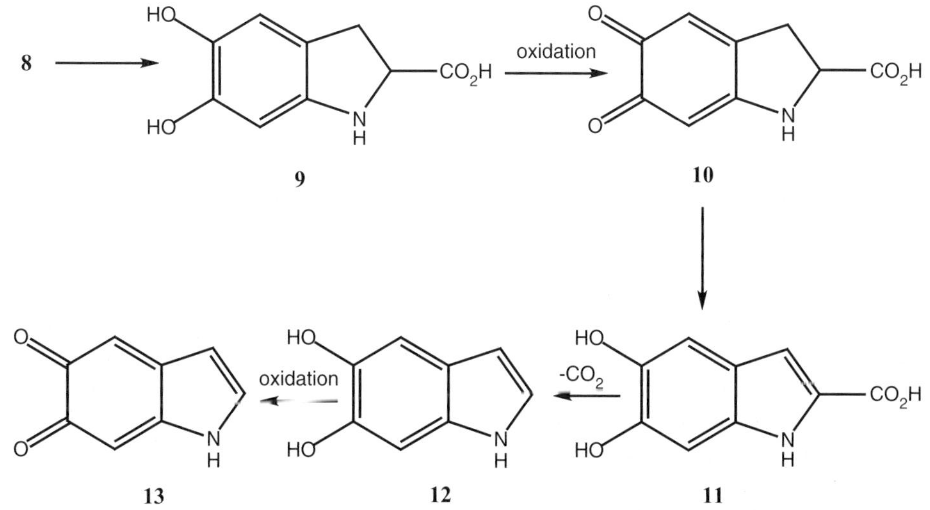

Scheme 11.2   *Biosynthesis of precursors to eumelanin.*

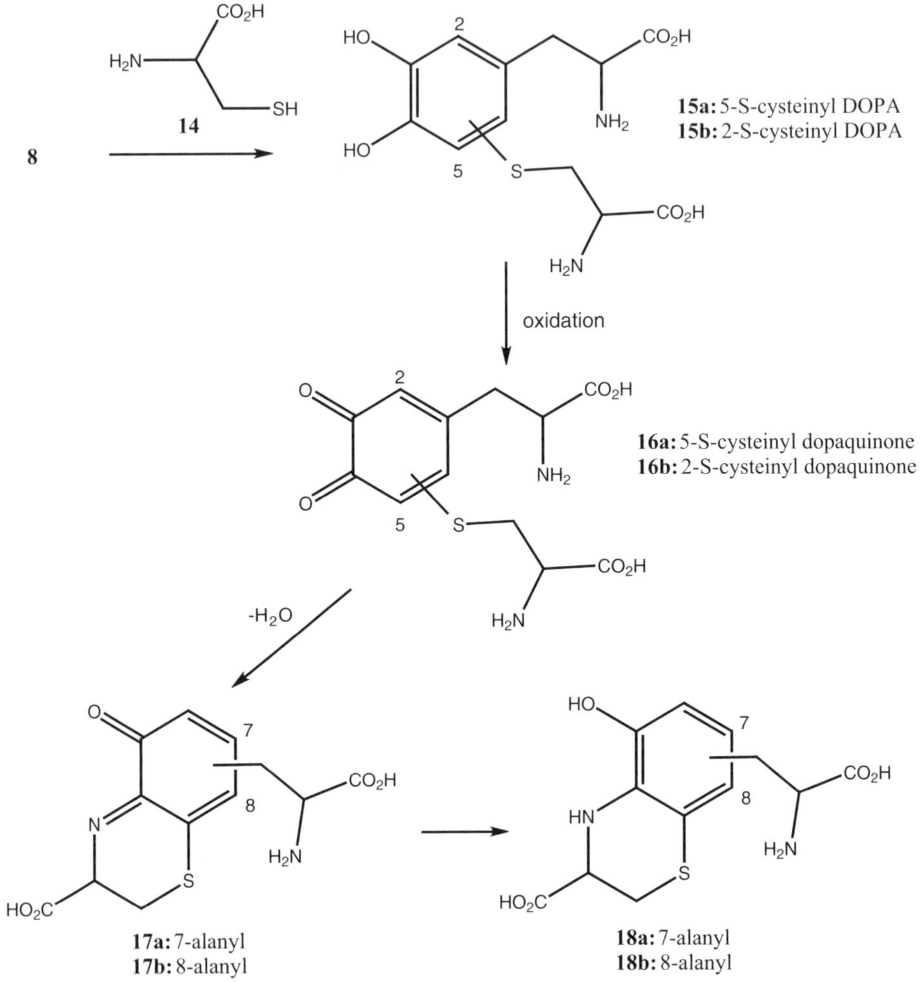

Scheme 11.3 *Biosynthesis of precursors to pheomelanin.*

melanin, which is transferred into adjacent keranocytes. In this way, hair acquires its colour as it grows, which lasts throughout its lifetime. A melanocyte produces essentially two types of melanin [32]: *eumelanin* and the less prevalent *pheomelanin* [33]. The ability of nature to produce its variety of colours using only two species contrasts with the extensive range of materials used in synthetic hair dyeing. The pigments appear to be complex, irregular polymers, whose structures remain incompletely characterised. Eumelanin granules are ovoid or spherical, fairly uniform in shape and have sharply defined edges. Their colour varies from reddish-brown to black. Pheomelanin granules are smaller, partly oval and partly rod-shaped, with colour varying from blond to red. Generally, hair contains a mixture of the two; the more eumelanin, the darker the hair. Japanese black hair contains virtually only eumelanin, and this pigment also predominates in Mongolian, Ethiopian and dark Caucasian hair. Celtic red hair is rich in pheomelanin, while Scandinavian blond hair contains mainly eumelanin at low levels. The variety of natural hair colours arises from not only the concentrations of the two pigments but also the size and shape of the granules, their distribution patterns [34] and their crystal structures [35]. However, the interrelationships between these various features remain incompletely understood [36].

The biosynthetic routes to the two melanin types are closely related. The amino acid tyrosine (**6**), the common starting material, is initially hydroxylated to form 3,4-dihydroxyphenylalanine (DOPA, **7**), which is oxidised to dopaquinone (**8**) in reactions mediated by the enzyme tyrosinase (Scheme 11.1). Thereafter, the pathways diverge. However, the control mechanisms that lead to the divergence have not been established with clarity [37,38].

Scheme 11.2 illustrates the route that ultimately generates eumelanin. Dopaquinone (**8**) is converted, via leucodopachrome (**9**), dopachrome (**10**), 5,6-dihydroxyindole-2-carboxylic acid (DHICA, **11**) and 5,6-dihydroxyindole (DHI, **12**), to the reactive intermediate indolequinone (**13**), from which an oxidative polymerisation leads to eumelanin [39–48].

Scheme 11.3 shows the route towards pheomelanin, whereby dopaquinone (8) [49] reacts with the amino acid cystine (**14**) to give cysteinyl DOPA derivatives (**15a/15b**) [50–52], which are converted via intermediates (**16a/16b** and **17a/17b**) to dihydrobenzothiazines (**18a/18b**). Details beyond this stage are uncertain. The partial structures **19** and **20** represent a current view approximating the general structure of pheomelanin [38].

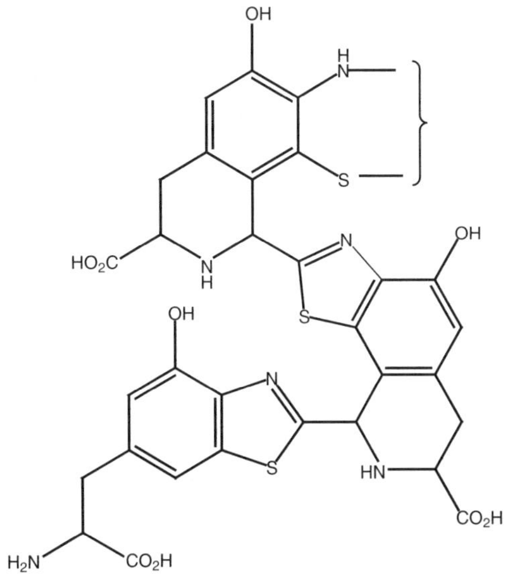

19

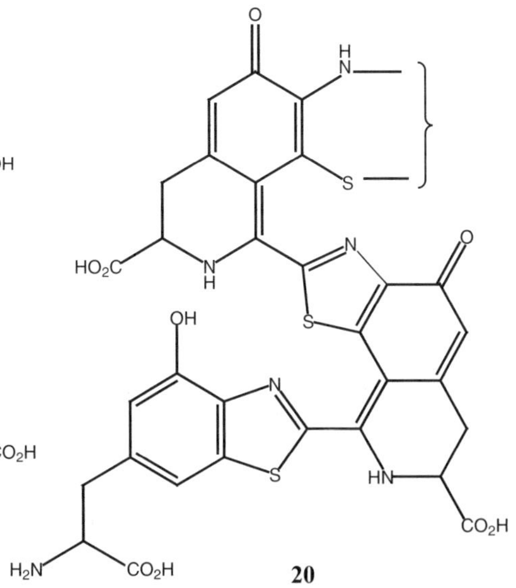

20

Cysteinyl DOPA and its metabolites are found in the hair of both eumelanic and pheomelanic subjects [53]. Every individual therefore appears to have the capability to produce either type. A study identifying a copolymer of eumelanin and pheomelanin suggests the possibility of interaction between the pathways [54].

11.4   Physical Chemistry of Hair Dyeing

The elucidation of the diffusion pathways followed by dye molecules and precursors as they penetrate into the hair fibre remains an area of great interest. It has been demonstrated that water-soluble molecules may access all components of hair, although the diffusion pathways and their final locations may differ with molecular structure [55]. The study reveals that molecules are found during the coloration process both in layers within cuticle cells and within the cell membrane complex. It is thus proposed that there may be alternative, complementary pathways into the hair fibre. Molecules enter the cuticle through its outer edge and from there migrate into different layers of the cuticle, which becomes their final location; alternatively, depending on factors such as their molecular size, shape and polarity, they may continue further to reach the cortex via the cell membrane complex. Another study supports these conclusions, and proposes that for 'small and active molecules' there is a transport system within which the cell membrane complex – and particularly the δ-layer, with its ability to swell [56] – provides the continuous pathway by which molecules are conducted by capillary forces towards the cortex [57].

Hair dyeing is commonly conducted in the pH range 9–10. At alkaline pH, significant swelling is observed, due largely to ionisation of diacidic amino acid residues. The isoionic point (the pH at which a protein has an equivalent number of positive and negative charges) and the isoelectric point (the pH at which a protein does not migrate in an electric field) are relevant parameters in hair-dyeing mechanisms. The isoionic point of unaltered hair has been determined as pH 5.8 ± 1 [58], while a separate study using electrophoretic mobility reported the isoelectric point of a single hair sample as being pH 3.67 [59]. Under alkaline conditions, hair acquires a negative charge and so has a stronger affinity for cationic than for anionic dyes. This behaviour is particularly relevant in the cell membrane complex and the endocuticle, both of which are involved in diffusion of dyes into the fibre and in both of which acidic amino acids predominate over basic amino acids. Another relevant feature is the effect of oxidative bleaching of hair, which significantly increases anionic character through formation of cysteic acid.

A critical factor governing the ability of a dye to penetrate into the hair fibre is its molecular size [60]. A model proposed for keratin fibres as a solid constructed from 'micelles' [61], although lacking rigorous justification, provided early recognition of the influence of molecular size on dye diffusion. The concept of a critical molecular size for penetration was reinforced on the basis of microscopic investigations of cross-sections of human hair treated with dyes and their precursors [62]. It was proposed initially that dyes with molecular diameters >6 Å (assuming spherical shape) were prevented from penetrating into hair. From a complementary study, hair was likened to a sieve with a hole size of ~14.8 Å [63]. The differing results from these two studies may be due to differences in the hair samples and dyeing conditions used. A subsequent, more definitive study introduced the concept of 'the longest diagonal line of the smallest shadow of the projected figure' (S_{LD}) as a measurement of molecular size that also takes some account of molecular shape [64]. The correlation between permeation distances into hair measured for selected fluorescent dyes with S_{LD} values calculated by molecular modelling suggested that molecules with S_{LD} values <10 Å migrated much more easily into hair. A recent reevaluation of the original experimental data, correlated with results from a more sophisticated molecular modelling calculation methodology (XED98), has led to a revised size descriptor, L_D (the longest dimension of the smallest cross-section of the optimum parallelepiped enclosing the molecule), as a measure of the diffusivity of the dye into the fibre. On this basis, a size limit of ~9.5 Å has been proposed for nonionic dyes, with slightly larger limits for ionic dyes [65]. In practical terms, the pore sizes and the consequent effect on dye diffusion will also be modified by fibre swelling, which is in turn influenced by features of the dyeing process and the presence of other hair dye formulation ingredients.

11.5   Toxicology of Hair Dyes

Concerns about the human safety profile of some hair dye product ingredients, especially certain precursors used in permanent oxidative hair colouring, have been raised throughout the years [66]. Because of their widespread use in regular and direct contact with humans, the general nature of the chemical structures of some ingredients, especially aromatic amines, and a few signals in the epidemiological literature, the human safety profile of hair dye components has been studied extensively, and some original ingredients are now prohibited [67]. In certain individuals, hair coloration may cause skin irritation or allergic reactions with various levels of severity; this applies to both oxidative and nonoxidative processes. Symptoms can include redness, itching, sores and a burning sensation. To prevent or limit allergic reactions, manufacturers have introduced risk-management measures [68]. It is commonly recommended that a patch test be conducted on a small area of skin before using a particular product, and that the client should not use the product if irritation develops. Two useful publications describe the results of extensive international epidemiological studies on the link between hair coloration and more serious human conditions [69,70]. The authors, who draw attention to inconsistencies in other studies, indicate that the personal use of hair coloration may play a role in the risk of certain lymphomas. The increased risk is described as moderate, and as being much more significant among women who used dyes before 1980, when the compositions were more likely to include potentially carcinogenic components. In parallel with this, the industry continues to provide reassurances over the formulation ingredients in current use [71]. For example, the European Commission has requested that the safety profile of all hair dye ingredients currently in use in the EU be updated according to recent standards and be reviewed by appropriate EU scientific committees, in an ongoing process designed to provide reassurance on the safety of oxidative hair dye ingredients in particular. Nevertheless, a change to the situation cannot be excluded in the future as stricter legislation, regulations and controls on the use of chemicals emerge. In Europe, the safety of hair dyes is controlled by a Cosmetic Directive, including two annexes which date from 1976 but are still currently valid [72]. A scientific committee of the European Commission (Scientific Committee on Cosmetic Products and Non-food Products Intended for Consumers, SCCNFP) advises the EU on the safety of hair dyes. In the USA, the legal responsibility rests with the Food and Drug Administration (FDA) [73]. In 1976, the Cosmetics, Toiletry and Fragrance Association (CTFA) established a Cosmetic Ingredient Review (CIR), an independent expert panel for the safety of cosmetic ingredients. The

CIR published safety assessments of cosmetic ingredients, which were qualified as *safe as used* or *safe under specific use restrictions*. In Japan, the Ministry of Health, Labour and Welfare (MHLW) regulates the safety of hair dyes, considering them 'quasidrugs', subject to approval based on evidence of their safety. Other East Asian countries, including Korea, Taiwan and China, have their own regulatory schemes for hair dyes, which are similar to those practised in Japan. Discussion of the detail of the legislation dealing with hair dye use is beyond the scope of this chapter, due to its complexity, variability from country to country and dynamically changing nature.

11.6   Oxidative Hair Coloration

The chemistry of permanent hair-dyeing technology is based on a 150-year-old observation by Hofmann [74], that *p*-phenylenediamine produces brown colours on a variety of substrates when exposed to oxidising agents. The first hair-colouring process based on this observation was patented 20 years later [75]. Perhaps surprisingly in view of subsequent immense advances in chemistry and concerns about the safety profile of some of the precursors used, this still forms the basis of by far the most common hair-coloration process.

The process requires three main components. The first is an *o*- or *p*-substituted (hydroxy or amino) aromatic amine, referred to as the *primary intermediate*, *oxidation base* or *developer*. Primary intermediates include *p*-phenylenediamine, *p*-aminophenol and their derivatives. The second component, the coupler, is commonly an aromatic compound with electron-donating groups arranged *m*- to each other; these include *m*-phenylenediamines, resorcinol and their derivatives, and also certain naphthols. Couplers alone do not produce significant oxidation colours, but rather modify the colour produced when used with the primary intermediates and an oxidant. Couplers are classified into three groups, according to the colour obtained with the primary intermediates: yellow-green (including resorcinol and its derivatives), red (mainly phenols and naphthols) and blue (including *m*-phenylenediamines). The third component is the oxidant, almost exclusively hydrogen peroxide, which is used with an alkali, usually ammonia. The oxidant serves two main purposes: to oxidise the primary intermediates and to lighten the natural hair colour. Ammonia also assists in swelling of the hair fibre, thus facilitating dye penetration. A decreased level of bleaching ensues when a 'no-ammonia' alkaliser, typically monoethanolamine, is used. Permanent hair-colouring preparations are generally marketed as a two-component kit. The first component contains the dye precursors (primary intermediates and couplers) and ammonia. A mixture of precursors is used – sometimes as few as three, usually five or six, and at most ten to twelve – in the combinations required to produce the desired range of colours. The second component is a stabilised hydrogen peroxide solution. The two are mixed immediately before use to give the pH of ~9.5 required for the coloration process. The mixture is initially applied near the hair roots for 20–40 minutes, to allow exposure to undyed new growth, and is then applied to the rest of the hair.

The dye precursors are small molecules which penetrate readily and deeply into the hair fibre. The first stage in colour formation involves oxidation of the primary intermediate with alkaline hydrogen peroxide to form reactive electrophilic species. In the case of

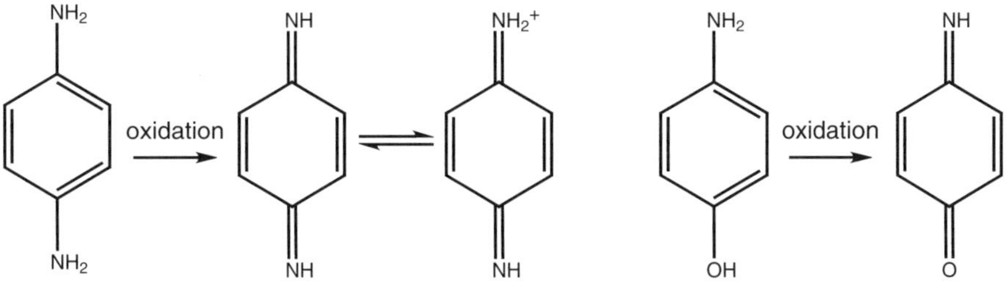

Scheme 11.4   *Generation of the reactive species from oxidation of* p-*phenylenediamine (21) and* p-*aminophenol (23).*

p-phenylenediamine (**21**), this is p-benzoquinonediimine (**22**) in equilibrium with its con-jugate acid (**22a**) [76,77] (Scheme 11.4), while p-benzoquinonimine (**24**) is formed from p-aminophenol (**23**) [7,78]. These species undergo an electrophilic substitution reaction with the coupler (**25**) preferentially at a position *para* to an electron-releasing amino or hydroxy group, forming leuco compounds (**26**), which oxidise further to dinuclear indo dyes (**27**) (Scheme 11.5). In the case where R = H, reaction of **26** with a further molecule

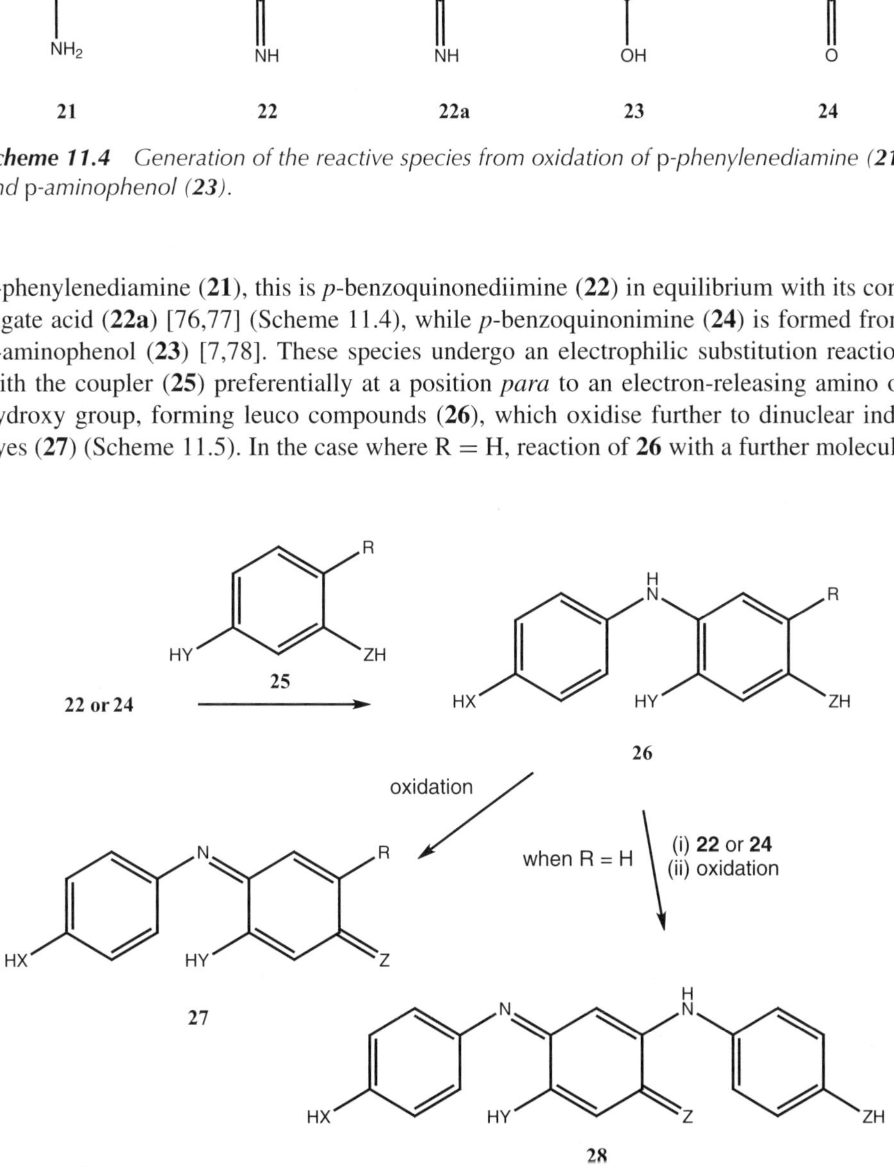

Scheme 11.5   *Pathway proposed for the formation of dinuclear (27) and trinuclear indo dyes (28) in the oxidative coupling involved in permanent hair dyeing. X, Y and Z may be independently O or NH.*

of the reactive species may lead to formation of trinuclear indo dyes (**28**). However, the presence of a blocking group (e.g. R = CH₃) in a position *para* to the electron-releasing groups prevents reaction beyond the dinuclear dye. The possibility of reaction to generate even larger structures has not been fully clarified. In the absence of couplers, primary intermediates may form brown to black oxidation colours, but, since they are much less reactive than couplers, there is almost no self-coupling in their presence. The coloured molecules generated in this oxidative process are much larger than the precursors, and so they become permanently entrapped within the hair fibre.

In formulated products containing mixtures of primary intermediates and couplers, the particular hair colour developed depends on competing reactions inside the fibre between the various precursors, influenced by such factors as concentrations, diffusion rates, redox potentials and pH [79]. Predictions based on kinetics have proved consistent with the colours obtained, at least qualitatively, implying that the relative reactivities of reactants are more important than their relative concentrations. With a higher redox potential, *p*-aminophenol (**23**) and its derivatives are oxidised before *p*-phenylenediamine (**21**) and its derivatives. Coupler selection is critical in hair dye formulation. For example, if an auburn shade is required from a composition containing primary intermediates **21** and **23**, the most reactive coupler is chosen so as to assure formation of a red dye from **23**, which undergoes oxidation first and then reacts with the coupler, before the formation of dyes from coupling with **21**. The pH may also be used as a tool to adjust tone or depth through its influence on the rate of colour formation. In a typical oxidation dye composition with *p*-phenylenediamine (**21**) and three couplers – resorcinol, *m*-aminophenol and *m*-phenylenediamine at pH 9.0–9.5 – the couplers react at similar rates. Reducing the pH by one unit increases the rate of the reaction with *m*-phenylenediamine relative to the phenols, while an increase of one unit enhances the relative rate of reaction with resorcinol.

Over the years, countless numbers of dye precursors have been examined for their ability to fulfil the requirements of oxidative hair dyeing. These studies have generally been aimed at addressing issues associated with toxicology, colouristic performance and fastness properties, particularly light fastness [67]. While this research intensity has produced a few individual successes, most of the compounds in current commercial dye formulations are the same ones that have been used for decades. The most important primary intermediates in current use include *p*-phenylenediamine, 2-chloro-*p*-phenylenediamine, *N,N*-bis(2-hydroxyethyl)-*p*-phenylenediamine, *N*-phenyl-*p*-phenylenediamine and *o*- and *p*-aminophenol. The most important couplers include pyrogallol, hydroquinone, 2,3-naphthalenediol, resorcinol, 4-chlororesorcinol, 2-methylresorcinol, 1-naphthol, *m*-aminophenol, 5-amino-2-methylphenol, 5-hydroxyethylamino-2-cresol, *m*-phenylenediamine, 2,4-diaminophenoxyethanol and 2,4-diaminophenetole. The range of additional compounds that have been claimed in patents is documented comprehensively in our review article [14]. Substituent variation in derivatives of the traditional primary intermediates and couplers has been investigated exhaustively. Numerous heterocyclic compounds have also been extensively considered as alternatives to the carbocyclic analogues. A successful outcome of this particular strategy is demonstrated by the commercial launch by L'Oreal of its Rubilane range of primary intermediates based on derivatives of dihydropyrazolone **29** [80–84]. These are claimed to provide colouristic advantages such as improved intensity, chromaticity and aesthetic qualities, with good resistance to shampooing, light, perspiration and permanent waving, and to offer the advantage of operating

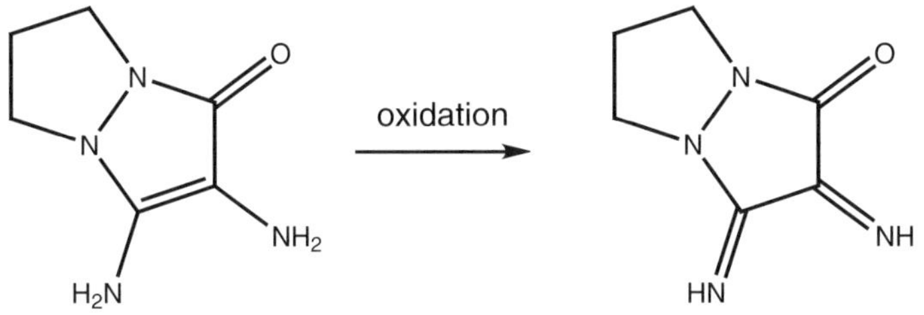

Scheme 11.6   *Oxidation of dihydropyrazolone (**29**) to the quinonediimine analogue (**30**).*

at neutral pH. They undergo oxidation to generate the quinonediimine analogue **30** as the reactive species (Scheme 11.6).

A number of approaches have been explored in attempts to enhance certain features of the oxidative hair-dyeing process. Cationic derivatives, derived from nonionic dye precursors, are claimed to provide enhanced depth of penetration into the fibre and reduced sensitivity of the colour formed to photodegradation [85–88]. Patents have been issued on alternative colour-development chemistry aimed at avoiding the use of hydrogen peroxide and alkali, which can degrade the hair fibre, rendering it coarse and fragile. For example, electrochemical methods and enzymatic oxidation systems have been claimed to oxidise dye precursors [89,90]. Although such systems are capable of oxidising the primary intermediates, they do not have the ability to remove natural hair colour. However, certain fluorescent compounds have been investigated for their potential to lighten hair on the basis of their light-emission properties [91].

Autooxidative hair dyeing involves air oxidation of dye precursors, without an additional oxidant [4]. In contrast to traditional oxidative dyeing, the hair is not lightened and so the system is often targeted at colouring grey hair. The chemistry of the process, which employs polyphenols, is exemplified in Scheme 11.7. Here, a coloured structure (**34**) is generated oxidatively, in the presence of ammonia, from the dye precursor 1,2,4-trihydroxybenzene (**31**), via diphenylquinone (**33**), which is formed by self-coupling of the semiquinone radical **32**, to provide a medium-brown hair colour [92,93]. This technology has niche commercial importance, especially in Asia and in products aimed mainly at the male market in Europe and the USA.

11.7   Alternative Approaches to Permanent Hair Dyeing

A number of processes have been investigated in attempts to provide alternatives to traditional oxidative hair dyeing. The general aim is to develop safe dyeing systems that give intense coloration and good fastness properties, while avoiding hair damage from the use of oxidising agents and strongly alkaline conditions. Like the oxidative process, these methods also mostly involve chemical reactions that generate colour *in situ*. An overview of the approaches is presented here, while the reader will find more comprehensive coverage in our review [14].

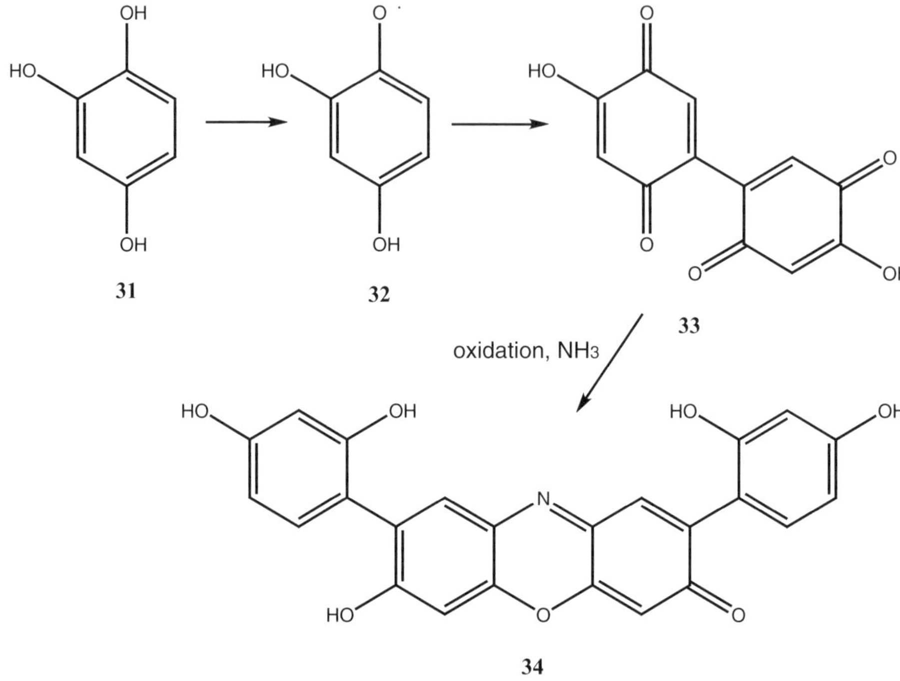

Scheme 11.7   *The autooxidation of 1,2,4-trihydroxybenzene (31).*

Azo dyes and pigments are the most important chemical classes of industrial colorants, providing bright, intense colours and good technical performance. They are synthesised by the two-stage process of diazotisation and azo coupling, at ambient temperatures in water; because they use commodity starting materials, they are inexpensive. Therefore, a process based on the synthesis of insoluble azo colorants within the hair fibre appears to be an attractive concept. Early patents describe a process involving treating hair for 20 minutes with an alkaline solution of a coupling component, similar to the compounds used as couplers in oxidative hair coloration, followed by treatment with a stabilised diazonium salt solution, and then allowing 20 minutes for *in situ* azo coupling to occur [94–96]. However, this process has not been used commercially. One negative issue is the potential explosion risk during storage of diazonium salts. Similarly, processes have been investigated that form methine or azomethine dyes based on *in situ* reactions between amines, for example *p*-phenylenediamines, mono-, di- or triethanolamine, and carbonyl compounds, such as phthalaldehyde derivatives.

In a related process, illustrated by example in Scheme 11.8, a cationic species, 3-ethyl-2-methylbenzothiazolium iodide (**35**), is reacted with 4-dimethylaminobenzaldehyde (**36**), leading *in situ* to methine dye (**37**) through base-catalysed Knoevenagel condensation. Methine dyes of the type formed in this process have been proposed as semipermanent hair dyes, providing bright, intense colours [97,98]. However, they tend to show inferior fastness properties, which may explain why permanent hair-dyeing processes based on this chemistry have not reached the market.

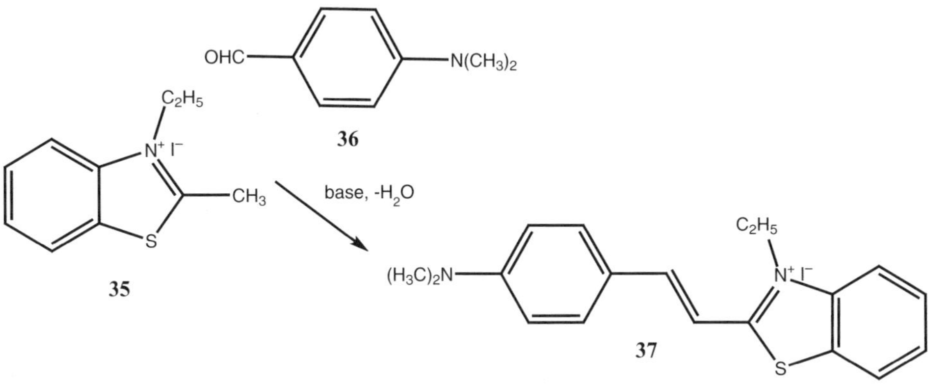

Scheme 11.8 In situ *reaction of 3-ethyl-2-methylbenzothiazolium iodide (35) with 4-dimethylaminobenzaldehyde (36) by Knoevenagel condensation.*

Vat dyes are a highly important dye class used for the coloration of cellulosic textile fibres. They are applied to the fibre via a reduced (*leuco*) form, which is later reoxidised within the fibre. Although this chemistry is not directly applicable to hair coloration, a process has been reported that uses solubilised vat dyes, which are water-soluble sulphate esters of leuco vat dyes with the generalised structure **38** [99,100]. The method involves application of the dye to hair, followed by acid hydrolysis to give the leuco vat dye (**39**). This is then oxidised within the hair fibre to generate the insoluble vat dye (**40**), which is characterised structurally by the presence of at least two carbonyl groups linked through a conjugated system (Scheme 11.9), giving the excellent fastness properties that are characteristic of vat dyes. The use of hydrogen peroxide as the oxidising agent means that the process is potentially compatible with conventional oxidative hair colouring.

Indo dyes are the coloured species obtained from the reaction between oxidised primary intermediates and couplers in conventional oxidative coupling. As exemplified in Scheme 11.10, an indo dye, such as the benzoquinoneimine **41**, may be reduced by alkaline sodium dithionite and then neutralised to give a leuco form: the stable, colourless diphenylamine derivative **42**. It has been shown that compounds of this type may be applied to hair and oxidised to generate colour *in situ* by regenerating the indo dye [101–108].

A hair-dyeing process is reported using indican (**43**), an indigo precursor extractable from leaves of *Indigofera tinctoria* [109,110]. A formulation containing the precursor and

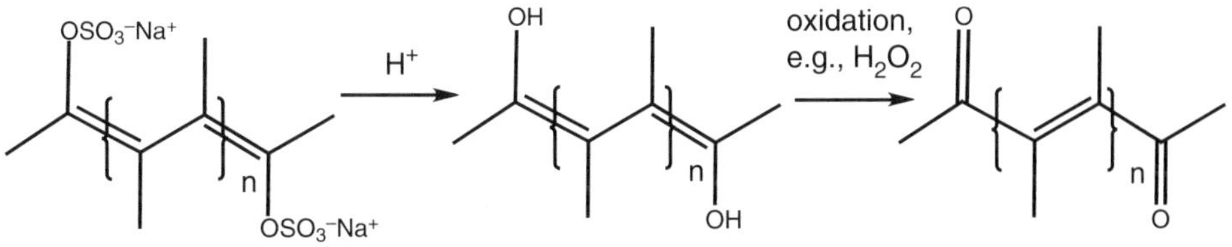

Scheme 11.9 *Solubilised vat dyeing.*

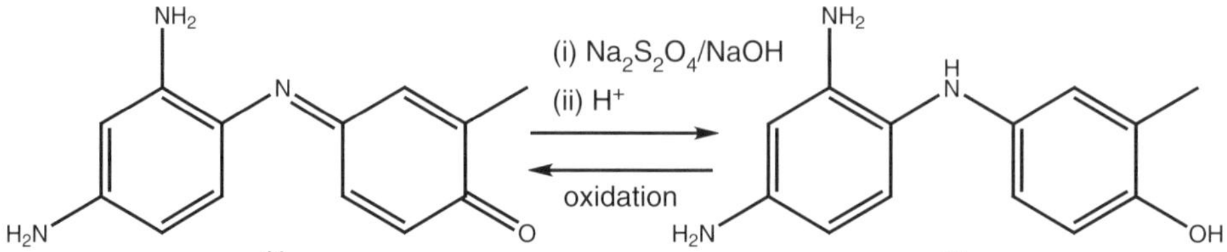

Scheme 11.10   *Hair-dyeing process with leuco forms of indo dyes.*

a separate formulation containing a β-glucosidase are applied to hair at 40–45 °C for 30 minutes. Enzymatic hydrolysis leads to indoxyl (**44**), which oxidises rapidly in air to form indigo (**4**) within the hair (Scheme 11.11).

In 1995, Ciba introduced the concept of *latent pigments*; these are precursors that are soluble in the application medium and which generate the pigment *in situ* following a physical or chemical treatment [111–113]. These products were initially developed for the coloration of plastics, although the patents claimed the potential to apply the technology to a variety of porous materials, including human hair [114,115]. However, pigment formation from these compounds generally required severe conditions that were inappropriate for hair coloration. Nevertheless, the design of a precursor which would penetrate the hair and, in a subsequent treatment, generate the pigment within the fibre represents an interesting concept. An example has been reported in which hair is treated with the cationic derivative **45** [116,117], which undergoes base-catalysed hydrolysis at room temperature to give indigo (Scheme 11.12) [118,119]. Pigments generated within hair are likely to form as nanoparticles. It has been postulated that the application of nanotechnology in hair dates

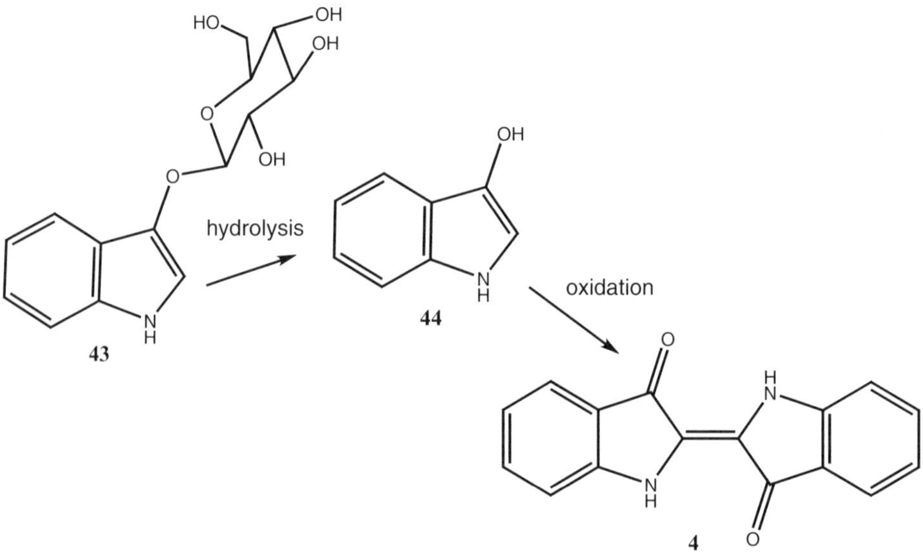

Scheme 11.11   *Formation of indigo from indican with a β-glucosidase enzyme.*

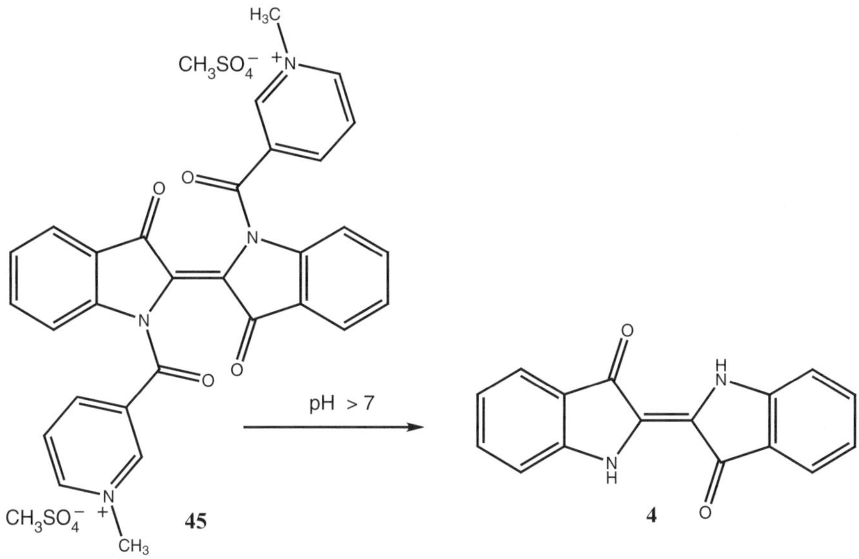

Scheme 11.12   *Latent pigment technology applied to hair coloration.*

back to the ancient Greeks and Romans, when mixtures of lead oxide and calcium hydroxide were used to produce a black colour on hair. A study has shown that this is due to formation of lead sulphide with a crystal size of ~5 nm, the sulphur being derived from the keratin amino acids [120]. More recent developments in nanotechnology are providing new possibilities for permanent hair dyeing. While the toxicity of carbon nanotubes is in question and preliminary studies suggest serious risk to human health, chemically functionalised and physically modified carbon nanotubes are claimed to impart a thin black coating that confers a smooth feel on the hair and provides a volumising effect. The high surface area of the nanoparticles enhances interaction with the hair and produces a long-lasting black colour [121].

Reactive dyes form a covalent bond between a C-atom of the dye and an O-, N- or S-atom of hydroxy, amino or thiol groups on textile fibres [122–130]. This dye class is most important industrially for cellulosic fibres, providing a wide range of colours with excellent wash fastness. Specific reactive dye types have also been developed for protein textile fibres, such as wool and silk [131,132], as discussed elsewhere in this book. It has been anticipated that a process linking dye molecules covalently to human hair might provide, in addition to superior colouristics and wash fastness, improved resistance towards other hair treatments and to perspiration, and that it might simplify the dyeing of new hair growth, since previously-dyed hair would provide fewer dye binding sites. The key structural feature of a reactive dye is the fibre-reactive group. In textile dyes, reactive groups may be categorised as reacting either by nucleophilic substitution, such as the halotriazinyl ring system, or by nucleophilic addition, such as the vinylsulphone (VS) group, generated *in situ* from a β-sulphatoethylsulphone. Early attempts at reactive dyeing of hair were judged unsuccessful because only minimal coloration could be achieved under the mild conditions necessary. However, processes have been claimed for reactive hair coloration [133] based on reaction with the thiolate group ($^-$S$^-$) on the hair protein, the process requiring pretreatment with a thiol derivative, such as thioglycolic acid, to generate the $^-$S$^-$ ion in such a way that it reacts with the dye at low temperatures [4]. However, hair-colouring products based on reactive dyes have not yet reached a commercial outcome, possibly due to inadequate light fastness and the prohibitive cost of the requirement to assess their toxicological profile in order to achieve approval and registration as hair dyes.

A process that mimics melanin formation synthetically inside the hair fibre would provide obvious attractive market advantages, notably the ability to designate a product as 'natural' and the potential to provide colours with natural appearance. However, the challenges are immense, since natural hair colour is determined not only by the melanin structure but also by the sizes of granules and their distribution throughout the hair, as described in Section 11.3. An overview of the developments leading to such a process is given here, with further detail presented in our review [14]. The approaches have been aimed primarily at synthetic imitation of the biosynthetic pathways leading to the natural pigments, focussed mostly on eumelanin, by oxidation of intermediates as identified in Schemes 11.1, 11.2 and 11.3. An early patent claims that a preparation consisting of tyrosine (**6**) or DOPA (**7**) with a tyrosinase gives a light-brown hair colour, with repeated dyeing being necessary to build up colour [134]. Subsequently, improved processes were reported using DOPA in combination with hydroquinones [135], diamines or aminophenols [136], and with hydrogen peroxide oxidation replacing the tyrosinase. A range of DOPA derivatives have been explored, with a particular focus on DHI (**12**), the closest isolable intermediate in melanin biosynthesis. Hair may be dyed with DHI using hydrogen peroxide as the oxidising agent, giving brown to black colours [137–140]. The process may be enhanced further using metal-ion promoters, notably salts of Ce(III). However, DHI is a difficult material to work with because it is unstable in air and its synthesis is elaborate, although research has made some progress towards more efficient synthetic routes [141]. One approach uses 5,6-diacetoxyindole (DAI; **46**), which polymerises less easily in air, in a process in which it is deacetylated in alkali to form DHI, just before application to hair [142]. Comparable approaches towards synthetic pheomelanin generation have been restricted by difficulties in synthesising the intermediates. Nevertheless, an oxidative hair-dyeing process using dihydrobenzothiazines (**47**), related structurally to intermediates in pheomelanin biosynthesis, has been reported to generate colours ranging from blond to brown [143]. In these processes, it is likely that the melanins are formed dispersed throughout the fibre, rather than being deposited in granules as in the natural process, so that the nuances in colour are not imitated. The colour range may be extended by introducing other reactants with the melanin precursors, such as oxidative permanent hair-dyeing components and certain sulphur-containing nucleophiles.

46   **47**

11.8   Nonoxidative Hair Dyeing

Nonoxidative hair-dyeing processes are often referred to as *direct* dyeing because the dyes are applied directly and the molecules remain chemically unchanged within the hair after application, in contrast with the oxidative dyeing processes. These products yield nonpermanent colour effects on hair, in that they resist a few washings, while slowly and evenly fading. The dyes are classified, according to the duration of the colour on the hair, as either *semipermanent* or *temporary*.

The term 'semipermanent' refers to products that provide hair colour lasting through roughly four to six washings. While permanent oxidative hair dyeing had become firmly established by around the mid-1920s, the first successful semipermanent hair dye products were not introduced until around 1950. This occurred at a time when preconceptions and attitudes towards hair colouring appear to have changed. It was no longer seen only as a means to conceal the greying process but also as a way of changing one's appearance in order to become more attractive or fashionable. Early semipermanent hair dyeing was based mainly on low-molecular-weight direct dyes, especially a series of nitro dyes that had been patented for use in permanent hair-dyeing formulations, but 'the inventors did not recognise their departure from the oxidation base type of product' [144]. It was recognised eventually that these dye molecules simply add to the overall colour in the hair, rather than participating in the oxidative coupling reactions in which colour is formed. The early semipermanent hair dyes, which were restricted to yellows through to reds, were used mainly to add gold and red tones to blond and brown hair. The initial absence of suitable purple and blue dyes restricted the range of shades that could be produced. Additional original uses of semipermanent dyes were to cover early signs of greying (through *tone-on-tone* colouring, where the aim is to tint the hair to its original colour) and to modify naturally grey hair to a more desirable silver-grey by removing yellowish tones. Other applications that have emerged in more recent years include the use of high dye concentrations that give intense coloration in order to brighten or highlight the original hair colour. This application is aimed mainly at the younger market, and also to colour hair that has been heavily damaged by bleaching or perms, and which therefore may not tolerate rigorous oxidative dyeing. Semipermanent dyes currently offer the advantages of ease of use, a lack of requirement for mixing and reversibility, since an undesirable result can be removed by repeated shampooing.

Nitro dyes constitute by far the most important chemical type of semipermanent hair dye, providing colours ranging through yellow, orange and red-brown to violet. Structurally, the range may be subdivided into nitroanilines, nitrodiphenylamines, nitrophenylenedi-amines, and nitroaminophenols, together with the related ethers [145]. Of these types, nitroaminophenols provide most of the yellow to orange dyes, with nitrophenylenediamine dyes dominating the orange through red to purple-violet colours. Nitro dyes cannot provide pure blues, a feature which accounts largely for the importance of anthraquinone dyes. A few azo dyes are also used, but there are issues of compatibility in formulations with other dye classes, in terms of solubility, diffusion and photochemical characteristics, for example. Some other dye types, including acid dyes and basic dyes – more common in temporary hair-colouring products – are occasionally incorporated into semipermanent formulations. In contrast to permanent oxidative dyeing, which involves *in situ* chemical reactions, semipermanent dyeing relies only on the diffusion of small, coloured, nonionic molecules into the hair cortex. The rate of diffusion increases with an increase in pH and temperature [146]. The principal interactions between the dye and fibre molecules are relatively weak dipolar and van der Waals forces. Larger dye molecules, characteristic of many anthraquinone dyes, have a higher affinity for hair than do smaller molecules, due to a stronger set of dye–hair interactions, and thus wash out more slowly [147].

Simple nitroanilines provide a few yellow semipermanent hair dyes, including dyes **48a–d** [148]. The *N*-hydroxyethyl substituent is particularly widely encountered in the structures of all chemical classes of semipermanent hair dye. Compared with the *N*-ethyl substituent, this group increases the water solubility without giving rise to a significant change in hue, and modifies the water–lipid partition coefficient, a property that has an important impact on dye uptake by the hair. The substituent at the 4 position in this series of dyes may play a part in optimising this partition coefficient and in enhancing certain technical features, such as shampoo resistance, again with minimal effect on the hue [149].

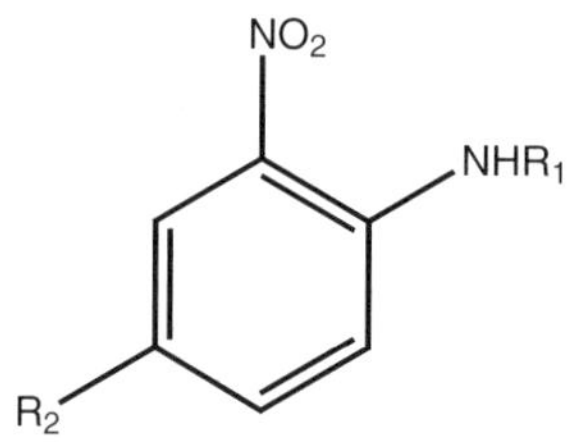

$$NO_2$$

$$NHR_1$$

$$R_2$$

48

a: $R_1 = CH_2CH_2OH$; $R_2 = H$

b: $R_1 = CH_2CH_2OH$; $R_2 = Cl$

c: $R_1 = CH_2CH_2OH$; $R_2 = CH_3$

d: $R_1 = CH_2CHOHCH_2OH$; $R_2 = CF_3$

Nitrophenylenediamine dyes are derived structurally from the simple isomers **49–51** and offer a wide colour range, from yellow through to violet. The range of nitrophenylene-diamine dyes currently used as semipermanent hair dyes is shown grouped according to colour, as determined by the absorption maximum in solution (in ethanol), in structures **52–54** [150]. The colour of a particular dye in this series is dependent on the substituent pattern, with dyes based on isomers **50** and **51** being bathochromic compared to those based on isomer **49**. The presence of *N*-alkyl, *N*-aminoalkyl and *N*-hydroxyalkyl substituents shifts the colour bathochromically, compared with unsubstituted derivatives, leading ultimately to purple-violet colours. Dyes **52a–c** and **53** provide greenish-yellows, dyes **54a–h** are reds and dyes **54i–s** are violets. The synthesis of many of these dyes involves relatively inexpensive starting materials, a particularly important factor as these nitro dyes generally find little use in other applications, and thus the limited quantities produced rarely result in economies of scale.

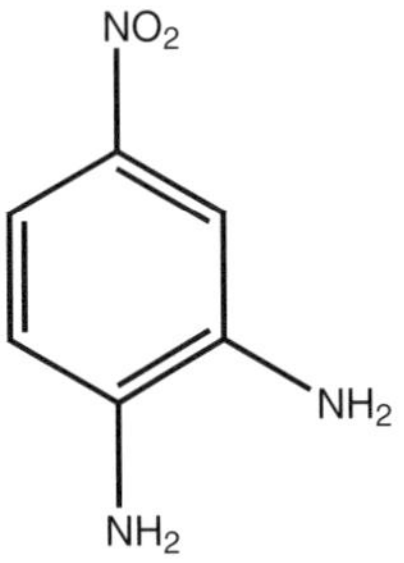

49

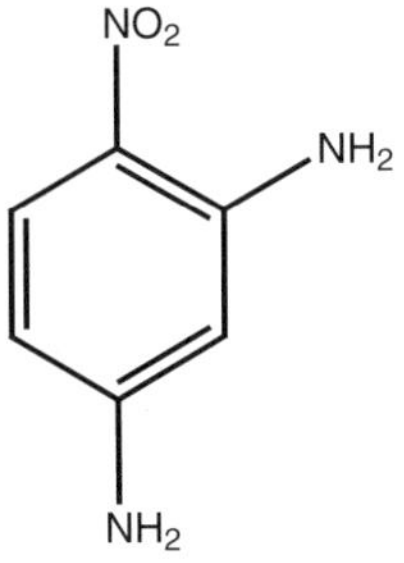

50

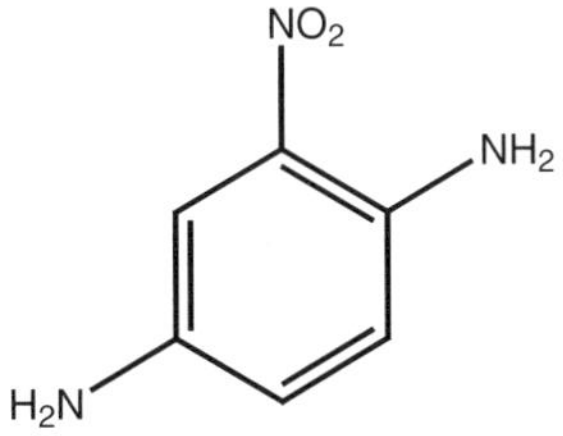

51

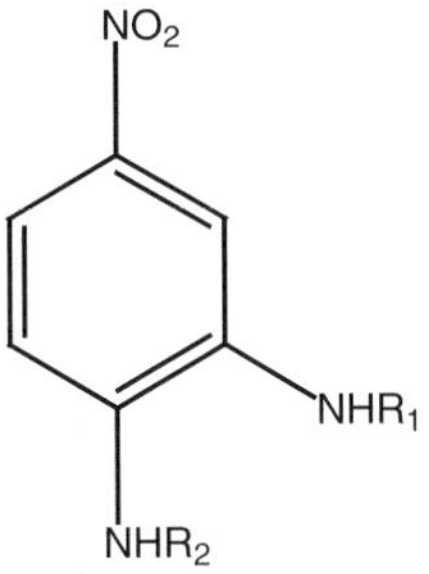

52

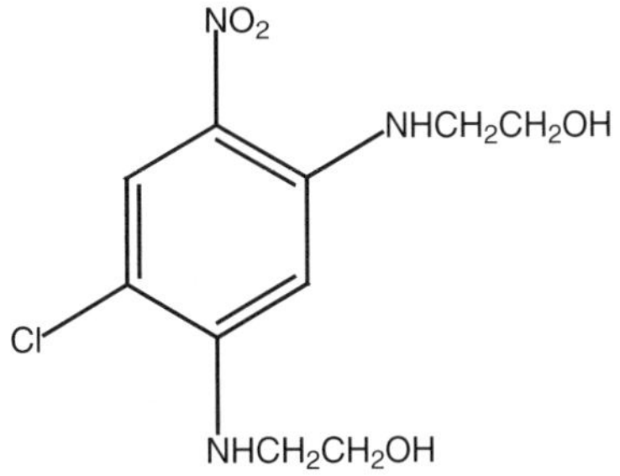

53

a: R₁=H; R₂=CH₂CH₂OH

b: R₁=CH₂CH₂OH; R₂=CH₂CH₂OH

c: R₁=H; R₂=C(CH₂OH)₃

54

a: R_1=H; R_2=H; R_3=H; R_4=CH$_3$; R_5=H; R_6=H

b: R_1=H; R_2=H; R_3=H; R_4=CH$_2$CH$_2$OH; R_5=CH$_3$; R_6=H

c: R_1=H; R_2=H; R_3=H; R_4=CH$_2$CH$_2$OH; R_5=Cl; R_6=H

d: R_1=H; R_2=H; R_3=H; R_4=CH$_2$CH$_2$NH$_2$; R_5=H; R_6=CH$_3$

e: R_1=H; R_2=CH$_3$; R_3=H; R_4=H; R_5=H; R_6=H

f: R_1=H; R_2=H; R_3=H; R_4=CH$_2$CH$_2$OH; R_5=H; R_6=H

g: R_1=H; R_2=CH$_2$CH$_2$OH; R_3=H; R_4=H; R_5=H; R_6=H

h: R_1=H; R_2=H; R_3=H; R_4=CH$_2$CHOHCH$_2$OH; R_5=Cl; R6=H

i: R_1=H; R_2=CH$_3$; R_3=CH$_3$; R_4=CH$_2$CH$_2$OH; R_5=H; R_6=H

j: R_1=H; R_2=CH$_2$CH$_2$OH; R_3=H; R_4=CH$_2$CH$_2$OH; R_5=H; R6=H

k: R_1=H; R_2=CH$_3$; R_3=CH$_3$; R_4=CH$_2$CHOHCH$_2$OH; R_5=H; R_6=H

l: R_1=H; R_2=H; R_3=CH$_2$CH$_2$OH; R_4=CH$_2$CH$_2$OH; R_5=H; R_6=H

m: R_1=H; R_2=CH$_2$CH$_2$NH$_2$; R_3=CH$_2$CH$_2$OH; R_4=CH$_2$CH$_2$OH; R_5=H; R6=H

n: R_1=H; R_2=CH$_2$CH$_2$OH; R_3=CH$_2$CH$_3$; R_4=CH$_2$CH$_2$OH; R_5=H; R_6=H

o: R_1=H; R_2=CH$_2$CHOHCH$_2$OH; R_3=H; R_4=CH$_2$CHOHCH$_2$OH; R_5=Cl; R_6=H

p: R_1=H; R_2=CH$_2$CH$_2$CH$_2$OH; R_3=CH$_2$CH$_2$OH; R_4=CH$_2$CH$_2$OH; R_5=H; R_6=H

q: R_1=H; R_2=CH$_2$CH$_2$OH; R_3=CH$_2$CH$_2$OH; R_4=CH$_2$CH$_2$OH; R_5=H; R_6=H

r: R_1=H; R_2=CH$_3$; R_3=CH$_2$CH$_2$OH; R_4=CH$_2$CH$_2$OH; R_5=H; R_6=H

s: R_1=CH$_2$CH$_2$OH; R_2=CH$_2$CH$_2$OH; R_3=H; R_4=H; R_5=H; R_6=H

Nitroaminophenols, and related ether derivatives, represent an important class of semiper-manent hair dye, offering colours from yellow to orange. They provide a more extensive range of dyes than the other nitro dye types in these shade areas. Dyes **55a–c** and **56a–b** are greenish yellow, **56c**, **57**, **58a–c**, **59a–b** and **60** are mid to red-shade yellow and **58d–f** are orange. A feature that restricts the use of the phenolic derivatives is the colour change due to ionisation at pH 9–10.

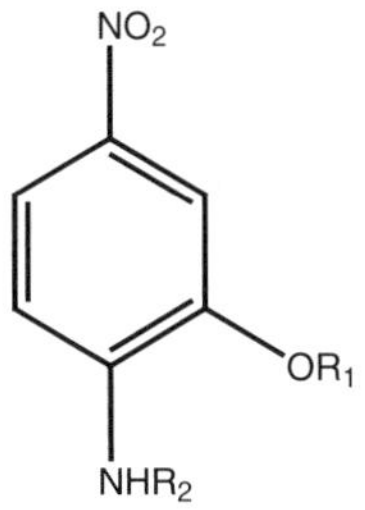

55

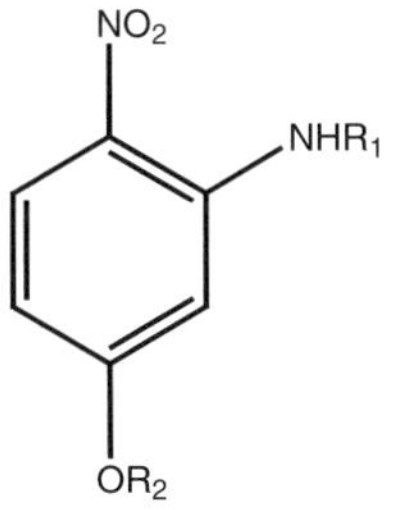

56

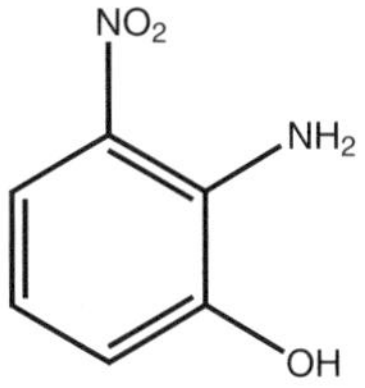

57

a: R_1=CH_3; R_2=CH_2CH_2OH
b: R_1=H; R_2=H
c: R_1=CH_2CH_2OH; R_2=CH_2CH_2OH

a: R_1=CH_3; R_2=CH_2CH_2OH
b: R_1=CH_3; R_2=$CH_2CHOHCH_2OH$
c: R_1=$CH_2CH_2NH_2$; R_2=CH_3

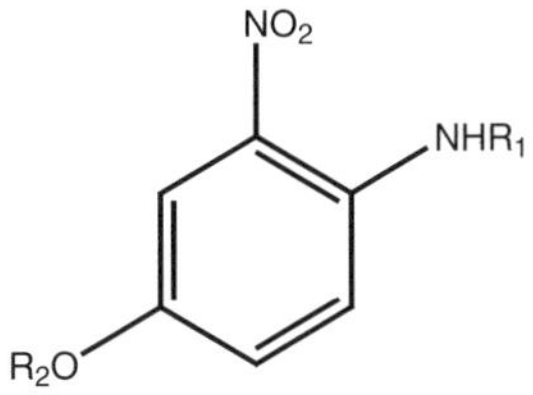

58

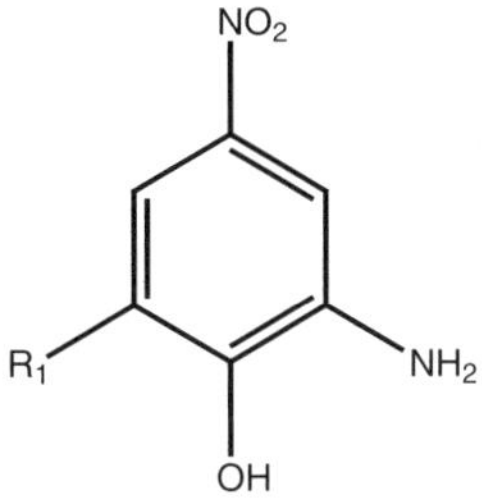

59

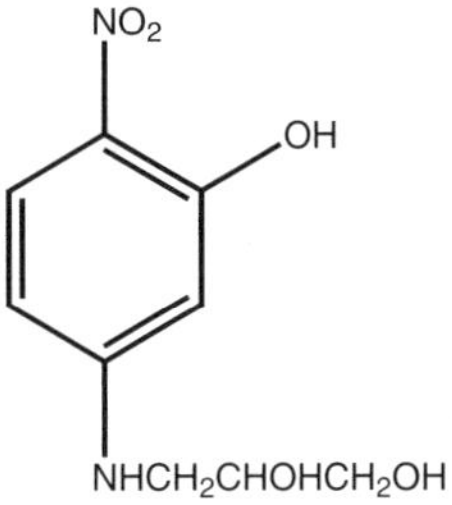

60

a: R_1=CH_2CH_2OH; R_2=$CH_2CHOHCH_2OH$
b: R_1=CH_2CH_2OH; R_2=CH_3
c: R_1=H; R_2=H
d: R_1=CH_2CH_2OH; R_2=H
e: R_1=$CH_2CH_2CH_2OH$; R_2=H
f: R_1=CH_2CH_2OH; R_2=CH_2CH_2OH

a: R_1=H
b: R_1=Cl

A range of other nitro dyes are used in semipermanent dye formulations. The dinitrophe-nols **61a–c** and **62** add useful greenish-yellow shades. Other types include diphenylamines **63a–c** and heterocyclic derivative **64**. A few dyes, such as the anionic dye **64d** and some larger-molecular-sized nonionic dyes in this series may be used in products for specific hair types. For example, they are useful for damaged hair, which is more porous and thus facilitates the diffusion of molecules both in and out of the fibre, but which may retain larger molecules.

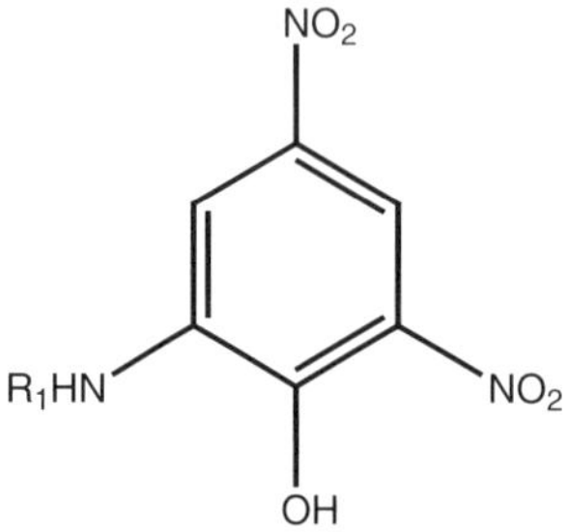

61

a: R_1=H

b: R_1=CH$_2$CH$_2$OH

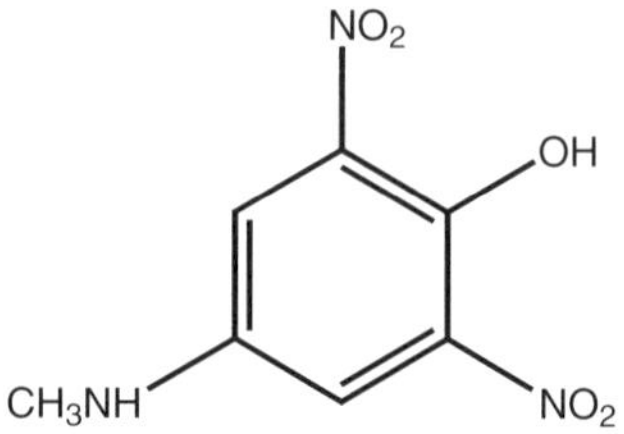

62

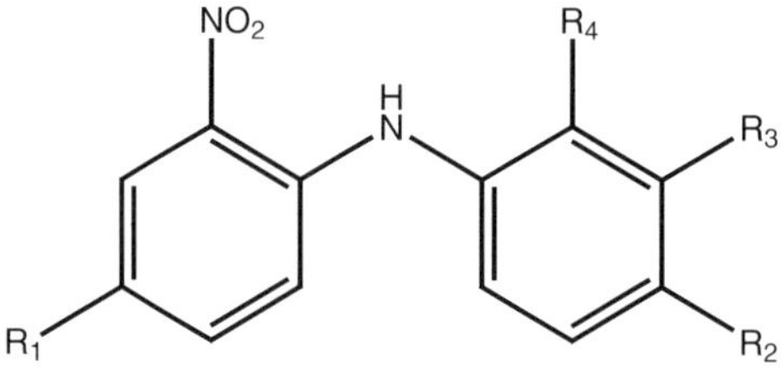

63

64

a: R_1=NH$_2$; R_2=H; R_3=H; R_4=H

b: R_1=H; R_2=OH; R_3=H; R_4=H

c: R_1=H; R_2=H; R_3=H; R_4=OH

d: R_1=NO$_2$; R_2=H; R_3=NHC$_6$H$_5$; R_4=SO$_3$Na

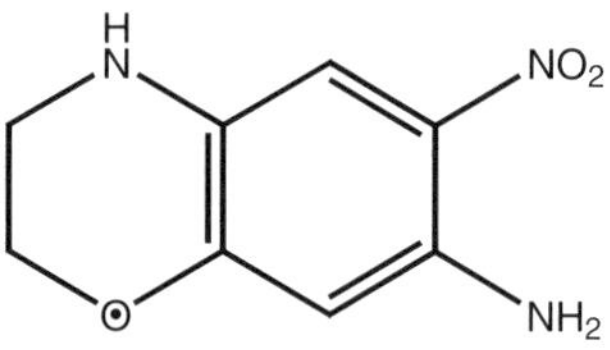

Anthraquinone dyes are important in completing the semipermanent hair colour palette as they contribute violet and blue colours, since nitro dyes cannot provide compounds absorbing at long wavelengths [151]. Appropriate dyes are generally selected from the range used for dyeing synthetic textile fibres, which minimises the cost, although a few have been specifically designed for the application, building in particular attributes such as enhanced hydrophilicity. The anthraquinone dye molecules are often larger than nitro dyes, but are able to penetrate into the hair fibre because of their planarity and the use of nonbulky substituents. Anthraquinone dyes are typical donor/acceptor types, in which substituent effects can be used to provide a wide range of colours. A notable structural feature is hydrogen bonding between α-substituents and the carbonyl group, which enhances light fastness. Dyes **65a–b** are red-violet, **65c–d** are violet, **65e–f** are violet-blue and **65g–j** are blue.

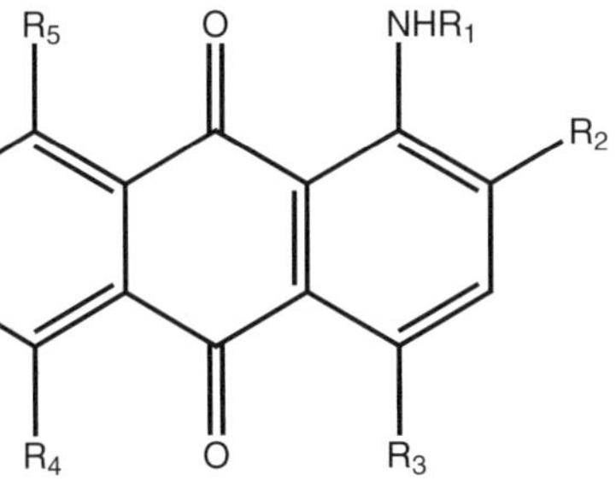

65

a: R_1=H; R_2=H; R_3=OH; R_4=H; R_5=H
b: R_1=H; R_2=OCH$_3$; R_3=NH$_2$; R_4=H; R_5=H
c: R_1=H; R_2=H; R_3=NH$_2$; R_4=H; R_5=H
d: R_1=H; R_2=H; R_3=NHCH$_3$; R_4=H; R_5=H
e: R_1=H; R_2=H; R_3=NH$_2$; R_4=NH$_2$; R_5=NH$_2$
f: R_1=*p*-CH$_3$C$_6$H$_5$; R_2=H; R_3=OH; R_4=H; R_5=H
g: R_1=CH$_2$CH$_2$OH; R_2=H; R_3=CH$_3$; R_4=H; R_5=H
h: R_1=CH$_2$CH$_2$OH; R_2=H; R_3=CH$_2$CH$_2$OH; R_4=OH; R_5=OH
i: R_1=CH$_2$CH$_2$OH; R_2=H; R_3=CH$_2$CH$_2$OH; R_4=H; R_5=H
j: R_1=*p*-CH$_3$C$_6$H$_5$; R_2=H; R_3=*p*-CH$_3$C$_6$H$_5$; R_4=H; R_5=H

Although azo dyes constitute the most important chemical class for most coloration applications, only a few nonionic derivatives have found use in semipermanent hair dyeing. They give bright, intense yellow to red colours [152]. Dyes **66a–c** and **67** are yellow, **66d–e** and **68** are orange, while dyes **69** and **70** are red.

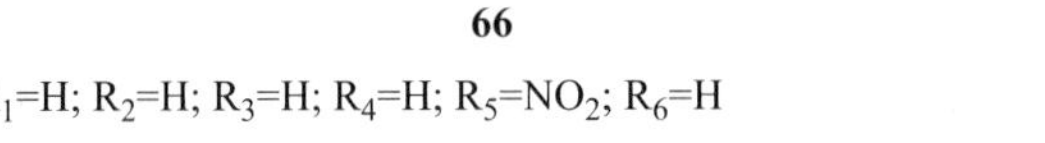

66

a: R_1=H; R_2=H; R_3=H; R_4=H; R_5=NO$_2$; R_6=H
b: R_1=CH$_2$CH$_2$OH; R_2=CH$_2$CH$_2$OH; R_3=H; R_4=NH$_2$; R_5=H; R_6=H
c: R_1=CH$_2$CH$_2$OH; R_2=CH$_2$CH$_2$OH; R_3=CH$_3$; R_4=NH$_2$; R_5=H; R_6=H
d: R_1=CH$_2$CH$_2$OH; R_2=CH$_2$CH$_2$OH; R_3=CH$_3$; R_4=NO$_2$; R_5=H; R_6=H
e: R_1=CH$_2$CH$_2$OH; R_2=CH$_2$CH$_2$OH; R_3=Cl; R_4=NO$_2$; R_5=Cl; R_6=Cl

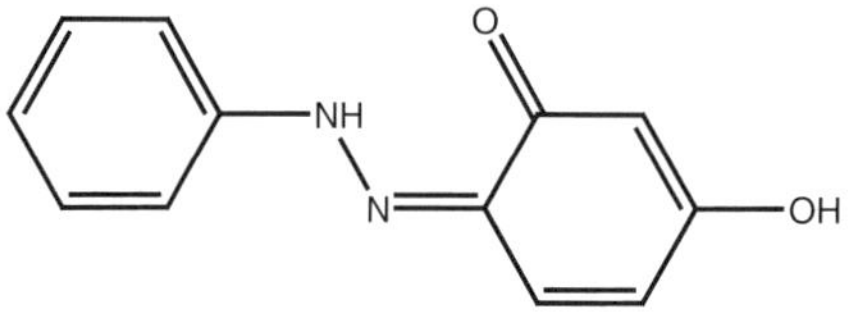

67

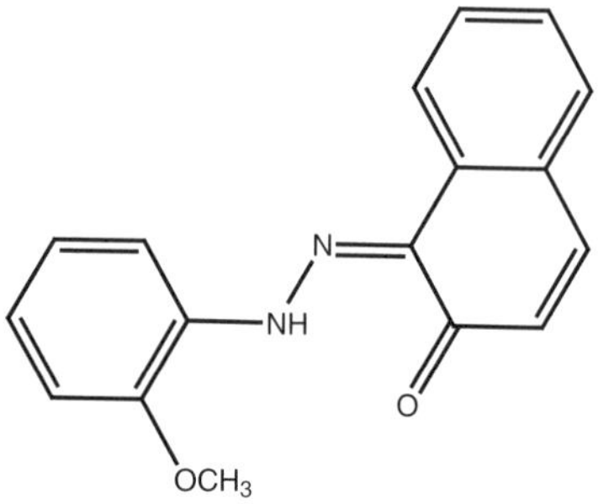

68

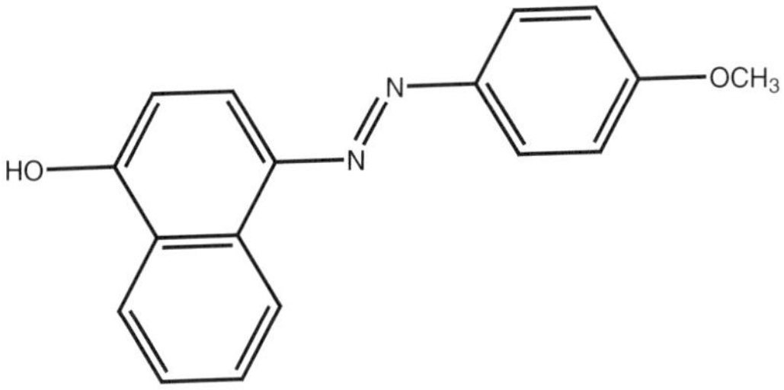

69

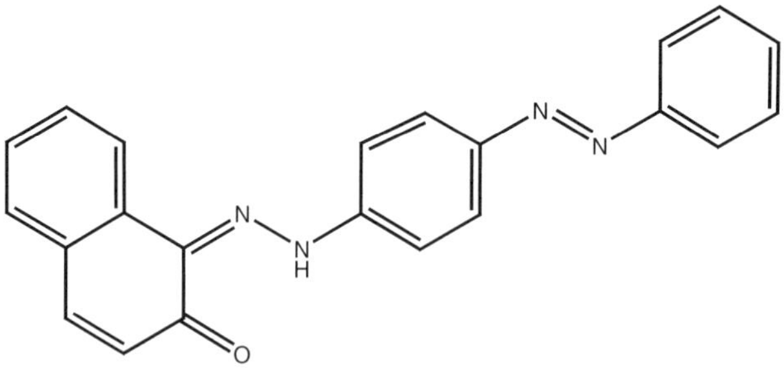

70

A significant recent development in hair coloration is the introduction of a range of cationic semipermanent hair dyes containing a quaternary heterocyclic nitrogen atom, in which the positive charge is delocalised throughout the chromophore. The first dyes developed had issues with stability in aqueous solution [153], but subsequent developments could colour hair deeply in a simple process using mild conditions and provided excellent fastness to light and shampooing. These include the azomethine yellow dye **71**, the azo orange **72a** and the azo red **72b** [97,98,154].

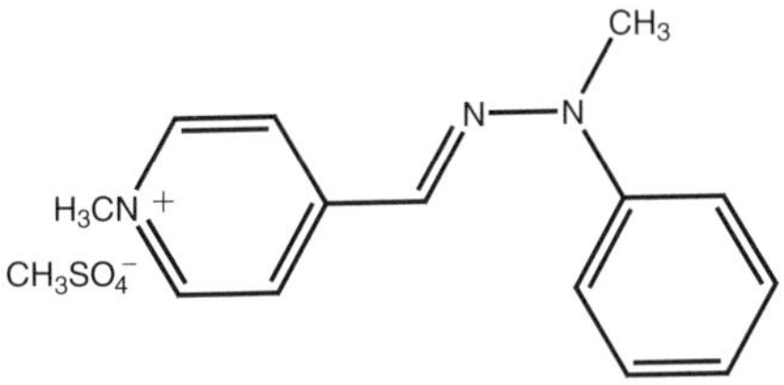

71

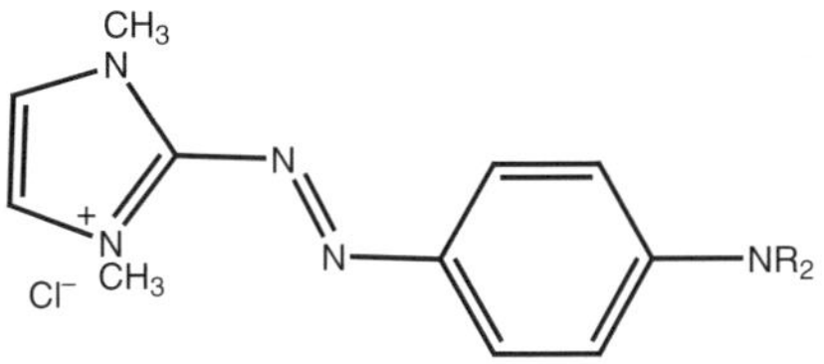

72

a: R=H

b: R=CH$_3$

Semipermanent dyes are usually applied to natural, unbleached hair after shampooing. After 10–30 minutes, the hair is rinsed. The formulations use mixtures of dyes, blended to the desired shade [13,155]. Browns, reds, and blacks are most popular, as the primary function is tone-on-tone colouring based on the range of natural hair colours. It is common to use several dyes of similar colour but with different molecular sizes, in order to provide even coloration. Larger dye molecules tend to be retained by the more damaged tip of the hair but do not penetrate so readily into the roots, while smaller molecules penetrate the entire hair fibre but wash out of the more porous tip. These differences are magnified on hair that is severely damaged. This feature is also important in minimising shade change during colour loss by shampooing. Dye selection requires a balance of physical characteristics that are often in conflict with one another, such as water solubility, partition coefficient and affinity for the fibre; these are determined largely by the substituents in the molecular structures. This feature explains why such a variety of substituent patterns is encountered, especially in commercial nitro dyes. Dye selection also requires consideration of the effect of other hair treatments, such as shampoo, fixative and permanent-wave lotions. It is desirable to avoid certain azo dyes which can be degraded by reducing agents in permanent-wave lotions, causing a change in colour and potentially leading to formation of allergenic diamines. It is also important to avoid dyes with acid–base pH indicator properties, which might produce undesirable colour changes. In addition to these technical features, cost is inevitably an important factor.

Semipermanent dyeing systems are typically supplied as lotions, although they are also marketed as aerosol foams, mousses and in shampoo form. The viscosity of the lotion is controlled with thickening agents, often used with surfactants; hair-conditioning agents may also be included. An important feature of semipermanent products is the ability to swell the hair fibre temporarily, in order to facilitate dye penetration under the mild application conditions. Most products incorporate an alkanolamine swelling agent, giving a pH in the range 8.5–9.5. These are selected carefully to avoid amine-exchange reactions and reduction of nitro groups, which might cause colour changes during storage. Ammonia is rarely used, due to its unpleasant odour and because many dyes are unstable in its presence. Water is the main solvent used in semipermanent formulations, although water-miscible organic solvents may be incorporated in order to promote dyeing by assisting diffusion into the fibre.

Temporary hair coloration, which is much older than semipermanent hair coloration, generally lasts only from one shampoo to the next. Originally, temporary hair dyes were aimed mainly at women with greying hair, offering masking or toning effects. The products, because they were applied to hair and then dried without rinsing, were often referred to as *colour rinses*. A current use of temporary hair dyeing is to produce a colour change for a single event; these are thus referred to as 'party' colours. They may also be used after permanent or semipermanent hair dyeing, or after bleaching, to provide an immediate shade correction. The dyes colour less intensely than permanent or semipermanent hair dyes, but they are simple and appealing in use. They involve minimal commitment or health risk.

Temporary hair-colouring products are formulated exclusively with dyes that have been approved by the US FDA for use in food, drugs and cosmetics (FD&C) or drugs and cosmetics (but not food) (D&C). Typical representative examples of dyes used are C.I. Acid Yellow 3 (Quinoline Yellow, C.I. Food Yellow 13), described in the Colour Index as a mixture of sulphonated derivatives **73a** and **73b**, C.I. Acid Orange 7 (**74**, Orange II), C.I. Food Red 1 (**75**), C.I. Acid Violet 43 (**76**), C.I. Acid Blue 9 (**77**; C.I. Food Blue 2) and C.I. Acid Green 25 (**78**). Their nontoxicity means that they do not require a preliminary patch test in order to check for allergenic response. Temporary hair colour formulation incorporates some apparently contradictory constraints: the colour should be removed easily by shampooing, but should be colour fast towards rain and perspiration; the dyes should have low substantivity for damaged hair yet achieve good coverage of white hair; the dyes should have sufficient resistance to friction so as to avoid staining clothing. Dyes **73–78** are anionic and belong to the acid application class. They show good water solubility and have little affinity for hair. Cationic surface-active agents may be used in conjunction with these dyes. Anion–cation complexes are formed, which increase affinity, decrease solubility and facilitate dye adsorption on to the hair fibre. The formation of complexes offers the additional advantage of colouring hair evenly, an effect rarely achieved in other hair-dyeing systems. Pigments may be used in temporary hair colour formulations, but the effect produced on hair is not particularly natural or appealing and suffers from flaking and removal by abrasion. Basic (cationic) dyes with higher substantivity have found a successful function as *colour refreshers*; that is, treatments to permanently and semipermanently dye hair which may be employed when the hair is washed. The mechanism of temporary hair dyeing is similar to that involved in the semipermanent process, except that, because of their larger molecular size, the dyes are deposited only on the surface of or marginally into the cuticle, a phenomenon usually referred to as 'ring dyeing' [156].

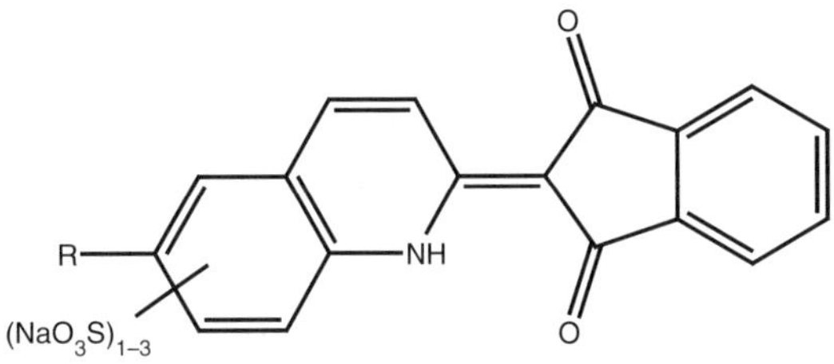

73: (a) R=H; (b) R=CH$_3$

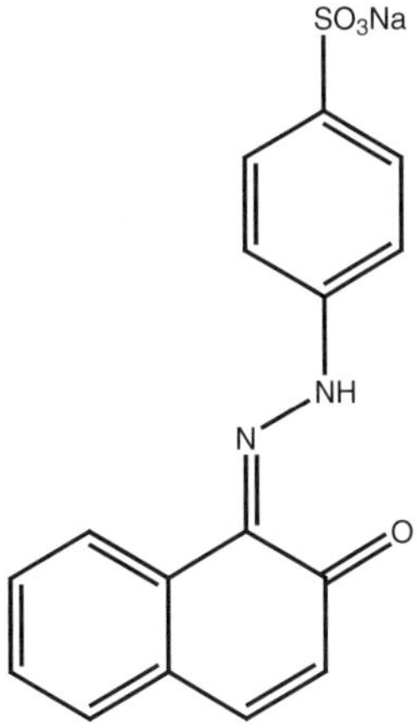

74

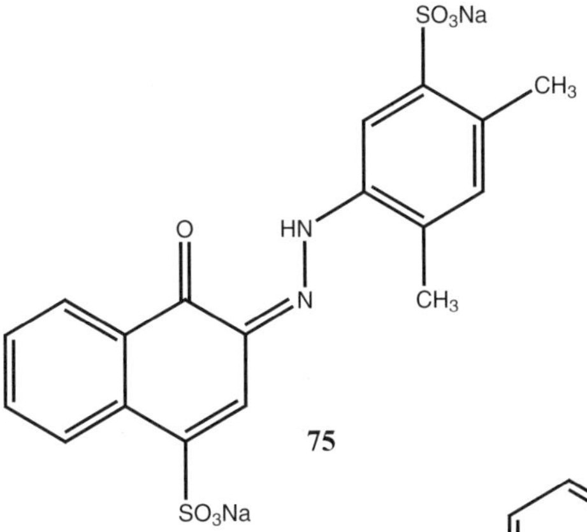

75

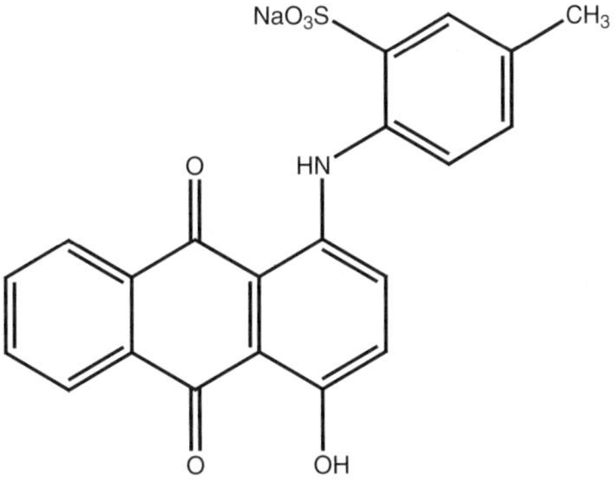

76

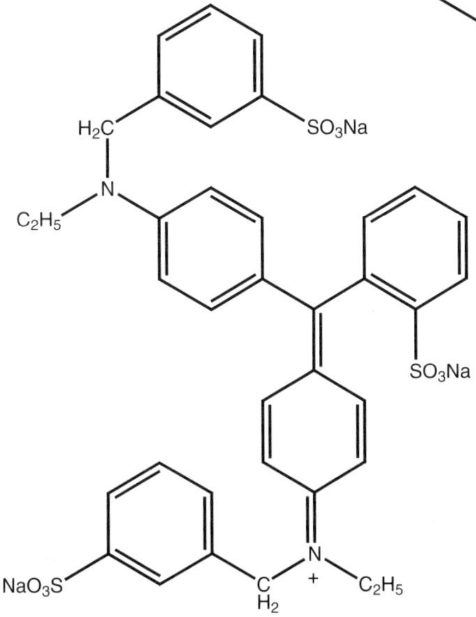

77

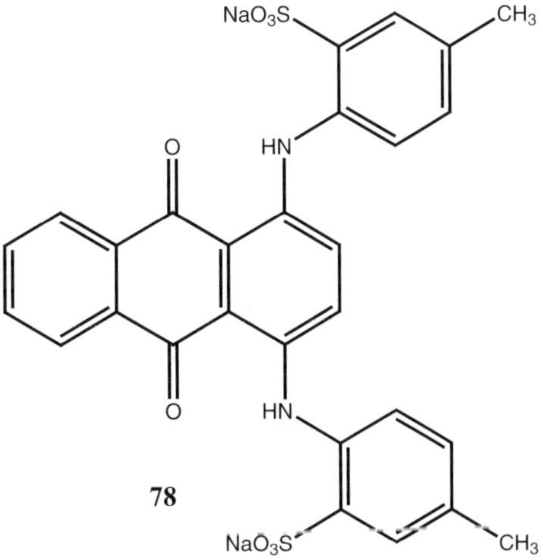

78

Temporary hair dyes are used in a variety of forms, including powders, capsules, rinses, lotions, shampoos, mousses, sprays and as coloured fixatives. The selection of formulation ingredients that facilitate application is often more important than selection of the dyes. The products may contain thickeners, surfactants, weak organic acids (to provide an appropriate acidic application medium) and water-miscible organic solvents (to increase the substantivity of the dyes for hair). Temporary dyeing is often carried out in association with hair styling, so the reagents for this purpose should be compatible with the colouring process. These coloured fixatives consist of dyes dispersed in plasticisers and incorporated into the styling aid, as the vehicle for the coloration. After application, usually in the form of a gel or aerosol mousse, the hair fibres become surrounded by a sheath of a coloured translucent film. Temporary hair dyes formulated as a coloured hair spray contain the dye and a polymer, which forms a coloured film after application to dry hair. This is a quick and easy way to add highlights to hair that has been styled, although often with nonuniform results.

11.9   Conclusion

Hair dye manufacture is currently a massive international industry, and remains substantially located in Western Europe and the USA, thus far resisting the movement of traditional dye and pigment manufacture towards China and India which has taken place in recent decades. Indeed, the emergence of these and other global economies has created growth in the demand for these products. The past few decades have witnessed a significant growth in our understanding of the chemical and physical properties of hair structure, of the mechanisms involved in the hair-dyeing process and of the interrelationships between these two features. Enhanced knowledge of the composition of the hair fibre, in particular the chemical structure of its outer boundary, and an emerging understanding of dye diffusion pathways and the factors influencing the process kinetics are assisting the development of tailored dye precursors and of new technologies commonly aimed at achieving more intense and durable permanent hair coloration. In addition, numerous studies have continued to address the perennial concerns over toxicological issues associated with the commercial products and processes, especially in permanent hair dyeing based on the oxidative process derived from Hofmann's 150-year-old discovery. There have been some interesting individual new developments, which have been commercialised, such as the dihydrazopyrazolone oxidation precursors and some cationic semipermanent dyes. Nevertheless, it seems likely that the permanent oxidative process, in an optimised form, and the current range of semipermanent dyes, as described throughout this chapter, will remain the dominant products into the foreseeable future. The expense associated with the approval process for new formulation ingredients, enforced by social pressures and legislation, is certain to limit the introduction of new dyes and precursors to those for which there is unequivocal evidence for a commercial investment return. However, there remains scope for research into new technologies for hair coloration. The rapidly developing science of genetics and an emerging understanding of the molecular basis of hair pigmentation may be key elements in the development of systems that encourage natural, biotechnological or semisynthetic hair repigmentation. While there has been some preliminary success in this area, delivering colouring agents to hair follicles [157–159], further development is essential before the technology is ready for the consumer. A growing understanding of the principles of the greying process, and the possibility of its inhibition or prevention, may well accelerate this development [160]. The explosive growth in nanotechnologies is certain to continue to attract the interest of hair colour chemists, who will want for example to exploit the potential of photonics to make use of materials which manipulate light to create bright colours without using traditional colouring materials, as one way of addressing potential environmental and toxicological concerns.

References

[1] R H Thompson, *Naturally Occuring Quinones* (New York, USA: Academic Press, 1957).
[2] Global Industry Analysts, *Hair Care Products* (Global Industry Analysts, 2008).
[3] C Chavigny, *Parfums Cosmétiques Actualités*, **180** (2004) 54.
[4] J F Corbett, in *The Chemistry of Synthetic Dyes*, **Vol. 5**, ed. K Venkataraman (New York, USA: Academic Press, 1971), 475.
[5] C K Dweck, *International Journal of Cosmetic Science*, **24** (2002) 287.
[6] J F Corbett, *Hair Colorants: Chemistry and Toxicology* (Weymouth, UK: Micelle Press, 1998).
[7] C Zviak, *The Science of Hair Care* (New York, USA and Basel, Switzerland: Marcel Dekker, 1986).
[8] J F Corbett, in *Colorants for Non-textile Applications*, ed. H S Freeman, A T Peters (Amsterdam, The Netherlands: Elsevier Science BV, 2000), 456.
[9] K C Brown, *Cosmetic Science and Technology*, **17** (1997) 191.
[10] J F Corbett, *Journal of the Society of Dyers and Colourists*, **92** (1976) 285.
[11] J F Corbett, *Review of Progress in Coloration and Related Topics*, **15** (1985) 52.
[12] J F Corbett, *Review of Progress in Coloration and Related Topics*, **4** (1973) 3.
[13] J S Anderson, *Journal of the Society of Dyers and Colourists*, **116** (2000) 193.
[14] O J X Morel and R M Christie, *Chemical Reviews*, **111** (2011) 2537.
[15] W Montagna and R A Ellis, *The Biology of Hair Growth* (New York, USA: Academic Press, 1958).
[16] J A Swift, *Journal of Cosmetic Science*, **50** (1999) 23.
[17] J A Swift, *Int. Journal of Cosmetic Science*, **13** (1991) 143.
[18] M S C Birbeck and E H Mercer, *Journal of Biophysical and Biochemical Cytology*, **3** (1957) 223.
[19] M S C Birbeck and E H Mercer, *Journal of Biophysical and Biochemical Cytology*, **3** (1957) 203.
[20] M S C Birbeck and E H Mercer, *Journal of Biophysical and Biochemical Cytology*, **3** (1957) 215.
[21] J D Leeder and J A Rippon, *Journal of the Society of Dyers and Colourists*, **101** (1985) 11.
[22] J D Leeder, *Wool Science Review*, **63** (1986) 3.
[23] A P Negri, H J Cornell and D E Rivett, *Textile Research Journal*, **63** (1993) 109.
[24] A P Negri, D A Rankin, W G Nelson and D E Rivett, *Textile Research Journal*, **66** (1996) 491.

[25] L N Jones and D E Rivett, *Micron*, **28** (1997) 469.

[26] L N Jones, *Biochimica et Biophysica Acta*, **412** (1975) 91.

[27] L N Jones, *Biochimica et Biophysica Acta*, **446** (1976) 515.

[28] J A Swift and A W Holmes, *Textile Research Journal*, **35** (1965) 1014.

[29] M Feughelman, *Mechanical Properties and Structure of Alpa-Keratin Fibres*, 1st Edn (Sydney, Australia: UNSW Press, 1997).

[30] B Rigg, in *Colour Physics for Industry*, 2nd Edn, ed. R McDonald (Bradford, UK: Society of Dyers and Colourists, 1997), 95.

[31] F Gjesdal, *Acta Pathologica Microbiologica Scandinavica*, **47** (1959) 1.

[32] P A Riley, in *The Physiology and Pathophysiology of Skin*, **Vol. 3**, ed. A Jarret (London, UK: Academic Press, 1974), 1102.

[33] J A Swift, *Chemistry of Natural Protein Fibres* (New York, USA and London, UK: Plenum Press, 1977).

[34] L J Wolfram and L Albrecht, *Journal of the Society of Cosmetic Chemists*, **38** (1987) 179.

[35] C E Orfanos and H Ruska, *Archiv fr Klinische und Experimentelle Dermatologie*, **231** (1968) 279.

[36] I Castanet and J P Ortonne, in *Formation and Structure of Human Hair*, ed. P Jollès, H Zahn and H Höcker (Basel, Switzerland: Birkhäuser Verlag, 1997), 209.

[37] G Prota, M Scherillo and R A Nicolaus, *Rendiconto dell'Accademia delle Scienze Fisiche e Matematiche* (Napoli, Italy: Sezione della Società Reale di Napoli, 1968), 2.

[38] G Prota, *Journal of Investigative Dermatology*, **75** (1980) 122.

[39] H S Mason, in *Advances in Biology of Skin: The Pigmentary System*, **Vol. 8**, ed. W Montagna and F Hu (New York, USA: Pergamon, 1966), 293.

[40] H S Mason, *Journal of Biological Chemistry*, **172** (1948) 83.

[41] H S Mason, *Advances in Enzymology*, **14** (1955) 105.

[42] H S Raper, *Journal of the Chemical Society*, (1938) 125.

[43] H S Raper XVI, *Biochemical Journal*, **21** (1927) 89.

[44] R A Nicolaus, M Piatelli and E Fattorusso, *Tetrahedron*, **20** (1964) 1163.

[45] M Piatelli and R A Nicolaus, *Tetrahedron*, **15** (1961) 66.

[46] R A Nicolaus, *Rassegna di Medicina Sperimentale*, **IX** (1962) 1.

[47] G A Swan, *Zeitschrift fr Naturforschung B*, (1976) 152.

[48] S Ito, *Biochimica et Biophysica Acta*, **883** (1986) 155.

[49] G Prota, *Melanins and Melanogenesis* (San Diego, USA: Academic Press, 1992).

[50] G Prota, S Crescenzi, G Misuraca and R A Nicolaus, *Experientia*, **26** (1970) 1058.

[51] S Crescenzi, G Misucara, E Novellino and G Prota, *Chimica e Industria*, **57** (1975) 392.

[52] S Ito and G Prota, *Experientia*, **33** (1977) 1118.

[53] H Rorsman, G Agrup, C Hansson, A-M Rosengren and E Rosengren, in *Pigment Cell*, **Vol. 4**, ed. S N Klaus (Basel, Switzerland: Karger, 1979), 244.

[54] S Ito, E Novellino, F Chioccara, G Misucara and G Prota, *Experientia*, **36** (1980) 822.

[55] C L Gummer, *Journal of Cosmetic Science*, **52** (2001) 265.

[56] L Kreplak, C Merigoux, F Briki, D Flot and J Doucet, *Biochimica et Biophysica Acta, Protein Struct. Mol. Enzymol.*, **1547** (2001) 268.

[57] A Kelch, S Wessel, T Will, U Hintze, R Wepf and R Wiesendanger, *Journal of Microscopy*, **200** (2001) 179.

[58] C R Robbins, *Chemical and Physical Behavior of Human Hair*, 4th Edn (New York, USA: Springer-Verlag, 2002), 249.

[59] V A Wilkerson, *Journal of Biological Chemistry*, **112** (1935) 329.

[60] D L Underwood, *Journal of the Society of Cosmetic Chemists*, **12** (1961) 155.

[61] J Speakman, *Proceedings of the Royal Society, London*, **132** (1931) 167.

[62] H Wilmsmann, *Journal of the Society of Cosmetic Chemists*, **12** (1961) 490.

[63] A W Holmes, *Journal of the Society of Cosmetic Chemists*, **15** (1964) 595.

[64] M Sakai, S Nagase, T Okada, N Satoh and K Tsujii, *Bulletin of the Chemical Society of Japan*, **73** (2000) 2169.

[65] O Morel, R M Christie, A Greaves and K M Morgan, *Coloration Technology*, **124** (2008) 301.

[66] C Flower, *Chemist and Druggist*, **July** (2002) 27.

[67] J F Corbett, *Dyes and Pigments*, **41** (1999) 127.

[68] M Krasteva, A Cristaudo, B Hall, D Orton, E Rudzki, B Santucci, H Toutain and J Wilkinson, *European Journal of Dermatology*, **12** (2002) 322.

[69] S de Sanjosé, Y Benavente, A Nieters, L Foretova, M Maynadié, P L Cocco, A Staines, M Vornanen, P Bofetta, N Becker, T Alvaro and P Brennan, *American Journal of Epidemiology*, **164** (2006) 47.

[70] Y Zhang, S de Sanjosé, P M Bracci, L M Morton, P Brennan, P Hartge, P Bofetta, N Becker, M Maynadié, L Foretova, P L Cocco, A Staines, T Holford, E A Holly, A Nieters, Y Benavente, L Bernstein, S H Zahm and T Zheng, *American Journal of Epidemiology*, **167** (2008) 1321.

[71] G J Nohynek, R Fautz, F Benech-Kieffer and H Toutain, *Food and Chemical Toxicology*, **42** (2004) 517.

[72] Council directive 76/768/EEC in *Official Journal of the European Communities*; *L 262*, 27September 1976.

[73] Cosmetic Ingredient Review, http://www.cir-safety.org/, last accessed 1 February 2013.

[74] A W Hofmann, *Jahr. Chem.*, (1863) 42.

[75] P Monnet, French patent 158558 (1883).

[76] J F Corbett, *Journal of the Chemical Society*, **B** (1969) 207.

[77] J F Corbett, *Journal of the Chemical Society*, **B** (1969) 213.

[78] K C Brown and J F Corbett, *Journal of the Society of Cosmetic Chemists*, **30** (1979) 191.

[79] G M Wis-Surel, *Int. Journal of Cosmetic Science*, **21** (1999) 327.

[80] L Vidal and A Fadli, US patent 7285137 (2007).

[81] J-B Saunier, US patent 7485156 (2009).

[82] J-B Saunier, US patent 7488355 (2009).

[83] J-B Saunier, US Patent 7488356 (2009).

[84] L Hercouet, US patent 7582121 (2009).

[85] A C Chan, Y G Pan and D L Chang, US patent 5139532 (1992).

[86] A C Chan, Y G Pan and D L Chang, US patent 5198584 (1993).

[87] A R Genet and A Lagrange, US patent 6340371 (2002).

[88] A R Genet and A Lagrange, US patent 6497730 (2002).

[89] J B Bartolone, P Prem and K Jacobs-Dube, US patent 6994733 (2006).

[90] E Zeffren and J F Sullivan, US patent 3957424 (1976).

[91] A Greaves, N Daubresse and X Radisson, US patent 7303589 (2007).

[92] J F Corbett, *Journal of the Chemical Society*, **C** (1970) 2101.

[93] H Musso, *Planta Medica*, **8** (1960) 431.

[94] P Berth and R Maul, US patent 3582253 (1971).

[95] A Bühler, A Fasciati and W Hungerbühler, US patent 4162893 (1979).

[96] A Bühler, A Fasciati and W Hungerbühler, US patent 4165967 (1979).

[97] P Möckli, US patent 5733343 (1998).

[98] P Möckli, US patent 6843256 (2005).

[99] M Bader and C Sunder, US patent 1448251 (1923).

[100] D M Lewis, US patent 5364415 (1994).

[101] G Kalopissis, A Bugaut and F Estradier, US patent 4054147 (1977).

[102] G Kalopissis, A Bugaut and F Estradier, US patent 4042627 (1977).

[103] G Kalopissis, A Bugaut and F Estradier, US patent 4008999 (1977).

[104] G Kalopissis, A Bugaut and F Estradier, US patent 4008043 (1977).

[105] G Kalopissis, A Bugaut and F Estradier, US patent 4288622 (1981).

[106] G Kalopissis, A Bugaut and F Estradier, US patent 4233241 (1980).

[107] G Kalopissis, A Bugaut and F Estradier, US patent 4222958 (1980).

[108] G Kalopissis, A Bugaut and F Estradier, US patent 4112229 (1978).

[109] K Taguchi, T Tokano, Y Yamaoka and K Furuse, US patent 6849096 (2005).

[110] K Taguchi, T Tokano, Y Yamaoka and K Furuse, US patent 6656229 (2003).

[111] A Iqbal, A Bize, P Bujard, H Dubas, A Hafner, V Hall-Goulle, Z Hao, G A Johnson, G De Keyzer, B Maire, U Schädeli, E Tinguely, H Wolleb and J Zambounis, *COLORCHEM'98, Rybitví, Czech Republic* (1998), L29.

[112] Z Hao, A Iqbal, H Dubas, U Schädeli and J Zambounis, in *Colour Science '98, Vol. I: Dye and Pigment Chemistry*, 1st Edn, ed. J Griffiths (Leeds, UK: University of Leeds, 1998).

[113] J S Zambounis, Z Hao and A Iqbal, *Nature*, **388** (1997) 131.

[114] H-T Schacht and G Moegle, US patent 6495250 (2002).

[115] J Zambounis, P Verhoustraeten, H Dubas, Z Hao and P Bujard, US patent 6767620 (2004).

[116] J-I Setsune, H Wakemoti, T Matsueda, T Matsuura, H Tajima and T Kitao, *Journal of the Chemical Society, Perkin Transactions I*, (1984) 2305.

[117] T Kitao, J Setsune, S Ishihara and R Yamamoto, EU patent 0085392 (1983).

[118] A Greaves, S Kratchenko and A Lagrange, EP patent 1426036 (2003).

[119] A Lagrange, S Kratchenko and A Greaves, US patent 7326255 (2008).

[120] P Walter, E Welcomme, P Hallégot, N J Zaluzec, C Deeb, J Castaing, P Veyssiere, R Bréniaux, J-L Léveque and G Tsoucaris, *Nano Letters*, **6** (2006) 2215.

[121] X Huang, R K Kobos and G Xu, US patent 7276088 (2007).

[122] I D Rattee, in *The Chemistry of Synthetic Dyes*, **Vol. VIII**, ed. K Venkataraman (New York, USA: Academic Press, 1972), 1.

[123] W E Beech, *Fiber-reactive Dyes* (London, UK: Logos-Press, 1970).

[124] K Carr, in *Advances in Color Chemistry Series*, **Vol. 3**, ed. A T Peters and H S Freeman (London, UK: Blackie, 1995), 87.

[125] A Hunter and A H M Renfrew, *Reactive Dyes for Textile Fibers* (Bradford, UK: Society of Dyers and Colourists, 1999).

[126] A H M Renfrew and J A Taylor, *Review of Progress in Coloration and Related Topics*, **20** (1990) 1.

[127] P Rys and H Zollinger, in *The Theory of Coloration*, ed. A Johnson (Bradford, UK: Dyers' Company Publication Trust, 1989), 428.

[128] E Siegel, D Schündehütte and D Hildebrand, in *The Chemistry of Synthetic Dyes*, **Vol. 6**, ed. K Venkataraman (New York, USA: Academic Pressk, 1972), 327.

[129] J A Taylor, *Review of Progress in Coloration and Related Topics*, **30** (2000) 93.

[130] I D Rattee, *Review of Progress in Coloration and Related Topics*, **14** (1984) 50.

[131] A Shansky, *American Perfumer and Cosmetics*, **81** (1966) 23.

[132] J C Brown, *Journal of the Society of Cosmetic Chemists*, **18** (1967) 225.

[133] R E Randebrock, US patent 3415606 (1968).

[134] S M Peck, US patent 2539202 (1951).

[135] P F Rosmarin and M Pantzer, US patent 2875769 (1959).

[136] M Pantzer and M Feier, US patent 3698852 (1965).

[137] R Charle and C Pigerol, US patent 2934396 (1960).

[138] J F Grollier and D Garoche, US patent 4804385 (1989).

[139] J F Grollier and D Garoche, US patent 4808190 (1989).

[140] J F Grollier and D Garoche, US patent 4888027 (1989).

[141] R J Beer, H G Khorana and A Robertson, *Journal of the Chemical Society* (1948) 2223.

[142] J R Seemuller, R Charle and C Pigerol, US patent 3194734 (1965).

[143] G Prota and G Wenke, US patent 5374288 (1994).

[144] R Charle and G Sag, *Manufacturing Chemist and Aerosol News* (1967) 33.

[145] R Raue and J F Corbett, in *Ullmann's Encyclopedia of Industrial Chemistry* (Weinheim, Germany: Wiley-VCH Verlag GmbH, 2002).

[146] S K Han, Y K Kamath and H-D Weigmann, *Proceedings of the 7th International Wool Textile Research Conference, Tokyo* (1985) 269.

[147] M Wong, *Journal of the Society of Cosmetic Chemists*, **23** (1962) 165.

[148] S E Rae, MSc Dissertation, Heriot-Watt University (1992).

[149] E Konrad, T Clausen, H-J Braun and H Mager, US patent 4835314 (1989).

[150] R M Christie, *Colour Chemistry* (Cambridge, UK: Royal Society of Chemistry, 2001).

[151] H-S Bien, J Stawitz and K Wunderlich, in *Ullmann's Encyclopedia of Industrial Chemistry* (Weinheim, Germany: Wiley-VCH Verlag GmbH, 2002).

[152] H Zollinger, *Color Chemistry*, 3rd Edn (Zurich, Switzerland and Weinheim, Germany: Wiley-VCH & HCA, 2003).

[153] P Möckli, US patent 5708151 (1998).

[154] P Möckli, US patent 5888252 (1999).

[155] K C Brown, *Cosmetic Science and Technology Series*, **17** (1997) 191.

[156] J T Guthrie, A Kazlauciunas, L Rongong and S Rush, *Dyes and Pigments* (1995) 23.

[157] L Li and V Lishko, US patent 5965157 (1999).

[158] L Li and V K Lishko, US patent 5641508 (1997).

[159] M Zhao, US patent 6372489 (2002).

[160] J M Wood, H Decker, H Hartmann, B Chavan, H Rokos, J D Spencer, S Hasse, M J Thornton, M Shalbaf, R Paus and K U Schallreuter, *FASEB Journal*, **23** (2009) 2065.

12

Wool Printing[1]

Peter J. Broadbent[1] and Muriel L.A. Rigout[2]
[1] Colour Chemistry Consultant, UK
[2] School of Materials, University of Manchester, UK

12.1   Introduction

The following chapter will focus on the printing of wool fabrics and will provide an updated review of the accomplished article so meticulously written by Bell [1]. It is not the authors' intention to rewrite this article, but to update and make amendments to the original as required. In addition, this chapter will also review the more recent advances in wool printing, such as the development of the digital printing process, that have taken place over the past 2 decades.

Traditionally, the proportion of wool printed, relative to production, is low in comparison to cotton. The reasons for this are almost certainly commercial rather than technical: factors such as base fabric cost, sampling costs and shorter run lengths all combine to make printed wool an expensive product. On the other hand, other factors should in principle make wool printing today a more attractive proposition than before: reliable methods of preparation for printing remove much of the uncertainty associated with this stage of the processing; the increasing popularity of 'green' policies has resulted in a trend towards natural fibres; and increased legislation on the flammability of textiles has renewed interest in printed wool furnishing fabrics.

---

[1] Some information in this chapter is based on that published in Chapter 10 of *Wool Dyeing*, ed. David M. Lewis, published SDC, 1992 and is used with permission from the Society of Dyers and Colourists.

The Coloration of Wool and other Keratin Fibres, First Edition. Edited by David M. Lewis and John A. Rippon.
© 2013 SDC (Society of Dyers and Colourists). Published 2013 by John Wiley & Sons, Ltd.

12.2 Preparation for Printing

Whereas a good pretreatment of any textile fabric destined for printing is generally considered desirable (e.g. the degumming of silk or the mercerisation of cotton), in the case of wool it is absolutely essential; it can never be emphasised too strongly that without efficient preparation of the goods prior to printing, full colour yields, levelness and brightness will not be achieved. Although oxidative bleaching can lead to a slight improvement, the resulting print will still exhibit poor yields and skitteriness [2]. Before discussing the methods currently available for preparation for printing, however, the processing steps that are necessary beforehand should be considered. These are described in considerable detail by Heiz [3], but may be summarised as follows:

- *Setting:* Depending on the quality of the goods, it is usually necessary to relax and stabilise the cloth before scouring; this is generally carried out by crabbing, either discontinuously, or continuously on the Konticrab (Hemmer). Alternatively, open-width scouring may be used to impart a degree of set to the cloth. The plain-weave challis and the worsted twill, both widely used as print substrates, are good examples of the kinds of cloth that require setting prior to scouring.
- *Scouring:* The objective of scouring is to remove processing oils, dirt and, with many lightweight fabrics, size, where sizing of the warp has been necessary. This can involve the use of enzymes where starch-based sizes have been employed, although nowadays there is a more widespread use of sizes based on poly(vinyl alcohol) (PVA). Efficient scouring is of course essential for level application of the subsequent preparation for printing. Both continuous and batchwise scouring methods are used.
- *Milling:* Depending again on the quality of the goods, a light milling is sometimes employed to impart the desired finish to the cloth.

Worsted finishing techniques, together with the machinery involved, are discussed by Bearpark, Marriott and Park [4], though certain cloths such as single jersey require special handling in their preparation and finishing.

12.2.1 Oxidative Processes

Chlorination, the traditional method of preparing wool for printing, has been used for decades. Only in isolated cases may it, or alternative preparation methods, be dispensed with. Chlorination, like other oxidative procedures, also imparts a degree of dimensional stability to the fabric, which is important for retention of print definition (particularly with fine-line designs) during washing-off of the printed goods; any milling or distortion of the fabric surfaces must be avoided.

All the most commonly used procedures for preparing wool for printing are based on chlorination. Other oxidative processes, such as treatment with permonosulphuric acid [5], have also been used in the past.

The reaction of chlorine with the wool fibre is markedly pH-dependent: under strongly acid conditions, a rapid reaction takes place, accompanied by little fibre yellowing; under alkaline conditions, the reaction is slower and the yellowing is more significant. Chlorination of wool for printing is normally carried out under acid conditions and in order to understand its effect, the chemical reactions involved should be considered.

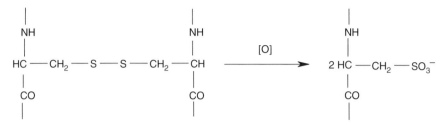

Scheme 12.1 *Oxidation of cystine linkages to form cysteic acid residues.*

The literature relating to the chemistry of wool chlorination is vast and has been comprehensively reviewed [6–9]. Bell and Lewis [10] have summarised the primary reactions, the most important of which are:

- *Cystine oxidation:* Cystine residues are rapidly oxidised to cysteic acid residues (Scheme 12.1).

- *Peptide bond cleavage:* Chlorine cleaves peptides and proteins at tyrosine residues. (Scheme 12.2).

Acid chlorination, therefore, performs two functions:

- It creates a number of strongly anionic groups, such as RSO_3^-, $RCOO^-$.
- It breaks disulphide and peptide crosslinks, making the wool more readily accessible.

These two factors of increased polarity and accessibility result in a fibre surface that is both highly attractive to and readily swollen by water. Furthermore, significant amounts of the anionic peptides at the wool surface are water-soluble, and wool protein material may actually be dissolved from the surface; under extreme chlorination conditions, the outer scales of the fibre may be completely removed (a detailed description of these physical changes is given by Makinson [11]). Prechlorination thus results in a much more hydrophilic fibre surface [12], allowing even distribution of print paste.

Wool chlorination for printing was traditionally carried out using a solution of sodium hypochlorite, the pH being adjusted to 1.5–2.0 by continuously metering in hydrochloric or sulphuric acid. Although this was practised until the end of the 1940s [13], it is now seldom used.

12.2.1.1 Chlorination with Dichloroisocyanuric Acid

The replacement of acid/hypochlorite chlorination by the use of the sodium salt of dichloroisocyanuric acid (DCCA) represented a major advance in chlorination technology. This compound acts as a chlorine donor, releasing a chlorinating species in a controllable manner. The advantages of DCCA over the older acid/hypochlorite methods have been

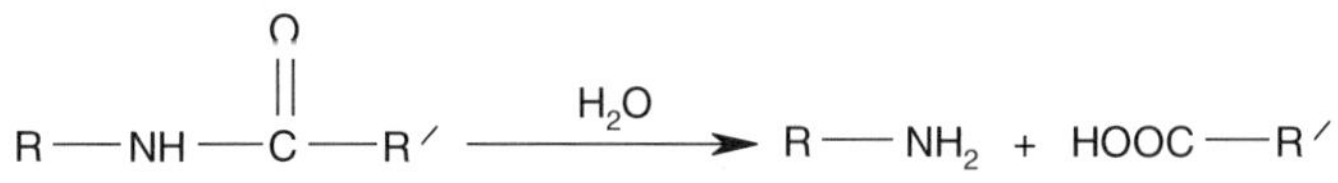

$$R-NH-\overset{\overset{\displaystyle O}{\|}}{C}-R' \xrightarrow{\ \ H_2O\ \ } R-NH_2 + HOOC-R'$$

Scheme 12.2 *Peptide bond cleavage.*

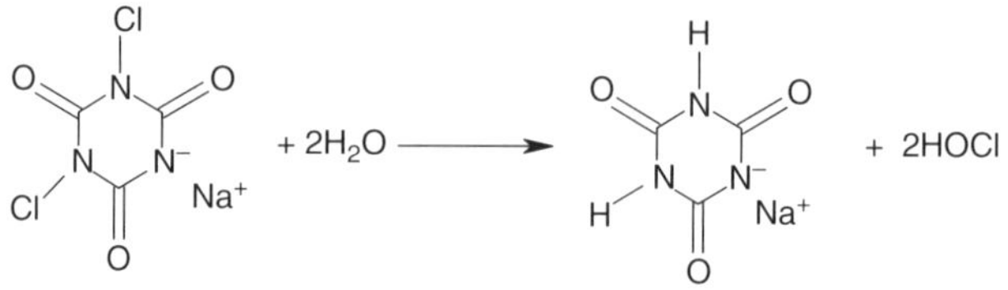

Scheme 12.3 *Hydrolysis of dichloroisocyanuric acid in aqueous solution to form hypochlorous acid.*

described by Reincke [2,14]. Commercial examples of this product include Basolan DC (BASF) and Fi-Chlor Clearon (Chlor Chem Ltd), both of which are soluble salts containing 60% of available chlorine. Hydrolysis of this compound in aqueous solution results in production of hypochlorous acid (Scheme 12.3). Hypochlorous acid is one of three reactive species present in aqueous chlorinating solutions: chlorine, hypochlorous acid and hypochlorite anions; these all exist in equilibrium, the predominant species depending on solution pH [6].

12.2.1.2   *Kroy Technology*

Kroy gaseous chlorination technology [17] was jointly developed by Kroy Unshrinkable Wools Ltd of Canada and the International Wool Secretariat Development Centre in Ilkley, UK. This technology, originally introduced for the chlorination of wool tops, has been adapted for the continuous chlorination of fabric as a prepare for print [18].

Chlorine gas, supplied by cylinders of liquid chlorine, and water are precisely metered using a special injector system. The chlorine is fed to the machine in amounts calculated on the basis of throughput and the feed is manually controlled directly from the chlorinator unit. The chlorine gas dissolves in water, resulting in a mixture of hypochlorous and hydrochloric acid (Scheme 12.4).

12.2.2   Polymer Treatments

The use of shrink-resist resins provides an alternative, but much less commonly used, approach to the problem of preparation for printing. There are two polymer treatment routes to consider: top treatment and fabric treatment.

12.2.2.1   *Top Treatment: Chlorination–Hercosett*

Continuous chlorination–Hercosett systems for the shrink-resist treatment of wool tops are well established and currently operating in many plants throughout the world. Hercosett is described as a water-soluble, cationic, crosslinking polyamide-epichlorohydrin polymer, and the chemistry, mechanism and operation of the process have been reviewed in detail [19–21]. Residual azetidinium cations and protonated tertiary amine sites in the polymer impart to the treated wool an increased affinity for anionic dyes, and this, together with

$$Cl_2 + H_2O \longrightarrow HOCl + HCl$$

Scheme 12.4 *Formation of hypochlorous and hydrochloric acids from chlorine gas.*

the prechlorination (albeit at a reduced level), provides a good preparation for printing. Furthermore, since considerably more processing takes place before the top becomes a woven cloth, any unevenness in treatment is blended out and a consistent, level prepare is achieved. However, both these factors can act disadvantageously. Due to the cationic nature of the polymer, care should be exercised in washing off the printed goods to avoid staining of white grounds, since the treated wool exhibits a high substantivity for any loose anionic dye in the wash liquor. Moreover, since a reduced level of chlorination is employed, the cloth wets out less readily, resulting in rather less print penetration.

This method of preparation for printing is encountered only occasionally – notably, in the printing of single jersey fabrics, where machine washability is required.

12.2.2.2   Fabric Treatments

Synthetic polymers have been developed which impart machine washability to wool fabrics without the necessity for a prechlorination step. It has been demonstrated that the water-soluble polymers Securlana (Cognis), formerly called Nopcolan SHR3, (Henkel) and Synthappret BAP (BAY) exhibit a high substantivity for the more hydrophobic types of dye, and this property may be put to advantage by using the treatment as a preparation for printing [3,10,21,22]. It has been shown that polymer application by a pad–batch (cold)/wash-off procedure imparts a much softer handle than the more usual pad–dry/heat-cure methods [23].

As no chlorination is involved, the fabric is not so easily wettable, resulting in reduced penetration, and steaming *must* be carried out under optimum conditions. The system is restricted to the use of selected acid milling and metal-complex dyes; poor yields are obtained with reactive dyes and dyes of large molecular size. This precludes, for example, the production of shades such as very bright turquoise and emerald green, based on copper phthalocyanine dyes. There are some advantages over chlorination procedures, however, such as reduced yellowing during steaming [21] and easier removal of unfixed dye during washing-off [24].

Although this method has been used commercially for the preparation of wool/cotton blends for dyeing and printing, it has failed to find commercial application for the preparation of 100% wool fabrics for printing.

12.2.3   Plasma Treatments

Since the 1970s, a number of researchers have investigated the possibility of using plasma treatments to modify the physical properties of wool fibres, and this approach still attracts much interest as a potential alternative to chlorination. Plasma processes offer many advantages over the traditional wet chemical processes employed to modify the physical properties of wool fibres [25]:

- They are environmentally friendly, as they do not use any of the reagents or solvents used in conventional wet-processing techniques and so do not produce high volumes of waste or toxic byproducts.
- They are uniform and reproducible.
- They use a variety of nonpolymerising reagent gases.
- They allow selective modification of the fibre surface with minimal bulk property change.

Plasma technology is a surface-specific process that results from the interaction of a plasma with the surface of a fibre to produce modifications in the outer part of the fibre to a depth of 5 nm, depending on the reagent gas employed [26]. A plasma is a material, usually a gas, in which a certain proportion of the particles are ionised to form charged species. A plasma can be composed of positive ions, free electrons, photons, neutral atoms and molecules in ground and excited states [27]. In low-temperature plasma-treatment processes, ionisation of the reagent gas to form a plasma is brought about using an electrical-discharge system, such as corona or dielectric barrier discharge. The wool substrate is placed in a reaction chamber containing the plasma generated from the ionisation of a high-flow feed gas; the feed gas consists mainly of an inert carrier gas, such as helium, and a small amount of a nonpolymerising plasma-generating gas, such as oxygen, nitrogen or air. Processing times can vary between 30 seconds and 30 minutes depending on the reagent gas employed and the degree of surface modification required. The efficiency of the plasma-treatment process is reliant on [28]:

- the nature of the gas employed;
- system pressure;
- discharge power;
- processing time;

Plasma-treatment processes can impart many desirable properties to the wool substrate, such as antifelting effects, degreasing, improved dye uptake and increased wettability through changes brought about in the fibre surface morphology. The underlying mechanisms for these changes are not yet fully understood, due to the varied procedures and treatment conditions that have been employed by different research groups. The theories put forward for the observed changes in the physical characteristics of the modified wool substrate after plasma treatment are related to fibre surface modifications such as erosion, etching, the creation of microcraters, the rounding of scale edges and ablation, which causes holes in the fibre surface [29]. Naebe *et al.* [30] studied the uptake of sulphonated dyes on wool that had been plasma treated with helium for 30 seconds and concluded that the improvement in dye uptake arose from the removal of the covalently bound lipids of the F-layer from the fibre surface during the plasma-treatment process, revealing the underlying hydrophilic protein layer. In addition, the plasma-treatment process also results in the oxidation of cystine disulphide bonds in the exposed protein layer to form cysteic acid residues and thus a further increase in the hydrophilic character of the wool fibre.

The Technoplasma technology was developed in Switzerland and is used to pretreat fabric in open width before printing (as an absorbable organohalogen (AOX)-free alternative to chlorination); a similar technique has been used in Russia for a number of years [31].

12.2.4   Other Methods of Preparation for Printing

Alternative methods of preparing wool for printing have been developed over the years, in order to avoid the need for chlorination. Lewis *et al.* [21] have described a 'sulphitolysis' process, which used sodium sulphite and an anionic surfactant. Brady, Rivett and Stapleton [32,33] have investigated the use of tetraethylenepentamine (TEP) as a means of increasing colour yields when printing with selected milling or metal-complex dyes. More recently,

Perachem has patented a process that very effectively removes surface lipids to produce a white wool which does not show the yellowing problems normally associated with the traditionally employed chlorination processes [34]. This process involves the continuous or batch treatment of the wool substrate with a peroxide/surfactant system. Typical treatment times are 10–15 seconds for continuous lines and 5–10 minutes for batch processes. In addition, this delipidisation process eliminates the environmental concerns arising from the levels of AOX residues, up to 1000 ppm, retained by the substrate after the conventional chlorination processes [35].

12.3 Direct Printing

Direct printing, or 'print-on', represents the most widely used and the most straightforward printing style used on wool today. Printing is carried out on goods that are:

- often bleached, where large areas of white ground are to remain;
- sometimes dyed to a pale background;
- sometimes even dyed to a medium shade; these are overprinted with deep colours to obtain subtle shadow effects.

12.3.1 Machinery

Developments in textile machinery are really outside the scope of this chapter. Suffice to say that the choice of machinery for wool printing is partially governed by run length and, to a lesser extent, by design, but more by the machines available in the printworks, where wool printing may form only a small part of the production.

Although rotary screen printing is sometimes chosen due to design considerations (e.g. warp-wise stripes or very open blotch areas), automatic flat screen printing remains the most popular method, due to the shorter run lengths processed on wool. The relative merits of flat and rotary screen printing have been enumerated in several publications [36–38].

12.3.2 Dye Selection and Print Recipes

Dyes for wool printing are usually selected according to the following criteria:

- price;
- brilliance of shade;
- good coverage properties;
- good solubility and print paste stability;
- good build-up properties;
- good washing-off properties;
- satisfactory wet and light fastness.

All these criteria are important to both printer and customer. Sometimes the satisfaction of one can be achieved only at the expense of another; for example, some very brilliant shades can only be attained by using dyes with poor fastness properties.

12.3.2.1   Nonreactive Dyes

The most widely used palettes of dyes are the acid milling and metal-complex types. These are also the ranges most favoured by the wool printer; perhaps the most important reason is the possibility of achieving most of the required print colours with two-component dye mixtures [13] (for reasons which will be explained later).

The development of metal-complex dyes over the last few decades has been reviewed by Beffa and Back [39] and by Lewis [40]. The sulphonated 1:2 metal-complex dyes offer the advantages of high tinctorial value, economy and fastness, which, coupled with their good solubility, make them a highly suitable choice for the printer requiring full, if not particularly bright, shades. The requirement for bright colours is satisfied for the most part by the acid milling dyes, many of which also possess good fastness properties. A wool printer's palette will comprise both types, and the major dye manufacturers offer a selection of these dyes as being particularly suitable for wool printing; these are listed in detail elsewhere [16], but, in summary, are taken from the following ranges:

- Acidol, Acidol M, Neutrichrome, Irgalan, Alizarine (Classic Dyestuffs Inc.).
- Ortolan, Palatin Fast (BASF).
- Isolan S, Isolan K (DyStar).
- Supranol (BAY).
- Polar, Erionyl, Lanacron S, Lanaset (Huntsman).
- Coomassie (Simpsons (UK) Ltd).
- Sandolan, Sandolan Milling, Lanasyn, Lanasyn S (Clariant).

12.3.2.2   Reactive Dyes

One tends to think of reactive dyes for wool printing only where very high wet-fastness standards are required – and there has indeed been a trend in recent years towards the printing of wool to machine-washable standards. Lewis [41,42] has reviewed the use of reactive dyes in wool dyeing, but their use in wool printing is not so well documented. Examples of reactive dyes for wool printing are listed in Table 12.1.

Reactive dyes offer several advantages [13,24] apart from the obvious one of good wet fastness (though for furnishing end uses, many of the metal-complex and milling dyes widely used in printing possess better light fastness):

- They possess better solubility than acid dyes, and can usually be sprinkled directly into the print paste as solids without the use of dye solvents; since most printers use 'standard' full-strength print pastes, however, this is of limited value.

Table 12.1   *Some reactive dyes suitable for wool printing.*

Dye Name	Reactive Group	Manufacturer
Levafix E-A, P-A	Difluoromonochloropyrimidine	Bayer
Lanasol	α-bromoacrylamido	Huntsman
Remazol	Sulphatoethyl sulphone	DyStar
Procion P, H-E	Monochloro-*s*-triazine	Atul
Drimarene R, K	Difluoromonochloropyrimidine	Clariant

- Shorter steaming times are required, which is of obvious advantage in continuous steaming.

On the other hand, in order to realise the full wet-fastness potential of reactive dyes, it is necessary to wash off with ammonia (this will be discussed later) at high temperatures, which can be both unpleasant from the point of view of health and safety and, moreover, technically problematic where continuous washing with short dwell times is employed.

Hofstetter points out [13] that printing with reactive dyes leads to certain problems, such as a tendency to unlevelness in large blotches in some shade areas; this arises from the necessity of using a trichromat, whereas with metal-complex or acid milling dyes the same shade can be obtained with a single dye, or at the most, two. Depending on the quality of the goods and the pretreatment, dye selection is therefore of paramount importance in minimising this tendency (this problem is presumably also compounded by the poorer levelling and migration properties of reactive dyes [41]). The use of computer match prediction and rationalisation of dye stocks now means that a printer will in any case formulate most shades with a three-component recipe. Also, as mentioned earlier, reactive dyes will not give satisfactory results with material prepared with the polymers Securlana and Synthappret BAP.

12.3.2.3 Pigments

The use of pigments is not generally encountered in wool printing (except in special cases, which will be discussed later), due to considerable modification of the fabric handle; however, Brady and Hine [43] have suggested that careful selection of binders and crosslinking agents can minimise this effect.

12.3.2.4 Recipes

Typical recipes for wool printing [13,16,24] are summarised in Table 12.2.

The most generally favoured *thickeners* are guar gum derivatives (mixed sometimes with crystal gum or starch ethers), the solids content of which depends on the type of design to be printed (blotch, fine line and other styles all have different requirements).

Table 12.2 Recipes for the printing of wool with reactive and nonreactive dyes.

Chemical	Acid Milling/Metal-Complex Dye (g dm^{-3})	Reactive Dye (g dm^{-3})
Dye	x	x
Urea	50–100	50
Thiodiglycol	50	–
Wetting agent	5–10	5–10
Antifoam	1–5	1 5
Acid or acid donor	10–30	10–30
Thickening (10–12%)	500	500
Water to bulk	1000	1000

Urea is an essential ingredient in wool print pastes; it aids solution of the dye in the print paste, and acts as a humectant, promoting wool swelling and dye penetration during steaming.

Dye solvents, such as thiodiethylene glycol or butylcarbinol, are usually employed to facilitate solution of the acid milling dyes.

A *wetting agent* is necessary if chlorination of the goods has been less than optimum, and also for end uses where complete penetration is required (such as scarves and shawls); antifrosting agents are sometimes also employed.

An *antifoam* is generally necessary in machine printing.

Wool is generally printed under acid conditions using either a non-volatile *acid* (such as citric acid) or an *acid donor* (such as ammonium tartrate or ammonium sulphate); the Remazols are fixed under neutral to slightly alkaline conditions, using sodium acetate at about $40 \, \text{g} \, \text{kg}^{-1}$.

Quite often, the *oxidising agent* sodium chlorate is added to print pastes to counteract the reducing effect of the wool on the dye; under the alkaline conditions employed for printing with Remazols, sodium *m*-nitrobenzene-sulphonate is effective.

Other possible additives include *glycerol*, often used to prevent screen blocking, and *printing oils*, to ensure easy running of the squeegee.

12.3.3   Steaming

After printing and drying, the dye is deposited on the fibre surface, within a thickener film, in a highly aggregated form. A steaming operation is thus necessary to swell the thickener film and dissolve the dye, swell the wool fibre to allow penetration of the dye and elevate the temperature of the fibre to effect fixation of the dye. Typical steaming times are 10–15 minutes for reactive dyes and 30–45 minutes for metal-complex and acid milling dyes.

Without a doubt, the most important condition during saturated steam treatment is that there is sufficient humidity present in the substrate; this must be uniform and constant throughout the treatment in order to prevent differences in colour development. The mechanisms involved in atmospheric steaming have been described in some detail [44]. At this point, however, it is worth considering the various factors that influence the efficiency of steaming of printed wool goods.

As soon as the fabric enters the steaming chamber, it is heated quickly to 100 °C, due to rapid condensation of the steam and consequent liberation of the latent heat of condensation ($2270 \, \text{J} \, \text{g}^{-1}$). Depending on the specific heat and initial moisture content of the substrate, only about 200 g of water is deposited per 1000 g of textile (assuming the print paste contains the appropriate humectants) [45,46]; this of course represents a very low liquor ratio, resulting in unfavourable conditions for dye fixation. If, instead of saturated steam at 100 °C, superheated steam at 110 °C is used, this moisture uptake drops to 75 g per 1000 g, compounding the problem even further and leading to reduced colour yields. It is thus vital to avoid overheating the steam. Wool is particularly problematic in that it has a much higher heat of water absorption than almost any other textile fibre (with the exception of viscose); thus, considerable heat is released when the moisture is absorbed. Fell and Postle [47] have shown that the moisture content of the wool *before* it enters the steaming chamber plays an important role in determining the temperature attained in the steamer due to the heat of absorption (Table 12.3).

Table 12.3   *Relationship between moisture content of wool samples and temperature attained in the steaming chamber.*

Initial Moisture Content (%)	Temperature Attained (°C)
0	125
2.5	115
5	107
10	102

It is therefore clear that the wool should be allowed to attain its natural regain moisture content before it enters the steamer. The reconditioning of several thousand metres of overdried wool is a formidable task and indeed can be a problem on the dryers of modern printing machines running at speed; it is much better to ensure that the goods are not overdried in the first place.

The criteria for effective steaming may be summarised as follows:

- The goods should always be at room temperature, and should have been allowed to condition.
- The print paste should contain the necessary humectants, such as urea or glycerol.
- The steam supplied to the steaming chamber should be truly saturated, and should not have any residual superheating.
- The heat liberated by condensation and absorption should be removed from the steaming chamber, so that the steam is maintained in the desired condition (this is especially important in the first part of the steamer, where the heat generated is at its maximum).

The last two criteria are functions of the steaming equipment, and their implementation in festoon steamers – the Stork HS-III Universal and the Arioli Vaporlitermotex – has been fully described by Schaub and Reina [44,48,49]. Perhaps the most important feature of these steamers is the humidifying system, based on the Venturi injector principle. This maintains the correct moisture content by injection of a fine water mist, which represents a far more economical method than using large quantities of fresh steam and a far safer method than spray damping the goods before they enter the steamer.

Steamers of the Star type (e.g. Dupuis or Sanderson equipment) are very rarely encountered, generally only with small hand printers.

A final point to mention with regard to steaming is the yellowing of chlorinated fabrics during the steaming process. Bell and Lewis [10] have shown that chlorinated wool yellows approximately twice as fast as untreated wool. Although it would seem an obvious solution to minimise steaming times, this is not a practical proposition, as the yellowness index of the steamed wool reaches a maximum after about 15–20 minutes – and even reactive dyes must be steamed for 15 minutes. As mentioned earlier, goods prepared by nonchlorination methods behave more or less as untreated wool, and are significantly less prone to yellowing during steaming.

12.3.4   Washing and Aftertreatment

The general problems arising in the washing of printed goods have been dealt with in some detail by Winkler [50] and Hofstetter [51], and apply equally to wool as they do to other fibres. The aim of washing-off is to remove thickeners, chemicals and unfixed dye, mainly in order to attain the desired level of wet fastness. This has to be achieved without detriment (such as surface felting) to the goods, and without staining of unprinted or pale-coloured areas. Several factors influence the efficiency of washing-off and the wet fastness attained:

- The washing process must be appropriate to the class of dye used; prints with acid milling and metal-complex dyes are washed at about 30–40 °C, while reactive dyes must be washed at 70–80 °C with ammonia at pH 8.5.
- The washing-off properties of the thickener must be satisfactory; incompletely removed thickener will give rise to fastness and handle problems.
- The dye concentration used and the amount of print paste applied to the fabric must not exceed a given maximum, established by conducting preliminary trials. Exceeding this maximum will result in very little increase in depth of shade, but will greatly increase the amount of unfixed dye to be removed by washing.
- The use of optimum steaming conditions ensures a high degree of fixation and leaves a minimum of unfixed dye to be washed out.

Since printed wool has usually been either prechlorinated or pretreated with shrink-resist polymers, surface felting during washing is generally not a problem. Wool prints are washed-off both on winches (either singly or several in series) and on open-width washing ranges, and sometimes using a combination of both methods. The winch offers the advantage of the extended washing times required for wool together with a certain amount of mechanical action. Where the open-width washer is used, several passes are often necessary to compensate for the relatively short dwell time. The highest possible liquor ratio is advisable to prevent staining of white grounds during washing. The use of auxiliaries with affinity for any unfixed dye in the washing liquor both helps prevent staining of white grounds and results in improved fastness properties [52].

Interest during recent years in machine-washable wool prints has placed greater demands on performance standards. Reactive dyes come first to mind when considering machine-wash colour fastness but, in order to realise the full fastness properties of these dyes, washing must be carried out at 80 °C, in the presence of ammonia. Bell has demonstrated [24] that if a maximum washing temperature of only 60 °C is employed, a marked deterioration in wet fastness is observed. As mentioned earlier, problems experienced in continuous washing, where it is difficult to maintain the ammonia level in the line, lead to incomplete removal of unfixed dye and corresponding deterioration of wet-fastness properties.

For prints with nonreactive dyes with normal fastness requirements, washing may be carried out at 30–40 °C, in the presence of a washing-off auxiliary. Machine-wash colour fastness may be attained with a wide selection of these dyes, provided that washing is carried out at 50–60 °C in the presence of ammonia (pH 8.5) and an appropriate auxiliary, and the goods are aftertreated with a cationic dye fixing agent [13,24].

The use of cationic aftertreatments is a way of dramatically improving the wet-fastness properties, in particular the perspiration fastness, of a print. Since these products function by forming a large, insoluble complex with small amounts of unfixed dye, they should only be employed after a thorough wash; otherwise, a marked deterioration in rub fastness occurs. These products, mostly based on condensates of formaldehyde and dicyandiamide, have been widely used in the cotton industry for some time [51].

12.4   Discharge Printing

In discharge printing, a predyed fabric is printed with a paste containing a reducing (discharge) agent and a dye (the illuminating colour) that is resistant to reduction. The ground shade is simultaneously destroyed and replaced by the 'illuminating' colour in the printed areas. Production of intricate patterns, usually strongly contrasting grounds, with great clarity, sharpness and fit, has become the hallmark of this style, which can be produced by direct printing methods only with great difficulty, if at all.

Little has been published on the discharge printing of wool (or, for that matter, silk), perhaps reflecting the difficulties involved in this process. The current state of the art has been described by Koch [53], Bell *et al.* [54], Hofstetter [13] and Berry and Ferguson [55], and is summarised in this section.

12.4.1   Ground Shades

Discharge printing involves the actual chemical destruction of the original dye in the printed areas. The ground-shade dyes are invariably azo dyes, and their dischargeability ratings according to the *Colour Index* system should be not less than 4–5. The selection of these dyes does not generally present a problem, many acid, metal-complex and reactive dyes being dischargeable to white [16]. Koch [53], however, points out the importance of carrying out preliminary tests under works conditions, since, in a trichromatic dyeing, not all the components are reduced at the same rate. The most readily reduced component is discharged first; thus, under unfavourable conditions, it could happen that a less readily reduced component will be inadequately discharged, even if it is only present in small quantities.

12.4.2   Discharge Agents

Several reducing agents are available; these are listed in Table 12.4. The most commonly used for the discharge printing of wool are the formaldehyde sulphoxylates, the mode of action of which has been investigated in depth [56–58]. A simplified representation of the decomposition is given in Scheme 12.5 [55].

Sodium formaldehyde sulphoxylate [59] is a water-soluble product that is easy to handle and gives good white discharges. It is stable in alkaline print pastes but not in acidic conditions; it develops its full redox potential when the print is steamed [53], and unless great care is exercised, excessive fibre damage and shrinkage occurs. For this reason, it is rarely used on its own; pH adjustment results in reduced stability.

Calcium formaldehyde sulphoxylate [60] is water-insoluble; although stable, it can lead to problems of screen blockage and unsatisfactory penetration of the discharge. BASF has subsequently developed a 30% dispersion of the compound; greater stability, fewer problems with haloing, less fibre damage and less shrinkage than with the sodium salt are claimed. The calcium salt is quite often used in combination with the sodium salt.

Table 12.4   *Reducing agents for use in discharge printing.*

C.I. Reducing Agent	Chemical Name	Structure	Trade Name
2	Sodium formaldehyde sulphoxylate	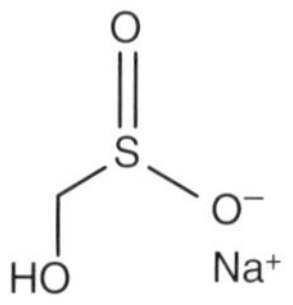 (structure **1**)	Rongalit C (BASF), Formosul (BOC)
6	Zinc formaldehyde sulphoxylate	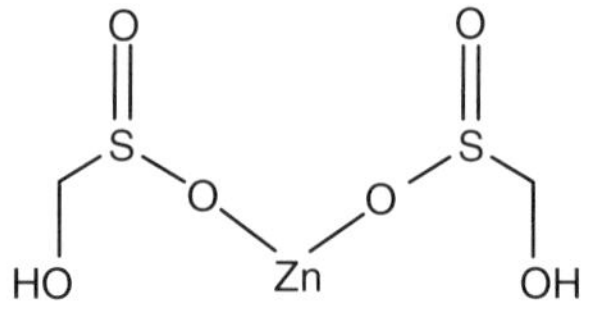 (structure **2**)	Decrolin (BASF), Arostit ZET (S)
12	Calcium formaldehyde sulphoxylate	analogous to **2**	Rongalit H (BASF), Rongalit H Liquid
11	Thiourea dioxide	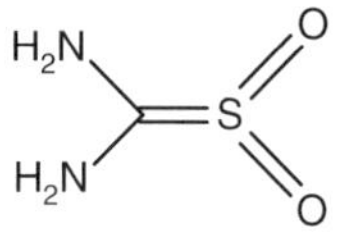 (structure **3**)	Reducing Agent F (Degussa), formerly Manofast
	Tin(ii) chloride (tin salt)	$SnCl_2 \ldots 2H_2O$ **4**	

$$HOCH_2SO_2Na \longrightarrow HCHO + HSO_2^- + Na^+ \xrightarrow{H_2O} HSO_3^- + 2H$$

$$Ar-N{=}N-Ar' + 4H \longrightarrow ArNH_2 + Ar'NH_2$$

Scheme 12.5   *Mode of action of sodium formaldehyde sulphoxylate in discharge printing.*

Zinc formaldehyde sulphoxylate [61] is stable in weakly acidic print pastes and discharges well in steaming; fibre damage and shrinkage are less of a problem than with the sodium salt. Although it is widely used, there can be problems in washing out the cleavage products of some reduced dyes, leading to after-yellowing of the discharge on exposure to light and air, and it is more aggressive towards the illuminating colours than is the sodium salt; it also gives rise to additional effluent problems.

Reaction schemes for the other reducing/discharge agents – thiourea dioxide and tin (II) chloride – have been given by Koch [53] and by Berry and Ferguson [55] (Schemes 12.6 and 12.7).

Thiourea dioxide seems to be an attractive alternative to the formaldehyde sulphoxylates [62–64], but it is rarely used due to its low solubility (37 g dm^{-3} at 20 °C).

Tin (II) chloride is not often encountered in wool discharge printing. Its lower redox potential permits the use of a wider range of illuminating dyes, but at the same time limits

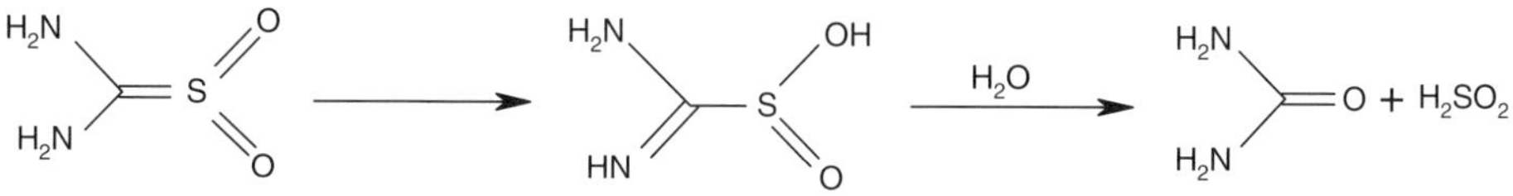

Thiourea dioxide

Scheme 12.6 *Mode of action of thiourea dioxide in discharge printing.*

$$Ar-N=N-Ar' + SnCl_2 + 4H_2O \longrightarrow ArNH_2 + Ar'NH_2 + SnO_2 + 4HCl$$

Scheme 12.7 *Mode of action of tin(II) chloride in discharge printing.*

the number of ground shades which are dischargeable; after-browning of white discharges is also a problem. Effluent disposal – of ever-increasing importance – and corrosion of the steamer due to production of hydrochloric acid (see Scheme 12.7) are further problems to be reckoned with. Hydrochloric acid production also precludes the use of thiodiglycol as a dye solvent, as these two compounds are unfortunately able to react to produce mustard gas – a most undesirable product!

12.4.3 Illuminating Dyes

The real problem in discharge printing lies in the selection of dyes for coloured discharge that are resistant to the reducing agent. The already short list is diminishing all the time, as manufacturers delete uneconomical colours from their ranges. It is usually necessary to resort to dyes of poor wet and light fastness, particularly where brilliant shades, so characteristic of this printing style, are demanded. Dye selection for coloured discharge [13,16,65] usually comprises C.I. Direct Yellow 28, C.I. Acid Yellow 3, C.I. Acid Red 52, C.I. Direct Blue 106 and C.I. Basic Blue 3. Other colours known to be discharge-resistant, and therefore of use as illuminating colours, are C.I. Acid Red 315, C.I. Acid Violet 90 and C.I. Acid Blue 61 : 1. Speciality colours for discharge are often based on mixtures of these dyes; perhaps the best three-colour mixture is made up of C.I. Acid Yellow 3, C.I. Acid Violet 90 and C.I. Acid Blue 61 : 1.

The use of pigments is sometimes encountered; although advances in binder technology mean that a relatively soft handle can be obtained, this approach is nevertheless limited to the printing of small motifs or outlines. The pigments must, of course, be resistant to the reducing agent.

Bell and Lewis [10] and Hofstetter [13] have reported the use of solubilised vat dyes; although not particularly deep or brilliant, nor indeed even offering a full shade gamut, these dyes do possess good fastness properties. Unfortunately, strongly acidic conditions are required for liberation of the leuco vat dye, which is subsequently oxidised to the insoluble vat pigment; these conditions can affect the ground shade.

12.4.4 Printing and Fixation

Recipes for the discharge printing of wool are numerous and diverse; most are based on the printer's own know-how and experience, on local works conditions and on effluent regulations. Detailed recipes from the major dye manufacturers are given elsewhere [13,16,65]. Although the recipe given in Table 12.5 may be regarded as fairly typical, it is very difficult

Table 12.5   *Recipe for the discharge printing of wool.*

Chemical	Quantity ($g\,dm^{-3}$)
Dye	x
Urea	30–50
Thiodiethylene glycol	30–50
Water	y
Sodium/zinc/calcium formaldehyde sulphoxylate	30–180
Thickening	500
Sodium *m*-nitrobenzenesulphonate (Revatol S, Ludigol, Matexil PA-L)	5–20
Ammonium chloride	5–20
Zinc oxide 1 : 1	20–50
Water to bulk to	1000

to generalise, and the importance of carrying out trials to establish optimum recipes with given printing, drying and steaming conditions must be strongly emphasised.

The reducing agent is used at the minimum level required to bring about discharge of the ground shade; this will vary enormously, depending on the ground shade. Use of higher levels will lead to both haloing and over-reduction of the print colour.

The sodium *m*-nitrobenzenesulphonate is used to prevent unwanted reduction of the illuminating dye; although not necessary for all illuminating colours, its use can in certain cases have a profound effect on the storage stability of the print paste and on the stability of the print colours during steaming, and consequently on the reproducibility of the colour yields [54,65]. Use of too high a level will affect the dischargeability of the ground.

The print should be dried as quickly as possible, but not at too high a temperature. Long storage times at this stage should be avoided, since storage under moist conditions can lead to decomposition of pastes containing sodium formaldehyde sulphoxylate and urea.

Steaming is carried out for 10–20 minutes at 100–120 °C (or sometimes for two separate passages through the steamer), air-free steam being essential. The addition of small amounts of hydrazine to the boiler water is advantageous in 'binding' free oxygen in the steam. It is also important that the required steaming temperature be attained as quickly as possible, since the formaldehyde sulphoxylate begins to break down at about 50–60 °C. The criteria for effective steaming (discussed earlier in this chapter) are even more important here, since the discharge process is exothermic.

Washing-off is carried out under mild conditions, as the dyes used for the coloured discharge generally possess poor wet-fastness properties.

12.5   Resist Printing

The end results of resist printing are similar to those of discharge printing, but the processing route is different. The resist agent is printed and fixed on the white fabric, preventing the fixation of the ground shade, which is applied by an appropriate dyeing or printing technique.

A coloured resist is obtained by including in the print paste an illuminating dye that is able to fix in the presence of the resist agent. Resist techniques fall into two classes:

- *Reactive resist* followed by piece dyeing, usually from long liquor.
- *Mechanical/chemical resist*, where the resist agent is printed and the ground shade is applied by wet-on-wet printing with a blank screen.

Many attempts have been made over the years to produce multicoloured effects on wool and polyamide fibres, with varying degrees of success.

12.5.1   Chemical Resist Processes

Several methods have been suggested for rendering wool nondyeable with acid dyes, the first practical proposition being the use of a reactive resist agent resembling a colourless reactive dye. One such product was Sandospace R (Clariant) [66], which is a highly reactive water-soluble anionic product, believed to have Structure **1**. Bell and Lewis [10] have described the reaction scheme by which this compound is able to react with amino groups in the wool, to produce Structure **2**. The resist thus arises both from blocking of reactive amino sites on the fibre and by anionic repulsion.

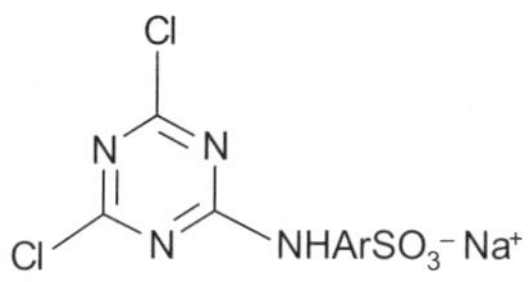

| **1** | **2** |

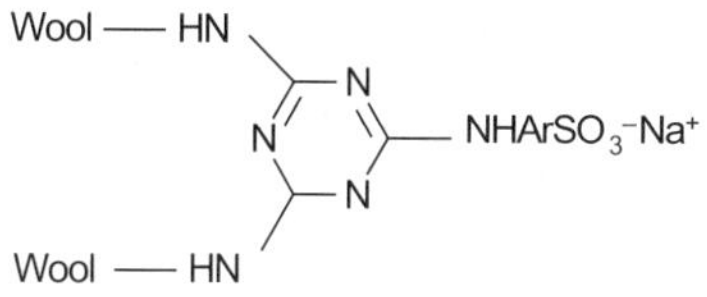

The use of this product to produce resist effects on wool and polyamide became quite well established [67–69], and it also found some application in the printing of wool garments [70]. When Sandospace R is applied to wool by padding or long-liquor dyeing, treatment levels corresponding to mass gains of 10–12% result in a resist effect of about 80–90% when the goods are dyed to a 2% shade [70,71]. At pale depths, a complete white resist is obtained. The alternative Sandospace S [72] is less reactive than Sandospace R, as it is probably a monofunctional chloro-*s*-triazine, and cannot be fixed adequately to wool to achieve the mass gains necessary for effective resists [71].

12.5.1.1   Sulphamic Acid Resist Printing

Most of the techniques described thus far suffer from one main drawback: the inability to achieve a perfect white resist except in pale depths of ground shade. Although tone-on-tone effects can be quite acceptable for certain end uses, such as the dyeing of carpet yarns, the objective as far as the printer is concerned is to find an alternative to discharge printing, where grounds are nearly always dyed to a full shade. These drawbacks have led to the development of a more effective resist system for wool.

The use of sulphamic acid as an anionic dye-resist agent for wool was patented in 1955 by Sandoz [73]. Other workers have since reported the dye-resist effectiveness of this chemical when applied by padding techniques [74,75]. Its use in wool printing was first described by Bell *et al.* [54]; the resist was found to be most effective towards reactive dyes.

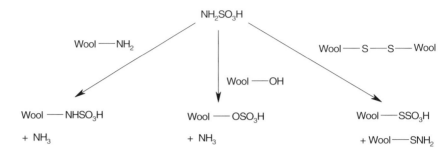

Scheme 12.8   *Sulphation of wool with sulphamic acid.*

Elliot *et al.* [74] have suggested that the most probable reactive sites on the wool for sulphation with sulphamic acid are basic amino groups, serine hydroxyl groups and the cystine linkage (Scheme 12.8).

The reactions between sulphamic acid and wool and the optimum reaction conditions have been studied in some depth [76,77]. For maximum reaction with sulphamic acid, urea is essential in the pretreatment liquor/print paste (confirming the results of Lewin *et al.* [75]), and the fabric should be cured at 150–160 °C for 4–5 minutes; under these conditions, a maximum uptake of 8.5% bound sulphamic acid is achieved. No further uptake is achieved by increasing either the curing time or the temperature. Omission of the urea results in a drop in bound sulphamic acid to 5%. Cameron *et al.* [76] showed that about 1% of the bound sulphamic acid was accommodated on the cystine linkages and between 0.6 and 1.5% on basic amino groups; by difference, about 5.9–6.8% had reacted with alcoholic amino acids. The reactions are complex, no one reaction going to completion under the conditions employed.

The conditions employed in printing are very similar (Table 12.6). A coloured resist is achieved by including an illuminating dye in the print paste; only reactive dyes, selected from the Lanasol and Drimarene R and K ranges, fix covalently to the fibre under these conditions; these may be added directly to the white resist paste as solids. Fixation is now, of course, a two-stage process, involving:

- dry heat fixation, 4–5 minutes at 150–160 °C, to fix the sulphamic acid, followed by;
- steaming, 30 minutes at 100–102 °C, to fix the illuminating dye.

Table 12.6   *Recipe for white resist print paste.*

Chemical	Quantity (g dm^{-3})
Sulphamic acid	150–200
Urea	150–200
Thickener 301 Extra RF 12% (Grünau) or Indalca PA40 Cesalpinia)	500
Alcopol 650 (Huntsman) (non-ionic wetting agent)	2
Antifoam	1
Water to bulk to	1000

Fixation of the reactive dye is lower than would be expected in direct printing, being adversely affected by the print-paste pH. However, this does increase during the baking and steaming [55]:

- as printed: pH 2.4;
- after baking: pH 4.5;
- after steaming: pH 5.2.

The steaming and baking steps are not interchangeable; hydrolysis of the sulphamic acid occurs if steaming is carried out first, even though solutions of sulphamic acid are very stable at room temperature [78]. Steaming of the sulphamated wool does result in a small but significant deleterious effect on the level of resist [79], but this cannot be avoided.

Thorough washing-off, including an ammonia wash at 70 °C and pH 8.5, is necessary to remove unfixed dye and to prevent cross-staining of unprinted areas during subsequent dyeing.

Overdyeing is carried out from long liquor with reactive dyes; an almost perfect resist is obtained with selected dyes from the Lanasol (Huntsman), Novacron F (Huntsman), Procion H-E (Atul) and Remazol (DyStar) ranges [16].

Since reactive dyes are used both for the coloured resist and the ground-shade dyeing, fastness properties of the prints are good. There is a certain amount of fibre damage, but this is not as great as that which occurs in discharge printing.

A disadvantage of the process is that the very brilliant shades associated with the rhodamine pinks and basic blues of discharge printing (albeit of very poor light fastness) are not attainable. A further practical problem is that in the dyeing of the ground shades it is necessary to know the extent of the printed area, which is virtually nondyeable, in order to calculate the mass of wool on which the dye recipe is based. This is not insurmountable, but the ground-shade recipe will obviously vary from design to design (which is not the case in discharge printing), and requires laboratory matching for each design.

12.5.2   Mechanical/Chemical Resist Processes

Mechanical resist printing techniques have been practised for centuries – in fact, for about 3500 years! The classical 'batik' style of wax resist printing is thought to date back to the Egyptian civilisations of 1500 BC. But the discussion of such styles, unique though they are (they are practised by only a very few specialist printers), is outside the scope of the present review (although their use on wool/cotton blends has been described [79]).

Matsumin High Zitt FR and FRC were offered by the Matsui Shikiso Chemical Co., Kyoto, Japan; these products were cationic in character and were described as polyamine resins, the FR being used for white resist and the FRC for coloured resist. The process involved printing the white or coloured resist, then applying the ground shade by wet-on-wet overprinting with a blank screen. Selected levelling dyes were recommended for the coloured resist, and acid milling and 1 : 2 metal-complex dyes for the overprint. After drying and steaming, the goods were simply washed and the ion-pair complex formed by the cationic resin and the anionic dye was broken down and removed, leaving a white or coloured resist. Unfortunately, due to the high solids content of the product, severe problems were encountered with screen blockage, and the process was really only suitable for hand printing.

12.5.2.1   Thiotan WS

A more promising development was the introduction of the Thiotan WS process from Sandoz [13,80]. This product too was based on a mechanical reserving action, and was also believed to possess a degree of cationicity. Wet-on-wet printing was again involved, and the process was aimed at overcoming some of the problems associated with the earlier Japanese processes.

Thiotan WS incorporates a thickener and is employed at about 600–750 g kg^{-1} in a white resist, and 600 g kg^{-1} in coloured resist. A typical recipe for white resist contains simply Thiotan WS, urea and antifoam. The recipe for coloured resist contains in addition dye, Lyocol BC and citric acid. Some minor variations may be necessary, depending on the working conditions. The recommended selection of dyes for the coloured resist is composed largely of Sandolan N, Milling N, Brilliant N and Fast P types, together with one or two Sandolan E dyes. In general, the wet- and light-fastness properties of these prints can be expected to be superior to those offered by classical discharge prints. The overprinting is carried out with a conventional wool print paste, using mostly selected Lanasyn S and Sandolan Milling N dyes, which are known to be reserved by the Thiotan. Steaming and washing-off complete the process, the reserved ground-shade dye being washed out with the Thiotan.

The process certainly has a high degree of simplicity and certain novel effects are obtainable, such as the reserving of several overprinting colours printed alongside each other. Due to the high solids content of the print paste, excessive mechanical handling of the printed goods must be avoided in order to prevent dusting and flaking.

12.5.3   Reactive-Under-Reactive Resist

This chemical resist process for chlorinated wool is based on the technology already used for resist printing on cellulosic fibres. The illuminating colours are generally monochloro-*s*-triazine reactive dyes and are printed from a mildly alkaline (pH 7.6 with sodium acetate) print paste containing sodium sulphite; the ground shade is printed wet-on-wet with vinyl-sulphone (VS) dyes, again from a mildly alkaline paste (pH 7 with caustic soda). After moderate drying, the print is steamed for 20 minutes at 102 °C and washed-off in ammonia at 50 and 70 °C. The 'ground-shade' print is resisted in the printed areas by the presence of the sodium sulphite, which reacts rapidly with the VS group, thereby preventing fixation of the dye. The monochloro-*s*-triazine illuminating colour is able to fix under these conditions.

12.6   Digital Printing

Digital printing offers many advantages over the traditional methods employed to print wool fabrics and provides an economical route to the delivery of printed substrates despite shorter run length, thus providing real opportunities to expand the traditionally small wool printing market. Benefits of digital printing include reduced production costs and lower environmental impact. These benefits are achieved through lower labour costs, the elimination of screen manufacturing and storage or disposal costs, lower water usage, lower emissions, reduced energy costs and the elimination of colour wastage from screen washing and colour reservoirs [81].

Disadvantages include much slower production speeds (350–500 m h^{-1}, compared to up to 4200 m h^{-1} for rotary screen printers) and the current high cost of digital inks (approximately £100 per litre).

12.6.1   Machinery

In recent years there has been a digital revolution in the textile printing arena that has seen the introduction of high-definition ink-jet printers that are capable of printing text and graphics in both monochrome and full colour. Ink-jet printers are used in conjunction with computer-aided design (CAD) software to produce high-quality prints on a wide range of textile substrates. CAD systems are employed to transfer artwork to a computer screen, where it can be edited to ensure the precise colour shades required and to determine optimum fitting of the design repeats and fall-ons. It is not possible at this time to expand further on the intricacies of such CAD systems, but detailed descriptions have been published by Dawson [82–84] and Loser *et al.* [85].

The past decade has seen a major evolution in the textile ink-jet printing sector, which has resulted in machinery manufacturers developing and marketing textile ink-jet printers that offer improved print quality, higher resolution and increased speed at lower capital cost. A range of textile ink-jet printers are now available which provide sufficient scope for sampling, small-production and even full-scale-production operations; the equipment offers printing speeds of 26–150 m^2 min^{-1} and print definitions of 50–720 dpi. The performance capabilities and benefits of some of these latest textile digital printers have been reviewed elsewhere, by a number of authors [86–89].

12.6.2   Ink Formulation

Dyes suitable for the ink-jet printing of wool should possess the same characteristics and properties as those described earlier for use in conventional print pastes for wool, but in addition they should exhibit high solubility and excellent storage stability. The dyes are specially purified prior to ink formulation to ensure that any particulates or electrolytes have been removed, as these can detrimentally affect the storage stability of the ink. The ink formulations are specifically tailored to meet the pH, viscosity, surface-tension and nonfoaming characteristics required for each type of printhead, in order to ensure that maximum ink storage stability, operability and firing performance are achieved [90–92]. It is not possible to add all the chemicals required to ensure acceptable colour-yield and fastness properties of the subsequent prints to the ink, as this might have a detrimental effect on the physical properties of the ink, resulting in ink-stability and performance issues [93]. Care is required in the selection of any additives, such as co-solvents or humectants, that are incorporated in the ink, as many contain hydroxyl groups that could lead to storage stability problems, especially when using reactive dyes [94]. Ink sets generally contain between four and eight different colours. The printers that utilise only four colours employ the cyan, magenta, yellow and black inks that are normally used in the graphics industry; however, these four inks are capable of reproducing only 70% of the gamut of colours available and so additional coloured inks, selected from bright orange, golden yellow, red, blue and green, are often included to extend the available range [83,94]. Digital printing applies much lower ink volumes, typically 20 g m^{-2}, to the fabric surface than conventional printing processes (125–200 g m²) and digital inks thus require a higher dye concentration to ensure that acceptable colour yields are achieved for the resulting prints [82,94].

12.6.2.1 Nonreactive Dye-Based Inks

Ink formulators generally choose acid and metal-complex acid dyes as their preferred colorants when formulating digital-printing inks for wool, because such dyes offer the low molecular weight, small size and high solubility that enables stable ink solutions to be prepared. In addition, such dyes enable the production of prints that exhibit high brightness [83,91,93].

12.6.2.2 Reactive Dye-Based Inks

Ink formulators do not specifically produce a range of reactive dye-based inks for wool substrates, but it is possible to print wool fabrics with the reactive dye-based digital inks that have been formulated for the ink-jet printing of cellulosic substrates. Ink formulations containing the monochloro-*s*-triazine-type reactive dyes have been developed which yield stable ink formulations that are compatible with most types of commercially employed ink-jet printheads [82,93]. Ink manufacturers often select their reactive dyes from the Procion P and Procion PX ranges of monochloro-*s*-triazine reactive dyes, as these offer good build-up, compatibility and wash-off characteristics [82]. Reactive dye-based inks offer many advantages for the digital printing of wool substrates, including a full colour gamut, brilliant shades and excellent fastness properties for the resulting prints [82,95,96].

Figure 12.1 illustrates some typical prints obtained on untreated, Kroy-chlorinated and Perachem-delipidised wool fabrics when ink-jet printed with monochloro-*s*-trazine-based inks and subsequently steamed at 100 °C for 30 minutes; the print on the delipidised wool fabric exhibit the highest colour yield and a degree of reactive dye fixation of almost 100%.

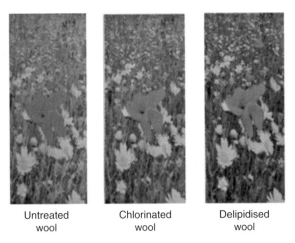

| Untreated wool | Chlorinated wool | Delipidised wool |

Figure 12.1 *The effect of the wool pretreatment process on the colour yield of wool ink-jet printed with monochloro-s-triazine reactive dyes. See colour plate section for a full-colour version of this image.*

Table 12.7   *A typical pretreatment applied to wool fabrics when digitally printing with acid or metal-complex acid dye-based inks.*

Chemical	Quantity ($g\,dm^{-3}$)
Guar gum or locust bean gum	150
Urea	100
Ammonium tartrate solution (1 part water to 2 parts ammonium tartrate)	50
Water to	1000

12.6.3   Fabric Pretreatment

In addition to the normal fabric preparation treatments that wool undergoes prior to conventional printing and dyeing processes, wool fabrics that are to be digitally printed require further processing before any ink-jet printing step [97,98]. This additional pretreatment process is necessary to ensure that all the chemicals and auxiliaries required to fix the dyes and maintain the integrity of the printed image are present, as they cannot be incorporated directly in the digital ink as would be the case in a traditional print paste, the pretreatment solution being applied to the fabric using a pad (75–80% wet pickup) or coating process prior to printing. The pretreated fabric is finally dried at temperatures below 100 °C prior to ink-jet printing. Pretreatment processes are necessary because:

- Chemicals can affect the stability of the ink and decrease storage stability.
- Chemicals and salts can corrode the printhead nozzles.
- Viscosity modifiers can unduly affect the rheological properties of the ink and make it unjettable.
- Large salt concentrations can have a detrimental effect on the solubility of the dye.

Most fabric pretreatment solutions will contain a thickening agent to control the degree of ink penetration into the fabric and its natural tendency to spread laterally along the substrate surface [82]. Excessive ink penetration decreases the observed colour yield of the subsequent print, while undue wicking results in a loss of print definition [92,97].

A typical pretreatment solution that can be applied to wool fabrics when they are being ink-jet printed with acid or metal-complex acid dye-based inks is given in Table 12.7, while Table 12.8 shows a fabric pretreatment suitable for use with reactive dye-based inks.

In an attempt to prevent the yellowing that occurs on steaming chlorinated wool fabrics, Broadbent and Lewis have proposed a wool pretreatment solution that can be applied to

Table 12.8   *A typical pretreatment applied to wool fabrics when digitally printing with reactive dye-based inks.*

Chemical	Quantity ($g\,dm^{-3}$)
Sodium alginate, medium viscosity	100
Urea	100
Water	1000

Table 12.9   *A bisulphite-based pretreatment for the digital printing of untreated wool fabrics [99].*

Chemical	Quantity (g dm^{-3})
Sodio-carboxymethylcellulose	20
Urea	300
or Formamide	100
Sodium bisulphite	10
Alcopol O 60 (Huntsman)	5–10
Water to	1000

untreated wool fabrics prior to digital printing [99]. In this process, the untreated wool fabric is padded with an aqueous solution containing a thickener, sodium bisulphite, an anionic surfactant and a hydrogen-bond breaker such as urea or formamide. A typical recipe is given in Table 12.9. The effect of the different pretreatment agents on the fixation of difluoromonochloropyrimidine reactive dye-based inks is demonstrated in Figure 12.2, the ink-jet-printed wool substrates being steamed at 104 °C for 10 minutes to covalently fix the reactive dyes. Wool fabrics pretreated with pad liquors containing an anionic surfactant,

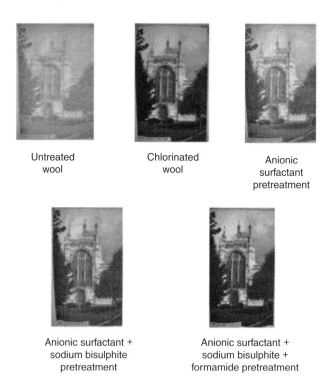

Untreated wool

Chlorinated wool

Anionic surfactant pretreatment

Anionic surfactant + sodium bisulphite pretreatment

Anionic surfactant + sodium bisulphite + formamide pretreatment

Figure 12.2   *The effect of wool pretreatment conditions on the colour yield of wool ink-jet printed with difluoromonochloropyrimidine reactive dyes. See colour plate section for a full-colour version of this image.*

sodium bisulphite and formamide exhibit the highest colour yields and almost-complete reactive dye fixation. The subsequent prints show no evidence of any yellowing of the unprinted white areas of the wool substrate following steaming, in contrast to those prints produced on the Kroy-chlorinated wool substrates, which exhibit some yellowing of the unprinted white areas.

12.6.4 Fixation

12.6.4.1 Steaming

Fixation of the ink-jet-printed wool fabric is usually carried out immediately after printing, without the need for any additional intermediate drying stage, as the water present in the low volumes of ink printed down is readily absorbed by the substrate and therefore the printed fabric is sufficiently dry on exiting the printer. Fixation of both acid and reactive dye-based prints is usually carried out by steaming with atmospheric steam at 102 °C; the acid-dye prints being steamed for 30–45 minutes and the reactive-dye prints for 10–30 minutes.

12.6.4.2 Print-Batch

This process involves interleaving wool fabrics that have been digitally printed with reactive dye-based inks between a dry and a wet cotton fabric, placing the composite sandwich on a polythene film and rolling it into a batch. The rolled batch is stored at 25 °C for 4–24 hours to enable dye diffusion and fixation to take place. The dry cotton fabric is placed against the face side of the printed wool fabric, while a wet cotton fabric that has been preimpregnated with a 1% w/w solution of sodium bisulphite is placed against the reverse side. This approach has been shown to successfully fix both difluoromonochloropyrimidine [96] and *p*-phenolsulphonate-modified difluoro-monochloropyrimidine (**3**) [100] dyes to wool substrates following digital printing. This process modification increases both the stability and the water-solubility of the difluoromonochloropyrimidine dye.

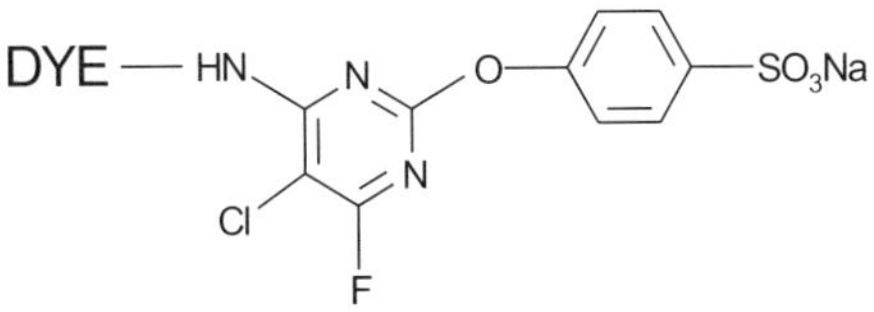

3

An example of a typical print produced by this process when using difluoromonochloro-pyrimidine-based inks is shown in Figure 12.3; this process again produces almost complete reactive-dye fixation.

12.6.5 Wash-Off

After fixation, the printed fabrics must be thoroughly washed-off to remove any unfixed dye, thickener and other auxiliaries that have been applied to aid the fixation process. The printed sample is initially washed-off with copious volumes of cold water until no further

Figure 12.3   *Print-batch development of wool ink-jet printed with difluoromonochloropyrimidine reactive dyes. See colour plate section for a full-colour version of this image.*

dye is removed, then given a hot wash, and finally a further cold-water wash. Sodium carbonate (1 g dm^{-3}) may be added to the first cold-water wash to prevent any protonation of the wool substrate that might lead to subsequent staining problems during removal of all the residual unfixed dye. A more in-depth account of the wash-off process was given in Section 12.3.4.

12.7   Wool Blends

Blends of wool with other fibres are often encountered amongst printed goods, for several reasons:

- A cheaper article can be produced by introducing a synthetic fibre.
- A wool-like handle and appearance can be imparted to a synthetic fibre by blending it with small amounts of wool.
- A synthetic fibre may be introduced on technical grounds (e.g. to impart improved wearing properties).
- Highly exclusive articles can be produced by blending wool with the so-called 'noble fibres' (silk or cashmere).
- Traditional blends, such as half-wool unions with cellulosic fibres, have been around for many years.

Where the other fibre possesses similar dyeing properties to wool, for example silk and cashmere, problems are not usually encountered, the blend being treated as 100% wool. This is likewise the case where small amounts of a synthetic fibre, with similar dyeing properties, have been introduced. For example, it is not uncommon to find blends of wool with 5–10% polyamide, particularly in the challis types, to impart greater fabric strength; this is often encountered in discharge-printed goods, the object being to compensate for the weakening of the wool during the discharge printing process [13]. This section, however, is concerned with the more frequently encountered blends, the most important of which are wool/polyester, wool/cotton and, to a lesser extent, wool/acrylic.

12.7.1   Wool/Polyester

Blend ratios in wool/polyester vary enormously – from 5 to 45% of wool. The lower levels of wool are usually used to improve handle and appearance, and the blend is printed as polyester with disperse dyes, often by heat-transfer printing. With any blend containing upwards of 20% wool, however, both fibre types must be considered; otherwise, poor coverage of the wool will occur, giving an undesirable skittery appearance [52].

It is usually necessary to prechlorinate the substrate (adjusting the level of chlorination for the proportion of wool in the blend) and print with a mixture of disperse and anionic dyes – either reactive, acid milling or metal-complex; most of the major dye manufacturers recommend matched pairs of dyes for this purpose [16], available either as readymade mixtures or as separate dyes. Fixation is usually a two-stage process, involving atmospheric steaming to fix the anionic dye on the wool, followed by high-temperature steaming at 175–180 °C to fix the disperse dye on the polyester. Surprisingly, however, in view of the earlier discussion on steaming, successful results have been achieved by reversing these steps [16,101], when reactive dyes are used for the wool component. The main problem in the printing of wool/polyester arises from the unfixed disperse dye – both in staining of the ground during washing-off and in poor wet fastness of the print, due to the small disperse dye molecules remaining trapped in the wool. Careful selection of auxiliaries in washing-off can lead to dramatic improvements [46,101].

Over the years, a few interesting variations have been proposed for the printing of wool/polyester blends. For example, it was demonstrated by Hoechst [102–104] that sodium formate brought about fixation of selected reactive dyes on wool and cellulosic fibres under high-temperature steam conditions (175 °C). The absence of free alkali meant that wool yellowing and fibre damage were kept to a minimum. Thus, mixtures of disperse (Samaron) and reactive (Remazol) dyes were fixed in a one-stage steaming process. It was necessary for the wool component of the blend to be chlorinated, and, if the wool content exceeded 30%, to carry out an atmospheric steam before the high-temperature steam. Washing-off was to be carried out, as always, with care, to avoid staining of grounds.

The Procilene PC dyes were developed by ICI for the printing of polyester/cellulosic blends. These were mixtures of Procion T (reactive dyes containing phosphonic acid) and Dispersol PC disperse dyes; they were designed to hydrolyse to water-soluble carboxylate derivatives during washing-off under strongly alkaline conditions [105], thereby alleviating staining problems associated with the use of conventional disperse dyes. Efforts were made to adapt this system to the printing of wool/polyester blends [106–108]; however, since very low levels of fixation (about 30%) of the Procion T dyes were obtained on wool [10], and since the severe wash-off conditions could not be employed on wool, there was little advantage in using the system.

The use of reactive disperse dyes was investigated by Brady and Cookson [108]. Although the range has now been withdrawn, the Procinyl dyes from ICI were shown to give satisfactory results when printed on 60 : 40 wool/polyester from a conventional print paste.

12.7.2   Wool/Cotton

For a comprehensive review of available and currently used methods for dyeing and finishing wool/cotton blends, the reader is referred to Chapter 10 and to Cookson [109].

Although chlorination is often used as a prepare for printing, it must be carried out under carefully controlled conditions; overchlorinated goods will suffer some impairment of handle under the printing conditions employed. As mentioned earlier, the polymer fabric pretreatment, used primarily to impart machine washability, also acts as a prepare for print, and this method has been used by one manufacturer of printed wool/cotton fabrics.

Prior to the introduction of reactive dyes, wool/cotton unions were printed with direct or acid dyes [110], with the accompanying problems of backstaining during washing-off and the necessity of a cationic aftertreatment to attain even moderate fastness. The direct printing of the blend is now more readily accomplished with reactive dyes, monochlorodifluoropyrimidine, VS and monochlorotriazine types all being suitable [13,109,110]. The dyes are printed from an alkaline paste with the minimum amount of sodium bicarbonate necessary for fixation on the cotton component of the blend (usually about $10–20 \, g \, kg^{-1}$). It is, however, essential to carry out preliminary tests on the material: if the wool/cotton is not an intimate blend, but composed of a cotton warp and wool weft, severe damage to the wool will occur, although it is reported [13] that reactive dyes of the difluoromonochloropyrimidine type require only $5–8 \, g \, kg^{-1}$ sodium bicarbonate, which allows such a construction to be printed without undue damage to the wool. The fabric pH should also always be checked: an acid pH will lead to partial neutralisation of the alkali and a decreased fixation.

Discharge printing of wool/cotton blends is nearly always carried out on grounds pad–batch dyed with reactive dyes: usually VS or dichloro-*s*-triazine types, but the use of difluoromonochloropyrimidine types is also reported. Sodium formaldehyde sulphoxylate is used in preference to the zinc salt, for optimum discharge on both the wool and the cotton [110], and coloured discharge is carried out with selected vat dyes in the presence of sodium and/or potassium carbonate at about $40 \, g \, kg^{-1}$. It will be appreciated that this is a strongly alkaline printing paste, which can only be tolerated because of the presence of the cotton; it can therefore be applied only to intimate blends, and never to the cotton warp/woollen weft type of construction.

12.7.3   Wool/Acrylic

A certain amount of interest has been shown in this blend, possibly because it offers a means of producing a cheaper article using a synthetic fibre with a wool-like handle. It is the most difficult blend to print, however, demanding the use of incompatible dyes: an acid dye for the wool and a basic dye for the acrylic fibre. A novel approach has been reported by Robert [111], where the material is first treated with Sandospace R (about 5–10% on mass fabric (o.m.f.)), either in a long-liquor exhaustion process or by a pad–dry–steam–wash-off process. The pretreated material is then printed as if it is 100% acrylic fibre, with basic dyes chosen for good light-fastness properties; prints of acceptable wet fastness have been obtained.

12.8   Cold Print Batch

In the early 1970s, the International Wool Secretariat (IWS) developed a new method for printing wool which required neither prechlorination nor steaming to fix the dyes

[112,113]. This was developed from earlier experience with the pad–batch (cold) dyeing method and was based on the use of sodium metabisulphite and high levels of urea to accelerate the cold fixation of highly reactive dyes, of the dichloro-*s*-triazine (Procion MX) [114] and difluoromonochloropyrimidine (Drimalan F, Drimarene R and K, Verofix, Levafix E-A) types.

Primarily designed for wool piece goods, the method involved printing, interleaving the wet fabric with polyethylene film and batching for up to 24 hours. Fixation levels of about 80% were achieved [112]. Steaming was not necessary, the process being completed by an ammonia wash at pH 8.5 to remove unfixed dye. Problems such as marking-off of the wet print and insufficient moisture take-up (leading to poor fixation, particularly on lightweight fabrics) prevented the method from becoming commercially adopted.

The process was later adapted to the printing of wool knitwear [70] in both garment and panel form. As such, it has enjoyed a certain amount of success, the handling techniques being much simpler than those required for the batching of wet piece goods. The garments are printed on a former, and batching is carried out by gradually building up a stack of garments separated from each other by polythene film. After batching, the goods are washed-off in a side paddle. This operation requires that the garments be given a shrink-resist treatment prior to printing, in order to avoid felting and loss of print definition. Either chlorination or the chlorination–Hercosett process can be used; in both cases, fixation of the dye is enhanced. Full details of the garment-printing method are given elsewhere [70].

12.9   Transfer Printing

Heat transfer printing (HTP) is based on the ability of certain disperse dyes to sublime from a printed paper at temperatures of 180–200 °C and is the very essence of simplicity. It is therefore no wonder that it was once regarded as something of a philosopher's stone.

Considerable efforts have been devoted to adapting the technique to wool, with the IWS and the Commonwealth Scientific and Research Organisation (CSIRO) being especially active in this field. The particular problems associated with the transfer printing of wool are:

- The anionic dyes used for the dyeing and printing of wool are not sublimable, and therefore not suitable for an HTP process.
- Untreated wool has no affinity for the disperse dyes used in HTP.

Most of the methods that have been developed are aimed at overcoming these problems. Some of the more important ones are described in this section.

12.9.1   Wet or 'Migration' Transfer Printing

The concept of wet transfer printing was first introduced by Dawson International in the late 1960s, and was the subject of several patents and publications [115,116]. It was based on the use of transfer papers printed with Lanasol reactive (α-bromoacrylamido) dyes, which were brought into contact with the substrate, prepadded with a thickened acid liquor. Through the application of heat and pressure, transfer (or 'migration') and fixation of the dyes took place, and a wash-off completed the process. The original Fastran process was based on the printing of garments, and radio-frequency (RF) heating was employed to enable a stack of garments to be printed simultaneously.

Dawson's extended the wet transfer concept to the continuous printing of piece goods on the Arcamatic Calender (SA Monk); collaboration between Transprints and Tootal's resulted in the development of the Dewprint machine, a more specialised form of calender. The general principles of the process remained the same:

- The use of Lanasol reactive dyes in the form of Aquatran W papers.
- Continuous prepadding of the goods with a thickened liquor.
- Contact between goods and paper for 30–60 seconds at 106 °C, with the application of pressure.
- Washing-off to remove unfixed dye.

Optimisation of the process with respect to pH, addition of urea and auxiliaries was reported. It was found to be necessary to work with prechlorinated wool for maximum dye fixation [117], although improved results on unchlorinated grounds could be obtained by using papers printed with more highly reactive dyes.

12.9.2 Sublimation Transfer Printing

The wool fibre has no affinity for the disperse dyes used in conventional HTP. Sublimation transfer techniques developed for wool were therefore based on modification of the fibre, modification of the dye or a combination of both.

Methods used to try and modify the fibre and increase its affinity for disperse dyes included:

- Pre-impregnation with simple polar compounds such as urea, thiodiglycol or lactic acid [118].
- Using tertiary and quaternary organic bases and salts to cleave the disulphide crosslinks, thereby opening up the structure of the wool [119].
- Pad application of surfactants [119] and acrylic monomers [120].

All these methods were impractical for one reason or another. The most promising approach was the Keratrans process, jointly developed by the IWS and CSIRO (Australia). This combined a pretreatment with moderate levels of an anionic surfactant and a chromium salt, with the use of specially synthesised, sublimable metallisable disperse dyes [122,123]. The dyes, which were available from Croda Colours as the Sublichrome range, possessed the general Structures **4–7**, giving yellow/orange, orange, red and blue shades, respectively.

4

5

6

7

The optimum pretreatment comprised a padding and drying operation with:

- 5% Alcopol O 60 (sodium di-*iso*-octyl sulphosuccinate, Huntsman);
- 5% urea;
- 2% lactic acid;
- 0.75% chromium(III) chloride hexahydrate.

An example of a typical print obtained by this process is presented in Figure 12.4, while Figure 12.5 demonstrates the difference in the hues of the metal-complex dyes formed on changing the metal salt present in the wool pretreatment solution.

The surfactant imparts an affinity for disperse dyes to the fibre surface, chromium(III) chloride promotes the formation of the chromium–dye complex during transfer printing and lactic acid is essential for pad-liquor stability. Transfer printing is carried out for 30 seconds

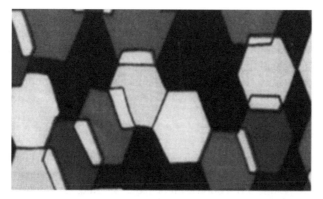

Figure 12.4   *Wool fabric pretreated with a chromium III salt and dry transfer printed with metallisable disperse dyes. See colour plate section for a full-colour version of this image.*

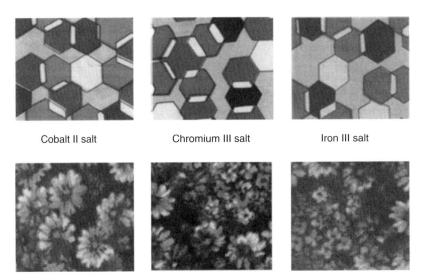

Cobalt II salt   Chromium III salt   Iron III salt

Figure 12.5 *The effect of metal salt type on the hue obtained for wool substrates dry transfer printed with metallisable disperse dyes. See colour plate section for a full-colour version of this image.*

at 195–200 °C; as the metal-complex dye is formed only on the fibre surface, steaming for 30 minutes at 100 °C is necessary for full colour development and wash fastness (the urea plays an important part in this step). The dye–metal complexes formed during the process have been characterised and shown to be predominantly 1 : 2 complexes, with only traces of the 1 : 1 complex present.

The main drawbacks to the Keratrans process on 100% wool are:

- The restricted shade gamut (that of a metal-complex dye range).
- The necessity for a separate steaming step.

Bell *et al.* [54] have described the application of the technique to wool-rich wool/polyester blends, using mixtures of the Sublichrome dyes and standard sublimable disperse dyes. A much improved shade range with good fastness was obtained, but steaming was still necessary (although conditions were not so critical) and, moreover, only a brief hot-water wash was necessary, since the pretreatment chemicals were not completely absorbed (as they are in the case of 100% wool).

12.9.3   Benzoylated Wool

Perhaps the most fundamental approach to making wool transfer-printable with disperse dyes is to chemically incorporate bulky aryl residues; that is, to impart to it polyester-like properties. Pretreatment of cotton with benzoyl chloride, as a preparation for transfer printing, formed the basis of the Shikibo-Uni process [124]. Lewis and Pailthorpe [125] demonstrated that, by treating wool with benzoic anhydride from *N,N*-dimethylformamide, machine-washability, heat-settability and outstanding dyeability with disperse dyes could

As printed   Washed sample (IS02)

Benzoylated
wool

Untreated
wool

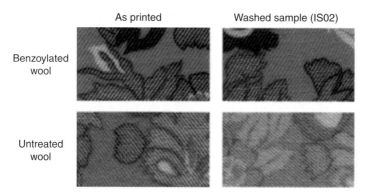

Figure 12.6   *Disperse dye transfer prints on benzoylated and untreated wool. See colour plate section for a full-colour version of this image.*

be achieved; the major practical drawback was the necessity to treat from a nonaqueous medium. An example of a typical print obtained on such a benzoylated wool substrate is given in Figure 12.6. These workers have also studied the application of reactive hydrophobes from an aqueous medium, with some promising results [71,126]. In particular, long-liquor treatment with 6-anilino-2,4-dichloro-*s*-triazine (ANEX) imparted to the wool an affinity for disperse dyes equivalent to that of benzoylated wool of similar mass gain (around 12–13%); shrink resistance of the benzoylated wool was vastly superior, however.

12.10   Novel Effects

Finally, some mention should be made of certain novel printing effects used on wool and wool blends. While not of great commercial importance, they are nonetheless interesting, and exploit well-known properties of the fibre for their effect.

12.10.1   Burn-Out (*devorée*) Printing

So-called burn-out or *devorée* techniques are practised on various fibre blends [127], the principle being to print a chemical/auxiliary which destroys one component of the blend in the printed areas. This effect [128] may be achieved on wool/polyester-blend fabrics by printing with sodium hydroxide, then pressure steaming at 120 °C for 15 minutes. The wool component is destroyed in the printed areas, and removed during subsequent washing-off.

12.10.2   Sculptured Effects

The felting properties of the wool fibre may be exploited in the production of unmilled patterns on milled ground structures, giving a relief effect. The process is aimed primarily at garments, especially woollen spun lambswool and Shetland types, for which milling is a standard finish, but it may also be applied successfully to knitted fabrics.

A pattern is printed on an untreated, unfinished garment, with a print paste containing a shrink-resist polymer, such as Synthappret BAP or Securlana, in the presence of sodium bicarbonate or sodium sulphite, respectively. After printing, the garment or fabric is dried, then baked at 150 °C to cure the resin, and finally scoured and milled in the usual way. The printed areas resist the milling action, resulting in a sculptured effect. Subtle colour effects may be achieved by including in the print paste very small amounts of pigment; the resin acts as a pigment binder, and the unmilled areas are faintly tinted.

References

[1] V A Bell, in *Wool Dyeing*, ed. D M Lewis (Bradford, UK: Dyers' Company Publication Trust, 1992).

[2] K Reincke, *Textil Praxis*, **25** (1970) 419.

[3] H Heiz, *Textilveredlung*, **13** (1978) 205.

[4] I Bearpark, F W Marriott and J Park, *A Practical Introduction to the Dyeing and Finishing of Wool Fabrics* (Bradford, UK: Dyers' Company Publication Trust, 1986).

[5] Stevensons (Dyers) Ltd, British patent 716806 (1952).

[6] H Zahn, *Textilveredlung*, **17** (1982) 421.

[7] J A McLaren and B Milligan, *Wool Science: The Chemical Reactivity of the Wool Fibre* (Marrickville, NSW, Australia: Science Press, 1981).

[8] W S Simpson, WRONZ Report No. 47 (1976).

[9] N Fair and B S Gupta, *Progress in Polymer Science*, **11** (1985) 167.

[10] V A Bell and D M Lewis, *Textile Research Journal*, **53** (1983) 125.

[11] K R Makinson, *Shrinkproofing of Wool* (New York, USA and Basel, Switzerland: Marcel Dekker, 1979).

[12] H D Feldtman and J R McPhee, *Textile Research Journal*, **34** (1964) 634.

[13] R Hofstetter, *Textilveredlung*, **21** (1986) 141.

[14] K Reincke, *Textilveredlung*, **3** (1968) 555.

[15] BASF, *Technisches Merkblatt* (1980).

[16] V A Bell, *Review of Printing* (Sydney, Australia: International Wool Secretariat, 1986).

[17] K M Byrne, P Smith, J Jackson and J Lewis, *Textile Horizons*, **4** (1984) 25.

[18] IWS, Dyeing Technical Information Bulletin No. 19 (1984).

[19] J Lewis, *Wool Science Review*, **54** (1977) 2.

[20] J Lewis, *Wool Science Review*, **55** (1978) 23.

[21] V A Bell, P R Brady, K M Byrne, P G Cookson, D M Lewis and M T Pailthorpe, *AATCC National Technology Conference, Book of Papers*, (1983) 289.

[22] P R Brady and P G Cookson, *Proceedings of the 6th International Wool Textile Research Conference, Pretoria*, **5** (1980) 517.

[23] D M Lewis, *Textile Research Journal*, **52** (1982) 580.

[24] V A Bell, *Textilveredlung*, **21** (1986) 148.

[25] C-W Kan and C-W M Yuen, *Textile Progress*, **39**(3) (2007) 121.

[26] D Biniaś, A Wlochowicz and W Biniaś, *Fibres and Textiles in Eastern Europe*, **12** (2004) 58.

[27] P A Sturrock, *Plasma Physics: An Introduction to the Theory of Astrophysical, Geophysical and Laboratory Plasmas* (Cambridge, UK: Cambridge University Press, 1994).

[28] C Kan and C M Yuen, *Plasma Process Polymers*, **3** (2006) 627.

[29] M Naebe, P G Cookson, J A Rippon and X G Wang, *Textile Research Journal*, **80** (2010) 611.

[30] M Naebe, P G Cookson, J Rippon, R P Brady, X Wang, N Brack and G van Riessen, *Textile Research Journal*, **80** (2010) 312.

[31] K M Byrne and J. Godau, *Proceedings of the 9th International Wool Textile Research Conference, Biella*, **IV** (1995) 415.

[32] P R Brady, D E Rivett and I W Stapleton, *Proceedings of the 7th International Wool Textile Research Conference, Tokyo*, **4** (1985) 421.

[33] P R Brady, *Proceedings of the 7th International Wool Textile Research Conference, Tokyo*, **5** (1985) 161.

[34] Perachem Ltd, PCT/GB 2006/002955 (2006).

[35] B M Müller, *Review of Progress in Coloration and Related Topics*, **22** (1992) 14.

[36] U Meyer, *Melliand Textilber*, **63** (1982) 51.

[37] H B Elsässer, *Textil Praxis*, **40** (1985) 63.

[38] H Ellis, *Textile Horizons*, **5** (1985) 37.

[39] F Beffa and G Back, *Review of Progress in Coloration and Related Topics*, **14** (1984) 33.

[40] D M Lewis, *Review of Progress in Coloration and Related Topics*, **8** (1977) 10.

[41] D M Lewis, *Journal of the Society of Dyers and Colourists*, **98** (1982) 165.

[42] D M Lewis, *Melliand Textilber*, **67** (1986) 717.

[43] P R Brady and R J Hine, *American Dyestuff Reporter*, **69** (1980) 23.

[44] R Reina, *AATCC 1981 National Technology Conference, Book of Papers*, (1981) 238.

[45] A Schaub, *Textilveredlung*, **19** (1984) 351.

[46] S Glander, SVCC Druckerei-Tagung, Näfels (1980).

[47] K T Fell and R Postle, *Textile Research Journal*, **40** (1970) 683.

[48] J H W Schaub, *Textilveredlung*, **20** (1985) 58.

[49] R Reina, *Textilveredlung*, **20** (1985) 60.

[50] J Winkler, *International Textile Bulletin*, **64** (1979) 25.

[51] R Hofstetter, *Textilveredlung*, **17** (1982) 517.

[52] R Koch, *Bayer Farben Review*, **35** (1983) 20.

[53] R Koch, *Textil Praxis*, **37** (1982) 1301.

[54] V A Bell, D M Lewis, P R Brady, P G Cookson and M T Pailthorpe, *Schriftenreihe des Deutschen Wollforschungsinstitut*, **93** (1984) 102.

[55] C Berry and J Ferguson, in *Textile Printing*, ed. L W C Miles (Bradford, UK: Dyers' Company Publications Trust, 1981).

[56] R J Hannay and W Furness, *Journal of the Society of Dyers and Colourists*, **69** (1953) 596.

[57] R J Hannay, *Journal of the Society of Dyers and Colourists*, **76** (1960) 11.

[58] A Jansen and W Kuppers, *Melliand Textilber*, **35** (1954) 880.

[59] BASF, Technical Information M1059e (1983).

[60] BASF, Data Sheet 03.85, Unternehmensbereich Textilchemie (1985).

[61] BASF, Technical Information M1379e (1985).

[62] M Weiss, *Canadian Textile Journal*, **97** (1980) 47.

[63] M Weiss, *American Dyestuff Reporter*, **67** (1978) 35.

[64] M Weiss, *American Dyestuff Reporter*, **67** (1978) 72.

[65] Sandoz Information Booklet 9128.00.83, Recommendations for Textile Printing.

[66] Sandoz, British patent 1410552.

[67] J Frauenknecht and D Schwer, *Textilveredlung*, **5** (1970) 912.

[68] Sandoz, Technical Information Bulletin No. 1570 (1971).

[69] P Koltai and W Lindermann, *Melliand Textilber*, **55** (1974) 365.

[70] J H Mills, *Wool Science Review*, **54** (1977) 46.

[71] V A Bell, D M Lewis and M T Pailthorpe, *Journal of the Society of Dyers and Colourists*, **100** (1984) 223.

[72] Sandoz technical information bulletin, Sandospace S: fibre reactive resist for polyamide substrates (1979).

[73] Sandoz, US patent 2726133 (1955).

[74] R L Elliot, R S Asquith and M A Hobson, *Journal of the Society of Dyers and Colourists*, **79** (1963) 188.

[75] M Lewin, P K Isaacs and B Schafer, *Proceedings of the 5th International Wool Textile Research Conference, Aachen*, **5** (1975) 73.

[76] B A Cameron, D M Lewis and M T Pailthorpe, *Proceedings of the 7th International Wool Textile Research Conference, Tokyo*, **5** (1985) 79.

[77] V A Bell, D M Lewis and M T Pailthorpe, *Proceedings of the 7th International Wool Textile Research Conference, Tokyo*, **5** (1985) 89.

[78] J M Notley, *Journal of Applied Chemical Biology*, **23** (1973) 717.

[79] H D Pleasance, *Textile Horizons*, **5** (1985) 21.

[80] Sandoz and Thiotan WS, Technical Information Bulletin 2285.00.85.

[81] J Provost, *The Colourist*, **3** (2011) 4.

[82] T L Dawson, in *Textile Printing*, 2nd Edn. ed. L W C Miles (Bradford, UK: Dyers' Company Publication Trust, 2003).

[83] T L Dawson, in *Digital Printing of Textiles*, ed. H Ujiie (Cambridge, UK: Woodhead, 2006).

[84] T L Dawson, in *Digital Printing of Textiles*, ed. H Ujiie (Cambridge, UK: Woodhead, 2006).

[85] E Loser and H-P Tobler, in *Digital Printing of Textiles*, ed. H Ujiie (Cambridge, UK: Woodhead, 2006).

[86] B Glover, in *Textile Ink Jet Printing*, ed. T L Dawson and B Glover (Bradford, UK: Society of Dyers and Colourists, 2004).

[87] M Raymond, in *Digital Printing of Textiles*, ed. H Ujiie (Cambridge, UK: Woodhead, 2006).

[88] L Caccia and M Nespeca, in *Digital Printing of Textiles*, ed. H Ujiie (Cambridge, UK: Woodhead, 2006).

[89] H Kobayashi, in *Digital Printing of Textiles*, ed. H Ujiie (Cambridge, UK: Woodhead, 2006).

[90] U Hees, M Freche, J Provost, M Kluge and J Weiser, in *Textile Ink Jet Printing*, ed. T L Dawson and B Glover (Bradford, UK: Society of Dyers and Colourists, 2004).

[91] H Noguchi and K Shirota, in *Digital Printing of Textiles*, ed. H Ujiie (Cambridge, UK: Woodhead, 2006).

[92] K H Blank, J M Chassagne and W Reddig, in *Textile Ink Jet Printing*, ed. T L Dawson and B Glover (Bradford, UK: Society of Dyers and Colourists, 2004).

[93] A Lavery, J R Provost, R Docherty and J Watkinson, in *Colour Science '98, Vol. 2: Textile Dyeing and Printing*, ed. S M Burkinshaw (Leeds, UK: University of Leeds 1999), 1.

[94] J R Provost, in *Colour Science '98, Vol. 2: Textile Dyeing and Printing*, ed. S M Burkinshaw (Leeds, UK: University of Leeds 1999).

[95] D M Lewis, *Textiles-UK Conference, London*, (2010).

[96] D M Lewis, *Digital Textile*, **6** (2010) 11.

[97] C Hawkyard, in *Digital Printing of Textiles*, ed. H Ujiie (Cambridge, UK: Woodhead, 2006).

[98] Y K Kim, in *Digital Printing of Textiles*, ed. H Ujiie (Cambridge, UK: Woodhead, 2006).

[99] P J Broadbent and D M Lewis, *Proceedings of the SDC International Conference, Belfast*, (2008).

[100] M Clark, K Yang and D M Lewis, *Coloration Technology*, **125** (2009) 184.

[101] R Koch, *AATCC National Technology Conference, Book of Papers, Atlanta*, (1986) 169.

[102] H J Weyer, *Journal of the Society of Dyers and Colourists*, **97** (1981) 50.

[103] K Speier and K Roth, *International Textile Bulletin*, **401** (1977) 333.

[104] German patent 2405057 (1974).

[105] B Glover, *Review of Progress in Coloration and Related Topics*, **8** (1977) 36.

[106] IWS Nominee Company, *International Dyer*, **165** (1981) 259.

[107] IWS Nominee Company, British patent 1583261.

[108] P R Brady and P G Cookson, *Journal of the Society of Dyers and Colourists*, **95** (1979) 302.

[109] P G Cookson, *Wool Science Review*, **62** (1986) 3.

[110] H Putze and G Dillmann, *Textilveredlung*, **15** (1980) 457.

[111] G Robert, *Textilveredlung*, **10** (1975) 431.

[112] D M Lewis and I Seltzer, *Journal of the Society of Dyers and Colourists*, **88** (1972) 327.

[113] J F Graham and I Seltzer, *Textilveredlung*, **9** (1974) 551.

[114] ICI, Technical Information D1575 (1978).

[115] Ciba Geigy, British patent 1227271.

[116] Joseph Dawson (Holdings), British patent 1284824.

[117] V A Bell and J F Graham, *Textilveredlung*, **14** (1979) 326.

[118] V A Bell and D M Lewis, *Journal of the Society of Dyers and Colourists*, **94** (1978) 507.

[119] I Rusznak, L Trezi and S Csanyi, *Textilveredlung*, **16** (1981) 172.

[120] V A Bell, D M Lewis, P R Brady and P G Cookson, *Textile Chemist and Colorist*, **11** (1979) 100.

[121] P R Brady and P G Cookson, *Journal of the Society of Dyers and Colourists*, **97** (1981) 159.

[122] P R Brady, P G Cookson, K W Fincher and D M Lewis, *Journal of the Society of Dyers and Colourists*, **96** (1980) 188.

[123] V A Bell, D M Lewis, P R Brady, P G Cookson and K W Fincher, *Journal of the Society of Dyers and Colourists*, **97** (1981) 128.

[124] Shikibo Ltd, German patent OLS 2608083.

[125] D M Lewis and M T Pailthorpe, *Journal of the Society of Dyers and Colourists*, **99** (1983) 354.

[126] D M Lewis and M T Pailthorpe, *Journal of the Society of Dyers and Colourists*, **100** (1984) 56.

[127] U Perkuhn, *Textil Praxis*, **41** (1986) 566.

[128] IWS, Technical Information Bulletin AP 123 (1982).

Index

The Coloration of Wool and other Keratin Fibres, First Edition. Edited by David M. Lewis and John A. Rippon.
© 2013 SDC (Society of Dyers and Colourists). Published 2013 by John Wiley & Sons, Ltd.